169 Topics in Current Chemistry

Electron Transfer I

Editor: J. Mattay

With contributions by
M. Baumgarten, C. S. Foote, W. Kaim,
R. Memming, K. Mizuno, K. Müllen,
Y. Otsuji, W. Rettig, M. Schmittel

With 79 Figures and 18 Tables

Springer-Verlag
Berlin Heidelberg GmbH

This series presents critical reviews of the present position and future trends in modern chemical research. It is adressed to all research and industrial chemists who wish to keep abreast of advances in their subject.

As a rule, contributions are specially commisioned. The editors and publishers will, however, always be pleased to receive suggestions and supplementary information. Papers are accepted for "Topics in Current Chemistry" in English.

ISBN 978-3-662-14915-7   ISBN 978-3-540-48225-3 (eBook)
DOI 10.1007/978-3-540-48225-3

Library of Congress Catalog Card Number 74-644622

© Springer-Verlag Berlin Heidelberg 1994
Originally published by Springer-Verlag Berlin Heidelberg New York in 1994
Softcover reprint of the hardcover 1st edition 1994

Typesetting: Macmillan India Ltd., Bangalore-25

SPIN: 10101078   51/3020 - 5 4 3 2 1 0 - Printed on acid-free paper

Guest Editor

Prof. Dr. *Jochen Mattay*
Organisch-Chemisches Institut,
Westfälische Wilhelms-Universität Münster,
Orléansring 23, 48149 Münster, FRG

Editorial Board

Preface

In 1989 the first volume of the series Photoinduced Electron Transfer I - V (Topics in Current Chemistry, Vols. 156, 158, 159, 163 and 168) was published by Springer-Verlag. Meanwhile this area has been reviewed with increasing frequency in various books and journals, reflecting its importance which was already emphasized in the prefaces to the above mentioned series of Topics in Current Chemistry.

Needless to say, electron transfer not only plays a crucial role in excited state systems but also in organic,organo-metallic, biochemical and heterogeneous systems as well as in new materials. Therefore the new general series, Electron Transfer, has been designed to provide the reader with up-to-date chapters by leading scientists who have already contributed in terms of a long-term research to the topic of their own articles.

In this respect the present reviews reflect the interdisciplinary character of this new field of chemical research covering aspects of inorganic, organic and physical chemistry as well as material sciences. I have the pleasure in thanking the contributors and all coworkers of Springer-Verlag especially Dr. R. W. Stumpe for bringing this volume to a successful completion.

Münster, January 1994   Jochen Mattay

Attention
all "Topics in Current Chemistry" readers:

A file with the complete volume indexes Vols.22 (1972) through 168 (1993) in delimited ASCII format is available for downloading at no charge from the Springer EARN mailbox. Delimited ASCII format can be imported into most databanks.

The file has been compressed using the popular shareware program "PKZIP" (Trademark of PKware Inc., PKZIP is available from most BBS and shareware distributors).

This file is distributed without any expressed or implied warranty.

To receive this file send an e-mail message to:
SVSERV@DHDSPRI6.BITNET.
The message must be:"GET/CHEMISTRY/TCC_CONT.ZIP"

SVSERV is an automatic data distribution system. It responds to your message. The following commands are available:

HELP	returns a detailed instruction set for the use of SVSERV
DIR (name)	returns a list of files available in the directory "name",
INDEX (name)	same as "DIR",
CD <name>	changes to directory "name",
SEND <filename>	invokes a message with the file "filename",
GET <filename>	same as "SEND".

For more information send a message to:
INTERNET:STUMPE@SPINT. COMPUSERVE.COM

Table of Contents

Erratum

Topics in Current Chemistry, Vol. 165 (1993).

D. A. Tomalia, H. D. Durst: "Genealogically Directed Synthesis: Starburst/ Cascade Dendrimers and Hyperbranched Structures", pages 193 - 313.

The reference to the Copyright owner was inadvertently omitted for the following illustrations:

Schemes 19 (p. 252):
   Hawker CJ, Fréchet JMJ (1990) J Am Chem Soc 112: 7638
   Copyright 1990 American Chemical Society;
Figure 30a (p. 267):
   Newkome GR, Lin X (1991) Macromolecules 24: 1443
   Copyright 1991 American Chemical Society;
Figure 30b (p. 267):
   Newkome GR, Nayak A, Behera RK, Moorefield CN, Baker GR
   (1992) J Org Chem 57: 358
   Copyright 1992 American Chemical Society.

Permission to reproduce was granted by the American Chemical Society.

Topics in Current Chemistry, Vol. 165
Editor: E. Weber
© Springer-Verlag Berlin Heidelberg 1993

Radical Ions:
Where Organic Chemistry Meets Materials Sciences

Martin Baumgarten and Klaus Müllen

Max-Planck-Institut für Polymerforschung, Ackermannweg 10, 55128 Mainz, FRG

Table of Contents

Topics in Current Chemistry, Vol. 169
© Springer-Verlag Berlin Heidelberg 1994

Radical ions generated by electron transfer reactions are known as important intermediates in organic chemistry. On the other hand, their formation, recombination and transport in organic materials is responsible for a series of attractive physical properties. Radical ion formation is often accompanied by structural changes being well understood in small organic molecules, which also constitute repeating units of intensively studied macromolecules. Therefore, an approach is described herein to compare and combine the structural and energetic description of monomeric and oligomeric radical ions with that of partially oxidized or reduced polymeric materials.

Many optical and electrical properties of high-molecular-weight conjugated polymers closely correspond to those of oligomers containing only a few repeating units. These oligomers can be synthesized as monodisperse species, facilitating the spectroscopic description and enabling systematic studies of physical properties as a function of chain length (Sect. 2).

The mode of charge and spin distribution on conjugated chains is a central question for conducting polymers which are electrical insulators and semiconductors in the neutral, pristine state (Sect. 3). Both, intra- and interchain charge transport have to be considered in describing the overall conductivity. Electroactive polymers are applied as change storage materials, e.g. in rechargeable batteries, where the detailed charging mechanisms and minimization of Coulombic repulsion in highly charged states are crucial (Sect. 4). Electron transfer can also induce chemical reactions under formation or cleavage of σ-bonds (Sect. 5). While this is an unwanted side effect in the doping of conjugated polymers electrooxidation of suitable π-systems is a common method of producing electroactive and conducting polymers. Conductivity does not necessarily require polymeric materials, but is also obtained in radical ion salts and charge transfer complexes which are crystalline one dimensional conductors (Sect. 6). Their electrical conductivity can adequately be described as an electron hopping process between neighboring molecular layers. While the mobility of charge is important in processes like electrical conductivity or photo- and electroluminescence, localized radical states with as many unpaired electrons as possible are needed in magnetic materials (Sect. 7). Finally, the control of electron transfer processes in radical ion states can be used in molecular electronics (Sect. 8).

1 Structural and Energetic Aspects of Radical-Ion Formation

Organic materials made from conjugated polymers exhibit a series of attractive physical properties such as tuneable electrical conductivity [1, 2], photo-conductivity [3], charge-storage capacity [4], photoluminescence [5] or elec-troluminescence [6]. Remarkably enough, all these properties are tightly related to the formation, recombination and transport of radical ion sites. Radical ions – charged species with unpaired electrons – are, on the other hand, import-ant intermediates in organic chemistry [7, 8]. An attempt is therefore made herein to combine organic chemistry and materials sciences in a unified view upon the structures and energetics of radical ions.

Radical ions are created in solution by chemically or electrochemically induced electron transfer to or from a conjugated π-system. Even if these ions are thermodynamically stable they are only of limited persistence since they are susceptible to reactions with electrophiles and nucleophiles or undergo other processes like dimerization or electron-transfer induced bond cleavage [9, 10]. Pairs of radical anions and radical cations can also be formed by electron transfer between neutral donors and acceptors either in the ground state or upon photochemical excitation [11, 12].

It is significant that the most important experimental methods for radical ion characterization provide information amenable to interpretation in terms of molecular orbital theory [13]. Some typical examples follow:

1) Electron transfer under cyclic voltammetric control provides the half-wave potentials of radical anion and radical cation formation [14]. These data reflect the ease of an electron uptake or delivery and can be correlated with the energies of the newly formed singly occupied molecular orbitals. More-over, the difference of the redox potentials for radical cation and radical anion formation provides a measure of HOMO-LUMO gap, ΔE, which is highly relevant for many physical properties of the neutral compound [4, 14]. One anticipates from the conditions of the redox experiment that this energy gap depends upon the ion pairing and solvation energies of the charged species.
2) Optical absorption spectroscopy of the neutral molecules, therefore, provides an even better measure of the HOMO-LUMO gap ΔE for the undisturbed system [13]. The deep color often observed for the corre-sponding radical ions results from new long-wavelength absorptions due to transitions from the singly occupied molecular orbital (SOMO) to the LUMO (anion) or from the HOMO to the SOMO (cations), respectively (see Scheme 1).
3) The hyperfine coupling constants of protons attached to a π-center can be determined by EPR/ENDOR spectroscopic measurements [15, 16]; these coupling constants are proportional to the local spin densities at the π-centers as predicted by the corresponding atomic orbital coefficients of the

3

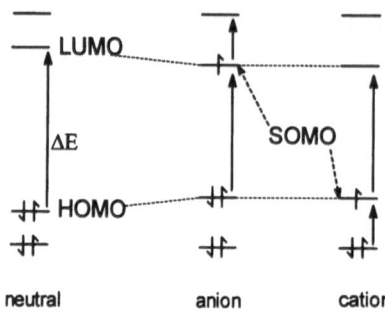

Scheme 1. Orbital scheme and optical transitions of low energy for a neutral p-system and the corresponding radical anion and radical cation

SOMO. Thereby spectroscopic experiment and elementary MO theory combine to a consistent description of the elctron-spin distribution.

4) X-ray crystallography of radical ions reveals that the geometry of the charged species does not necessarily correspond to that of the neutral compounds [17, 18]. Typical effects of the crystal lattice such as the relative spatial arrangement of the single molecules are also disclosed. These aspects will be important since e.g. electrical and magnetic properties of the solid state are strongly dependent on the spatial arrangements of the molecules.

Simple conjugated hydrocarbons like **1–5** can be invoked to demonstrate the occurrence of electron-transfer induced structural changes [19, 20]. What should be emphasized, before considering such structural changes in detail, is

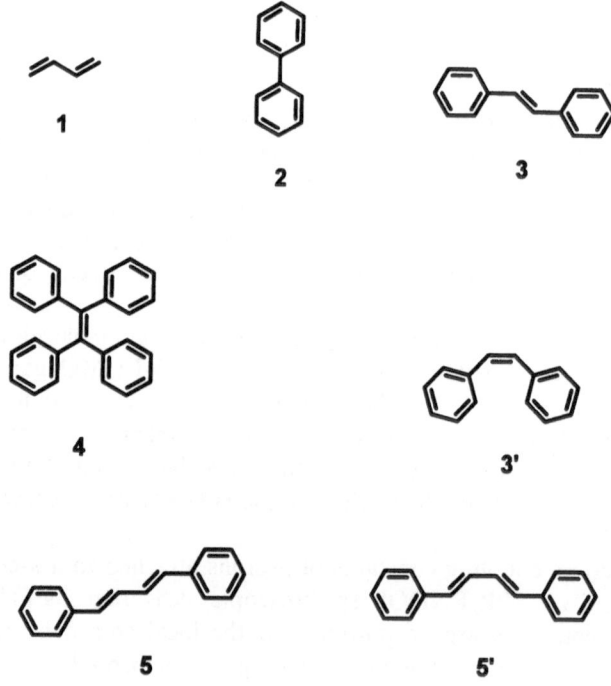

that the materials science of conjugated oligomers and polymers with extended π-systems faces completely analogous problems. Indeed, compounds **1–5** are invoked here since they constitute repeating units of intensively studied conjugated macromolecules such as polyacetylene (PAc, **6**), poly-*para*-phenylene (PPP, **7**), poly-*para*-phenylenevinylene (PPV, **8**) and poly-*para*-phenylenediphenylvinylene (**9**) (see Scheme 2) [21].

The radical anion of *cis*-stilbene **3'** rapidly transforms into that of the corresponding *trans*-isomer **3**[22]. In contrast, electron transfer to or from tetraphenylethene (**4**)[23] produces a significant torsion about the formal double bond. Similar configurational changes are observed for the dianthrylethene **10**[24]. Formation of the radical anion of biphenyl (**2**) [19] or of 1,4-diphenylbutadiene (**5**)[20] is accompanied by a conformational change: thus the radical anion **2**⁻• exhibits a tendency toward a lowering of the angle of torsion

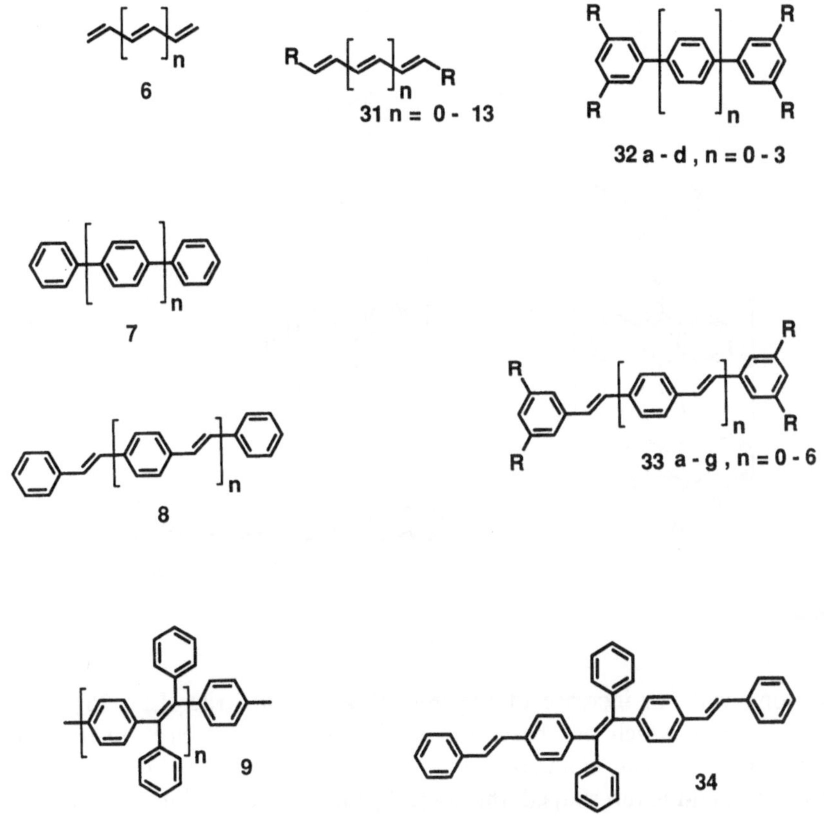

6

31 n = 0 - 13

32 a - d , n = 0 - 3

7

33 a - g , n = 0 - 6

8

9

34

Scheme 2.1.

5

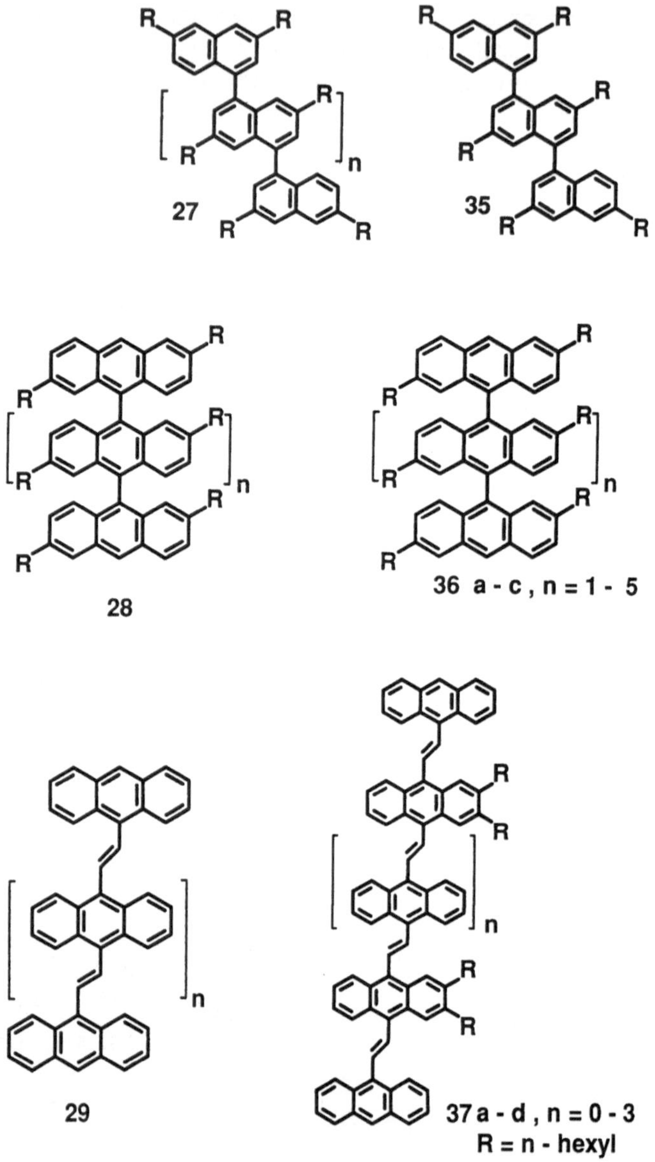

27

35

28

36 a - c , n = 1 - 5

29

37 a - d , n = 0 - 3
R = n - hexyl

Scheme 2.2.

going along with an increase of the inter-ring π-bond order [25, 26]. The classical example which has best been studied in terms of electron-transfer induced conformational changes is that of cyclooctatetraene (**11**): while the neutral compound is tub shaped, the corresponding radical anion and dianion are planar [7, 27]. There is evidence that electron transfer to or from conjugated polymers such as PP (**7**) and PPV (**8**), in a quite similar fashion, causes a

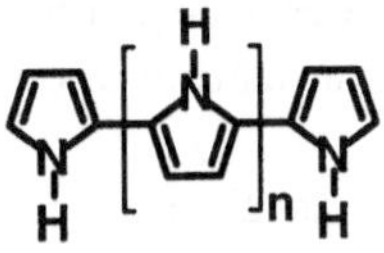

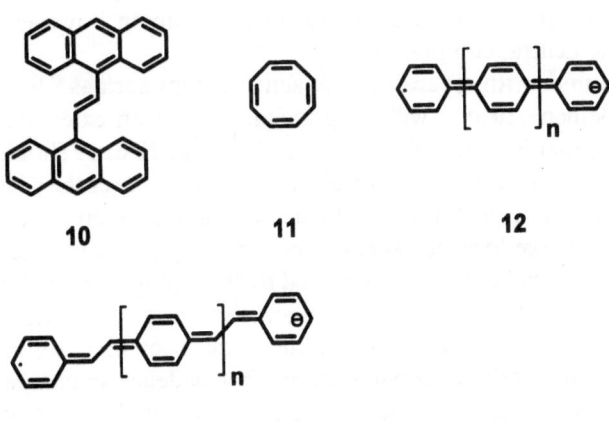

25

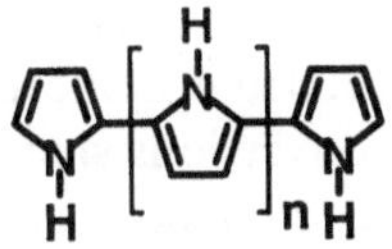

38 a - c , n = 1, 3, 5

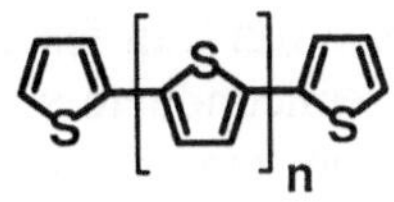

26

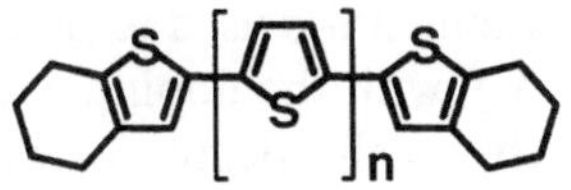

39 a - d , n = 1 - 4

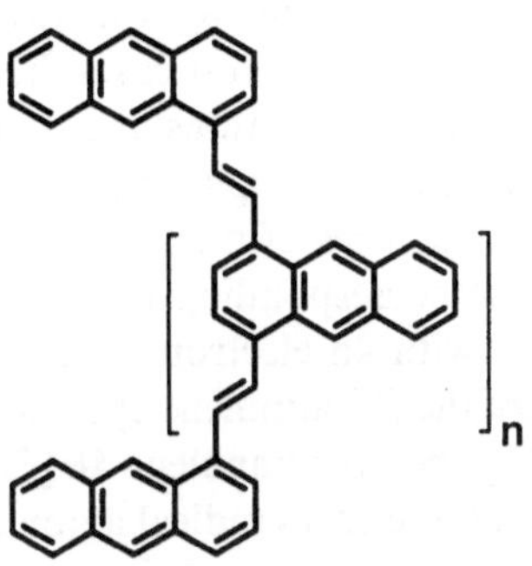

30

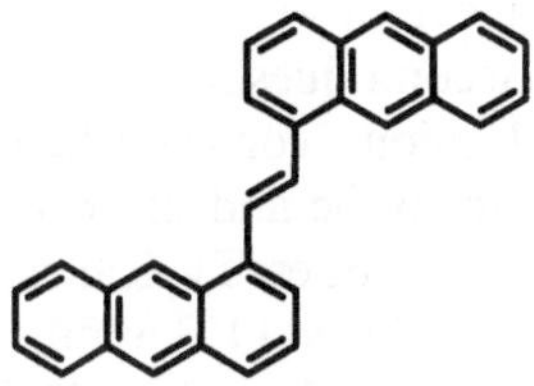

40

Scheme 2.3.

Scheme 2. Conjugated polymers and the corresponding oligomeric model compounds. R is *tert.* butyl unless otherwise indicated

10

11

12

13

flattening of the π-chains and a tendency toward "quinoidal" structures such as **12** and **13** [28]. Apart from conformational changes, one can also observe configurational interconversions since PAc (**6**) is described in various pieces of spectroscopic evidence to undergo a *cis/trans*-isomerization upon electrochemical oxidation [29].

The important role of conjugated polymers as electro- and photoactive materials is tightly bound to charges occurring on their π-chains. These materials are insulators in their neutral, pristine state and become electrically conducting upon doping, that is, upon partial oxidation or reduction [30]. The term doping is sometimes used to describe this electron transfer thus indicating the close analogy with the doping of inorganic semiconductors. An important question thereby is the degree of doping, i.e. the fate of the π-chain upon successive electron-transfer steps. The corresponding reaction for "small" organic radical anions and radical cations is their transformation into the dianions and dications, respectively [8, 31]. The first question then to be answered is that of spin pairing versus parallel spin alignment: while in most cases the two extra charges undergo spin pairing giving rise to singlet structures, the occurrence of orbital degeneracy as in coronene (**14**) can result in triplet-dianion structures [32]. The second question is whether the structures of mono-charged species, e.g. radical anions, correlate better with the corresponding neutral or diionic species. Monocyclic and bicyclic π-systems with 4n electrons often exist as π-bond localized species. Typical examples are the [8] annulene cyclooctatetraene (**11**) [33], the bridged [12] annulene **15** [34], the [16] annulene **16** [35] and the bicyclic 12π-system heptalene **17** [34]. In all these cases radical anion formation produces π-bond delocalized structures similar to those of the corresponding dianions [7, 36]: one electron per molecule is sufficient to induce a symmetry change of the ring. On the other hand, injection of charge can also induce a symmetry distortion. Thus, tetraphenylene (**18**), which can alternatively be considered as a cyclic oligophenylene or as a tetrabenzo-fused cyclooctatraene, possesses D_{2d} symmetry with a tub-shaped structure while the corresponding dianion has only C_2 symmetry with two different biphenyl moieties [19, 20, 37]. The geometric structure of the intermediate radical anion, however, corresponds to that of the neutral compound [38].

In compounds with separate redox-active groups such as 9,9'-bianthryl (**19**) or dianthrylethene **10** one will, in principle, except an extended conjugative interaction so that injection of an extra electron into **10** or **19** should give rise to radical anions with a complete delocalization of charge over the whole molecule. This, however, is not generally true. In fact, the radical anion of the bianthryl **19** can exist as a charge-localized species since the conjugative interaction between the two halves of the molecule is inhibited by the strong torsion about the inter-ring bond [39, 40]; in a similar fashion, an electron-transfer induced structural change (see above) can seriously inhibit the conjugation of the anthracenes in **10** [24]. For the present discussion it is crucial that evidence for charge localization in particular subunits also emerges from studies of the structurally related polymers.

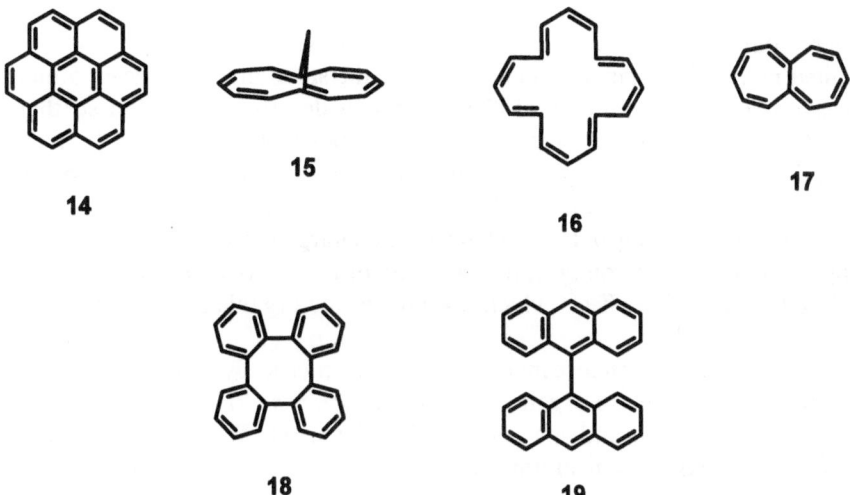

14

15

16

17

18

19

Spontaneous localization of charge in a π-system with extended π-conjugation can occur not only as a consequence of steric hindrance, but also as a result of the topology [41]. Thus, the dianion of dibenzotetracene **20** localizes the excess charge in the anthracene subunit [42, 43], and the radical anion of 1,3-distyrylbenzene (**21**) has the unpaired electron localized on one stilbene unit [44], although there is no significant steric inhibition of resonance. The immediate conclusion is, and this will be reconsidered in Sect. 3.2, that the electron-transfer behavior of a structurally related poly-*meta*-phenylenevinylene (**22**) should be closely related to that of the small model compound **21**.

20

21

22 a - d , n = 1 - 4

There is, indeed, evidence that excess charges occurring on a polymer chain can be confined to an effectively conjugated chain segment which extends over a limited number of repeating units [45]. For the small organic species, a major question concerns the mode of charge or spin density distribution so that a structural elucidation of charged conjugated polymers appears to be tightly bound to the "physical organic" criteria outlined above; in other words, it is totally adequate to relate the formation of charge-localized states on an electrically conducting polymer chain to the charge-induced changes of bond lengths, double-bond configuration and conformation as well as to counterion induced polarization effects which are known for organic model systems (see Sect. 3) [41, 45, 46].

A strong plea is therefore made herein to systematically increase the number of repeating units within organic radical ions (see Sect. 2) in order to close the gap between organic and macromolecular chemistry. When focusing on the properties of "large" radical anions and radical cations it must be added that materials properties such as electrical conductivity or magnetism of organic materials are solid state (bulk) properties and thus strongly depend upon the mode of *inter*molecular interactions (see Sects. 3 and 7). It will appear from the analysis below that the structural description of organic radical anions and radical cations can well be extended to include supramolecular structures. Crystalline radical-ion salts (see Sect. 6) which exhibit a strongly anisotropic electrical conductivity, can thereby demand particular attention. This is because their description, on the one hand, goes back to the knowledge of organic radical ions in solution and, on the other hand, extends to the charge-transport mechanism of conducting polymers.

In a similar fashion, further reduction of radical anions or oxidation of radical cations by successive electron transfer provides the molecular basis for charge storage in organic molecules, which is one key-step in the design of secondary battery electrodes (see Sect. 4). The initial question concerns the effective minimization of Coulombic repulsion. Finally, if in multi-charged derivatives of extended π-systems a spin pairing is prevented by topological or geometric reasons, one expects the formation of organic high-spin states, which is a crucial prerequisite for organic magnets (see Sect. 7).

Electron-transfer processes are often conducted in such a way that they affect only the π-electron system and leave the molecular σ-frame unobstructed. Thus, chemical reactions following an initial electron transfer are under many circumstances unwanted side effects. On the other hand, radical anions and

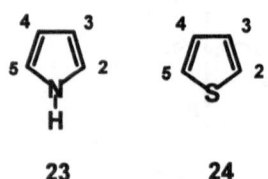

23   **24**

cations can undergo useful reactions which are not observed for the neutral compound (see Sect. 5). The electropolymerization of pyrrole (23) and thiophene (24) to yield polypyrrole (PPY, 25) and polythiophene (26), respectively, is a remarkable example of a radical-ion coupling [47].

The electrical conductivity of radical-cation salts can adequately be described by means of an electron-hopping process between neighboring molecular layers (see Sect. 6), and a related question arises for both the intramolecular and intermolecular transport of localized charge carriers in conjugated polymers (see Sect. 3). The energy profile of electron-transfer processes is also a major challenge of physical organic chemistry [46]. It is particularly important, thereby, to interpret the relevant rate constants in terms of structure. Along this line it becomes possible to tune low molecular organic and macromolecular redox systems for a particular electron-transfer behavior in the light of materials science (see Sect. 8).

These few examples are not meant to fully cover the materials science aspect of the organic chemistry of radical anions and radical cations. They point out, however, that the combination of both fields can be extremely fruitful for developing a reliable structural theory of conjugated oligomers and polymers and for tuning their physical properties.

2 Oligomers Versus Polymers

When dealing with the materials aspect of conventional polymers, mechanical properties such as, e.g., rubber elasticity depend upon the availability of high molecular-weight compounds [48, 49]. When considering conjugated polymers with extended π-systems, however, many optical and electrical properties of high molecular-weight polymers closely correspond to those of oligomers containing only a few numbers of repeating units [50, 51]. It should be emphasized, on the other hand, that before following the active physical function of conjugated π-systems one must create a well defined macroscopic state of the material, whereby the formation of thin films on a substrate or of mechanically robust, free-standing films is particularly important. In this respect polymers are advantageous over their oligomeric analogues [5, 6, 52].

Polymers exhibit certain disadvantages, however, when being subjected to the methods of radical ion characterization outline in Sect. 1:

1) Polymers exist as polydisperse mixtures with different molecular weights. The resulting physical properties such as, e.g., a redox potential cannot be ascribed to a specific length of the extended π-chain [53]. This and the large size of the molecules create an element of uncertainty in the physical description.

2) Conjugated polymers with their rigid π-systems often possess a low solubility even for low molecular weights, which seriously inhibits full structural

elucidation. It is therefore necessary to bring about sufficient solubility. This goal can be achieved by alkyl substitution which can be looked at as providing the molecule with its own solvation shell [54]. It is clear, on the other hand, the extensive substitution of a conjugated polymer such as **PPP** (**7**) [55] or **PPV** (**8**) [56] can create a steric inhibition of conjugation [57]. Interestingly enough, the phenyl-subsituted **PPV**-system **9** has sufficient solubility [21, 50]. This finding is related to the non-planar structure of the π-system which also inhibits the regular arrangement of different chains in a crystal lattice. Here again, however, solubility, as introduced by the substituents, goes along with a change of the electronic properties.

3) The most important drawback in the electronic description of conjugated polymers is the occurrence of structural defects such as sp^3-hybridized carbon centers in an extended π-chain [58, 59]. This leads to an interruption of π-conjugation which may determine the effective conjugation length within the polymer. The scientific literature is full of examples in which one does not pay proper attention to a complete structural elucidation and to the synthesis of defect-free, homogeneous polymeric materials; instead, the consideration of conjugated π-systems is often restricted to idealized, over-simplified or even partially decomposed structures.

The synthetic organic chemist, on the other hand, who is used to deal with structurally defined, but small molecules should be aware that avoiding structural defects in conjugated polymers is by no means a trivial task. A side reaction in any of the polymer forming steps does not lead to a separable side product, but produces an inherent structural defect of the macromolecule.

The polydispersity, the limited solubility, and the structural inhomogeneity of many conjugated polymers strongly call for an inclusion of related oligomers. If solubilized by suitable substitution, they can be synthesized as monodisperse species of a given size. This will not only facilitate the structural description by spectroscopic means, but will also allow one to systematically follow physical properties as a function of chain length [53]. By a plot physical properties such as a λ_{max}-value versus $1/n$, with n giving the number of repeating units, one can extrapolate toward the borderline value characteristic for the related polymer. It will, indeed, appear in the following that the careful study of oligomers and their corresponding radical ions can provide information on the polymer which is not available from a study of the polymer itself.

Alkyl and alkoxy groups attached to an oligomer chain can create some "disorganization", thus increasing the solubility, but they can also induce self assembly. As a consequence, one obtains well organized domains of parallel π-chains with a regular interchain distance [60]. Such a supramolecular architecture is crucial when investigating e.g. the charge-carrier transport mechanism of electrical conductivity (see Sect. 3).

Scheme 2 provides an overview over some conjugated polymers **6–9** and **25–30** together with their related oligomers **31–40**. For the latter, the number of repeating units is always specified. The simplest topology is that of polyacetylene **6** which is obtained by tetrabutoxytitanium- and triethylaluminum-catalyzed polymerization of acetylene [61]. Polyarylenes such as **7** and polyarylenevinylenes such as **8** are generally prepared by polycondensation of suitably functionalized precursors [62]. The insolubility and limited processability of many of these polymers have prompted researchers to refer to precursor routs [63, 64, 65]. Thereby the material is processed in the form of a soluble precursor polymer and finally transformed into the insoluble target structure with extended π-conjugation, e.g., by a thermally induced elimination process. It should also be noted that synthetic polyarylene and polyarylenevinylene chemistry has mostly been restricted to the "benzene"-type (**7**) and "styryl"-type (**8**) structures and has only recently been extended to include larger π-systems as building blocks. Polymers **27–30** are typical examples of this search for tailored conjugated polymers [39, 66–69]. A third class of conjugated polymers comprises e.g. polypyrrole (**25**) and polythiophene (**26**) which are formed by electropolymerization and deposited as solid films on an electrode [47, 70].

3 Electrical Conductivity and Charges on a Conjugated Chain

3.1 The Doping Process

Conjugated polymers such as polyacetylene (PA, **6**), poly-*para*-phenylene (PPP, **7**) or poly-*para*-phenylenevinylene (PPV, **8**) are electrical insulators. It has, therefore, been a major breakthrough when such species were subjected to redox reactions thus increasing the conductivity by up to 8 orders of magnitude [71]. The fabrication of conducting plastics has appeared as most promising in view of the combination of metal-like conductivity with the low specific weight and processability of polymers [51, 52]. The range of potential applications comprises e.g. rechargeable organic batteries [47, 72], electronic devices [2, 73] or antistatic materials [74]. It is clear that research on conducting polymers creates an interdisciplinary challenge in view of the necessary contributions of synthetic chemistry, physics, theory and processing science.

Before dealing with the electrical conductivity of doped polymers, a few comments on the electronic description of conjugated polymers are appropriate. The conjugative interaction of unsaturated building blocks with distinct electronic levels in a polymer chain gives rise to new electronic bands of the polymer, which are characteristic for the conjugated π-system considered. These bands are denoted as band structure, provided that the effective conjugation length is infinite, otherwise these electronic states are the HOMO and the LUMO of the conjugated segments. The band width of the newly formed energy levels depends on the strength of interaction between the individual molecular orbitals MOs of different molecules, on the topology and geometry of the polymer, and on the electronic correlation [75, 76]. This can even lead to a crossing of molecular orbital levels within the macromolecular structure. Thereby, the highest occupied electronic levels form the valence band and the lowest unoccupied electronic levels from the conduction band. The energy difference ΔE between both bands, the so-called band gap, corresponds to the HOMO-LUMO gap of a single molecule. This energy difference ΔE is relevant for many physical properties of conjugated polymers such as long-wavelength optical absorption, non-linear optical activity and intrinsic conductivity. The large band gap of most conjugated polymers ($\Delta E > 1.0$ eV) readily explains that they are electrical insulators or semiconductors at best. It is not surprising therefore that much attention has been paid to the design and synthesis of low-band gap polymers with intrinsic conductivity [77–82].

The electrical conductivity of a material is a macroscopic solid-state property since even in high molecular-weight polymers there is not just one conjugated chain which spans the distance between two electrodes. Then it is not valid to describe the conductivity by the electronic structure of a single chain only, because intra- and interchain charge transport are important. As with crystalline materials, some basic features of the microscopic charge-transport mechanism can be inferred from conductivity measurements [83]. The specific conductivity σ can be measured as the resistance R of a piece of material with length d and cross section F within a closed electrical circuit,

$$\sigma = 1/\rho = d/FR$$

Where ρ is the specific electrical resistance. R can easily be measured through the applied field (voltage) and the charge current (Ohm's law). At medium resistance R with conductivities of 10^{-4} to 10^{-10} S/cm), a simple two-point contact is sufficient, while at higher resistances in order to prevent leakage currents a guard ring configuration of electrodes is recommended [84]. Measurements of high conductivities as in organic metals are performed with a four-point contact to eliminate the resistance of the electrical contacts and wires leading to the sample [83]. For determining anisotropic conductivities of a sample the Montgomery technique can be applied [85]. The conductivity behavior of conjugated polymers may be compared to the one from ordinary metals and inorganic semiconductors. Depending on the energy of activation and mobility of the charge carriers (see below) the materials possess a very characteristic temperature dependence.

The conductivity of metals is described by the Drude-model. Therein the charge carriers are thought to be free moving electrons as in a gas. The current density J is proportional to the applied electric field, and the proportionality factor is the conductivity σ of the material ($J = \sigma E$). According to the Drude-model the electrical current can be expressed through an averaged drift velocity of electrons, which is time-independent in the case of a direct current (DC), but much lower than their averaged thermal velocity. This model allows the conductivity σ to be expressed by the mobility μ of the charge carriers (electrons):

$$\sigma = n\mu e$$

Where n is the charge-carrier concentration and e the elementary charge. The electrons do not move undisturbed, but are scattered by collision with other carriers and thermal lattice vibrations, so-called phonons, and structural defects. Upon lowering the temperature, collision probability and thermal lattice vibrations are reduced, and the mobility of the electron increases, while n is constant. This leads to a large increase of conductivity with decreasing temperature.

Semiconductors show a different temperature dependence [87]. Due to the energy gap between valence and conduction band the valence electrons need a thermal activation to enter the conduction band, thereby, leaving unoccupied states in the valence band (holes or defect electrons). Both charge carriers-electrons and holes-contribute to the conductivity, and the number of thermally activated charge carriers follows an Arrhenius law. With increasing temperature the number of charge carriers increases, but the phonon concentration also increases and tends to reduce the charge-carrier mobility. This leads to the observed temperature dependence of semiconductors, where conductivity increases from very low (1–10 K) to medium temperature (RT) and then decreases at even higher temperatures.

For conjugated polymers the gap between valence and conduction band normally is too high to be thermally overcome (> 1000 °C); therefore additional charges (charge carriers) are necessary, which are introduced by the doping process (see below). The temperature dependence of the conductivity is then expected to change with the doping level. At low doping levels a strong exponential increase of conductivity with increasing temperature is obtained as in semiconductors. At high doping levels (above 6% in PAc, **6**) very large conductivities, in some cases comparable to those of metals, are measured [61, 71]. The conductivity of highly conducting organic polymers lies in the order of 10^2–10^5 S/cm what may be compared to semiconductors such as doped germanium and indium-antimony (10–10^2 S/cm) and metals such as bismuth and iron (10^4–10^5 S/cm) [88]. Although these highly doped polymers exhibit a much smaller temperature dependence of the conductivity than the weakly doped polymers, the conductivity still decreases with decreasing temperature in contrast to the situation for metals. Accordingly, some activation barrier for the transport of charge is present even in highly conducting polymers. This activation barrier depends on the structure of the polymer and on the nature of the charge carriers. The charge carriers in organic materials have been identified as 1) neutral radicals or just charges without a spin, so-called solitons, 2) charged radicals so called polarons or 3) spinless doubly charged states, so-called bipolarons [89].

The soliton concept has been developed to explain charge transport in *trans*-PAc **6** [90], where the neutral radical centers are already created upon thermal, chemical, or electrochemical treatment of the synthetically produced *cis* form [29]. Within the charge-transfer description, **6** is unique in that it is the only polymer with a degenerate ground state, which can support neutral soliton formation. The solitonic energy states are located midgap in the center between valence and conduction band. A number of roughly 400 ppm of spins has been measured by EPR/ENDOR spectroscopy in undoped *trans*-PAc **6**, while *cis*-PAc contains no radicals at all [91]. The neutral soliton delocalizes over roughly 15 double bonds as measured through the hyperfine couplings by ENDOR spectroscopy, and it can move along a chain with very small activation energy [92, 93]. Evidence for this high soliton mobility has been found by detection of a motionally narrowed line width in the EPR studies. This mobility has been further confirmed by nuclear-spin relaxation measurements [94]. Upon light doping, some solitons loose their spin properties and become charged. These charged species are localized to the chain, but by interaction with another neutral soliton they can exchange their charge. The soliton model can perfectly explain charge transport in the low-doping regime of polyacetylene, but inevitably fails to explain the high, metal-like conductivity under high doping concentrations where polarons and bipolarons are created. Note, however, that this model cannot be applied to other conducting polymers with non-degenerate ground states.

Most conjugated polymers have non-degenerate ground states. In these cases the charging process leads to the formation of radical anions and radical cations (polarons) and diamagnetic dianions and dications (bipolarons), which are considered? (see Ref. 138–140 and Sect. 3.2, p 28) to be the charge carriers in the *low* and medium doping level, respectively [89, 90, 95]. Polarons are radical ions which can be envisaged as consisting of a charged and a neutral soliton, where the interaction between them leads to a splitting of the midgap solitonic level into two polaronic levels for the bonding and antibonding combination (see Scheme 3). Thereby, the lower state is split off from the valence band and the upper state is split off from the conduction band. The interaction of the charge with the nuclear lattice leads to a change in geometry, e.g. changes in bond length and torsional angles. The localization of the geometry changes and the charge distribution to a part of the conjugated chain is energetically favored

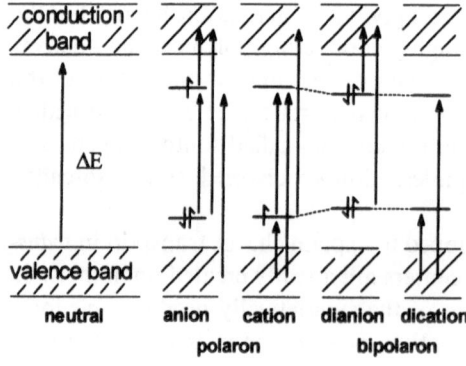

neutral — anion — cation — dianion — dication

polaron   bipolaron

Scheme 3. Band-structure and electronic transitions for conducting polymers in the neutral and charged state

over complete delocalization. Accordingly, within the terminology of organic chemistry the polaron has to be considered as an unpaired electron trapped by the induced lattice distortion and the counterion. In polymers containing aromatic subunits the lattice distortion is described as a partial transition from an aromatic to a quinoide structure as in **12** and **13** [96], which also induces a planarization of the π-system. The combined effects of localization of lattice distortion and of charge distribution describe the actual polaron length.

Polaron formation has found support from optical absorption spectroscopy and EPR spectroscopy. The creation of polaronic states on a polymer chain leads to drastic changes of their optical absorption spectra with new absorption bands of lower energy than in the neutral form (see Scheme 3) [90, 95]. In the case of oxidative doping (creating radical cations) the lower polaronic level is occupied, and the absorptions are due to the transition from the valence band to the polaronic states and from the polaronic states to the conduction band where two of these transitions are energetically equivalent. Additionally, there is a transition between the two polaronic levels. In the case of reductive doping, the upper polaronic level is singly occupied, and, again, three optical transitions of energy comparable to that for the cation should occur.

At low doping levels, the EPR signals increase in intensity with an increasing amount of the redox reagents. The signal intensity reaches a maximum at a certain doping level and then decreases again. This decrease has been attributed to spin pairing at higher doping levels while the conductivity still increases.

The fact that at higher doping levels the carriers of charge are nearly spinless has been attributed to the formation of bipolarons, which are believed to be formed by combination of two polarons or, as in the case of PAc **6**, by two charged solitons. The driving force for polaron combination, which of course has to overcompensate Coulombic repulsion is, again, the charge-induced distortion of the geometry of the neutral form. The structural changes going along with bipolaron formation are, therefore, predicted to be more significant than those of a polaron. Consequently, the electronic states appearing in the gap resulting from bipolaron formation should be more distant from the band edges than in the case of polaron formation (see Scheme 3). Bipolarons on a polymer chain will give rise to two optical transitions with lower energy than the band-gap transition which, in the case of oxidative doping, are described as the transitions between the valence band and the two bipolaronic levels and, in the case of reductive doping, between the two bipolaronic levels and the conduction band. At low and medium doping levels the electronic picture of the charges on a polymer chain, forming localized polaronic and bipolaronic states, is quite similar to the description of ion radicals and diionic states in organic π-molecules (see Schemes 1 and 3). At high doping levels ($> 10\%$) a direct overlap of these localized states occurs, leading to a new band structure with a much smaller band gap than in the neutral form [83, 90, 97, 98].

In an attempt to combine the description of the charge-carrier formation within the language of condensed-matter physics and of physical organic chemistry, the exposure of a mono-charged π-chain to additional doping reagent under removal or addition of a second electron raises the immediate question whether a second polaronic state is formed on the chain or whether two localized polarons combine to form a diionic or bipolaronic state, respectively. The possibility of bipolaron formation can thus be reduced to the question of whether or not there is an extra-stabilization of diionic states, and this question will be reconsidered in Sect. 3.2 [90, 95].

A major concern within the description of conjugated polymers, is the extension of excess charges on a conjugated π-chain. The extension of charged states combined with the lattice distortion as in polarons or bipolarons is described as the so-called effective conjugation length of a conjugated polymer. The stronger the lattice distortion, the smaller is the effective conjugation length, thus, reducing the mobility of the charge carriers. The effective conjugation length is then considered to serve as a direct measure of the ease of charge transport along a chain, which is one important aspect for the electrical conductivity of conjugated polymers. Section 3.2. will therefore be devoted to 1) the mode of charge distribution adopted upon charging of extended π-chains, 2) structural changes induced upon charging, and 3) whether there is an extra-stabilization of diionic states, i.e. going along with polaron or bipolaron formation. The key question is how the knowledge accumulated for well-defined oligomers can be utilized for a better understanding of the corresponding polymers. Another important aspect of electrical conductivity in conjugated polymers, namely the interchain charge transport will be dealt with in Sect. 3.3.

3.2 Distribution and Stabilization of Charge on a Conjugated Chain

The linear oligophenylenevinylene species **33a–g** are highly appropriate for the study of charge distribution in doped derivatives, because relatively large soluble oligomers are available and may be compared to the corresponding polymers [99]. The π-systems **33a–g** can be subjected to chemical or electrochemical reduction in an aprotic solvent. This reduction process can be controlled by EPR-[100] and absorption spectroscopy [101] as well as by cyclic voltammetry [102, 103]. By this parallel approach the resulting radical anions as well as the dianions obtained upon further reduction can reliably be identified. The following features are characteristic for the absorption spectra [Fig. 1]:

1) Radical anions are characterized by three and the dianions by two bands.
2) The absorption bands of the anions are shifted bathochromically with increasing chain length, but exhibit a constant spectral shape. [Fig. 1B].
3) A dianion is only formed after all the neutral compound has been transformed into the radical anion.

From the experimental λ_{max}-values of the oligomeric radical anions **33**$^{-\bullet}$ the absorptions of a monoanionic state of the corresponding PPV polymer can be determined by extrapolation [101]. The results show that even in the highest

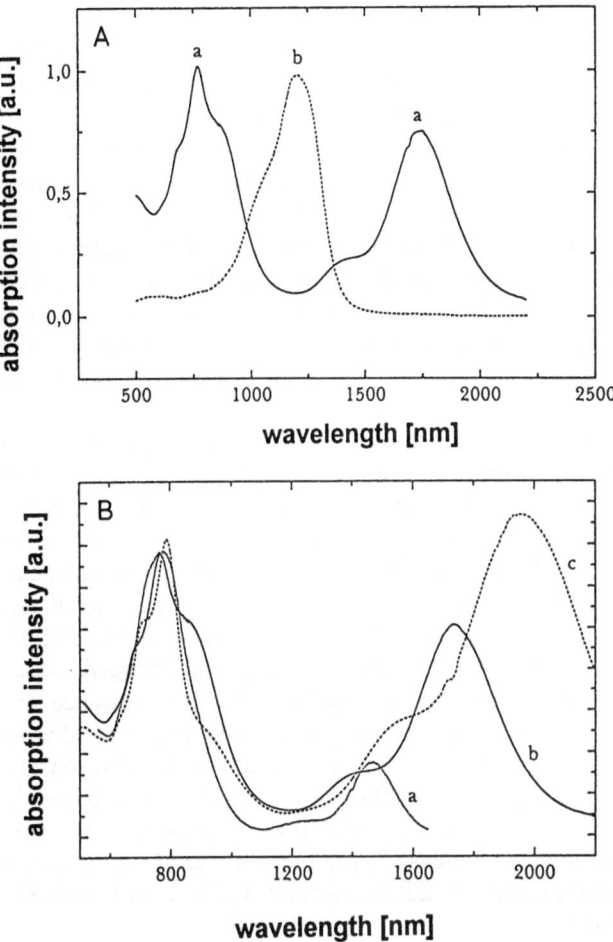

Fig 1. **A** Optical absorption spectra of Oligophenylenevinylene **33d** for the mono- (*a*) and dianion (*b*) in THF/K. **B** Optical absorption spectra for the monoradicals of **33c–e**

available oligomer, a limit of convergence is not yet attained. It can be concluded from the extrapolation, however, that the delocalization of an excess charge in a PPV chain affects about 8 to 10 styryl units. This entity would then have to be identified as the effectively conjugated segment to which the unpaired electron is confined in a defect-free polymer.

The conclusion is supported by vibrational FT-IR and FT-Raman spectra of **33a–g** in the neutral and doped state [104, 105]. A comparison of the intensities of the FT-IR bands from 1,3,5-trisubstituted benzene end group signals near 704 cm^{-1} with the intensities of the C–H out of plane motion of 1,2-*trans*-disubstituted ethylenes near 964 cm^{-1} shows a linear decrease of the intensity ratios with increasing chainlength approaching very small values for 8–10 units.

Upon doping with either AsF_5 or I_2, strong new infrared bands accompanied by weaker satellites are observed in the range of 1600–1000 cm^{-1}. A theory of the effective conjugation coordinate (ECC) has been developed to explain the results from IR and Raman spectroscopic measurements in terms of conjugation length [28b, 105]. The ECC theory is defined by a linear combination of the internal coordinates R, which are related to the stretching of the C–C and C=C bands. This approach includes the frequencies and intensities of the IR and Raman spectra, showing that most of the observations are due to the delocalization of π-electrons. The theory of ECC allows to determine an effective conjugation length of 7–9 units for **33**$^{-\bullet}$ [105] with a substantial contribution from quinoidal structures as **13**, which leads to a coplanar alignment of the subunits and an increase of the conjugation length in the doped state compared to the neutral precursor.

Further optical studies on the photoinduced absorption and photolumine-scence spectra of the phenylenevinylenes **33a–g** have been performed [106]. A strong photoinduced absorption signal due to a triplet-triplet transition with a monomolecular decay kinetic has been found. From the red shift of the peaks with increasing chain length, an effective conjugation length for the polymer can be estimated. No photoinduced absorption signals from charged excitations have been found. This may be due to poor interchain contact that prevents an *interchain* charge separation necessary for generating long-lived charged excitations. The high-energy emission peaks observed in photoluminescence are assigned to a radiative recombination, and the lower energy peaks are due to excitations of vibrational quanta, where excited and ground state geometry are coupled [106]. Photoluminescence and absorption spectra of oligomeric samples in solution are blue-shifted as compared to those in the solid state. This blue-shift is explained through an increase in the mean value of the torsional angle to a maximum value around 15–20° in solution, which is in agreement with theoretical calculations.

An independent approach toward the definition of an effective conjugation length rests on EPR and ENDOR spectroscopic investigations. Structurally defined oligomers of limited size such as **33a–g** offer the additional advantage of providing EPR hyperfine coupling constants for individual positions [100, 107]. Toward a reliable assignment of hyperfine coupling constants specifically deuterated derivatives have to be included. The radical anions of **33a–g** have been prepared in dimethoxyethane by reduction of the neutral compounds with potassium. Not unexpectedly, only the lower homologues give rise to resolved EPR spectra while the higher homologues show S-shaped EPR spectra without any hyperfine resolution so that the coupling constants have to be determined by ENDOR experiments. For systems with an odd number of benzene units the protons with the largest coupling constants occur at the two double bonds directly connected to the central ring; for the analogous systems with an even number of benzene moieties these coupling constants are due to protons at the two double bonds connected to the central stilbene moiety [100]. When focusing on the largest coupling constant one would expect from a MO theoretical point of view that the increasing size of the π-chains leads to an increasing distribution of spin density. Calculating the hyperfine coupling constants by an HMO-McLachlan procedure ($\lambda = 1.2$) and the McConnell equation ($Q = 2.5$ mT) one obtains a very good agreement with the experimental values for the lower homologues. For the larger π-systems, however, the experimental value of a_{max} is stabilized around 0.3 mT whereas the calculated value drops to 0.15 mT [107]. Thus, this very crude approach seems to indicate a localization of spin density in central parts of the molecules.

In dealing with the chain-length dependence of the spin delocalization one can also introduce the second moment of the hyperfine coupling $\langle a \rangle^2$ according to

$$\langle a \rangle^2 = \int v^2 F_{ENDOR}(v) d(v) / \int F_{ENDOR}(v) d(v)$$

where v is the radio frequency and $F_{ENDOR}(v)$ is the "upper half" of the ENDOR spectrum [107]. Plotting the experimentally determined width of the ENDOR spectrum, defined as $\langle a^2 \rangle^{1/2}$, versus $1/n$, where n is the number of benzene units, it is obvious that the ENDOR width decreases with increasing chain length. When extrapolating linearly to $1/n = 0$, however, the final ENDOR width does not equal to 0, as one would have expected for a completely delocalized spin. Instead there is, again, evidence for a finite extension of the spin density in an infinite PPV chain.

In the theoretical description of conjugated polymers such as polyphenylene 7 or polyphenylenevinylene 8 the geometric relaxation associated with the formation of charge on a chain is identified as quinoidal structures 12 and 13 [28, 95] . Indeed, one would have also anticipated from a MO theoretical picture that the π-bond order of formal single bonds is increased when going from the neutral to the mono-charged species. Further experimental evidence in favor of redox-induced structural changes can be obtained from the fact that the activation barrier for rotation about relevant bonds is increased so that, as a borderline case, one might even observe stable conformational isomers within the time scale of the spectroscopic experiment. This expectation is, indeed, born out from electron-transfer experiments for various oligophenylene and oligophenylenevinylene species. Typical examples are the *ortho*-terphenyl (**41**) [108–110], stilbene (**3**) and its higher homologues **33a–g** [111, 112], the cyclic

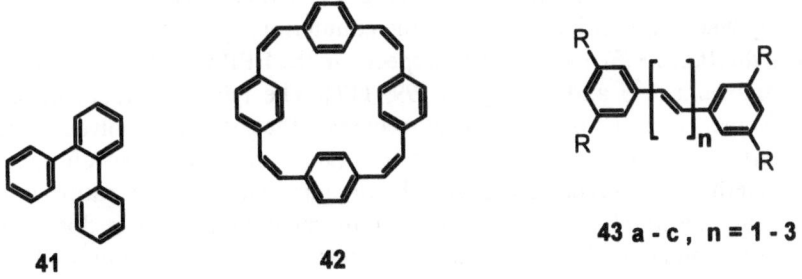

41   **42**

43 a - c , n = 1 - 3

phenylenevinylene **42**, a so-called paracyclophanetetraene [112–115], as well as the 1,n-diphenylpolyenes **3**, **5** and **43** [116–118].

For the radical anion of **41** the rotational barrier, which has been determined by EPR spectroscopy, is higher than in the neutral compound, but by 6 kcal/mol lower than in the corresponding dianion [15, 16, 108]. Along with an increase of the rotational barrier one encounters a decrease of the interring torsional angle. A "frozen" phenyl rotation in the radical anion and dianion of stilbene (**3**) can readily be inferred from EPR- and NMR spectra by symmetry arguments, i.e. by counting the number of hyperfine coupling constants or NMR chemical shifts in the slow exchange domain of the experiments. The cyclic phenylenevinylene species **42** exhibits a particularly interesting redox chemistry, since upon alkali metal reduction it does not only transform into a radical anion, but also into a diatropic dianion and a paratropic tetraanion [113–115]. While in the neutral compound the rotation of the benzene rings about the neighboring single bonds is extremely rapid, the activation barrier is 9 kcal/mol in the corresponding radical anion and 15 kcal/mol in the dianion. The tetraanion, finally, is an extremely rigid species since even at elevated temperature one does not observe a rotation of the benzene rings.

There is evidence from the ENDOR spectra of oligophenylenevinylene radical anions for the existence of non-interconverting stereoisomers such as **33a** and **33a'** [107]. The closely related 1.4-diphenylbutadiene (**5**) can form a s-*cis* **5'** and a s-*trans*-isomer **5** (see Fig. 2), and in the higher 1,n-diphenyloligoenes **43** the number of conformers increases drastically [99, 116–118]. For the neutral hydrocarbons neither the interconversion of these conformers by rotation about the formal single bonds, nor the rotation of the terminal phenyl rings could be detected on the NMR-time scale. Here, again, however, the stereodynamic situation is markedly different in the resulting ionic species. One must be aware in approaching this structural problem that the redox products of π-systems such as radical anions or radical cations as well as the more highly charged derivatives exist as ion pairs so that the excess charge is screened by the counterions. Paramagnetic or diamagnetic carbanions are known from extensive evidence to exist as equilibria of contact ion pairs and solvent separated ion pairs in solution [119]. The latter are generally favored by smaller counterions (with better eigensolvation), lower temperatures, larger organic ions or by strongly cation solvating solvents. Structural features such as the above mentioned charge localization in a particular subunit of an extended π-system or configurational and conformational changes can thus only be described adequately when accounting for role of the counterion. For the radical anion of diphenylbutadiene (**5**) one detects two species in the EPR which are identified as the stereoisomers **5** and **5'** (Fig. 2) [99, 117]. The relative amount of these stereoisomers appears to depend sensitively upon the counterion/solvent system (counterion: Li$^+$, K$^+$; solvent: tetrahydrofuran, 2-methyltetrahydrofuran, dimethoxyethane). Interaction between the organic anion and the cation might contribute to a stabilization of the s-*cis* arrangement if the cation is close to the carbons C-1 and C-4. This stabilization is expected to be most pronounced in an

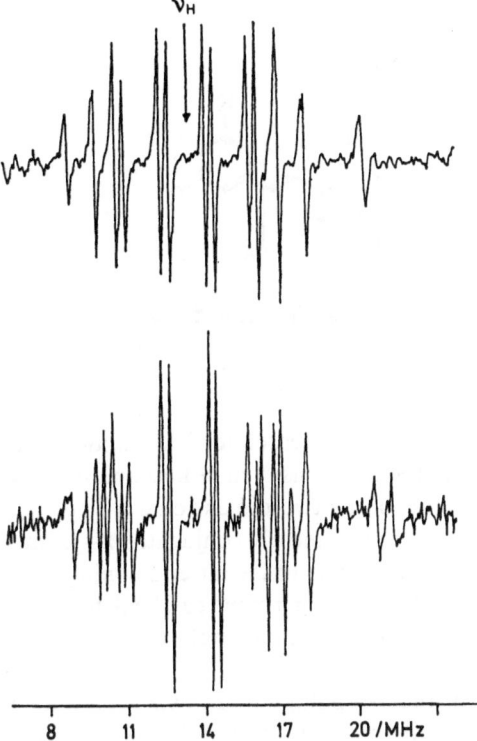

ν_H

8   11   14   17   20 /MHz

Fig 2. ENDOR spectra of the radical anion of Diphenylbutadiene (5) in (A) DME at T = 200 K and (B) in 2-Methyltetrahydrofurane at T = 200 K

ion pair with a tight interaction of the countercation and the π-system, and, indeed, **5′** is only observed under those experimental conditions which favor tight ion pairing.

Not surprisingly, the cation-anion interaction is much more important for the stereochemistry of the doubly charged anions. The number of ^{1}H- and ^{13}C-NMR signals obtained for the dilithium salt of dianion $5^{\prime 2-}$ points out that there is a single stereoisomer which can be identified as the s-*cis* species [99]. Support for this interpretation comes from an analysis of the crystal structure of dianion salt $5^{\prime 2-}/2Li^+$. Different from the dilithium salt the dipotassium salt exists as a mixture of two stereoisomers at low temperatures. This is not surprising, however, since in a contact ion pair the stabilization exerted by the larger K^+-ion is less pronounced than that by the Li^+-ion.

When interpreting the solution and solid-state ENDOR spectra of radical anions of oligophenylenevinylenes it is not perfectly clear to what extent the observed localization of the charge in a particular subunit can be attributed to counterion induced polarization or to structural changes such as a twisted conformation. It is therefore important to include rigid and planar structures with extended π-systems for which the series of *peri*-fused naphthalenes **44** appears as a suitable example [120, 121]. Such two-dimensional ladder-type

23

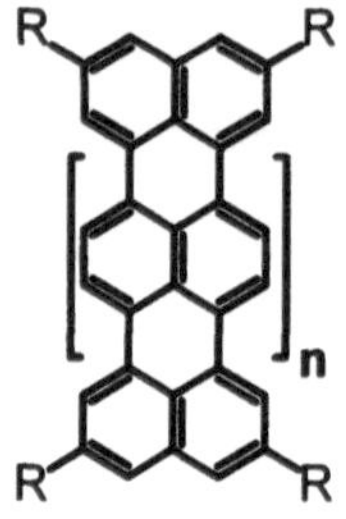

44 a - d , n = 0 - 3

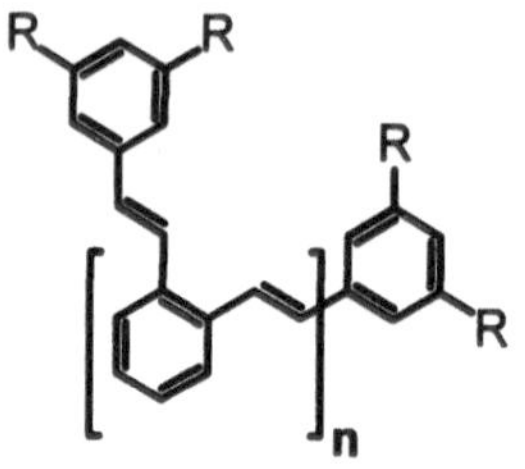

45 a - d , n = 1 - 4

structures have attracted particular interest in the search for low band-gap π-systems [76, 78, 122, 123]. It appears that the spin density in radical anions of **44** tends to be partially localized at both ends of the two-dimensional π-structure simultaneously which is, clearly, in contrast to the linearly conjugated π-systems like oligo-*para*-phenylenes and oligopyrroles. While in the latter a conjugation length comprising only a limited number of repeating π-units can be detected, it is not straightforward to assign an effective conjugation length to the ribbon-type system like **44**.

Another interesting consequence of the π-topology of an extended system can be demonstrated when comparing the electron-transfer behavior of the *para*-substituted phenylene species **33** with that of their corresponding ortho- [124] and *meta*-isomers [44], **45** and **22**, respectively. Surprisingly enough, 1,2-distyrylbenzene (**45a**) is a more powerful electron acceptor than the *para*-isomer [125]. The former transforms into a tetraanion salt upon reduction while the latter only gives rise to a dianion. This finding is unexpected in view of the relevant electrostatic effects and the close neighborhood of the olefinic bonds in **45a**. On the other hand, it reflects, again, the role of counterions in stabilizing the excess charge on a π-system because, similar to 1.4-diphenylbutadiene (**5**), 1.2-distyrylbenzene (**45a**) allows for structures in which charge-carrying centers are bridged by counterions [125, 126].

The 1.3-distyrylbenzene (**21**) and its higher homologues **22b–d** have been transformed into radical anions and the latter investigated by EPR and electron absorption spectroscopy [44]. Thereby, the *meta*-series **22a–d** appears to behave completely different from the *para*-isomers. The following findings are obvious from the study of **21** and **22**. (1) the absorptions at shorter wavelengths closely correspond to those of the stilbene radical anion, and no chain-length dependence can be observed; (2) the oligomeric radical anions exhibit an absorption at very low energies which from various pieces of experimental evidence are identified as charge-transfer bands. One concludes, and this is in full agreement with the results from MO theoretical calculations, that the unpaired electron is confined to a stilbene unit while the remaining part of the molecules remains uncharged. This spontaneous charge localization is not a consequence of steric hindrance, but follows from the role of the *meta*-phenylene unit as a conjugational barrier. The immediate consequence is that many spectroscopic properties of the related poly-*meta*-phenylenevinylene (**22**) should be analogous to those of the lower oligomers and that an eventually formed polaron in **22** is nothing else than a stilbene polaron **46** [44, 127]. In contrast, as can readily be deduced from a comparison of structures **46** and **13**, polaron formation in PPV (**7**) comprises a larger number of repeating units.

While the charge localization in the radical anion **22**⁻˙ is a consequence of the π-topology, one can also observe a clustering of the charge in particular subunits as a result of steric effects. Homologous series of oligo-para-phenylenes **32a–d** have been synthesized which obtain solubility by *tert*-butyl-substitution at the terminal phenyl rings [54, 99, 128]. The characterization of the radical anions by absorption spectroscopy provides a picture which is totally analogous to that obtained for the oligo-para-phenylenevinylenes **33** [101]. In particular, one observes a strong bathochromic shift with increasing chain length and a convergence of the λ_{max}-values towards a threshold value. In another series of oligo-para-phenylenes, **47a–e**, which owes its solubility to the presence of *ortho*-methyl substituents inside the chains, there is evidence from cyclic voltammetry

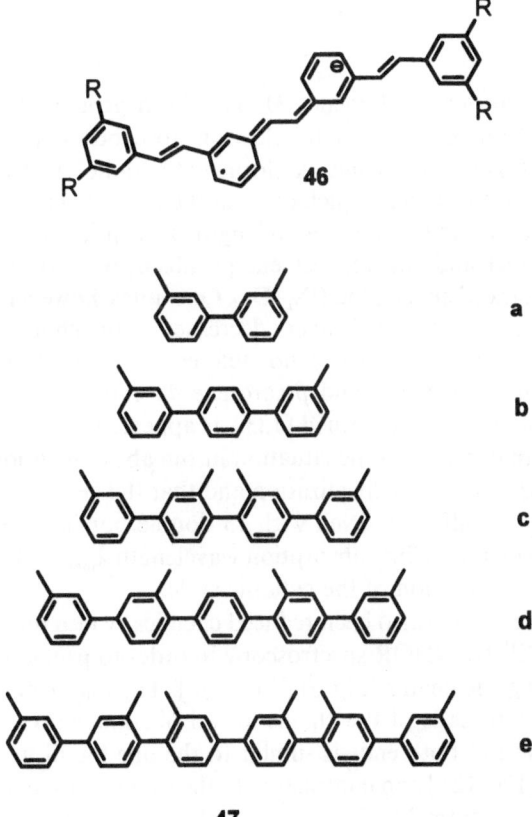

47

(see also Sect. 4) and from electron absorption spectroscopy that the radical anions tend to localize the spins in smaller subunits [54, 129]. Thus **47c** and **47e** behave as biphenyl radical anions, and **47d** as a terphenyl radical anion. The two classes of oligophenylenes demonstrate convincingly that oligomeric models must be carefully tuned to the structures of the polymers; the nature of solubilizing side groups must be chosen carefully since they can lead to drastic changes in the electronic properties.

Polyacetylene **6** is the best known conjugated polymer and yields the highest conductivity of organic materials in the doped state ($\sigma = 10^5$ S/cm) [2, 61, 83]. While oligoenes are an important component of natural products such as carotenoids (see e.g. **48**) [130, 131] and a large number of isoprenoid oligoenes has been studied, oligoenes with unsubstituted backbones are rare. Therefore the description of *tert*-butyl end-capped oligoenes **31**, which are stabilized and solubilized by terminal *tert*-butyl groups, immediately prompts a study of their electronic structure in the pristine and doped states [132, 133].

48

The radical cations of oligoenes **31** have been generated radiolytically in Freon matrices and investigated by electron absorption spectroscopy [133]. Even, when starting from all-*trans*- conformers the radiolytic treatment leads to formation of several rotamers which differ in their λ_{max}-values. Thereby, the all-*trans* conformer has the longest wavelength absorption for the low energy transition (D_0–D_1) and the highest energy absorption for the second even stronger band in the visible region (D_0–D_2). One notes, however, that the spread in λ_{max}-values for the different isomers decreases as the chains become longer. Not surprisingly, therefore, nearly no difference in the electron absorption spectra of the dodecaene **31e** and β-carotene **48** cation radicals, which both possess 11 double bonds, is obtained [133]. It appears that the methyl group in the β-carotene, in contrast to the situation in the above mentioned phenylenes **47**, does not lead to a charge localization and that the geometry of the charged state is only slightly affected. Even with 13 double bonds the absorptions are further red-shifted and no final absorption wavelength λ_{max} is obtained, pointing to a relative large extension of the cationic state.

Oligoenes **31a–c** have also been reduced chemically into their radical anions and studied by EPR/ENDOR spectroscopy in order to probe the electron spin distribution along the chains (Fig. 3) [134–136]. The major finding is the high spin density at both ends of the chains for all oligomers (**31a–c**). This partial charge localization at both ends is similar to the one found in the ladder-type rylenes (**44a–d**) [120, 121] and contrasts with the suggested spin distribution in a long oligoene chain from MNDO/AUHF calculations where the highest spin

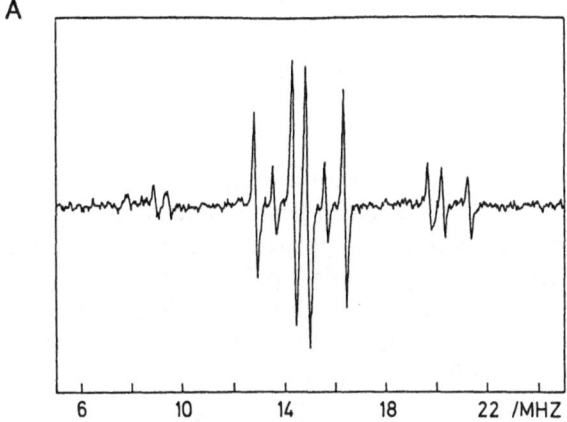

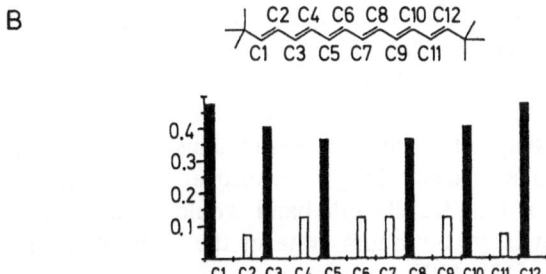

Fig 3. A ENDOR spectrum of the radical anion of dodecylhexaene (**31b**) in THF/K at 220 K. **B** The hyperfine coupling assignment

density is predicted to occur in the center of the molecule [133]. High spin densities in the central part of the molecules have also been suggested for the neutral radicals in *trans*-polyacetylene **6** (solitons) from the aforementioned ENDOR study of the solid material (Sect. 3.1) [92, 93], where a maximum extension of the soliton over 15 double bonds has been deduced. This delocalization length of the neutral radical state is astonishingly close to the data obtained for the cation radicals in the photophysical model studies mentioned above.

A good case can be made when comparing the spin delocalization in paramagnetic radical ions with the charge delocalization in spinless diamagnetic carbanions. The diphenyloligoenes **49a**, with an odd number of carbon centers, are ideally suited to study the charge delocalization of the spinless state after transforming them into the corresponding carbanions via deprotonation [137]. The carbanions **49b** have been characterized by NMR spectroscopy and the ^{13}C-NMR chemical shifts correlated with the corresponding π-charge density. The possible effect of countercations on the charge distribution has been drastically reduced by the use of crown ether. It appears that the charge density

49a   49b

localizes in the center of the polyenyl anion, with a decreasing proportion of charge extending to the end groups, as the chain length increases [137]. The width for the carbanion extension has been extrapolated from the experimental data to include 31 carbon centers in a hypothetical polymer, where it should reflect the negative soliton extension. After all, no substantial difference between the effective conjugation length of a neutral soliton in trans-PAc 6 and a carbanion (or charged soliton) in 49 has been found.

The results from optical absorption spectroscopy of long chain oligomers such as 31 and 33a–g have prompted researchers to challenge the polaron-bipolaron concept [138–140]. As the relatively small oligomers like 33a–g do not possess a continuum state the spectra of the radical anions and radical cations can be interpreted in a straightforward way as Frank-Condon allowed vertical transitions among molecular orbitals. It has been pointed out that such an assignment also holds for the polymer, thus, describing the PPV polymer as an array of oligomers with statistically varying lengths reflecting the local degree of ordering. The fact that the shape of the absorptions remains constant upon increasing chain length argues against a switching of the mechanism by which in the corresponding polymer the long-wavelength band would result from a transition between a localized inter-continuum state of the same electronic origin, but different geometries [138, 139].

It should be noted that this view does not exclude conformational and configurational changes upon radical anion or radical cation formation, which have clearly been established for small organic radical anions and radical cations (see Sect. 1). These are not evidenced, however, by the absorption spectra of the ionic species because the spectra reflect the electronic energy band scheme, after occurrence of the geometric relaxation.

The fact that optical absorptions of the dianions are only detected, after all neutral compound has been consumed, indicates that the disproportionation

$$R^{-\bullet} + R^{-\bullet} \rightleftharpoons R + R^{2-}$$

is an endothermic process providing evidence against bipolaron formation and excludes an extra-stabilization of a diionic state. Related arguments against bipolaron formation in the oligophenylenevinylene model compounds 33a–g are obtained if the species are subjected to electrochemical reduction under cyclic voltammetric control [102]. While the detailed analysis of results from cyclic voltammetry will be described later, the following findings are important here: The potentials of already existing redox states shift to less negative values with increasing chain length (thereby, the redox states degenerate pairwise with increasing chain-length), and the potential difference $\Delta\varepsilon$ of, e.g., the first and second electron transfer approaches a minimum. Remarkably enough, however, no two-electron transfers are observed as would have been predicted for energetically stabilized diionic states.

This approach, in which one refers to $\Delta\varepsilon$ as a measure of the coulombic repulsion between two extra charges, is in agreement with knowledge accumulated earlier for small benzenoid and non-benzenoid π-systems [14, 19, 20, 25, 26, 99]. In a number of benzenoid hydrocarbons radical anion and dianion formation are separated by about 0.5 eV as a consequence of electrostatic repulsion. As one anticipates, this potential difference becomes smaller when the π-system is increased because the charges can "better avoid each other". Even smaller potential differences are to be expected, if a strong shielding of the extra-charge occurs as a result of ion pairing, and if the dianion benefits from conjugational stabilization. Typical examples from organic chemistry are the cyclooctatetraene (11) and the related tetraene species 50 [7, 33, 141–144]. In 11 the first transfer of charge is quasi-reversible; the cyclic voltammogram shows one wave, which does *not* imply a thermodynamically favored second transfer under the chosen experimental conditions, because the second electron transfer is shifted negative by 120–150 mV with respect to the first one [143]. A change of counterion, on the other hand, can induce a thermodynamically stabilized second electron transfer, which is mainly dependent on the size of the counterion [145]. Compound 50 is closely related to 11 [142]. NMR-spectroscopic evidence has led to characterization of the dianion salt $50^{2-}/2Li^{+}$ as a strongly diatropic (aromatic) species, in which the perimeter-type behavior is a result of bishomoconjugation. 50^{2-} is thus a true analogue of the well known cyclooctatetraene dianion [141–146]. Cyclic voltammetric experiments characterize the reductive formation of dianion 50^{2-} as an ECE process (E: electron transfer, C: chemical step) in which the C-step corresponds to a structural change improving the conjugative interaction between the two diene subunits. As a result of this, a second electron transfer occurs at less negative potentials than the first one. It is important to note that a similar finding cannot be made for the oligophenylenevinylene reduction [147].

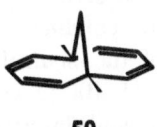

50

Up to now only charges on conjugated *hydrocarbons* have been considered, but a large number of conducting polymers is based on heteroaromatic subunits, mainly with thiophene 24 and pyrrole 23 [47, 148, 149]. These polymers often show lower band gaps in the pristine state and higher conductivities in the doped state than the corresponding benzene (or purely carbon based) analogues.

Toward an understanding of the conduction properties of polythiophenes (26) and polypyrroles (25) a large number of soluble oligomers has been prepared. Oligothiophenes, indeed, represent the most common model compounds for electrically conducting polymers [149]. Thereby, lower oligomers (n < 5) tend to undergo oxidative coupling upon doping [148, 150], due to the high reactivity of the 2-position, while an increasing chain length (n > 6) leads to a loss of solubility. A series of extended 2,5-conjugated thiophene oligomers **51a–d** (n = 3, 4, 5, 6) has been oxidized to the corresponding radial cations upon treatment with $FeCl_3$ in dilute methylenechloride solution [148]. Very sharp and intense visible/NIR bands for the radical cations have been observed, contrasting again with the frequently observed broad and unresolved bands of the doped polymer **26**. The spectrum of the cation radical **51a$^{+\bullet}$** is identical to that of **51d$^{+\bullet}$**, demonstrating the high reactivity of the trimeric cation radical **51a$^{+\bullet}$** which immediately dimerizes with a neutral molecule to the hexamer **51d** cation radical (see Sect. 5). This dimerization process is also observed, although to a smaller extent, for the tetrameric analogue **51c$^{+\bullet}$**. The strong tendency towards intermolecular coupling of the radical cations proves to be extensively useful in the electrochemical polymerization which will be further described in Sect. 5.

51 a - d , n = 1 - 4

The longest wavelength transition in the optical absorptions for the radical cations in the series **51b–d** are shifted bathochromically with increasing chain length (λ_{max} = 1068, 1265, and 1319 nm, for **51b–d**, respectively) [148]. They are assumed to stem from the transition of the valence band (highest doubly occupied MO, see Figs. 1 and 3) to the SOMO of the cation. The diamagnetic dicationic state has only been obtained for the hexamer and sharp absorptions have been detected at 1000 nm while in the polymer they are expected at 1240 nm from extrapolation. The conjugation length in the polymer is thus not reached within the hexamer **51d**. The series of thiophene has been transformed into thin films to measure the bulk conductivity of the semiconducting oligomers [150]. The conductivity of the neutral films strongly increases with increasing chain length ($\sigma = 10^{-11}$–10^{-6} S/cm for **51a–d**, respectively) and reaches a maximum in films of **51d**, while larger oligomers do not lead to further enhancement. For films of the doped hexamer **51d** a solid-state conductivity of σ = 1 S/cm has been measured.

Towards a full characterization of their neutral and charged states, oligomeric systems are often synthesized with substituents capping the reactive chain-end positions and alkyl chains at the 3-positions enhancing the solubility [149]. Using *tert*-butyl end-groups and dodecyl side chains oligomeric thiophenes have been kept soluble with up to 11 coupled thiophene units **52** [149]. For a chain with 11 thiophene units (**52a, b**) the effective conjugation length seems to have been reached, and the conductivity of the doped state increases up to 20 S/cm. The synthesis of oligomers has been further extended to twelve thiophene units

52 a , b   n = 1 , 2   R = tert. butyl
R' = n - alkyl

[151], with the result that in the doping process the mono- and dication formation can no longer be separately detected.

In a similar approach end-capped oligothiophenes (**39a–d**) have been synthesized [152–154], in which the saturated cap is also thought to increase the solubility. Spectroelectrochemical investigations of the oxidation of **39** in methylenechloride show a clear dependence of the optical absorption bands and of the first oxidation potential on chain length and, additionally, an unusual temperature dependence with new absorptions appearing at low temperatures. This temperature dependence can be explained by dimerization of the radical cations to π-radical dimers as judged from the drastically shift of the visible absorptions and the loss of EPR signal intensity. The tendency toward dimerization is enhanced with increasing chain length. The dimerization is not hindered, if alkyl substituents are introduced [152]. The extrapolation to infinite chain length for the energy of the two main absorption bands of the radical cations yields 2254 nm and 886 nm which are in fair agreement with the data obtained for polythiophene.

The series of compounds **39** has also been investigated by FTIR studies [155]. Upon formation of radical states by doping with iodine, broad electronic transitions in the 5500 cm^{-1} region are reported which shift to lower wavenumbers during further oxidation processes. This band is interpreted as an electronic transition from the occupied valence band to the polaron state in the band gap. The electrochemical and chemical oxidation do not differ in their characteristic IR-bands, while an additionally observed band at 1080 cm^{-1} in the electrochemically oxidized sample can be assigned to the perchlorate counterion used in the electrolyte [155]. The observed bands can be correlated with the much broader bands in the corresponding polymer, and a few signals only observed in the oligomers are assigned to the corresponding end groups.

A dialkylated derivative of the end-capped sexythienyl **39d** allows for a comparison of the oxidized with the reduced states. The reduction leads to radical anions and dianions with similar optical absorption features than observed for the dications, but they are shifted slightly bathochromically (0.1 eV) [153]. In contrast to the cation radicals the anions fail to show any tendency toward dimerization (see Sect. 5).

The oligomers **39** have further been functionalized through introduction of redoxactive units such as viologen and ferrocene via an alkyl side chain **53** [156, 157]. The functionalization is thought to be of interest for the physical organic changes of electronic properties, but may also induce the formation of new supramolecular structures of well defined oligomers. The additional redox centers behave as independent electroactive subunits. Upon doping of the thiophene chain a stronger tendency toward charge localization on the π-chain is obtained in the case of viologen functionalization as compared to the non-functionalized oligomers. This can be explained by a counterion complexation between the viologen and the thiophene backbone [156, 157].

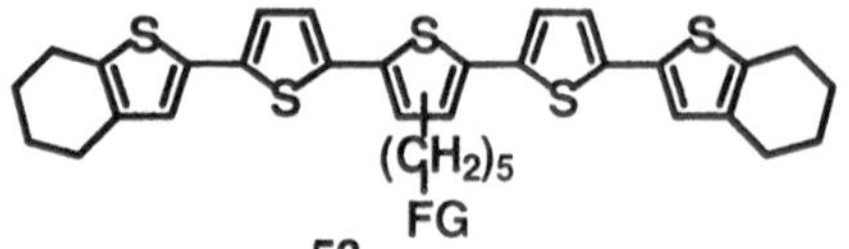

$(CH_2)_5$
FG
53

53 a) FG = viologen

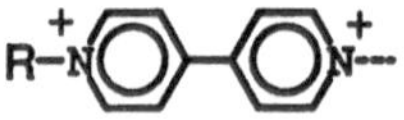

53 b) FG = ferrocene

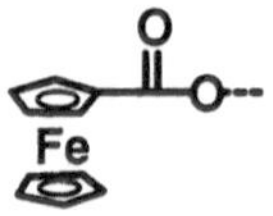

Despite the importance of polypyrrole (**25**) as conducting polymer, only a few papers deal with the synthesis and investigations of unsubstituted and well defined oligo-pyrroles **38** [158, 159]. This might be due to experimental problems of achieving a selective 2,5-coupling of the pyrrole building blocks and due to the considerable sensitivity of the oligomers towards oxygen. A recent synthesis has provided soluble oligo-pyrroles **38a–c** with n = 3,5,7 [158]. The first oxidation potential is shifted to less positive potentials with increasing chain length, while the absorption maxima of the neutral and the charged species are shifted bathochromically. For the oligomers **38b–c** the cyclic voltammograms show reversible first and second oxidation waves (e.g. − 0.28 and − 0.08 V for **38b**), while in the trimer **38a** some oligomerization of the reactive radical cation still occurs. The polymerization of the higher oligomers is only obtained at relatively high positive potentials where further oxidation processes occur (0.8 V). The polymer films produced from the pentamer and heptamer at this high potential are reversibly oxidized at − 0.57 V and show a maximum of absorption at 435 nm. Interestingly, the linear extrapolation to the hypothetical "defect free" polymer from the data for ter-, penta-, and heptapyrrole **38a–c** results in much lower oxidation potentials and absorption maxima (− 0.57 V and 435 nm, respectively) than obtained for the PPy **25** polymerized from the reactive pyrrole **23** under the same conditions (− 0.3 V and 405 nm) [159]. This clearly demonstrates the large number of defects in polymers obtained from pyrrole, while the agreement of the extrapolated data from the oligomers with the data of the polymers electropolymerized from higher oligomers is astonishingly good. The electrochemical analysis of well-defined pyrrole oligomers thus provides the borderline potentials for reversible charging of PPy and the electrochemical synthesis of a virtually defect-free, hypothetical PPy by further oxidation of higher oligomers [159, 160].

3.3 Intra-and Interchain Charge Transport

Due to their relatively large band gap, many electroactive polymers are semi-conductors at best. The electronic structure drastically changes upon doping because new electronic levels in the band gap become accessible. The discussion in the preceding Sect. 3.2. is, therefore, devoted to the influence of additional charges (upon oxidation or reduction) on the bonding properties of well-defined, soluble oligomeric model systems. The effective conjugation length in doped π-systems appears to critically depend upon the:

— structure of the subunits,
— topology,
— counterion/dopant,
— geometrical rearrangement upon charging, and
— attachment of solubilizing side groups.

The conductivity is a solid-state phenomenon, and, as pointed out already, conductivity is not a single chain phenomenon. The band-gap description of a conjugated chain is a one-dimensional model. Additionally, considerably inter-chain charge transport is necessary to describe a metal-like behavior in the highly doped three-dimensional sample and further transfer mechanisms across the polymer chains have to be discussed. What one actually needs to know in explaining conductivity of organic polymers is how the charge transport proceeds

1) along the chain,
2) between neighboring chains,
3) across defects and chain ends.

This complexity has led to a large number of theoretical considerations on the nature of the charge carriers and the mechanisms of transport [71, 83, 86, 90, 161–165].

In oligomeric model compounds which do not possess structural defects, three factors are relevant for the control of charge transport: the effective conjugation length, the interchain spacing and the change of energetic structure by polarization effects in the solid state [99, 166]. In a chemical approach toward structural organization, long-chain alkyl groups have been attached to the terminal 2(5)-positions of oligothiophenes (51) and films on Si substrates created by evaporation [167]. X-ray diffraction spectra point toward well organized domains with a high degree of crystallinity. The characteristic features, thereby, are

1) the segregation between the rigid core of conjugated π-chains and the flexible alkyl chains,
2) the formation of ordered layer structures with layer planes preferentially oriented perpendicular to the substrate.

In such self-assembled structures the conductivity σ as well as the charge carrier mobility μ can be measured both parallel and perpendicular to the substrate. These studies nicely document how the structural organization of the chains increases both the anisotropy of the conductivity and of the charge-carrier mobility.

It has been generally accepted that the description of charge transfer in doped conjugated polymers depends on the amount of doping and the number of chain-defects in the conducting polymers [58, 163a, 163b]. At low doping levels and under an applied field, these singly charged carriers (solitons or polarons) may move or hop along the chain together with their lattice distortion. The activation energy for this process depends on the energy needed to overcome lattice distortion and counterion attraction. This transfer process depends on the energy needed to overcome lattice distortion and counterion attraction. This transfer process along the chain is limited to a chain defect or a chain end. Thus, the overall length of conjugated centers within a chain or the "mean free path" along a chain becomes a polymer characteristic just like the "effective conjugation length" (ecl) of the radical state [96]. The activation energy needed to overcome the "structural inhomogeneity" should differ from the activation for moving along the mean free path length and has to be compared to the activation energy necessary for interchain hopping which is also needed at the chain end. The total chain length lies in the order of some 100 Å and is thus not macroscopically important.

The electron transfer through the saturated chain-defect or chain ends may be compared to ground state electron-transfer processes between organic π-radical ions which are bridged by a saturated spacer [46]. It appears that in the latter, the electron transfer sensitively depends on the distance between the neutral and charged radical π-units, the counterion and the reorganization energy of the system. While, for instance in anthracenes, which are bridged by short alkanediyl spacers such as ethane, propane and butane, the monoradical state is delocalized over both subunits within the timescale of the EPR experiment 10^{-7}–10^{-8} (deduced from hyperfine couplings) [46, 168], longer bridging units such as hexane inhibit the electron transfer, leading to localization of the unpaired electron on one subunit [168].

While standard preparations often lead to clustered polymers with many more interchain barriers to be passed by the charges, stretch alignment has successfully been used to obtain higher conductivities [61b]. This approach corresponds to the above mentioned self-assembly of conjugated oligothiophenes [167]. Upon stretch alignment an anisotropic conductivity is obtained, with higher conductivities along the stretch direction than perpendicular to it. This effect can easily be explained by the smaller activation energy along the "defect free" parts of the chain as compared to the interchain hopping at large distances. Additionally, the interchain hopping can only occur if there is an adjacent free path in the neighboring chain that is not blocked by a defect or another charge, otherwise the hopping process has to overcome coulombic repulsion.

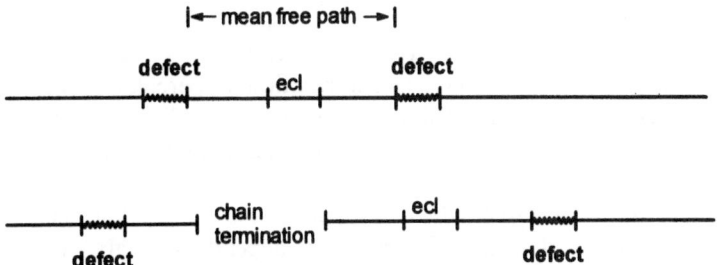

ecl = effective conjugation length

Scheme 4. Simplified picture of aligned polymer chains with undisturbed conjugation (mean free path), their defects, chain termination and segments of effective conjugation length (ecl)

The interchain hopping transport can be described by the probability $Q1(x)$ - where x is the effective conjugation or delocalization length - of finding a comparable mean free path on another chain weighted by the Frank-Condon factor FC. The conductivity σ can then be described as a function of the concentration C of polarons, multiplied by the integral over the interchain hopping probability $P_p(x)$ [96, 97].

$$P_p(x) = FC(x)Q1(x); \quad \text{and} \quad \sigma = C\int P_p(x)dx$$

This model is supported by experiments which show that the conductivity in conjugated polymers can be enhanced drastically by synthetic procedures which minimize the number of chain defects. Thereby, the mean free path as well as the probability of interchain coupling are enhanced [41, 96, 163]. At higher doping concentrations, when coulombic interaction between the polarons becomes important, the formation of bipolarons has been considered [95]. This can be accounted for by introducing the probability Q2 of finding a second polaron on the chain. However, at even higher polaron and bipolaron concentrations, as in the highly conducting polymers with 0.1-0.3 charges per unit [98], this model no longer holds, since the original electronic structure has turned into a new band structure and behaves like a disordered metal [97].

The transition between the low and high conductivity regimes occurs at a critical doping level, which has been obtained for nearly all conducting polymers. Upon reaching this doping level, the conductivity is suddenly enhanced. The transition between the two regimes has been described by percolation theory, which refers to the mixtures of two phases with large differences in conductivity from a statistical point of view [83, 86, 161, 163, 165]. Considering the polymers as amorphous solids with domains of different crystallinity and structures, the doping process should be inhomogeneous and occur at the borderline of morphological substructures. A charge transport can only occur across the edges of the particles. The partially doped polymeric material may then be considered as a mosaic of insulating and conducting domains. Some of the domains become conducting, while others are still insulating. Upon reaching a critical volume concentration of conducting subunits, such that a current can flow from one electrode to the other, a drastic increase in conductivity is observed. Within the percolation theory, a transformation (or percolation) of the averaged conductivity should occur at this critical volume ratio [169–171]. The critical volume concentration depends on the dimension, the degree of dispersion and the more detailed morphology.

This model explains the overall conductivity, but not the microscopic details. The description of inter- and intrachain transport depends strongly on the structure of the polymer, its total conjugation length (mean free path), the number of defects allowing an estimation of the mean free path of conjugated centers, its effective conjugation length and the interchain distances. The overall complexity of these processes makes it difficult to find a unified theoretical description, in which all molecular details can be included [83, 163–165]. It therefore appears that the measured conductivity σ in conjugated π-systems can best be described within a statistical approach leading to a variable-range hopping model [163–165, 169–174]. Within this theory, a fluctuation-induced tunnelling through potential barriers at low temperatures and an activation over the barrier at high temperature is included.

$$\sigma = \sigma_0 \exp - [T_1/(T_0 + T)]$$

The value of T_0 and T_1 are obtained from a computer fit of the temperature dependent conductivity data and expressed as geometrical factors containing the height and thickness of the barriers [83, 165, 175].

This kind of model can even be extended to a more microscopic view of the conduction phenomena, using separate terms for the

1) intrachain conductivity (σ_1) along defect-free chains,
2) charge transport across defects on a chain (σ_2), and
3) interchain transfer of charge carriers (σ_3), together with geometrical factors g for taking details of the sample into account [176]:

$$\sigma^{-1} = (g_1\sigma_1)^{-1} + (g_2\sigma_2 + g_3\sigma_3)^{-1}$$

The latter description is more general, and a larger set of experimental data can be explained. It is even possible to describe the conductivity within low doping concentrations at low temperatures where the conductivity drops nearly to zero and charge localization due to vanishing σ_3 dominates.

Conductivity and X-ray diffraction studies on highly conducting stretch-aligned polyacetylene samples can be fitted well by this expression, and it has been found that [175–178]:

• The anisotropic conductivity with σ_{par} measured parallel to the chains is at least one order of magnitude larger than σ_{per} measured perpendicular to the chain alignment,
• The σ_{par} and σ_{per} values both follow the same temperature behavior, supporting the same mechanism of conduction in both directions.

- the rate determining step seems to be the hopping process from charge carriers via defect sites on one chain or between chains,
- high crystallinity and hence small number of defects are necessary for high conductivity,
- there is an intercalation structure where rows of counterions and stacks of polymer chains alternate (see Sect. 6.1).

Section 3.3 clearly documents the limitations of a "molecular description" of electrical conductivity. Nevertheless, the molecular and supramolecular architecture of defect-free oligomers provides an easier access to a description of a much more complex situation in polymers. The conclusion to be drawn is that the intrinsic conductivity of conjugated polymers will be even higher than reached up to now, if synthetic procedures succeed in avoiding structural defects (see Sect. 9).

4 Charge Storage

4.1 The Cyclic Voltammetric Classification of Conducting and Redox Polymers

Electroactive polymers, apart from serving as electrical conductors, have attracted interest as materials for charge storage in rechargeable batteries. Two aspects are crucial when dealing with the latter property on a molecular basis: the detailed charging mechanism and the minimization of the Coulombic repulsion in highly charged states. In both cases cyclic voltammetry proves as a most revealing tool [4, 14].

Oligomers and polymers containing unsaturated redox-active building blocks may be classified according to the way in which the subunits are linked: the coupling of the building blocks can create an extended π-conjugation or it can give rise to electronically independent moieties [4]. This division closely corresponds to the two well known categories of electroactive polymers, namely *conducting polymers* [147, 179, 180] and *redox polymers* [181, 182]. Typical examples of conducting polymers with extended π-conjugation and the possibility of charge transport along the chain are polyacetylene (6) [71], polyphenylene (7) [183], and polyphenylenevinylene (8) [21, 50], as pure hydrocarbons on the one hand, and polypyrrole (25) [184], polythiophene (26) [185] and polyaniline (54) [186], on the other hand. It will be shown in Sect. 5 that 25, 26, and 54 are prepared by an electrochemical coupling process which leads directly to partially charged species [47, 70]. Conducting polymers exhibit a sharp peak in their cyclic voltammograms which can be assigned to energetically similar redox states. There is experimental and theoretical evidence that the broad current

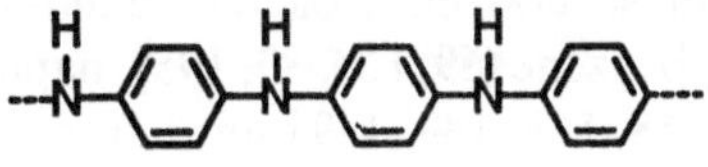

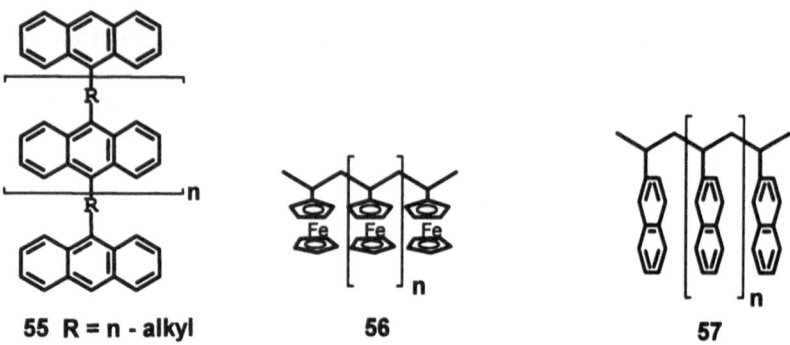

55 R = n - alkyl   **56**   **57**

plateau observed at higher voltages originates from states of increasing energy (see below) [187]. The redox polymers, on the other hand, contain independent monomeric π-units which are attached to a saturated backbone or incorporated into a main chain. Examples of main-chain structures are the poly-9,10-anthrylenealkylenes (**55**) [188, 189] while examples of side-chain polymers are polyvinylferrocene (**56**) [190, 191] or polyvinylnaphthalene (**57**) [192]. The most common application of redox polymers is their use as modified electrodes. These polymers, in which π-conjugation between the separate moieties is interrupted, are usually insulators even in the charged sate. Cyclic voltammetry experiments, in which the material is dissolved or, if not soluble, deposited on an electrode reveal that it is possible to charge the redox polymer at potentials very similar to that of the monomer with at least one electron per unit [190–192]. Additional charges are tranferred at the coresponding monomer potentials as well, and since the electroactive subunits constitute non-interacting redox-centers, the electron transfer is expected to follow the simple rules of statistics. In addition, for non-conjugated linkages with short saturated spacers one anticipates a splitting of cyclic voltammetric waves as a result of electrostatic interactions between the separate redox centers [181, 182].

Although several working models have been proposed, the charging mechanisms of both types of polymers are not yet fully understood. Relevant approaches are the concept of effectively conjugated segments [28, 104, 161, 193] and the bipolaron concept [45, 95, 96], which formulates multiple thermodynamically stable diionic states associated with geometric lattice relaxation of the solid system. Both concepts have been outlined already in the discussion of electrical conductivity (see Sect. 3). When extending these models to a description of multi-step redox sequences, an inclusion of oligomeric reference systems, again, appears helpful. Different homologous series of structurally defined soluble oligomers have been synthesized and subjected to cyclic voltammetric analysis in order to trace the continuous change of the charging mechanism as a function of oligomer structure. These species differ

1) in that they contain several electroactive building blocks, such as [14] annulene (**58**) [194], benzene (**59**) [54, 89, 195], naphthalene (**60**) [66, 196] or anthracene (**61**) [39, 68, 168, 188, 189] units, and

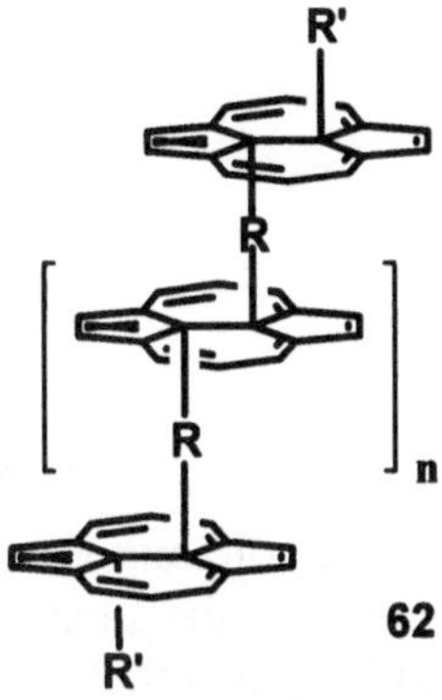

62

62 a : n = 0   R = (CH$_2$)$_3$

R' = CH$_3$CH CH$_2$CH$_3$

b : n = 1   R = (CH$_2$)$_3$

R' = CH$_3$CH CH$_2$CH$_3$

c : n = 1   R = (CH$_2$)$_4$

R' = CH$_3$

d : n = 2   R = (CH$_2$)$_6$

R' = CH$_3$

e : n = 3   R = (CH$_2$)$_6$

R' = CH$_3$

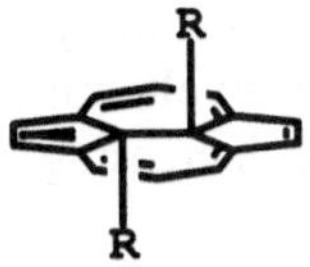

58

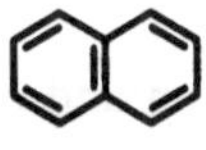

60

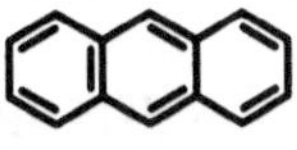

61

2) in that the modes of the linkage vary from saturated spacers to unsaturated spacers with different spacer lengths [66, 68, 168, 195]. Some typical results of relevant cyclic voltammetric studies are discussed below.

In the multi-layered [14] annulenes (62) (n = 0 − 3) the first reduction potential remains unaffected when going from the monomer to the trimer [194, 197]. The dimer 62a shows a potential difference of 110 mV between the first and second electron transfer which corresponds to an interaction energy of 1.8 kcal/mol after correction for an entropy term. The cyclic voltammogram of the trimer can be simulated as a two-electron transfer, followed by a third reduction step [4, 194]. The interaction energy between the first two electrons is less than 0.2 kcal/mol whereas it amounts to 1.8 kcal/mol between the second and third ones. The immediate conclusion is that the doubly charged species has two charges localized on the terminal layers. Upon increasing the spacer length by one methylene group (e.g. in 62c), an electrostatic interaction can no longer be observed both in reduction and oxidation experiments [194, 197]. Consequently, the first reduction and oxidation potentials of multi-layered higher oligomers are nearly the same independent of the number of repeating units in the stack. Related findings can be made for di-(9-anthryl) alkanes (63) and for the corresponding polymer 55 [168, 188, 189, 198]. One can detect an interaction between the mono-charged anthracene units for spacers shorter than butane-diyl; if, on the other hand, a long spacer group such as the undecanediyl is

63 R = n - alkyl

introduced the cyclic voltammograms of the oligomers show only one reduction peak, corresponding to the single charging of each anthracene without interaction. It should be added that the single charging of neighboring layers in **62** produces multiradical anion sites which can be detected EPR spectroscopically by zero-field splittings [168].

The oligophenylenevinylenes **33a–g** as models for PPV belong to the class of redox-systems with extended π-conjugation. Their cyclic voltammetric description reveals three most characteristic properties; with increasing chain length [99, 102, 103]:

1) the absolute values of the first reduction and oxidation potentials decrease;
2) the number of reduction and oxidation states increases; redox states of identical charges shift toward lower energies; and
3) the energy difference between the first and second redox steps approaches a minimum, as is true for all pairs of further redox steps.

The latter finding has already been used in Sect. 3.2 in the search for an extra-stabilization of diionic states. It has been concluded thereby, that the interaction energy of the charges in a dianionic state decreases to a minimum only when the oligomeric chain contains ca. 8 repeating units [99–107]. While this is relevant for describing the charging mechanism, it is also possible to determine the redox-activity of the oligomers and the related polymer. One can deduce that in the corresponding PPV polymer with the degree of polymerization n, n/4 charges can be stored which interact only by spin pairing [50, 199, 200]. As all these states are almost energetically degenerate, only one thermodynamic redox potential can be observed. According to the study of the oligomers the polymer may, in principle, be charged up to almost 100 mol% with the effect that for charge numbers greater than n/4, increasing electrostatic interactions gradually shift all the successive redox states to higher energy. Assuming that a PPV chain with a degree of polymerization n is made up of x weakly interacting effectively conjugated segments, each of which can be charged to a diionic state and comprises k monomeric units $(1 < k \ll n,\ x \cdot k = n)$ then it is clear that in the limiting case of $(n - k)$ charges on the chain any further redox step is energetically separated from the preceding one by at least the amount of the coulombic repulsion energy created by the new charge [47, 200].

It must be emphasized that such a detailed analysis of the charging mechanism and the redox-activity is not possible from a direct study of the polymer itself. Nevertheless, the assumption that upon increasing chain length the redox behavior of the oligomers approaches that of the related PPV polymer must be judged with care when considering the prevailing experimental conditions [201]: the cyclic voltammetric studies of the oligomers are performed in solution while PPV is an insoluble solid. It is therefore crucial to monitor the charging process of the oligomers and polymers in an immobilized phase. In the ideal case of a thin layer, reversible cyclic voltammograms of electro-active polymer films are supposed to exhibit symmetrical and mirror-image cathodic and anodic waves which situation certainly differs from the well-known waves observed in solution cyclic voltammetry [14, 47]. Another difference with respect to the situation of the solid redox systems is that current and scan-rate are proportional to each other. Complications are expected for increased layer thicknesses because the charging process will additionally depend not only on electron exchange between neighboring redox sites, but also on the diffusion of counterions which ensure electroneutrality [202–204]. For solid-state cyclic voltammetric measurements, the oligomers 33a–g have been cast on an electrode whereby, of course, the solvent and temperature applied in the cyclic voltammetric experiment must be chosen in such a way as to avoid further dissolution processes; at this time, the decreasing solubility of the higher oligomers becomes advantageous [187, 205]. In perfect analogy to solution experiments the solid-state cyclic voltammograms follow the characteristics 1), 2), and 3) which have been mentioned above for the charging of extended π-systems. The solid-state voltammograms, however, show significant hysteresis effects, which are most obvious for the first charging step. Upon going to higher oligomers the first cathodic peak potential is moved to more negative values, while the related anodic peak moves into the opposite direction. One readily concludes that in the solid state the charging process needs more energy and the discharging process needs less energy than in solution. The same findings have been made upon solid-state cyclic voltammetric studies for oligo-*para*-phenylenes [201, 205] and oligothiophenes [206–208]. It must also be noted that this behavior is seen even at very low scan rates and cannot be explained by slow heterogeneous electron transfer.

When rationalizing the observed hysteresis one can tentatively refer to three factors [201, 209, 210]: 1) the chain exists in a non-planar conformation and undergoes an energy consuming planarization upon charging; this process could be more demanding in the solid than in solution. The reverse arguments hold for the discharging which is accompanied by a relaxation into the twisted equilibrium structure. Such an interpretation is weakened by the fact that the same hysteresis effect is also obtained in cyclic voltammetric experiments for the rigid oligorylenes 44 which cannot undergo a charge-induced conformational change [211–213]; 2) the counterions, which diffuse into the film upon charging, can lead to a thermodynamic stabilization with new interchain interactions. The diffusion process also depends on the size of the counterions (3) the charging is accompanied by a structural change leading to increasing interactions between *different* chains. Up to now, the relative importance of these factors is not clear and must await further studies [213].

The reduction of the highest oligomer in the solid state gives rise to closely neighbored, but still separate waves, which are superimposed on a current contribution, which is nearly constant over the whole potential range [201, 213]. This finding is reminiscent of the current plateau which has frequently been observed in voltammograms of conducting polymers (see above) [190–192, 214, 215]. For all oligomers, however, the charging and discharging leads to a superposition of redox states over a broad potential range. The current plateau can therefore be identified as due to Faraday-type redox processes [205, 210, 216].

It has been mentioned already that in the series of oligophenylenes the substitution pattern leads to a dramatic change in the redox processes. For biphenyl (2) and the terphenyl 41 the radical monoanions and the dianion are formed in distinct reduction steps [54]. Not surprisingly, therefore, the interaction energy for the first and second electron transfer in 41 corresponds to that of the *tert*-butyl-substituted terphenyl 32b. In contrast, higher homologues 32c–e are charged with two electrons at nearly the same potentials [54]. This finding can be rationalized by the fact that the first electron induces a planarization of the resulting radical anion. Consequently, the transfer of this electron occurs at a more negative potential, and the second electron can now easily enter the flattened system at the same potential. Such a clustering of extended π-systems into biphenyl subunits is also born out by UV/Vis measurements [129].

It is interesting in this context to compare the redox behavior of oligo- and poly*phenylenes* with those of other *arylene* species in which benzene building blocks are replaced by larger acene species. The oligo- 1,4-naphthylenes 64 and oligo-1,5-naphthylenes 65 are closely related to the oligophenylenes except for the fact that the naphthalene *peri*-protons create additional steric hindrance of the π-conjugation [4, 66, 120, 121]. Several investigations of the geometry of the parent 1,1'-binaphthyl system 64a have been described in the literature [217, 218]. It has been concluded that the potential curve for rotation about the interring bond is rather shallow with two minima at 50° and 130°. It is very surprising now that the first oxidation potential of the oligonaphthylenes 64a–e at + 1.31 V is *independent* of chain length [4, 120, 211]. From the pentamer onwards the single steps corresponding to mono- and dication formation are no longer resolved, and even the potential difference between the tri- and tetra-cation formation slowly approaches a minimum. Clearly, the invariance of the

64 a - d , n = 0 - 3   65 a - d , n = 0 - 3

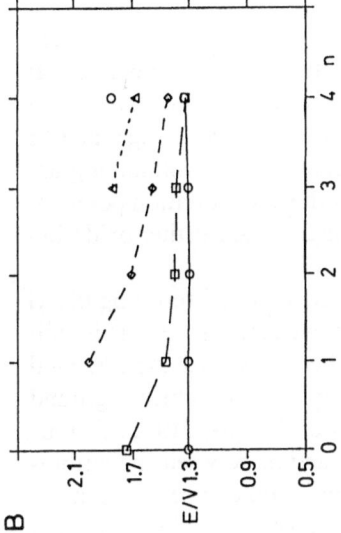

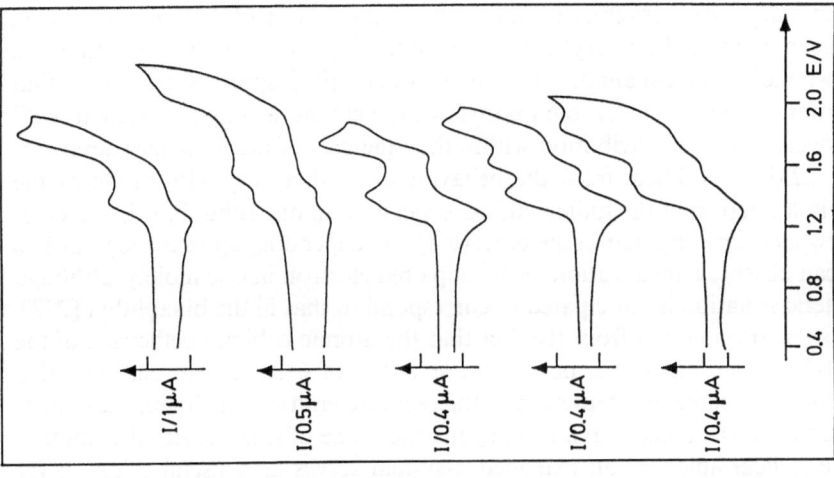

Fig. 4. A Cyclic voltammograms of 1,4-oligonaphthylenes (**64a–e**) oxidation in CH_2Cl_2 with tetrabutylammoniumhexafluorophosphate as electrolyte at $T = 0°C$. **B** Oxidation potentials versus number of repeating units

first redox step with the chain length is unexpected for extended species with conjugative interaction of the subunits. It is therefore relevant that the oxidation potential of the monomer is about 200 mV higher than that of the dimer so that some electronic influence of the second naphthalene unit is detectable. In a simple view, the oligonaphthylenes can therefore be described as being made up of weakly interacting binaphthyl units [4, 120]. From the cyclic voltammogram of the related polymer **27** [66] one can conclude that the first oxidation potential must lie in the same potential range as that of the oligomers. Further oxidation steps can no longer be resolved.

When replacing the naphthalene by an anthracene unit, the resulting biaryl and polyarylene species are predicted to possess an even stronger steric inhibition of the conjugative aryl-aryl interaction [219]. There is, indeed, experimental and theoretical evidence that 9,9'-bianthryl (**19**) adopts an equilibrium ground state structure with an interplanar angle not far from 90° [39, 219–222]. One concludes that there is only a small electronic interaction between the anthracene units, and this is supported by the small potential difference between the first and second electron transfer and by the invariance of the first reduction potential when going from the dimer to higher oligomers [39, 221].

An independent criterion for the degree of interaction of the arylene building blocks in oligo- and polyarylenes is the mode of spin density distribution in the corresponding radical anions. It appears from an EPR-spectroscopic study that the radical anion of 1,1'-binaphthyl (**64a**) acts as a single π-system with homogeneous spin distribution within the timescale of the EPR measurements [121, 223]. This differs from the behavior of 9,9'-bianthryl (**19**) in which the unpaired electron of the radical anion is localized in one subunit. It is therefore somewhat surprising that in the biperylenyl **66**, depending upon the ion pairing, one can observe a localization of the unpaired electron in one moiety, although the steric situation is anticipated to correspond to that in the binaphthyl [223]. An explanation comes from the fact that the atomic orbital coefficients of the SOMO for the bridgehead positions are larger for **61a** than for **66**; this value controls the degree of interaction of the separate entities which are, thus, more decoupled in the biperylenyl. Here again, the mode of spin density distribution in the radical anion of an extended π-system serves as a useful criterion for

66

probing the conjugative interaction between the repeating units and is consistent with the view originating from cyclic voltammetry.

The oligo-9,10-anthrylenevinylenes **37** [224] are, in a formal sense, related to the phenylenevinylenes **33a–g**. It is therefore surprising that according to cyclic voltammetry these *conjugated* hydrocarbon chains behave like independent redox systems as deduced from their charging with one electron per repeating unit at very close potentials. When all units are already carrying one charge, further step-by-step single electron transfers occur, leading to doubly charged repeating units. It is straightforward to ascribe the small conjugative interaction along the chain to the significant steric hindrance [224, 225]. Furthermore, there is evidence from related species such as e.g., **4** that electron transfer can induce a torsion about the formal double bond [23, 24].

It appears from the cyclic voltammetric evidence accumulated so far, that the charging mechanisms and the charge–storage activity of oligomers and polymers depend sensitively upon structure [4]. Furthermore, it is possible by recording the cyclic voltammetric behavior as a function of chain length to extrapolate reliably toward the corresponding polymers. The classification into two groups of polymers, namely conjugated polymers and redox polymers, is convincingly supported by the cyclic voltammetric results on oligomers. Borderline cases, obviously, arise when the conjugative interaction is accompanied by strong steric hindrance, such as in the methyl-substituted oligophenylenes **47** [54] oligonaphthylenes **64, 65** [4, 66, 120], and oligoanthrylenevinylenes **37** [68, 224], or when strong coulombic interactions occur as in the case of short saturated spacers between the electrophores.

4.2 Stabilizing Extreme Redox States

When dealing with the charge–storage activity of large π-systems, a good case can be made in comparing the linear oligonaphthylenes **64** and **65** with their isoelectronic two-dimensional analogues **44 a–d** [4, 120, 226]. Not surprisingly, the cyclic voltammograms of the latter fulfill all three criteria discussed above for conjugated π-systems. In addition, the following findings are noteworthy:

1) the values of the oxidation and redox potentials are much lower than those for linear π-systems;
2) the number of transferred electrons increases strongly; thus the quaterrylene **44c** exhibits a sequence of 8 separate redox states ranging from the trianion to the tetracation;
3) plotting the energy differences between the first oxidation and reduction steps as a function of the number of naphthalene repeating units, one does not observe a true convergence, although – different from **64** and **65** – a significant decrease is observed. By extrapolation one obtains a band gap ΔE of less than 1.0 eV for the polymer which is very low in comparison with all other model systems discussed above [4, 210–212].

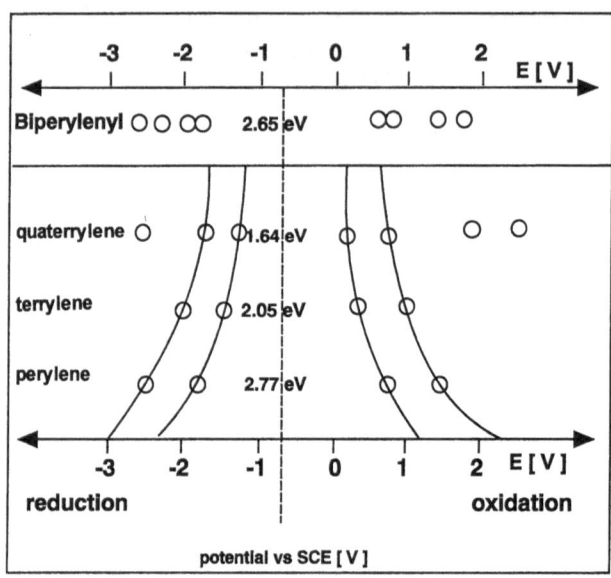

Scheme 5. Redox-potentials of rylenes **44a–c** and biperylene **66**

The 3,3′-biperylenyl system **66** can accept four excess charges both in reduction and oxidation, whereby the first electron transfer to or from **66** occurs at potentials being nearly identical with those of the monomeric perylene (see Scheme 5) [226]. There is thus no detectable facilitation of the charging step by the presence of the second unit as has been deduced for the oligo-1.4-naphthylenes **64** with the same steric situation. The second charging process of **66**, now, occurs at slightly higher potentials with the difference of 80 mV for reduction and 120 mV for oxidation. Furthermore, the second charging steps of both moieties exhibit only a small potential difference when compared to the second step in perylene itself. The tetracation is reversibly formed within a relatively small potential range. One is led to conclude that the biperylenyl **66** can store 8 charges in a potential range of 4.42 V in contrast to the isoelectronic monomeric quaterrylene **44c** which accepts 7 charges in a potential window of 5.2 V [226]. Consequently, the biperylenyl constitutes an attractive model species for materials with a high charge-storage activity and nicely reflects the role of the π-topology in minimizing the coulombic repulsion in high redox states.

In dealing with the charge-storage activity of standard organic π-systems, oligomers and polymers made up from benzenoid or polybenzenoid repeating units have so far been serving as active redox-materials. Thereby, the number of redox states can be increased by either increasing the size of the repeating unit (benzene, anthracene, perylene) or by increasing the chain length. On a molecular basis there are two other factors relevant for the redox activity of an organic hydrocarbon [227–229]: the screening of the charge by the counterions within the ion pairs and conjugative stabilization. The π-topology of the monomeric building block thereby plays an important role, and it is particularly rewarding to proceed from benzenoid to non-benzenoid species with cyclic π-conjugation. Thus while pyrene (**67**) forms a dianion upon reduction, the non-benzenoid analogue acepleiadylene (**68**) transforms into a tetraanion [8, 228, 230]. Likewise, the diatropic [18] annulene **69** does not only give rise to a dianion, but also to a tetraanion [227, 228]. Thereby, the compensation of the excess charge is, in part, due to the conjugational stabilization in the "aromatic" tetraanion which is again a $(4n + 2)$-π-system. Cyclooctatetraene (**11**) forms a stable dianion already at low potentials [27, 33], and an even more remarkable case is that of octalene (**70**) for which a stable tetraanion salt is known [231]. In the latter, the π-charge per carbon is close to 1/3. Although such an approach is in most cases severely limited for synthetic reasons it should be challenging to incorporate these active π-systems as building blocks into oligomers and polymers so that the redox-activity can be optimized by both the π-topology of the building block and the size of the macromolecule.

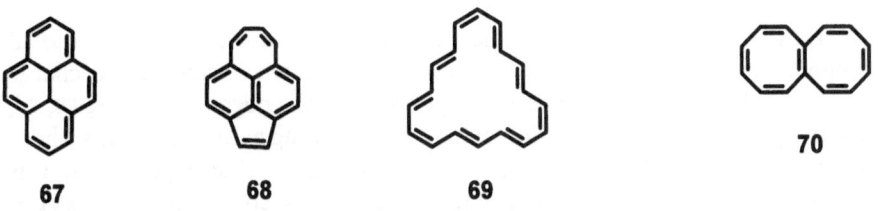

67 **68** **69** **70**

Relevant examples are the extended cyclooctatetraene species **71** and **72** which constitute non-benzenoid analogues of oligo- and polyphenylenevinylenes [232]. While in the oligophenylenevinylenes **33a–g** the difference between the first and second reduction, $\Delta\varepsilon$, approaches a minimum, the first wave in **71** and **72** is due to a chemically reversible, but electrochemically irreversible two-electron transfer. This situation closely corresponds to the situation for cyclooctatetraene itself, in the latter the small $\Delta\varepsilon$-value is ascribed to the stabilization of the dianion, and the electrochemical irreversibility is due to the flattening of the ring upon charging. The redox behavior of **71** and **72** is, thus, governed by the role of COT (**11**) as electrophore – inspite of the extended π-conjugation [33, 232–235]. Not surprisingly, therefore, the π-chains in **71** and **72** can be charged in such a way that each COT subunit transforms into a dianion without a significant charge density on the olefinic bridging groups. Like in COT the charging of each 8-membered ring is accompanied by a significant conformational change. Such a process has to overcompensate significant ring strain, and this energy contributes to the activation barrier of charge-shift processes along the chain. Thus, while for intermediate redox states any excess charge tends to be localized on the 8-membered ring, this charge can be shifted in a thermally activated process to the neighboring non-charged cyclooctatetraene.

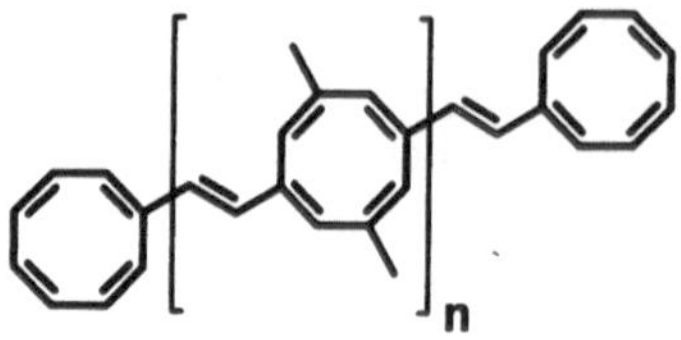

71

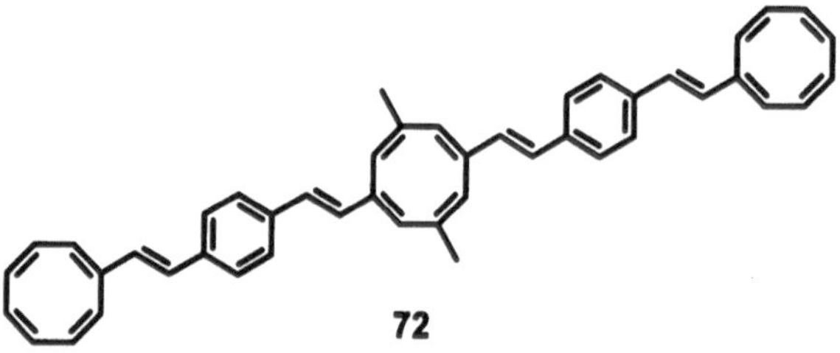

72

4.3 Toward Battery Elements

The above cyclic voltammetric studies provide the molecular basis for a rational design of organic battery cells. A major question when considering charge storage in electroactive oligomers and polymers is the degree of doping [4, 47, 98]. It is given by the mole fraction of the corresponding monomeric units whose charge is compensated by incorporated counterions. One expects that this information is obtained by elemental analysis or by coulommetric measurements. Care must be taken, however, in directly comparing the results of both approaches since and elemental analysis is affected by the presence of solvents or chemically different counterions, while coulommetry also includes the capacity of charging. Problems resulting from chemical reactions between the oxidized or reduced polymer and the solvent or the counterion will be considered in Sect. 5. The experimentally observed doping levels indicate that, e.g., the oxidation of polypyrrole (25) and polythiophene (26) charges every third or fourth ring [236–239]. The behavior of poly-*para*-phenylene (7) sensitively depends upon the nature of the solvent. When using liquid SO_2, a solvent having a low nucleophilic activity, the polymer can be reversibly oxidized to a doping level of 0.24 which corresponds to one excess charge on every fourth ring [102, 240, 241].

In the design of polymer electrodes for battery elements one has to consider the following factors [226]: 1) the number of charges transferred per molecule, 2) the reversibility of the charging and discharging steps, 3) the stability of the redox states over a number of cycles and 4) the conductivity of the polymer in the neutral and doped states. The measurement of charging/discharging cycles for the oxidation of a conjugated polymer such as 27 can be performed in cells with propylenecarbonate/LiBF$_4$ as electrolyte system and lithium as cathode [212]. The polymer is usually mixed with small amounts of graphite (for increasing the conductivity) and Teflon and then compressed to a pellet. When analyzing the charging/discharging cycle according to Faraday's law the measured capacity is

$$K_{meas} = I \cdot t = x \cdot F \cdot m / M_{monomer}$$

with

$\quad$ I = charging/discharging current
$\quad$ t = charging/discharging time
$\quad$ x = charges per repeating unit
$\quad$ F = Faraday constant (26.8 Ah/mol)
$\quad$ m = weighted mass of polymer
$\quad$ M = molar mass of monomer unit.

Taking into account both the anodic and cathodic reaction, the theoretical capacity of a polymer is calculated according to [2]

$$K_{theor} = x \cdot F / (M_{monomer} + x \cdot M_{LiBF_4})$$

For the oxidative charging of polymer 27 one calculates a theoretical capacity (x = 1 e/monomer, $M_{monomer}$ = 0.246 g/mol) of 78 A·h/kg; the experimentally determined value is 72 A·h/kg corresponding to 92% within the first cycle [212].

$\quad$ The design of such battery cells is confronted with many problems such as the limited capacity in the first cycle (low currents must be used to obtain conducting material) or the instability over many cycles as a result of, e.g., swelling and dissolution processes.

$\quad$ Another feature of the above discussion on charge–storage capacity of organic π-systems is that it has been restricted to *hydrocarbons*. On the other hand, polypyrrole (26) is often used as active material in battery electrodes

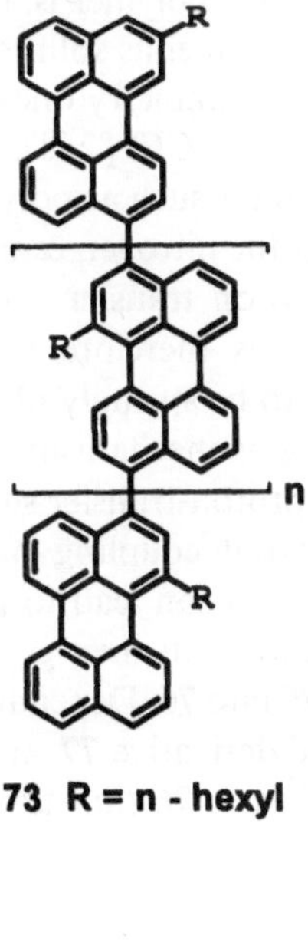

73 R = n - hexyl

whereby the crucial chemical process is its oxidation to a polycation [72b]. This requires a lithium counter electrode which in some cases can pose transport problems due to the high capacity. While polypyrrole is not susceptible to reduction, the above hydrocarbons such as biperylenyl as well as the related polymer **73** [243], which has also been prepared, undergo successive reduction *and* oxidation steps so that it is, in principle, possible to construct a cell in which both the anode and cathode are made up of organic material.

5 Reactions of Radical Ions

5.1 Dimerization and Polymerization

It appears from the description of radical ions in Sects. 1 and 3 that redox reactions can significantly change the chemical and physical properties of conjugated π-systems. Whether the extended π-species are treated within molecular orbital theory or within band-structure theory, the inherent assumption in these concepts is that an electron transfer is reversible and does not promote subsequent chemical reactions. While inspection of cyclic voltammetric waves and the spectroscopic characterization of the redox species provide reliable criteria for the reversibility of an electron transfer and the maintenance of an intact σ-frame, it is generally accepted that electron transfer, depending on the nature of the substrate and on the experimental conditions, can also initiate chemical reactions under formation or cleavage of σ-bonds [244, 245].

A simple chemical reaction following the charging process is the transfer of a proton. This can nicely be demonstrated for the electrochemical oxidation of polyaniline (**54**) [186, 246, 247]. This polymer is, in most cases, prepared by the anodic oxidation of aniline (**74**) in aqueous solution. When the oxidation of the material is monitored by cyclic voltammetry one detects at least two reversible waves between − 0.2 and + 1.2 V (SCE) [248]. The difference observed with respect to other conjugated polymers such as polypyrrole (**25**) or polythiophene (**26**) is mainly due to the fact that the nitrogen centers accept the greatest part of the positive charge and that proton transfer processes can occur during the charging process [247, 249]. It is therefore not surprising that the redox behavior of polyaniline appears to be strongly pH dependent. Different models have been proposed for the charging mechanism of polyaniline (**54**) which differ in the sequence of electron- and proton-transfer steps [248, 250–252]. Assuming a chain structure with head-to-tail coupling of the aniline repeating units according to structure **54**, oxidation can lead to poly-radical cation structures such as **75** [250]; further oxidation will then go along with a loss of protons according to transformation of **75** into **76**. Depending on the pH, the amine form **54** can also exist as a protonated derivative **77**, and protonation can, of course, also occur for the fully oxidized form **76** thus giving rise to **78** [248].

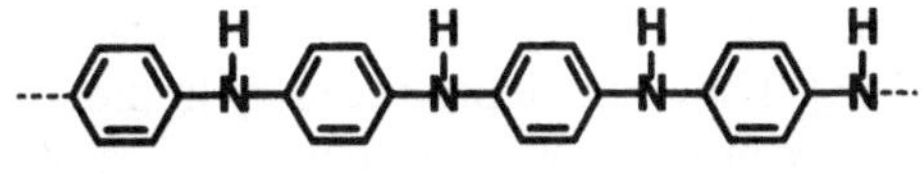

54

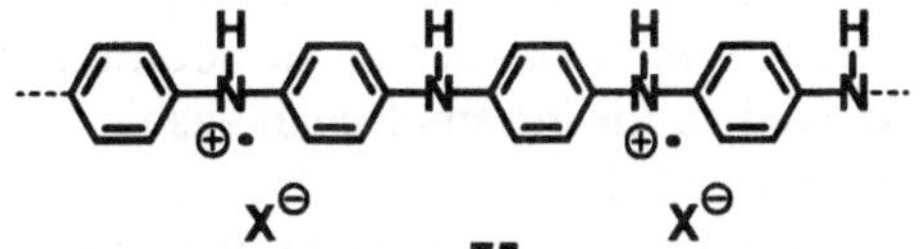

75

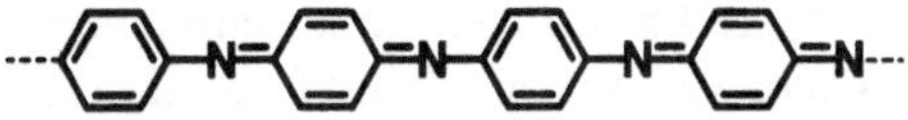

76

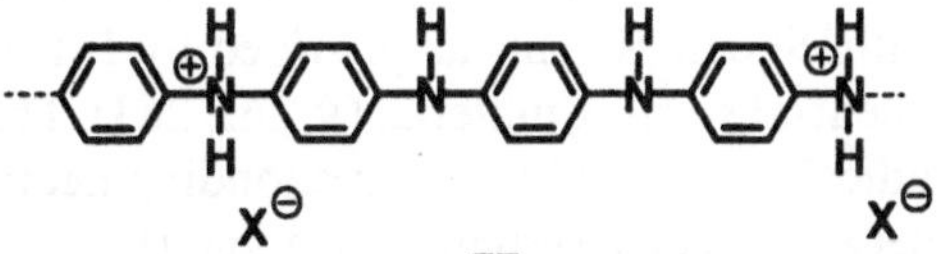

77

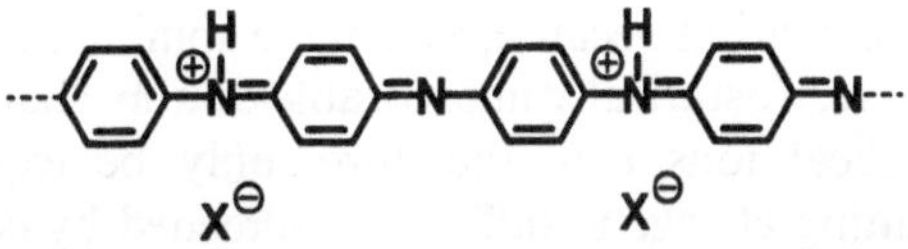

78

The problem of chemical reactions overlapping with electron-transfer steps is particularly severe in the doping of conjugated polymers which is achieved by treatment with oxidizing reagents such as halogens, antimony-III-halides or antimony-V-halides as well as reducing agents such as alkali metals [30]. A related doping process can also be achieved with electrochemical reactions [47]. What is not always fully recognized, e.g. in measurements of electrical conductivity of the doped species, is the occurrence of irreversible chemical modifications: Typical examples are (1) the coupling of oligophenylenes [253, 254], in particular, via the para-positions of the terminal phenyl units, (2) the cross-linking of polyene chains [83], or (3) the reactions of cationic species with nucleophilic counterions under formation of single bonds and interruption of the π-conjugations [221, 255].

While in many cases reactions of radical anions and radical cations are unwanted side effects, there are many examples in which electron-transfer induced reactions of unsaturated systems are of chemical and physical use. A brief consideration of the reactivity of neutral radicals is appropriate as an introductory case [256]. They can be produced by homolytic bond cleavage, by oxidation of anionic precursors or by reduction of cationic precursors [256, 257]. Neutral radicals possess a very high chemical reactivity, and it is not always possible to control the different reactions such as addition to double bonds, hydrogen abstraction, dimerization or disproportionation. For chemical applications, radical transformations such as the addition to double bonds are, therefore, often conducted in an intramolecular fashion in order to suppress intermolecular side reactions [256].

The stabilization of a radical ion by a chemical reaction is controlled by the presence of a spin *and* of a charge. To begin with, the interaction of two radical ions must not necessarily produce a π-bond: two radical cations, e.g. of 9.10-dimethylanthracene (**79**) are known to form a diamagnetic dimer in which the singlet ground state is achieved not by σ-bond formation [258, 259, 263], but by a face-to-face arrangement under formation of a so-called π-dimer. This dimerization is often reversible upon temperature variation. The resulting structure can consequently be compared with that of an anthracenophane such as **80** [260, 261]. A charged radical cation of a conjugated π-system will generally tend to interact with electron-rich components as provided by 1) the counteranions, 2) the solvent or 3) the neutral starting material [9, 262, 263]. The formation of a π-complex between radical cation and the corresponding neutral π-system (i.e., a 2:1 stoichiometry between redox system and dopant) is a crucial step in the formation of solid radical-cation salts which will be described in Sect. 6 [264]. In contrast to neutral radicals, radical ions are much less susceptible to dimerization under σ-bond formation because, even if the spins are lost upon formation of a new σ-bond, an electrostatically unfavorable doubly charged ion originates. Dimerization of radical ions can, therefore, only be expected to proceed smoothly if the resulting charge is sufficiently screened by conjugation or ion-pairing effects [9, 10, 47, 227–229]. The classical example of a reductive dimerization, i.e. of a coupling of radical anions, is that of ketones **81** whose ketyl derivatives **82** transform into the dimer **83** [265, 266]. It is interesting now to recognize cyclic hydrocarbons such as pentafulvenes (**84**) [267] and azulene (**85**) [268, 269] as keto analogues since the dimerization of their corresponding radical anions leads to very stable dimeric dianions, **86** and **87**, in which the charge is localized in a cyclopentadienide moiety [267–269]. A related example of technical importance is the hydrodimerization of acrylonitrile (**88**) in which the original radical anion transforms into the dimeric dianion **89** followed by

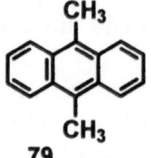

79

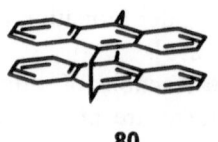

80

$$\underset{R}{\overset{R}{>}}=O^{-}$$

81

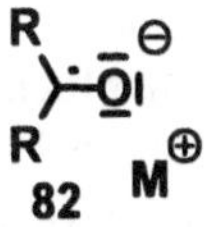

82 **83** 2 M⊕

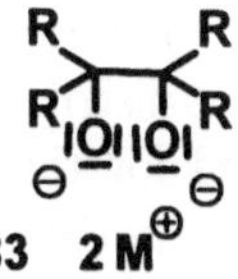

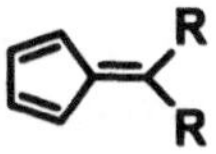

84 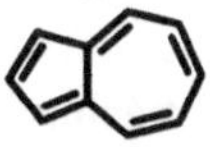 **85** 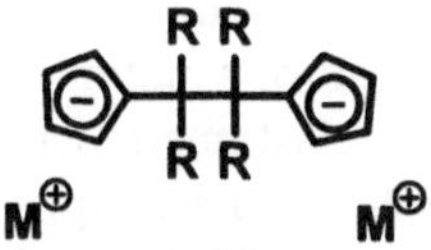 **86**

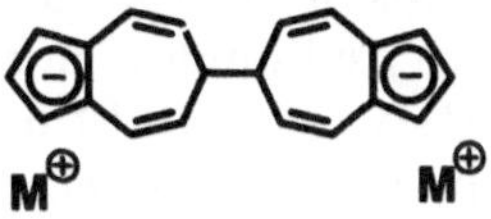

87 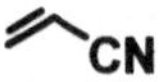 **88**

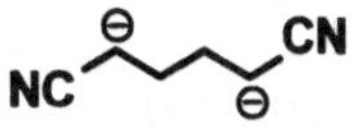

89 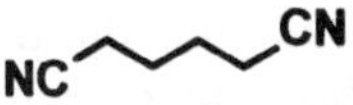 **90**

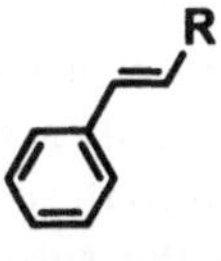

91 a : R = H
b : R = alkyl

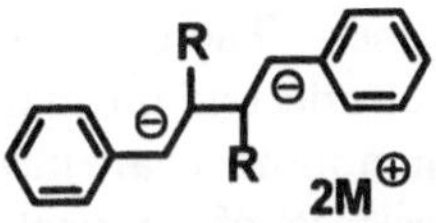

92 a : R = H
b : R = alkyl

protonation to adiponitrile **90** [270]. It should be a added, however, that the basic coupling step can also proceed as a Michael-type nucleophilic addition of a radical anion upon a neutral molecule [271].

If the (green) radical anion of naphthalene (**60**) as sodium salt is reacted with styrene (**91a**) in tetrahydrofuran solution the resulting styryl radical anions dimerize under the formation of the (red) dianion **92a** comprising two benzyl ions [272, 273]. The latter can then initiate the polymerization of styrene whereby the chain grows into two directions. If traces of a proton source can be avoided there occurs no termination reaction for the stable carbanionic chain ends [272, 274]. When all the monomer has been consumed, addition of the same or of another monomer can, again, start the chain growth. The growing polymers are therefore termed "living" [274, 275]. It is also important that the degree of polymerization for quantitative conversion can be determined by the molar ratio of monomer and initiator and that polymers with a narrow molecular-weight distribution are available. The above anionic polymerization can also be used for the synthesis of block copolymers for which the ABA-structure type is particularly important [276]. Commercial "SBC rubbers" contain a polybutadiene with 1,4-connection as central block and polystyrene as end blocks [277]. The combination of a glassy or crystalline polymer (e.g. polystyrene) and of a polymer with rubber elasticity (polybutadiene) provides thermoplastic elastomers with unique mechanical properties. The relative amount of 1,2- and 1,4-coupling of butadiene (1) in the anionic polymerization depends sensitively on the mode of ion pairing [278]. Hydrocarbon solvents, e.g. cyclohexane, favor the 1,4-isomer, but fail to dissolve the salt-like dianionic initiator **91a** [277]. The latter becomes soluble in hydrocarbons, however, if carrying long-chain alkyl groups. The necessary substituted initiator **92b** is then obtained by reductive dimerization of β-alkyl styrenes (**91b**) [276, 277, 279, 280].

If two identical radical cations or a radical cation and the neutral starting compound react under σ-bond formation, the cyclic π-conjugation is interrupted. Nevertheless, such a process is likely to occur when the cation is effectively stabilized, e.g. by heteroatoms such as nitrogen or sulfur, and if the neutral substrate is electron rich as is the case e.g. for the readily oxidizable heterocycles pyrrole (**23**) and thiophene (**24**) [2, 47, 62, 70]. Such a coupling process provides the basis for a commonly used method of polymer synthesis, i.e. electropolymerization, while the classical examples are the oxidative coupling of thiophene (**24**) or pyrrole (**23**), other hetero- and even homocyclic units such as azulene (**85**) [281, 282] can also be subjected to anodic oxidation giving rise to the corresponding polymers. The electropolymerization produces films which are deposited on the electrode, and the material is originally achieved in a doped i.e. conducting state. If n monomers are incorporated into the chain, 2n electrons are consumed under release of 2n protons, while the subsequent charging process of the originating film requires $n \cdot x$ electrons with $x < 1$ [283].

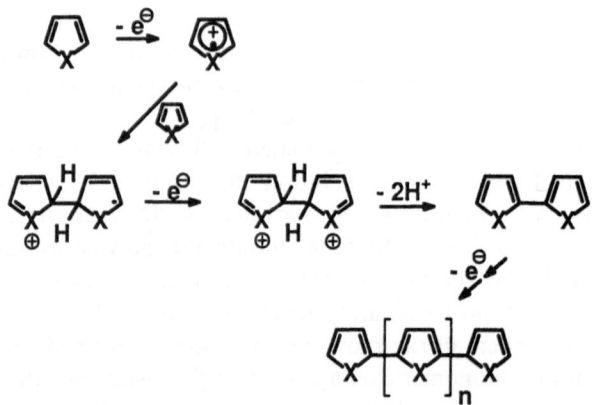

Scheme 6. The initial steps of electropolymerization

There has been a prolonged discussion on whether the initial coupling step implies the dimerization of two radical cations or the reaction between the radical cation and the neutral substrate [284–287].

While there seems to be a growing body of evidence in favor of the former mechanisms, it is clear that the coupling product can lose hydrogen in order to give a neutral dimer [288]. As the latter is oxidized with greater ease than the original monomer, it readily transforms into a radical cation which is then susceptible to further coupling with monomeric radical cation. The radical cations originally formed at the electrode can, in principle, enter three different processes: 1) polymerization, 2) diffusion into the solution – if the cation is stable; and 3) reaction with solvent or counterions – if the cation is reactive [288, 289]. The final polymer formation will, thus, depend sensitively upon the relative rate constants of these different processes. As a practical consequence, the polymerization rate, which increases with increasing oxidation potential of the substrate in the series pyrrole (23) < thiophene (24) < furan (93) < benzene (59), must be qualified by the role of nucleophilic solvents which compete with the radical cation for the growing chain [47]. The consequence for the formation of 25 is that, due to the low rate constant of radical-radical coupling, a relatively high concentration of substrates is needed for film formation [290, 291]. While electropolymerization appears as an extremely complex process, an additional challenge originates from a study of the deposition of the polymers on the electrode [288, 289]. As concerns chain growth and deposition, there has been evidence from particular cases that oligomers with about 3–4 repeating units are formed in solution which then deposit on the electrode.

93

It has been mentioned already that polypyrrole (25) and polythiophene (26) play an important role as electrical conductors and polymeric anodes in battery cells [2, 47, 226]. Since the charging and discharging of the conjugated polymer is accompanied by the incorporation and removal of counterions it is clear that the material can also act as a "carrier" of chemically different anions which influence the physical, chemical and physiological properties of the material [292]. With regard to the full structural elucidation of the polymers it must be added, however, that the electropolymerization process of pyrrole and thiophene does not provide a clean coupling of the heterocycles in the 2,5-positions. Instead, the 3- and 4-position can also be involved giving rise to further fusion processes under formation of complex polycyclic structures [47].

While the formation of poly-*para*-phenylene (7) from benzene in the presence of Cu^{2+}-salts is believed to involve radical cations similar to the polymerization of pyrrole [195, 253, 254], the formation of a related polymer, namely poly-*para*-phenylene sulfide (PPS, 94) according to Scheme 7 can readily be described as a

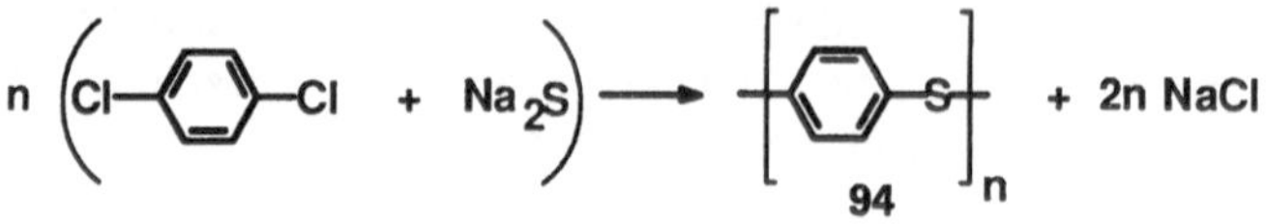

Scheme 7. Formation of polyphenylene sulfide

polycondensation based on a nucleophilic aromatic substitution [293]. It has been shown for the formation of PPS and of related aromatic polyethers, however, that the relationship between molecular weight and conversion does not exhibit the behavior expected for a normal polycondensation: one observes the formation of medium molecular weights already at the beginning of the reaction, so that the crucial assumption made for the mechanistic description of polycondensations, the equal reactivity of end groups, is no longer valid. It is known for various halobenzene compounds **95**, on the other hand, that the reaction with nucleophiles Nu⁻ under formation of product **96** (see Scheme 8) proceeds via an initial electron transfer under formation of radical anion **97** followed by elimination of the leaving group and recombination of the phenyl radical **98** with the nucleophile according to a so-called $S_{RN}1$ reaction [294]. Arguments have been proposed that a related electron transfer also occurs during formation of polyethers such as **94** and of polyether ketones. EPR spectroscopy points out that $S_{RN}1$ reactions do, indeed, play a role; one must be aware, however, that the detection of radical ion species is no convincing proof that $S_{RN}1$ is the mechanism of the *main* reaction.

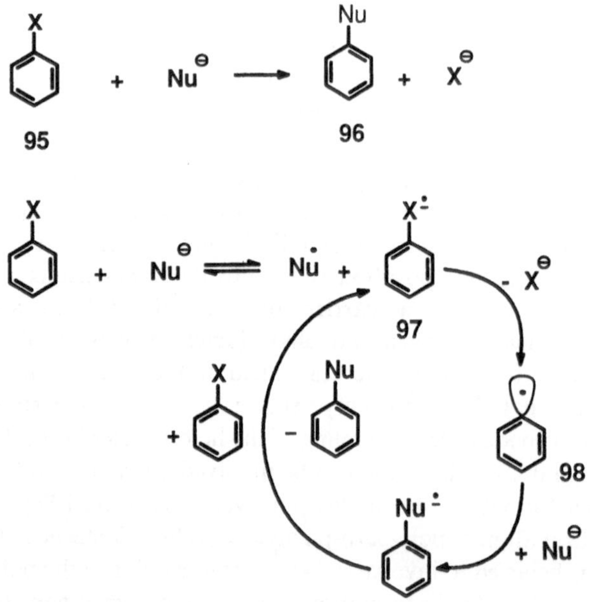

Scheme 8. The SRN_1 mechanisms

5.2 Skeletal Rearrangements

Electron transfer to or from a conjugated π-system can also induce pericyclic reactions leading to skeletal rearrangements. A typical example is the Diels-Alder cycloaddition occurring after radical-cation formation from either the diene or the dienophile [295–297]. The radical cation formation is in most cases achieved via photochemically induced electron transfer to an acceptor. The main structural aspect is that the cycloaddition product (see Scheme 9) contains a smaller π-system which is less efficient in charge stabilization than the starting material. Also, the original radical cations can enter uncontrollable polymerization reactions next to the desired cycloaddition, which feature limits the preparative scope of radical-type cycloaddition.

Electron transfer can also induce [2 + 2] cycloadditions if the olefinic double bonds are forced into close neighborhood [298]. Thus the perimeter-type polycycle **99** undergoes cycloaddition upon radical anion formation [298]. Interestingly enough, however, the resulting cyclobutane species **100** suffers from immediate cycloreversion under formation of acenaphthylene (**101**) and phenanthrene (**102**) [299, 300]. The driving forces of the latter reaction are the relief of strain and the chance for delocalizing the unpaired electron into a larger π-system.

In oligo- and polyphenylenevinylenes with *para*-phenylene subunits the double bonds are held far apart so that an intramolecular cycloaddition upon

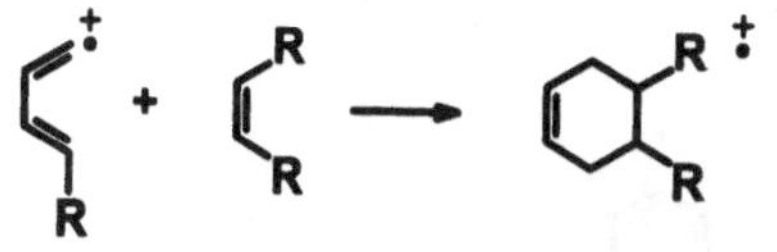

Scheme 9. Electron transfer induced Diels-Alder cycloaddition

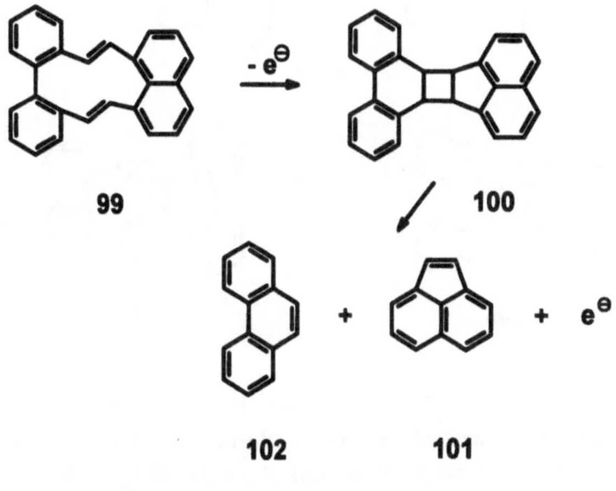

Scheme 9.1. [2 + 2] cycloaddition with formation of acenaphthylene **101** and phenanthrene **102**

photo-irradiation is not possible. 2,2'-distyrylbiphenyl (**103**), however, which is related to the "linear" 4,4'-distyrylbiphenyl (**104**) [99], is known to undergo a clean photochemical [2 + 2] cycloaddition giving rise to the cyclobutane species **105** [301–303]. Therefore, the PPV-analogous polymer **106** has been prepared [304]. It possesses an extended π-conjugation, but due to the existence of "kinks" within the chain, can undergo strictly intramolecular photochemical cycloaddition in analogy to model compound **103**. The important consequence of this process is that the π-conjugation can be interrupted by photochemical reaction and a photobleaching process achieved. It is therefore interesting that electrochemical and chemical dianion formation from **103** can also induce formation of the 9,10-single bond giving rise to a product **107** comprising two benzylanion subunits. The latter, upon subsequent oxidation, can again form a cyclobutane species. It is thus tempting to compare the electron-transfer induced bond formation with the photochemical process tested for the corresponding neutral compound [300].

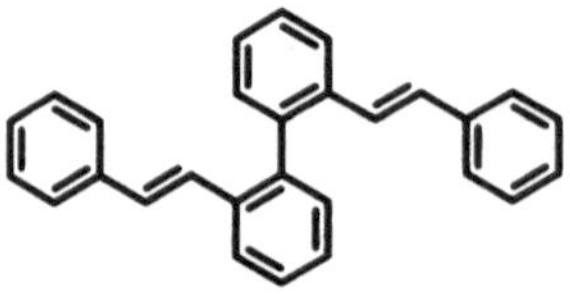

103

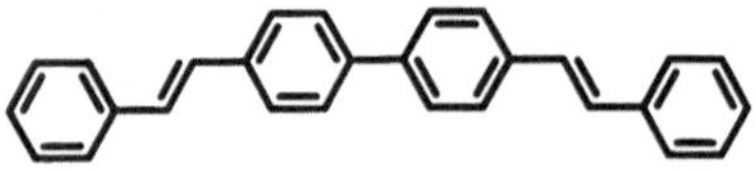

104

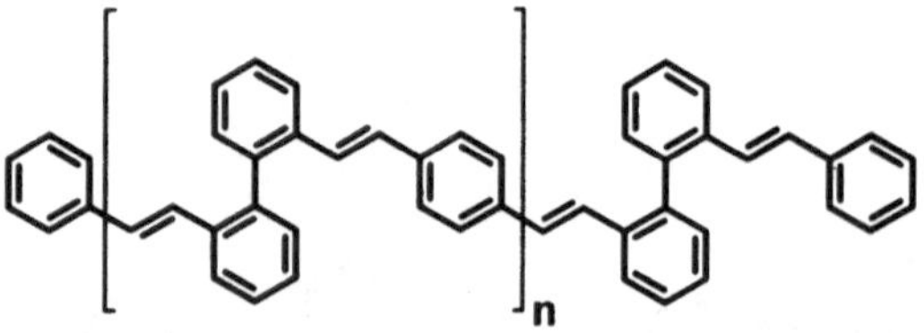

106

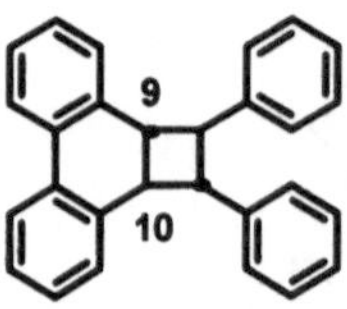

105

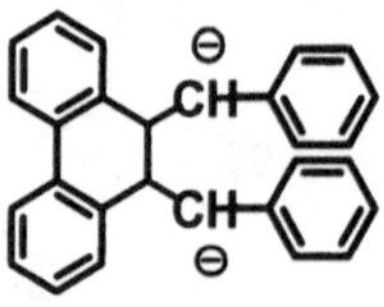

107

Oligomer and polymer structures with two-dimensional, so-called ladder-type π-conjugation have been shown to be of great relevance in material sciences since, apart from their inherent chemical stability, they seem to possess lower band gaps than their linear analogues. Attractive examples are the [n]acenes

108 and the perifused naphthalenes **109** [79, 306, 307]. While a repetitive Diels-Alder cycloaddition of bis-diene and bis-dienophile precursors, followed by a suitable dehydrogenation, seems to be the method of choice for the synthesis of **108**, there is no obvious synthesis of **109** via a "concerted" cycloaddition route. In such cases, a two-step approach has proven of special value [66, 120]; thereby, one, first, synthesizes an oligomeric and polymeric naphthylene chain such as **110** which is then subjected to a fusing reaction forming the extended "rylene" structure **109** [120, 308]. The well established model reaction is a transformation of 1,1′-binaphthyl (**111**) into perylene (**112**) which, according to various pieces of experimental evidence, appears to proceed via the dianion structure **113** followed by loss of hydrogen. It has been an important achievement that the ring-closure reaction of **111** can be extended to higher oligomers whereby the reaction is performed both on the reductive side (treatment with alkali metals) and on the oxidative side (treatment with copper-II-chloride and aluminum trichloride, so-called Kovacic conditions) [195, 253, 254]. The electrochemical oxidation of polymer **9** occurs at less positive potential when charge stabilizing substituents are introduced under formation of **114** [309]. The oxidation of **114** with $FeCl_3$ in CH_2Cl_2-solution at room temperature includes a clean irreversible dehydrocyclization to yield cationic derivatives of **115**.

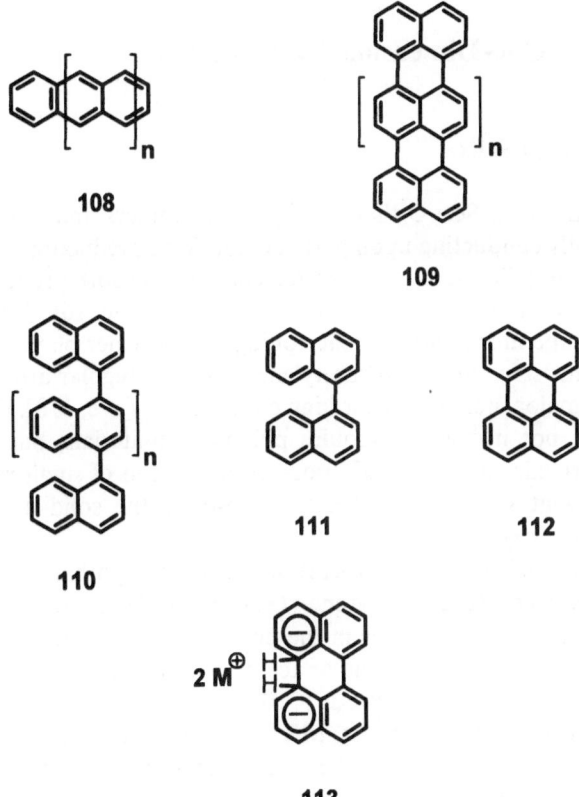

108

109

110

111

112

113

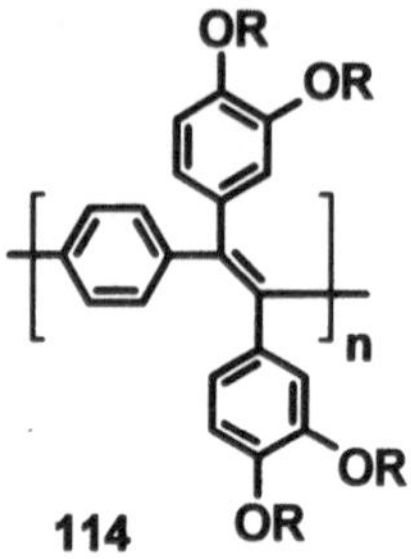

114

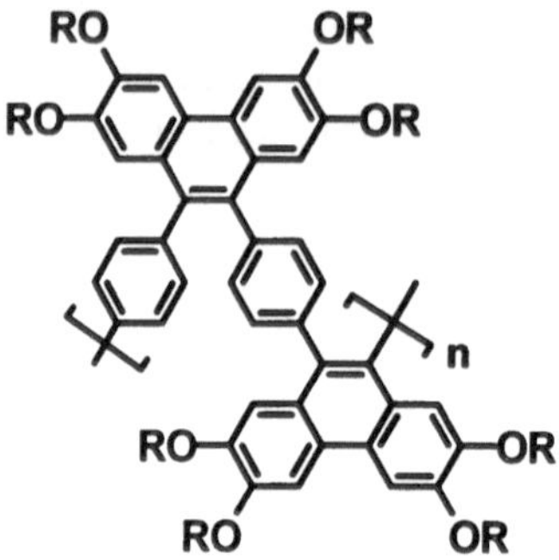

115

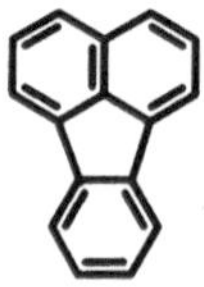

116    117

6 Crystalline One-Dimensional Conductors

6.1 Radical-Cation Salts

It was pointed out in Sect. 3 that conjugated oligomers and polymers can be made electrically conducting upon partial oxidation or reduction. This property is bound to the solid state, and several charge-transport processes must be considered in evaluating the over-all conductivity [83, 86, 161–165]. It is clear, therefore, that the morphology of the conjugated polymer plays a crucial role, and a crystalline structure of polyacetylene (6) with a regular array of π-chains seems to be a major prerequisite for high conductivity [175, 178]. Conductivity, however, does not necessarily require polymeric materials, but can also be obtained in organic solids formed upon crystallization of small radical-cation salts. The present section will, therefore, describe the solid-state aspects of radical ion chemistry.

When benzenoid organic hydrocarbons such as naphthalene (60), fluoranthene (116), perylene (112) or pyrene (117) are subjected to electrochemical oxidation at a platinum electrode in the presence of supporting electrolytes in solvents such as methylene chloride or acetonitrile, one frequently observes the deposition of crystals on the electrode [310]. When denoting the substrate as A and the supporting electrolyte as MX there are two nucleophilic species competing for the radical cation $A^{+\bullet}$, i.e., the neutral molecule A and the closed-shell counteranion X^-, and it is, indeed, the equilibrium constant of the

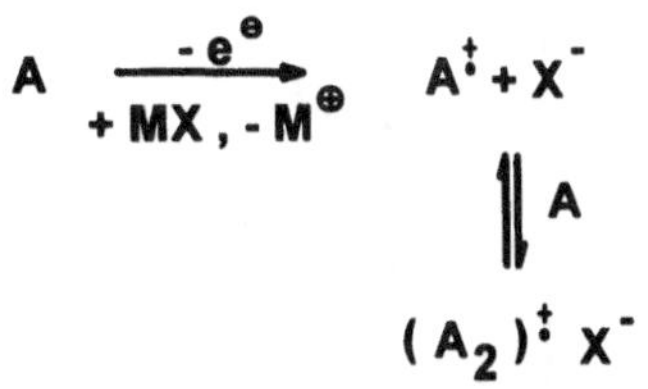

$$A \xrightarrow[+ MX, - M^{\oplus}]{- e^{\ominus}} A^{\dagger} + X^{-}$$

$$\Big\Updownarrow A$$

$$(A_2)^{\dagger} X^{-}$$

Scheme 10. The dimerization of hydrocarbons upon oxidation

dimerization reaction according to Scheme 10 which controls the crystals growth on the electrode. In fluoranthene (116) with a high equilibrium constant one always observes the stoichiometry A_2X of the resulting radical-cation salt [17, 18]. For other polybenzenoid species such as pyrene (117) and perylene (112) different stoichiometries are frequently observed, and the actual composition of the complex depends sensitively upon experimental conditions such as temperature, potential, current density, and supporting electrolyte [310–314].

A typical structural feature of the crystalline radical-cation salts is the formation of stacks. Thereby, the hydrocarbon units are packed in columns leaving space for channels in which the closed-shell counterions are placed. The planar hydrocarbons are oriented perpendicularly to the stacking axes with interplanar distances between 0.32–0.33 nm which are significantly smaller than the van-der-Waals radii of the carbon centers and allow for strong π-orbital overlap. It is also characteristic that the molecules are alternately rotated by 180°, as e.g. in the fluoranthene salt [17, 18]. Closed inspection shows also that a particularly strong interaction arises between π-centers which in the corresponding radical cation possess a high local spin density.

Many radical-cation salts exhibit high electrical conductivities which, in some cases, exceed 1000 S/cm. The conductivity is highly anisotropic, since charge transport occurs preferentially along the stacks [175–178, 310, 315]. The following structural principles seem to be particularly relevant: 1) incorporation of polarizable π-systems in segregated stacks, 2) constant interplanar distances allowing for significant π-orbital overlap, and 3) mixed valence states. While the first two aspects are immediately obvious, the importance of a mixed-valence state, that is, of only partial oxidation of the hydrocarbon, is deduced from inspection of Scheme 11 [86–88]. In the stack, according to the stoichiometry, only every second layer carries charge. If now a charge travels along the stack, e.g. by way of a hopping process, there will be no dication formation on a single layer. Such an electrostatically unfavorable process would necessarily occur for a 1:1 stoichiometry. More detailed studies of radical-cation salts of, e.g., fluoranthene or naphthalene show a metal-like behavior at high temperatures and a semiconducting behavior at temperatures below 180 K. Furthermore, the narrowest EPR line observed so far in a solid was found for the salt $(60)_2$ AsF$_6$ [315–317].

Conducting radical-cation salts have been obtained not only for benzenoid hydrocarbons, but also for their non-benzenoid analogues such as the pyrene

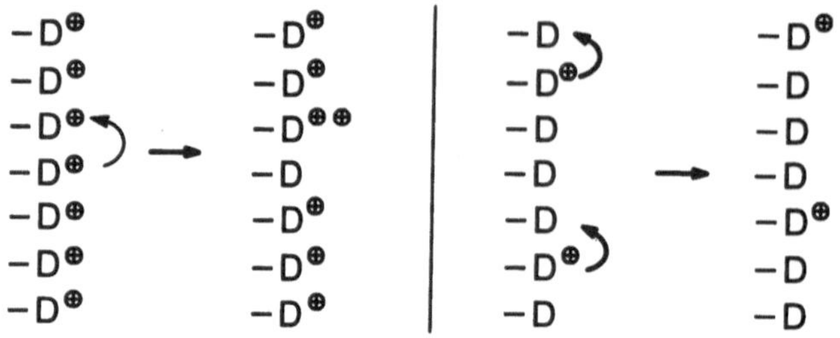

$$-D^\oplus \quad -D^\oplus \qquad\qquad -D \quad\quad -D^\oplus$$
$$-D^\oplus \quad -D^\oplus \qquad\qquad -D^\oplus \quad\quad -D$$
$$-D^\oplus \curvearrowright \quad -D^{\oplus\,\oplus} \qquad -D \quad\qquad -D$$
$$-D^\oplus \quad -D \qquad\qquad -D \quad\longrightarrow\quad -D$$
$$-D^\oplus \quad -D^\oplus \qquad\qquad -D \quad\quad -D^\oplus$$
$$-D^\oplus \quad -D^\oplus \qquad\qquad -D^\oplus \curvearrowright \quad -D$$
$$-D^\oplus \quad -D^\oplus \qquad\qquad -D \quad\quad -D$$

Scheme 11. Charge-transport in stack-type structures

isomer **118** [318]. Interestingly enough in view of the sometimes low persistence of doped conducting polymers, the salt $(118)_2$ $SbCl_6$ is stable for months even in the presence of oxygen. The non-benzenoid salt behaves significantly different from those mentioned above; its room temperature conductivity is about 10^{-1} S/cm, with a semiconductor-type temperature dependance, i.e., decreasing conductivity upon lowering the temperature. It can be concluded from EPR and conductivity measurements of $(118)_2$ $SbCl_6$ that the observed spin diffusion constants are not due to mobile charge carriers [318]. One, therefore, assumes that the EPR signal is not motionally narrowed, but exchange narrowed by the interaction of localized spins.

118

Although, at a first glance, it looks somewhat remote, to compare crystalline radical-cation salts with amorphous conjugated polymers, the intrastack interactions in radical-cation salts have been described as relevant models for the interchain interaction in conducting polymers [86]. The doping of conjugated polymer chains creates radical-cation sites which naturally stabilize themselves not only by an interaction with counterions, but also by the formation of complexes with neutral chain segments in their neighborhood. Accordingly, segments of the polymer chain are incorporated into stacks, and electroneutrality is, again, achieved by the alternation of polymer and counterion layers [178].

It should be noted that the stacking of isopyrene (**118**) can also be obtained chemically, i.e., by introducing saturated spacer groups between the layers [319, 320]. This approach gives rise to oligomeric and polymeric annulene multilayers **62** which can be doped by partial reduction processes. While the resulting species are not electrically conducting as a result of the large interplane distances, there is a chance of following the electron hopping between the layers as a function of the distance between the redox groups (see Sect. 8).

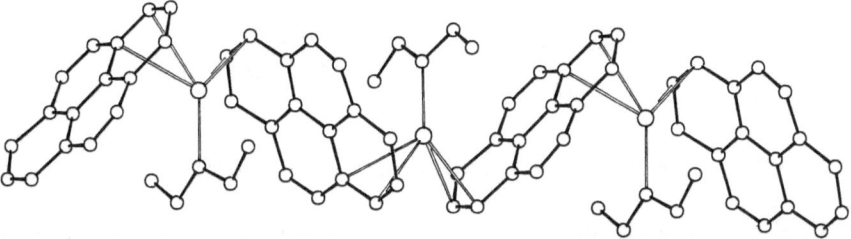

Fig 5. Crystal structure of the radical salt pyrene⁻˙(117)Na⁺˙ yielding a zigzag structure with solvent molecules

The question has been raised whether the conducting radical-*cation* salts have their counterpart in the corresponding radical-*anion* salts. This would allow for an interesting comparison in view of the mobility of the charge carriers [321, 322]. A priori, however, a stack-type structure seems to be less favorable in the case of radical anions because 1) the position of a neutral hydrocarbon on top of a negatively charged hydrocarbon creates additional Coulombic repulsion, and 2) there is a tendency toward a "capping" of the carbanion by the small alkali-metal counterion within an ion pair. Experimental investigations of the prevailing structural principles are rare since crystal structures have mostly been determined for diamagnetic carbanions such as e.g. the fluorenide alkali metal salt [321], and there are only a few radical anion-structures known. A particularly interesting case is that of the pyrenide sodium radical ion (**117⁻˙/Na⁺**) (Fig. 5) [322] for which the following findings are characteristic, 1) the stoichiometry is 1:1, 2) a stack-type structure is formed in which each sodium cation is interacting with two pyrene molecules and with one solvent molecule. According to Fig. 2 the resulting arrangement can best be described as polymeric contact-ion pair with a zigzag array of the π-system. An eventual charge transport along the stacking axis is thus severely inhibited by coulombic effects so that one will not expect an electrical conductivity.

6.2 Charge-Transfer Complexes

In the above radical-cation salts, the crystal contains partially oxidized donors, while the electroneutrality is achieved by the presence of closed shell anions. The structural requirements necessary for electrical conductivity in solid salts can also be met upon mixing of donors and acceptors; in the resulting charge-transfer (CT) complexes both the donor and acceptor exist in a partially oxidized and reduced state, respectively. Famous examples are the conducting CT complexes formed upon mixing of perylene (**112**) [323. 324] and iodine or of tetrathiafulvalene (TTF, **119**) as donor and 7,7,8,8-tetracyanoquinodimethane (TCNQ, **120**) as acceptor [325–327]; the crucial structural finding for the

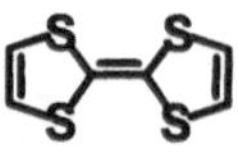

119

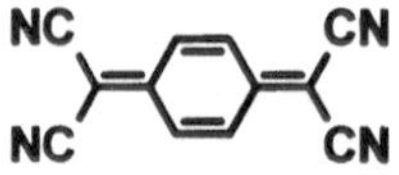

120

TTF · TCNQ complex is, again, the existence of segregated stacks of donors and acceptors.

TCNQ has been reported to form a large number of charge-transfer and radical-ion salts. Many of them form a chain-type alternating array of cations and TCNQ molecules or possess a donor-acceptor ratio different from $1:1$ [328, 329]. These species must be contrasted with the "simple" salts which are characterized by a $1:1$ stoichiometry and by segregated stack structures [263, 329–331]. These "simple" salts, in turn can be classified into two groups: the first group, with the potassium salt of TCNQ as a typical example, exhibits a semiconductor behavior, and the second group, with the TTF salt as a typical case, exhibits a metal-like conductivity [263]. The difference in conductivity between both groups by a factor of up to 10^6 is basically ascribed to differences in the degree of charge transfer from the donor to the TCNQ acceptor: for group 1 the amount of charge transfer is one, while for group 2 it is smaller than one. Similar to the situation of the above radical-cation salts, the TCNQ stacks of the latter group may be regarded, within a short time scale, as being made from both neutral and ionic TCNQ species, and only in such a mixed-valence stack is it possible to excite an electron from a charged TCNQ to a neighboring neutral TCNQ molecule without creating a strong coulombic energy [332, 333]. It was demonstrated by diffuse X-ray scattering that the degree of charge transfer in TTF · TCNQ amounts to 0.59 electron from the donor to the acceptor so that both stacks form real mixed-valence states [334].

The factors which influence the amount of charge transfer in complexes, e.g., of TCNQ have for long been a matter of dispute [263, 329–331, 335, 336]. In one approach one has considered the ionic binding of the solid materials. Thereby one assumes that the electrons in the crystal of a TCNQ complex with a donor are in equilibrium between the neutral and anionic electronic structure according to Scheme 12a [337].

The energy terms relevant for this crystal structure are the electrostatic Madelung energy E_M obtained for the ionic crystal and the energy of charge

a) $\text{TTF} + \text{TCNQ} \rightleftharpoons \text{TTF}^{\oplus} + \text{TCNQ}^{\ominus}$

b) $\text{TTF}^{\oplus} + \text{TTF} \rightleftharpoons (\text{TTF} \cdot \text{TTF})^{\oplus}$

c) $\text{TCNQ}^{\ominus} + \text{TCNQ} \rightleftharpoons (\text{TCNQ} \cdot \text{TCNQ})^{\ominus}$

Scheme 12. Charge-transfer and dimerization equilibria. A schematic description of a biselectrophore and degenerate electron hopping

transfer which is determined from the difference of the ionisation potential 1 of the donor and the electron affinity A of the acceptor. For the alkali metal salts of TCNQ one expects a large Madelung energy as a result of the small cations and the close packing in the crystals [338]. The ionisation potential of the donor, on the other hand, is extremely low so that the equilibrium according to Scheme 12a is shifted toward an ionic ground state with a complete charge transfer [263]. For TTF • TCNQ, in turn, the ionisation potential is higher and the E_M is small, so that the formation of mixed valence stacks is energetically favorable.

One must be aware that prior to the detection of a *segregated* stack structure of e.g. TTF • TCNQ the interaction of a neutral closed-shell donor with a neutral closed-shell acceptor would have been predicted to produce *mixed* stacks, thus optimizing electrostatic interactions [337]. While the above discussion of uniform segregated stack structures considers the relative contribution of different cohesive forces, there are also kinetic descriptions involving effects such as van-der-Waal's interaction, donor/acceptor interaction, specific interatomic contacts or molecular aggregation. It must be noted that in dilute acetonitrile solution only a small fraction of TTF and TCNQ is ionized. Similar to the dimerization of the hydrocarbon radical cations, the equlibria depicted in Scheme 12 must be referred to when describing the formation of the TTF • TCNQ complex in solution. It has been suggested that interchain inter-actions, e.g. between sulphur centers of the donor and nitrogen centers of the acceptor, are relevant to the formation of the segregated stacks [339, 340]. Here again, however, the situation in solution is of only limited relevance for the formation of the solid. Clearly, in solution one has a mixture of neutral and ionic species, and at some time between the initiation of aggregation and precipitation from solution of TTF • TCNQ a reorganization of charge from the neutral ionic mixture to the uniformly stacked structure must occur.

A class of acceptors related to TCNQ are the *N,N*-dicyanoquinodiimines (DCNQI, 121a) which are readily available and susceptible to chemical modifi-cation [341–343]. Another advantage of 121a is the fact that, due to the nearly linear N–CN-group, even tetra-substitution of the 6-membered ring does not cause a strong bending of the molecule. Mixing, e.g., a 2,5-dimethyl-*N,N*-dicyanoquinodiimin (121b) and CuI or oxidizing a mixture of DCNQI and $CuBr_2$ electrochemically one obtains black crystals of the composition DCNQI • 2Cu with a room temperature conductivity of 1000 S/cm [342].

121 a) R = H
     b) R = CH_3

Temperature dependent conductivity studies reveal a metallic character of the salt and the remarkably high conductivity of 500 000 S/cm at 3.5 K. It should be noted that the TCNQ copper salt is a semiconductor with a room-temperature conductivity of 2×10^{-6} S/cm. The crystal structure of the salt TCNQI $\cdot$ 2Cu reveals segregated columns formed from the quinone and the copper ion whereby the copper chains are surrounded by 4 quinone stacks [344].

When crystallizing TTF or tetramethyltetraselenofulven **122** with monovalent anions as PF_6^-, AsF_6^-, TaF_6^-, ClO_4^-, BF_4^-, and IO_4^- etc. in place of TCNQ some of the crystals are exhibiting superconductivity at low temperature and applied pressure [345]. The salts became known as Bechgaard salts and they are typically one-dimensional metals with high conductivity measured only along the stack direction. The first ambient pressure organic superconductor with a transition temperature $T_c = 1.4$ K was based on a tetramethyl-TTF with ClO_4^{2-} as counterion and found just 12 years ago [346]. The crystal structure shows short intra- and interstack chalcogen-chalcogen distances [347]. The next development in organic superconductors was a more effective donor with eight sulphur atoms, namely BEDT-TTF **123** and at least 20 superconductors with different anions and crystal phases are known up to date [348]. For the copper salts $(Cu[N(CN)]_2X$ and $Cu (NCS)_2$ of BEDT-TTF the highest conductivities for charge transfer complexes with transition temperature above 10 K have been found [349–351]. The crystal structure of these crystals show κ-phases, in which the donors are forming dimers setting up a two-dimensional network separated by inorganic copper complex layers.

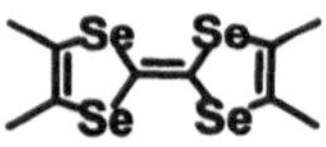

122 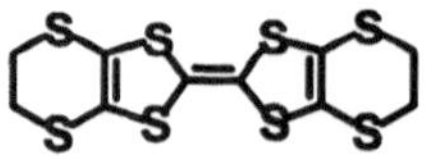   **123**

There has been a great number of suggestions for further stabilizing the metallic state of CT complexes. A model which seems particularly relevant to the description of organic radical ions (see Sect. 1) is based on the aromaticity of the ions [331]. According to this view the reduction of the non-aromatic quinone acceptors and the oxidation of the non-aromatic fulvalene donors afford ions with highly stabilized cyclic π-conjugation. On the other hand, an extension of this view to intrastack electron transfer, in which aromaticity would produce the driving force for the delocalization of electrons, seems to be inappropriate because, in a metallic CT salt, the electrons are already delocalized according to the band-structure description.

In a related fashion there have been attempts to increase the size of the donor and acceptor thus reducing the coulombic repulsion. The extended fulvalenes **124** and **125** would benefit from an appreciable conjugative stabilization upon cation formation [352–356]. The main approach within the search for new donors and acceptors tries to avoid the one-dimensional character of the charge

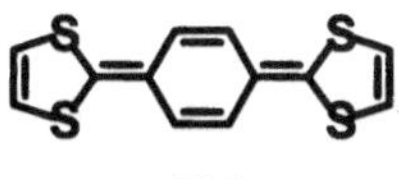

124 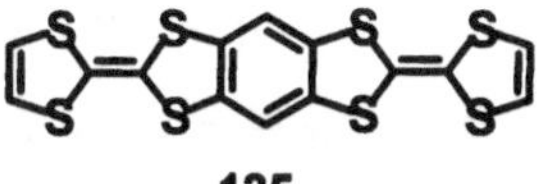   **125**

transport by creating increasing interstack interactions in the crystal. Another suggestion which readily emerges from the above discussion of radical ion chemistry is to fabricate organic metals by using neutral radicals, thus avoiding strong on-site coulombic repulsions and insulating closed-shell ions between the stacks. One should be aware, however, that the crystallization of radical-cation salts and charge-transfer complexes is a very complex process, and all models for the controlled design of crystalline electrical conductors have been of only limited success.

7 Organic High-Spin Molecules and Organic Ferromagnets

7.1 Electron Spin-Spin Interaction

The small organic π-systems described so far have been extended into larger conjugated oligomers and polymers or used as single redox units in radical salts. As their building blocks possess a closed-shell electronic structure, their charging with a single electron gives rise to a paramagnetic behavior, while higher charging leads to spin pairing as in the neutral state.

The field of organic high-spin molecules addresses the question of how to stabilize more than one unpaired electron in a molecule and how to align the spins in a parallel fashion with a ferromagnetic coupling [357]. Following the nomenclature of transition-metal complexes with their partially filled d-orbitals, organic molecules with two or more coupled spins are arbitrarily called high-spin molecules. They serve as models for studying the spin-coupling mechanisms which is an important issue towards "spin control", i.e. the understanding of detailed aspects of spin-spin interaction and its dependence on the molecular structure [358–360]. Some of the high-spin molecules may also be used as repeating units for high-spin polymers.

The magnetic exchange interaction between two local spins S_A and S_B can be described with the Heisenberg Hamiltonian H [361, 362]:

$$H = -2JS_AS_B = (E(T) - E(S))S_AS_B.$$

If the exchange interaction $J > 0$ the exchange coupling is ferromagnetic and the ground state configuration is high-spin with parallel aligned spins (triplet). If $J < 0$, the ground state configuration is low-spin with antiparallel spin orientation. If we consider a system of two spin momenta occupying localized orbitals ϕ_a and ϕ_b a simple molecular orbital (MO) and configuration interaction (CI) treatment lead to the equation [363, 364]:

$$E_T - E_S = J_{eff} = -2K_{ab} + (\varepsilon_a - \varepsilon_b)^2/(J_{aa} - J_{ab})$$
$$= -2K_{ab} + 2t^2/(U) = J_{pot} + J_{kin}$$

where K_{ab} is the exchange integral, ε_a and ε_b are the orbital energies, and J_{aa} and J_{ab} are the coulomb integrals and U is the difference between them.

The potential exchange integral J_{pot} stabilizes the triplet state ($K_{ab} > 0$), while the kinetic exchange J_{kin} is also positive and adds an antiferromagnetic contribution. When considering further excited state through extended CI treatment an additional polarization term becomes important that sometimes has been described as indirect exchange interaction J_{pol} [365–367]. Qualitatively this polarization is inducing unpaired spin density on formally diamagnetic parts of the molecule and for instance becomes especially important in radical centers linked unsaturated to a conjugated polymer. The total effect of this term can be either ferro- or antiferromagnetic, depending on the nature of the interacting orbitals.

There has been a great interest in the design of organic materials with ferromagnetic ordering. The idea is to combine the processability and the ease of production of organic materials with ferromagnetic properties common for some inorganic compounds. In order to prepare organic ferromagnets from high-spin molecules one has to meet two requirements [368]:

1) synthesis of stable high-spin organic molecules which can be used as building blocks e.g. for macromolecules,
2) organization of the high-spin molecules within a three-dimensional network, such that on the basis of the individual intermolecular spin couplings a spontaneous bulk magnetization results.

The first requirement seems easier to be satisfied than the second one, because there are already a number of molecules with high-spin ground states available [357, 360]. Even if a high-spin state is established for a monomeric unit, it must be verified whether the molecules can be extended to larger oligomers or polymers with retention of their individual intramolecular spin-coupling properties. For the design of new high-spin molecules one has to be aware of many discrepancies between theoretical predictions and experimental results. Clearly these shortcomings become particularly striking for cases in which the high- and low-spin alignment are close in energy (small absolute J).

The second requirement is much more demanding, because in organized media of persistent organic radicals, the electron spins have a strong tendency to either orient themselves randomly or to align themselves in an anti-parallel fashion, so that the total macroscopic spin is zero [357, 368]. Therefore, stabilization of a large number of spins leading to a magnetic moment within an organic/macromolecular system through long-range spin ordering is a formidable, but very difficult task. Several suggestions have been proposed toward the design of molecular ferromagnets, some of which can be summarized as follows [369, 370]:

1) Stacking of stable radicals with overlapping spin densities of different sign and different size, such that an overall magnetic moment is obtained. (Heitler-London spin exchange, ferromagnetic ordering) [371];
2) Synthesis of one- and two-dimensional polymers with high ground state spin multiplicity, which at least yield ferromagnetic domains [372–374];
3) Preparation of ion-radical salts of alternating donor ($D^{+•}$) and acceptor ($A^{-•}$) radicals, whereby the charge transfer stabilizes a high-spin triplet state of the D/A pair [369, 375, 376];
4) Use of antiferromagnetic coupled radical centers with a different number of spins per site, which can cancel only partially, leaving a magnetic moment (ferrimagnetic ordering) [377, 378].

The pros and cons of these suggestions have been discussed in depth [357, 360, 368]. Experimental evidence for high-spin states usually comes from the EPR fine structure and from magnetic susceptibility measurements. EPR is most often used for well defined radicals also in dilute solution. In the solid state the spin-spin interaction is usually referred to as zero-field splitting and represented by a traceless tensor D. The susceptibility data, on the other hand, are predictive for the bulk magnetization of the whole sample and typically measured for the powdered or crystalline material.

7.2 Intramolecular Spin Coupling

The dicarbene **126** and the Schlenk hydrocarbon **127** belong to the first organic molecules in which a quintet and a triplet ground state have been established [379, 380, 381] and they have frequently served as a structural basis for further theoretical and experimental studies on high-spin molecules [372, 373, 382, 383]. Both systems **126** and **127** contain the 1, 3-benzoquinodimethane moiety **128** for which no stable Kekulé structure can be drawn. This approach rests on the assumption of degenerate or nearly degenerate singly occupied orbitals with parallel electron spins according to Hund's rule. Within molecular π-systems degenerate orbitals can be achieved by 1) symmetry, 2) non-Kekulé structures and 3) orthogonal alignment of the π-systems [369, 376, 382, 384, 385]:

1) Orbital degeneracy occurs in molecules having at least a threefold rotational axis as in triphenylene (**129**) or possessing D_{2d} symmetry [382, 386]. Very

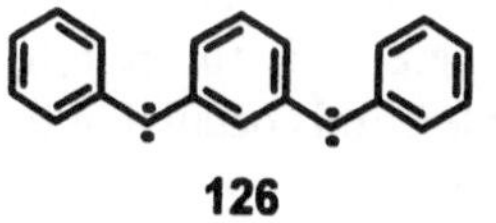

| **126** | **127 R = phenyl** | **128** |

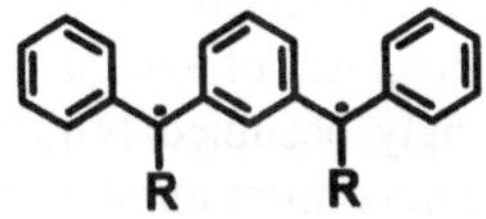

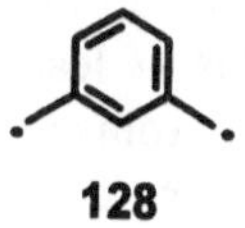

often the degeneracy is removed and the symmetry is broken through a Jahn-Teller distortion as in cyclobutadiene (130) [387]. The distortion of symmetry and the relatively high redox potential for transforming closed shell molecules of high symmetry into high-spin states has limited this approach to some biradicals such as derivatives of the antiaromatic benzene dication, while higher spin states (S > 1) within one molecule are not available [388]. These molecules, on the other hand, have attracted interest in the design of donor-acceptor radical stacks with ferromagnetic coupling [376].

2) As originally pointed out almost degenerate non-bonding molecular orbitals (NBMO's) are present in non-Kekulé molecules. Typical examples are trimethylenemethane (131) and 1, 3-benzoquinodimethane (128) [389, 390]. Although the NBMO's are not strictly degenerate, non-Kekulé molecules possess high-spin ground states due to the strong coulomb and exchange-correlation interaction between the unpaired electron [383, 390, 391].

3) Orthogonal alignment of simple aromatic subunits can be induced by steric hindrance, as in decachlorobiphenyl (132) [39, 222, 384, 392, 393], or by saturated sp^3-centers bridging the π-units [394, 395]. The aromatic subunits normally are diamagnetic in their neutral state and need additional charges as in 1) for achieving single occupancy of the degenerate states (dication or dianion).

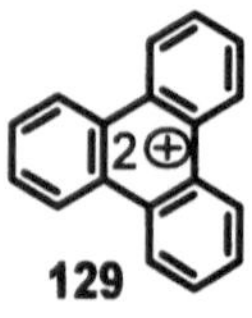

129

130

131

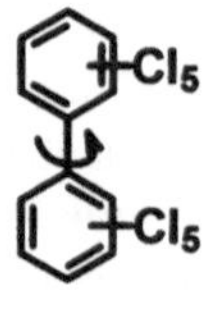

132

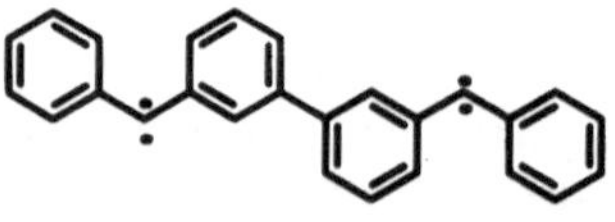

133 a

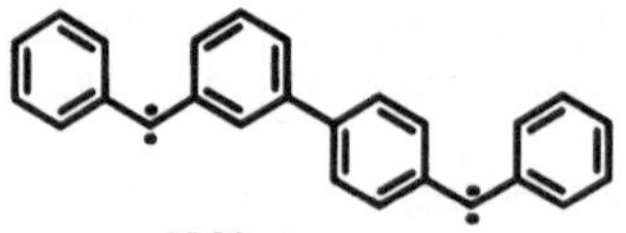

133b

Some rules have been put forward to design molecules with degenerate NBMO's. Following the proposal of Longuet-Higgins, an alternant hydrocarbon has at least N-2T singly occupied NBMO's, where N is the number of carbon atoms and T is the maximum number of double bonds occurring in any resonance structure [383, 396]. These rules have later been improved on the basis of a valence-bond approach [374, 397]. This concept predicts that the spin-quantum number is given by $S = (n^* - n)/2$; whereby (1) n^* and n are the numbers of starred and unstarred atoms of the alternant π-network, (2) as many stars as possible are created, and (3) starred atoms must not occur next to each other.

In some cases, experimental results have been in conflict with theoretical considerations; thus *meta, meta'*-biphenyldicarbene (**133a**) is predicted to have a quintet ground state from simple Hückel theory, while spin polarization leads to a singlet ground state, which, has indeed been observed [398]. For the *meta, para'*-biphenyldicarbene (**133b**) the quintet state is established. Hexamethylene (**134**), on the other hand, is foresaid to possess a singlet ground state by ab-initio calculations while a biradical has been detected even at low temperatures [399, 400–402]. Thus in order to establish the existence of high-spin ground states in large molecular systems like polymers, both, good numerical methods for the theoretical design and confirmation by experiments are needed.

The majority of organic high-spin molecules is based on the non-Kekulé structures, which are also represented in the classical arylmethyl **127** and dicarbene **126**. These molecules have served as structural basis for many theoretical and experimental studies [372, 373, 382, 383], and it has been shown that these structures are not limited to the one-dimensional extension. By use of 1,3,5-trimethylenephenyl **135** at least a two-dimensional structure is possible [403–407]. For poly-*meta*-arylmethyl **126** the spin multiplicity is predicted to be $S = n/2$, where n is the number of repeating units, and the exchange interaction between the radical centers in **136** is estimated to be 0.21 and 0.001 eV for the first and second neighbors, respectively [383]. The number of spins doubles when the methylene unit is replaced by a triplet carbene as in **137**. In the carbenes **126** and **137** two aspects are relevant for the description of the spin states: the meta-configuration of carbene centers favors the ferromagnetic interaction between different carbene units, and the orthogonally aligned n- and π-oribitals within each carbene moiety also couple ferromagnetically thus further stabilizing high-spin alignment [408, 409]. In diphenylcarbene **126** the ground state triplet shows a strong ferromagnetic exchange interaction J of

134   **135**

136 a :   m = 0
     b :   m = n
     R = phenyl

137 a :   m = 0
      b :   m = n

138 a : R = H
    b : R = H, N†
    c : R = ·/-, N·
    d : R = O·

139

140

0.20 eV. One spin is localized in the σ-orbital, while the other is delocalized over the phenylene unit. The carbenes, however, are problematic in some way because 1) they are very reactive and unstable at temperatures above 100 K; 2) the complete photolysis of all diazo precursors becomes difficult with increasing chain length, while a high spin multiplicity is only reached if all sites are intact, and 3) there are no polymerization reactions leading to this skeleton [410].

Of great synthetic interest, therefore, is the fact that high-spin alignment is also predicted when the carbon is exchanged by nitrogen to give poly-*meta*-aniline (138), which can be oxidized to the corresponding cation radical 138b, transformed into the neutral radical 138c or oxidized to the corresponding nitroxide 138d [411–413]. The predictions have been tested for the nitroxides 138d and cationic *meta*-anilines 138b [414, 415]. Nitroxides are known to constitute stable radicals, and consequently the synthesis of 1,3-di and trinitroxide 139 and 140 has led to di- and triradicals with triplet and quartet ground states, respectively [414, 416, 417]. The ferromagnetic exchange coupling is smaller than for the carbenes, which is explained by the reduced spin polarization from the NO·-group into the phenyl ring. After all it seems difficult to prepare even larger oligomers, because in the course of the synthesis the hydroxyamino group (NOH) has to be protected and finally converted into the nitroxide.

For the *meta*-anilines 138 only one relatively small cationic oligomer 1,3,5,-tris(diphenylamino)benzene 141 is available for comparison, where EPR spectroscopy yields a small zero field splitting D of D = 1.31 mT compared to the carbon analogue (D = 4.4 mT) [418]. 141 shows a linear increase of the EPR signal intensity I_{EPR} with decreasing temperature, obeying the Curie-Weiss law ($I_{EPR} \sim \chi = 1/T-\theta$). The negative Weiss constant of $\theta = -110$ K indicates an antiferromagnetic coupling, which is explained by strong intermolecular coupling even in frozen solution of methylene chloride. Polymers, on the other hand, have been prepared from *meta*-bromaniline (142) to yield poly-*meta*-aniline 138 [415, 419] and from 1,3,5 triaminobenzene (143) [420] which has been reacted with iodine to yield a three-dimensional polymer. Thereby, a clean conversion into the cationic or neutral radical forms has not been achieved. While the chemical constitution of these black insoluble polymers and the three-dimensional alignment are not clear, the samples show some ferromagnetic properties. It appears as a general problem within high-spin polymers that the total spin concentration is much lower than predicted from assuming an overall conversion of each subunit into radical forms (see below); for poly-*meta*-aniline (**138b**) the total spin concentration does not exceed 10^{20} spins/g which corresponds to only one spin per ca 70 aniline units in the iodine-doped sample [415, 419].

The *meta*-substitution of a central benzene ring has been further tested in many approaches. The *meta*-ketones **144**, for instance, which are precursors in the synthesis of the carbenes, have been shown to be reversibly reduced to the ketyl radicals with high-spin formation even at ambient temperatures [421]. The metaphenyldinitrenes **145** [379] and 1,3,5-tricyano-2,4,6-trinitrenes **146** [422, 423] have been shown to possess quintet and heptet ground states. Thereby, only the number of spin-carrying substituents on the same phenyl ring is enhanced, and the question is still open how the π-system can be enlarged still yielding ferromagnetic coupling between the spin centers. An elegant approach to answer this question has been tested for the nitrenes by an extension of phenylnitrenes bridged through acetylene (**147**) or diacetylene (**148**) [410, 424, 425]. One can then examine where to place the open-shell centers regiochemically and what kind and size of conjugated systems will allow for a

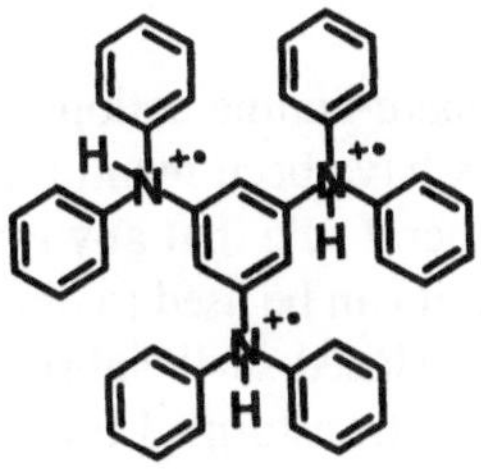

141

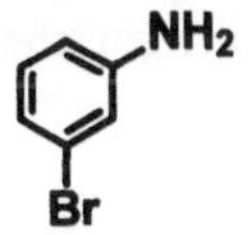

142

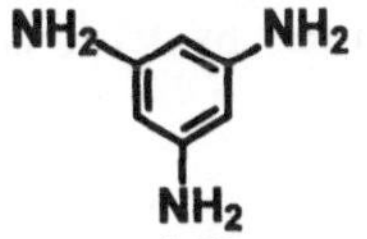

143

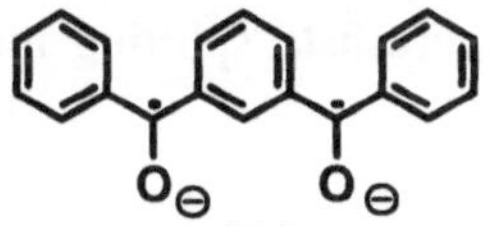

144

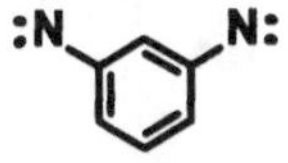

145

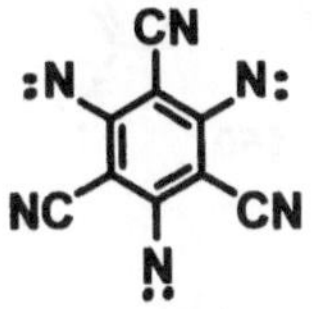

146

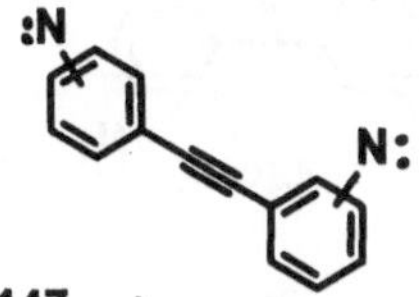

147 a) p,m
b) p,p
c) m,m

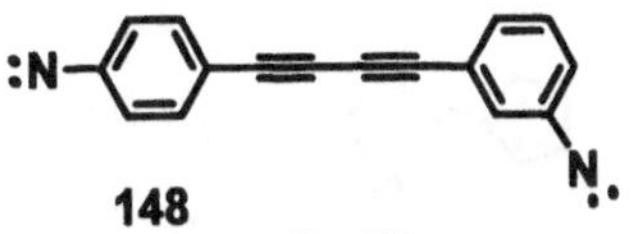

148

p,m

sufficient interaction. Only the *meta, para'*-isomers (**147a, 148**) show high-spin ground states, irrespectively of the distance between nitrenes, while the *meta, meta'*- and *para, para'*-isomers **147b, 147c** have singlet ground states as proven by disappearing EPR signals at low temperatures. Further studies on phenylnitrenes bridged by a carbonyl (**149**), 1,2-ethenediyl (**150**) or 1,1'-ethenediyl (**151**) [425] demonstrate that effective ferromagnetic interactions are also possible through these bridges as long as the nitrene substituents are properly located in a *meta, para'*- or *meta, meta'*-arrangement respectively.

Another approach toward high-spin molecules with non-Kekulé structures uses dimethylenecyclobutadienyl-units **152** which possess a triplet ground state [426–428]. A triplet ground state also prevails in the unsymmetric 3-methylenecyclobutanone (**153**), but the latter is much less stable [427]. Cyclobutanediyl may be used as coupling unit between cyclopentanediyls **154,** which also possess a triplet ground state, thus leading to a novel tetraradical **155** [358, 429]. Within a theoretical model the extension to higher spin systems as in **155** may be described as a combination of triplet centers that are connected by a ferromagnetic coupling unit [358, 430, 431]. This idea corresponds to the search for bridging units inducing ferromagnetic coupling between the radical centers as in the aforementioned nitrenes.

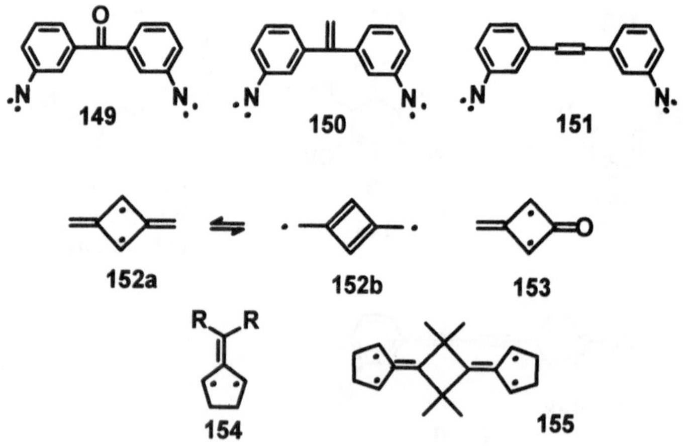

Some charged radical ions such as the poly-*meta*-aniline cations **138b** and the meta-diketyls **144** with non-Kekulé structures have been mentioned before [413, 415, 418, 419, 421]. It seems to hold as a general rule that any radical site connected in 1,3-position of a central phenylene unit can be used to obtain high-spin alignment. This is even true for aromatic closed-shell substituents as anthracene **61** or naphthalene **60**, which are easily transformed into a radical cation or radical anion [432]. Using anthracene cation radicals **156** or naphthalene anion radicals **157** as substituents, the EPR spectra of the triplet biradicals clearly follow a Curie law behavior dawn to 6.5 K, indicating a high-spin ground state. The zero-field splitting measured by EPR is larger for the napthylene radical sites than for the anthracene radicals, where the centers of maximum spin density in the latter have a larger distance [432]. The positive exchange interaction of these radicals is not limited to the *meta*-configuration, but can, again, be achieved in the 1,3,5-trisubstituted phenylene with naphthalene **158** and anthracene **159** radical substituents [433].

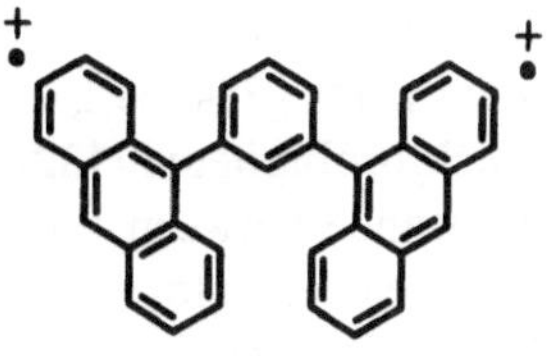

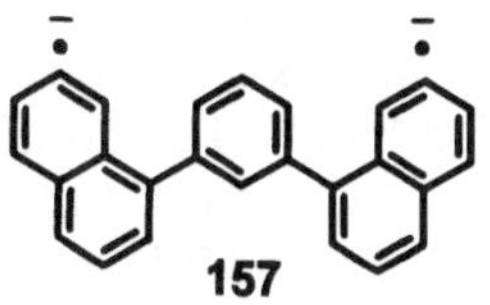

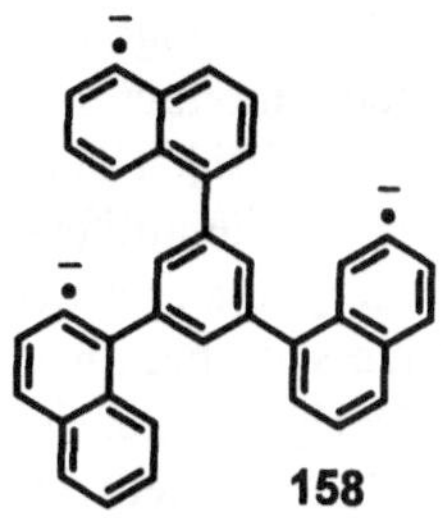

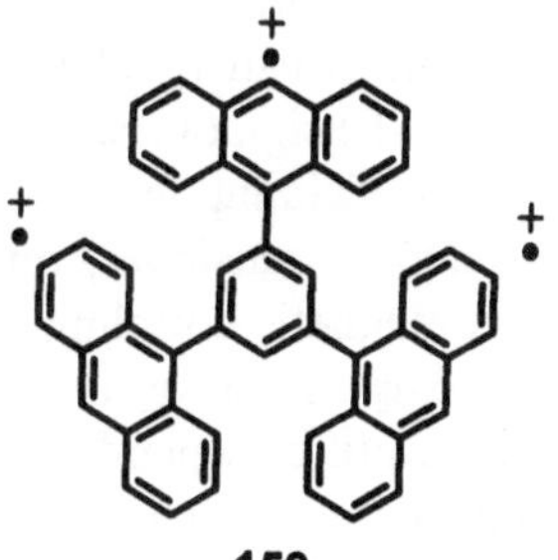

Even in the *meta*-phenylenevinylenes **22** the neighboring units can be charged to the biradical form. This is in line with the observation of spin localization in the monoradical, where the unpaired electron is distributed in one stilbene unit only. Upon further charging of e.g. **22b** the biradical formation is obtained [44, 434].

As has been pointed out already, antiaromatic systems with high symmetry may be used for biradical formation upon oxidation or reduction. The unsubstituted benzene seems to be too reactive for stabilizing the dicationic state, and the addition of stabilizing substituents helps in producing more stable triplet molecules. Stable triplet ground states have been observed for a number of five- and six- membered antiaromatics as the cyclopentadienyl (**160**) and pentachlorocyclopentadienyl cation (**161**) [376, 385], hexachlorobenzene dication **162** [435] and the dications of triphenylene (**129**), coronene (**14**), and the hexaaminobenzene derivative (**163**) [32, 376, 436–440]. The dicationic derivatives of **129** with ethylenediamines substituted at the outer benzene units (**164**) are stable in solution for several hours [385]. The substitution considerably lowers the oxidation potential and enables oxidation by strong donors as hexacyanocyclopropylene (**165**). This makes these molecules suitable for radical stack formation with alternating donor-acceptor molecules, following one of the above mentioned suggestions for the formation of organic ferromagnets [385].

The orthogonal alignment of aromatic π-molecules leads to nearly degenerate orbitals of the neighboring subunits within the corresponding oligomers and polymers; upon reduction or oxidation of the single subunits stable radicals are created, which may couple ferromagnetically. This approach has been used in perchloro-substituted bi- and oligophenylenes **166**, in which the chloro-substituents increase the torsion between the phenyl rings [384, 392, 441]. Thereby a stable dication biradical is formed for tetrachlorobiphenyl (**132**).

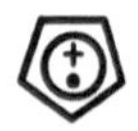

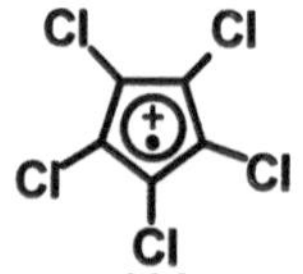

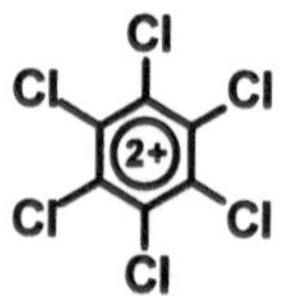

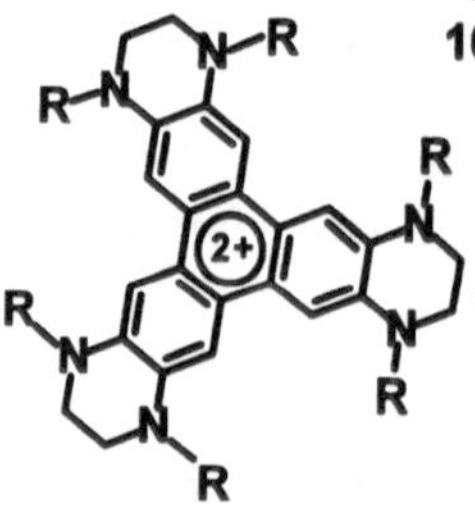

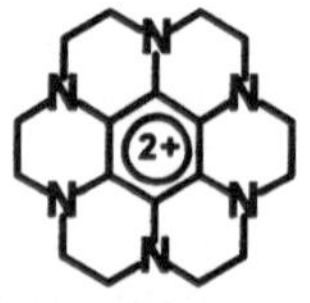

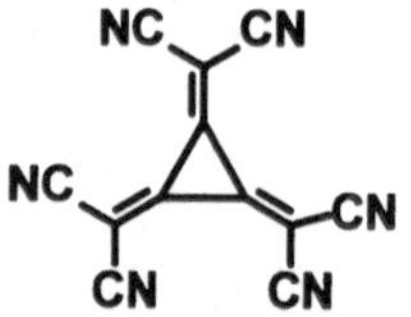

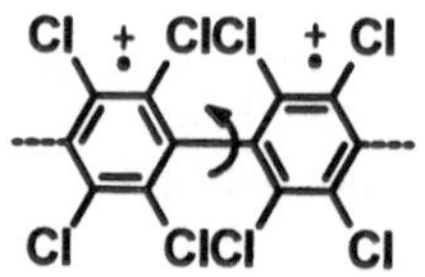

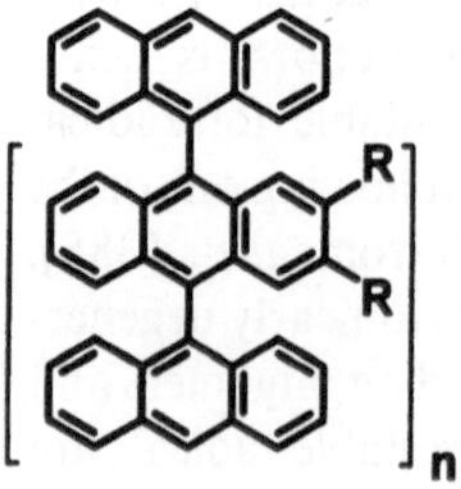

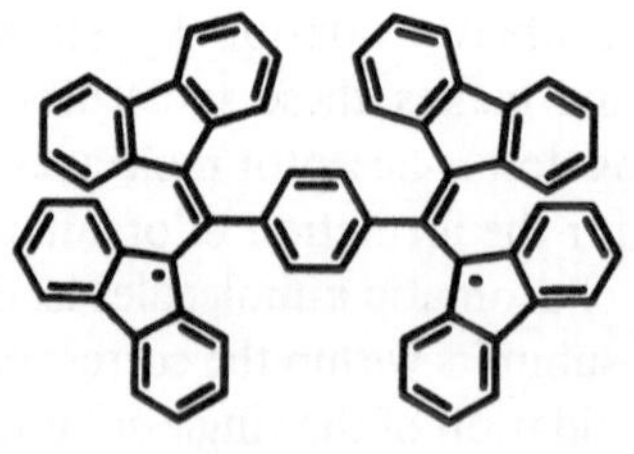

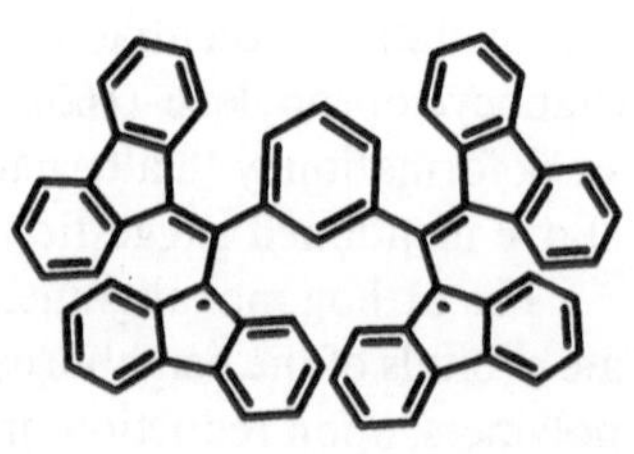

In oligoanthrylenes (**167**) the steric hindrance between the anthracenes is considerable, and no additional substituents are needed for a nearly orthogonal alignment of the anthracenes [39, 40, 219, 367, 442]. Therefore each of the anthracene subunits can be charged to the radical anion affording high-spin oligomers [39]. The sign of the direct exchange coupling through the π-orbitals in 9,9′ position yielding ferro- or antiferromagnetic interaction, has been calculated to depend sensitively on the angle of torsion [367]. The calculations predict a ferromagnetic coupling at angles of 84–90°, while antiferromagnetic exchange coupling should occur at lower angles leading to larger overlap of π-orbitals. Temperature dependent EPR measurements of bianthryl **19** have demonstrated that at low temperatures T = 18 K the signal intensity suddenly drops and the ground state becomes singlet, with a singlet-triplet energy

160

161

162

164

163

165

166

167

168

169

separation of only 60 cal/mol [40]. This is very similar to the results for *para*-phenylenebiphenylpropyl (*para*-tetrafulvenexylene) **168** [443], in which the triplet state is thermally excited. For the corresponding *meta*-isomer **169**, on the other hand, a stable triplet ground state has been observed [444]. This demonstrates that non-Kekulé structures are much better suited for establishing high-spin states than the orthogonal alignment. Up to now, there is no description of *ortho* and *para* linked aryl systems with orthogonal aligned π-units, which give rise to stable high-spin ground states. In the non Kekulé structures as **126**, the ferromagnetic coupling is partially due to a steric hindrance leading to a torsion between the aryl units [445]. Therefore the comparable biradicals with an additional methylene bridge leading to planarized phenyl rings loose the ferromagnetic coupling.

A different strategy for the design of high-spin polymers suggests the use of block copolymers of conjugated oligomers connected via bridges that enable ferromagnetic spin coupling. When charging the conjugated chains to the monoradical (polaron) the bridging unit should prevent the spin pairing leading to ferromagnetically coupled polarons within the polymer (see Scheme 13) [431].

This approach has been tested by use of *meta*-phenoxyl as bridging unit **170**. Although a significant number of spins have been produced and some high-spin ferromagnetically coupled domains are reported, the total number of spins is much lower than expected, and the magnetization of the sample is weak [429, 486]. Further investigations of the radical states are necessary because at low doping levels larger spin moments have been measured (with S > 2 components) than at high doping levels (S = 1/2).

It has not been possible up to now, to obtain real high-spin polymers with radical centers in the conjugated π-chains. This is partly due to the extreme difficulties of precisely generating all radical centers within a chain. As soon as a radical position is not intact the spin coupling is interrupted and confined to smaller units. Therefore, the highest spin states in organic polyradicals reported so far are S = 6–9 for oligoarylcarbenes of structure type **137b** [447] and S = 5 for 1,3,5-oligoarylmethyls of type **136b** [448].

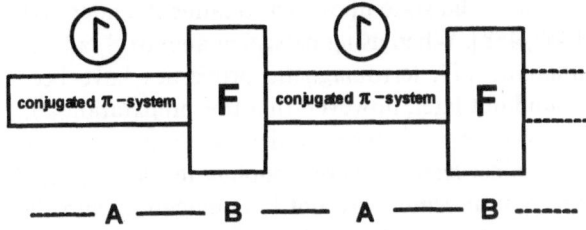

F = ferromagnetic coupling unit

Scheme 13. A schematic picture of a ferromagnetic coupling unit between two polaronic states according to the Fukotome model

7.3 Polymers with Radical Centers in the Side Chain

To overcome the problems faced with higher spin states in main-chain high-spin systems, the use of stable radical centers in the side chain of a polymer has been tested [358, 382, 449–456]. The main chain should then be a conjugated polymer like PAc 6, PPV 8, PPP 7 or polydiacetylene 171, to which are attached stable radical centers as nitroxides, phenoxyls (172), and galvinoxyls (173) [358, 382, 449–456]. Even when precursors of the radical or functional groups for subsequent conversion are used, the degree of polymerization may be low. This has been described for polymers from *para*-ethynylbenzaldehyde (174) [454] and from *para*-phenylacetylene with a hydroxyphenyl group 175 [382]. The total spin concentration is low (10-25% of theory), which already explains the preferential paramagnetic behavior of the samples. Nitroxides have been added to films of polyacetylene through a Diels-Alder reaction; the reaction is slow and the bulky spinlabel with a reactive ene component is introduced in low amounts, leading to typically 10^{-3} spins per CH unit [450–452].

The topochemical polymerization of phenyl substituted 1,3-butadiynes (176) can be performed in the crystalline solid state upon heat treatment, under high pressure or UV irradiation [457, 458]. When using persistent aminoxyl radicals at both ends of the 1,3-butadiyne 177, ferromagnetic properties have been reported [459]. It has been found out later, however, that this preparation and the magnetic properties could not be reproduced under standard conditions [460]. Two major arguments against ferromagnetic coupling in these systems have been put forward: 1) the radical centers are not in conjugation with the butadiyn chain and far apart from each other such that antiferromagnetic coupling is expected, and 2) the crystal structure does not appear to satisfy the empirical criterion for topochemical polymerization. Similar results have been reported for the 2,4-hexadyine-1,6-diol bridged diiminoxyls 178, where a high magnetization, first reported [461], could not be reproduced [462].

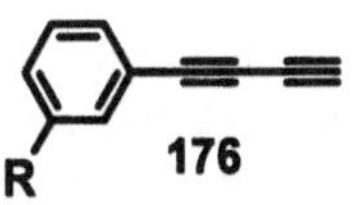

Following the original concept of topochemical polymerization and using only one *tert*-butylnitroxy-phenyl as pending group at the butadiyne **179** the polymerization through 1,4'-coupling in the crystal should allow for a ferromagnetic interaction between the spin centers, and the total spin should be (S = n/2) [463]. Surprisingly enough, the nitroxyls do not crystallize and only some mixed crystals – the nitroxide diluted with the precursor hydroxylamine molecules – undergo polymerization. Although the total spin concentration is relatively low $(1.59 \times 10^{19}$ spins/g), some "islands" with ferromagnetic interactions and S > 1 have been found, while the overall susceptibility of the sample points towards an antiferromagnetic behavior [463]. It is somewhat surprising, that the corresponding 1,3,5-hexatriyne with a 4-chloro-3-(*N-tert*-butyl-*N*-oxide) phenyl group **180** does undergo smooth polymerization in the crystal already at room temperature [464]. The magnetic susceptibility data of this compound follow a Curie-Weiss law with a positive $\Theta = +2.5$ K, and the exchange coupling observed by a least-square fit of the experimental data yields a positive exchange coupling of $J/k_B = +1.5$ K.

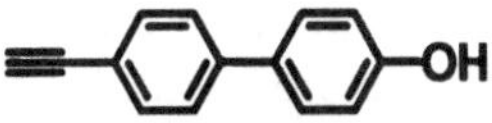

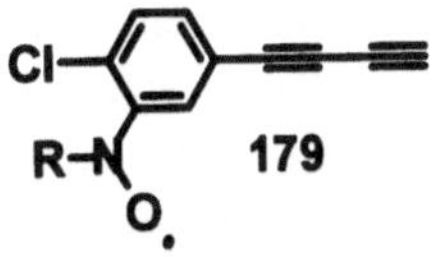

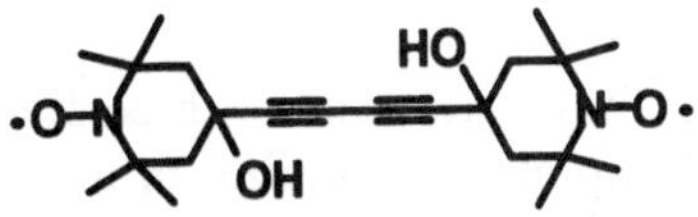

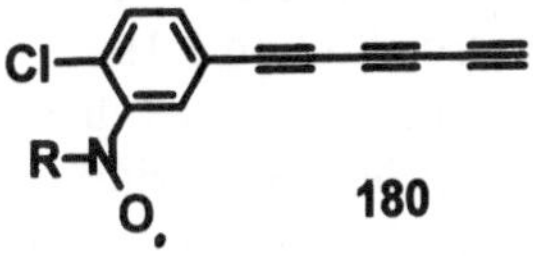

There have been many more efforts on polymer materials with ferromagnetically coupled units (magnetic polymers), leading to materials with some ferromagnetic domains and many spins [450]. However, no definitive evidence for three-dimensional magnetic ordering with spontaneous magnetization in a well characterized and reproducibly prepared polymer has been found. The main portions of all the samples reported so far behave as paramagnets with isolated spin centers.

7.4 Crystalline Radicals with Ferromagnetic Properties

The galvinoxyl **181** can be regarded as the first organic system exhibiting ferromagnetic coupling [465]. In the high-temperature crystalline phase it indicates ferromagnetic order with a positive Curie-Weiss temperature. Below

85 K there occurs a phase transition to an antiferromagnetic phase. When preparing mixed crystals of galvinoxyl and phenol in different ratios, the ferromagnetic coupling is preserved down to 2 K (6:1 and 4:1), and J becomes J = + 56.5 cm^{-1}. The magnetization curves follow closely the theoretical curve for S = 3. The galvinoxyls **181** are known to have atoms of different spin densitites, large negative and small positive spin density, which are coupled most strongly to positions of opposite spin density of different size in neighboring molecules [466]. Thereby the total spin is not cancelled, and an intermolecular ferromagnetic coupling results (Heitler London spin exchange). This outcome corresponds to the realization of the first suggestion for the design of organic ferromagnets (see Section 3.1).

A nitronyl-nitroxide **182** has been shown to exhibit three-dimensional ferromagnetic order. Surprisingly, the compound crystallizes in four different phases α, β, β$_h$ and γ, respectively, and all of them exhibit ferromagnetic coupling [467–470]. The crystal of y-**182** consists of a two-dimensional network with separated molecular stack [470]. The susceptibility data give a positive Curie Weiss temperature with Θ = 2.4 K, and from the field dependence of the magnetization an increase of the effective spin from S = 3/2 (4.9K) to S = 3 at 2.3 K has been reported. The Curie-Weiss temperature and the estimated spin values are largest for the y-phase compared with the series of crystals (α, β, β$_h$ and γ). The saturation behavior of the magnetization seems to be characteristic for these ferromagnetically coupled organic radicals where the exchange-correlated spin domain increases with decreasing temperature.

When using *meta*-phenylenebis (nitronyl-nitroxide) **183** or 1,3,5-phenylene-tris (nitronyl-nitroxide **184** stable bi- and triradicals with triplet and quartet ground state are formed [466]. Upon crystallization the high-spin ground state is maintained, but the antiferromagnetic intermolecular interaction dominates. In the triradical form the susceptibility exhibits a maximum at 16 K, then decreasing to nearly vanishing values.

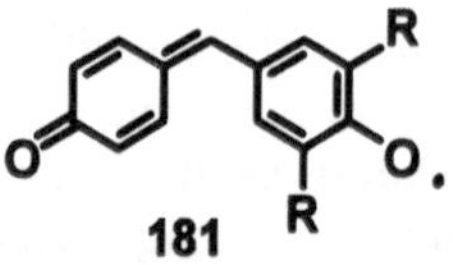

181

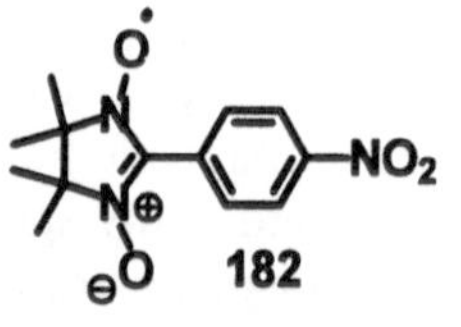

182

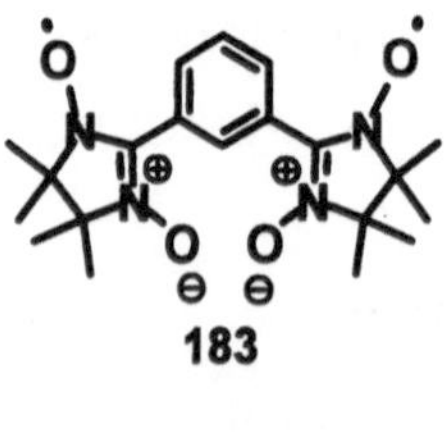

183

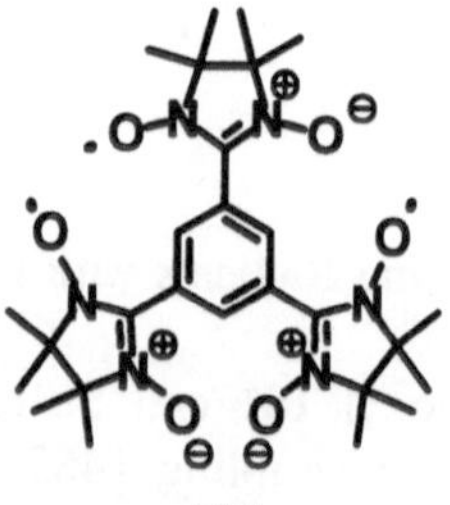

184

A number of antiaromatic biradicals mentioned before (129, 160–164) have been considered as building blocks for D/A charge-transfer stacks with ferromagnetic intrastack coupling [369, 376, 385]. Unfortunately, all the systems tested so far have shown antiferromagnetic interactions in the solid. This holds also if the D/A stacks, e.g. hexaethoxy substituted triphenyl 185 and TCNQ-F$_4$, are doped after crystallization [471].

While the progress in the search for purely organic materials with magnetic properties seems to be slow, the recent advances in organometallic donors stacked with organic acceptors as TCNQ (118) or TCNE (186) leading to ferromagnets with high Curie temperature is amazing [370, 378]. In complexes of decamethylferrocenium donors (187) and TCNE (186) acceptors the susceptibility obeys the Curie-Weiss expression with $\Theta = +30$ K for T > 60 K, suggesting dominant ferromagnetic interactions with an exchange coupling J of 19 cm^{-1} [472, 473]. Below the critical temperature T_c of 4.8 K, spontaneous magnetization has been observed. Ferromagnetic behavior also exists in other metallocene complexes with cyclopentadienyl ligands such as MnCp$_2$ and CrCp$_2$, where the critical temperature with field dependent magnetization is slightly higher for the manganocene ($T_c = 8.8$ K) than for the chromocene ($T_c = 3.7$ K) [378]. By far the highest T_c is obtained by starting from a bisbenzenovanadium complex. Upon addition of TCNE as acceptor there is evidence that the benzenes are lost, leading to a new vanadium complex in which the acceptor molecules set up the coordination sphere [474]. This preparation leads to a room temperature magnetic material ($T_c = 400$ K) that is attracted to a permanent magnet. Due to the insolubility of the material, further structural characterization is necessary for better insight into the molecular alignment.

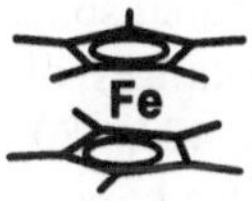

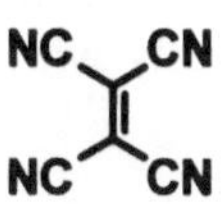

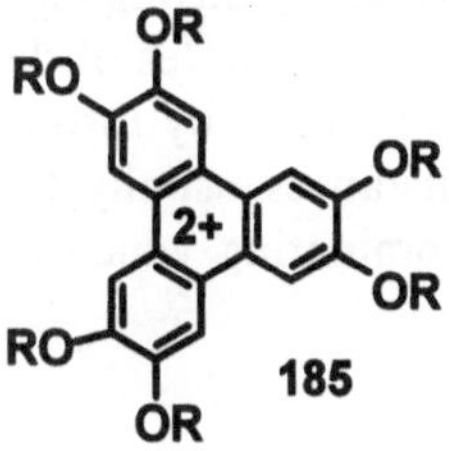

Within Sect. 7 it has been shown how important the design and synthesis of organic high-spin molecules becomes on the way to organic ferromagnets. Thereby, not only is the high-spin formation important, but also the stability of the radicals once formed, and the more complete transformation of accessible radical centers in higher oligomers and polymers. Ferri or Ferro-magnetic ordering has been obtained in crystalline radicals but they are difficult to obtain and there is no method to predict the crystalline ordering. Therefore, high-spin polymers with better processibility of the material are needed.

8 Electron-Transfer Processes and Devices

8.1 Ground State Electron Transfer

The nature of the spin-density distribution as a function of π-topology, conformation and ion pairing has been considered as the crucial piece of information in describing radical ions (see Sects. 1 and 3). The experimental detection of a spin-density distribution gains additional significance when a molecule is composed of two identical redox units **A** linked by a saturated spacer **L** according to the sequence **A-L-A** [46]. Upon addition or removal of an electron one faces the question whether the unpaired electron is localized in one subunit or whether it will undergo a degenerate electron transfer, a so-called self-exchange, between the identical redox systems **A** [475, 476]. If this process is rapid within the time scale of the prevailing experiment one observes an effective delocalization (see Scheme 14).

The difference between photoinduced electron-transfer and ground state electron transfer in the redox product **A-L-A**$^{+\bullet}$ is that in the latter one is dealing with an ion pair whose structure has to be reorganized upon electron transfer [477–483]. A great deal of experimental studies has been devoted to the question of how the self-exchange in **A-L-A**$^{+\bullet}$ can be controlled by the structure of the units **A** and **L**, i.e. by the nature of the redox unit **A**, of the spacer **L** and of the ion pairing [46]. A particularly important factor is the linking group **L** because it defines the distance and relative orientation of the redox centers **A** and provides a channel for the charge to be transferred. Concerning the role of **L** it is additionally interesting to emphasize the relative importance of through-bond and through-space mechanisms. Even with saturated spacers **L**, fast long-range electron transfer has been observed up to distances of 1.5 nm, which cannot be explained by through-space coupling [476, 479, 480, 484, 485, 486]. Therefore the energetic states of the bridging molecules have to be considered and an indirect exchange by through-bond coupling seems to become important. At short distances usually the direct coupling of the redox centers (π-systems) **A** is dominating which decreases exponentially with increasing distance.

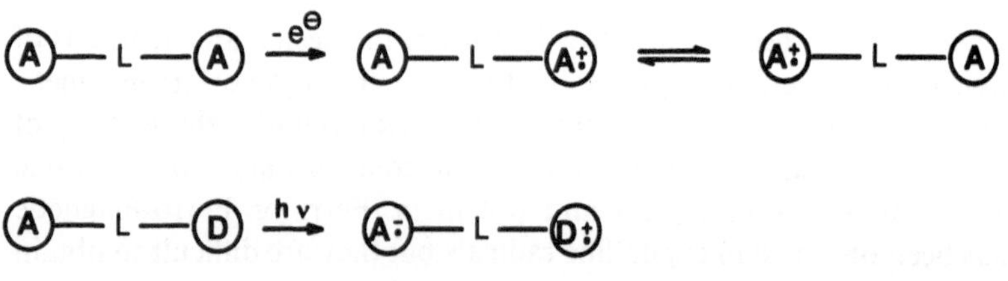

Scheme 14. Electron transfer in the ground state of a biselectrophoric monoradical **A**$^{+\bullet}$**A** bridged by spacer **L** and photoexcited electron transfer in donar-acceptor bridged molecules **A-L-D**

The rate of intramolecular electron-hopping processes has been treated in several reviews as a function of structural parameters [11, 20, 46, 475, 476, 486–488]. It seems appropriate here to point out a few aspects in which an electron transfer is used to switch between different states of a material [489]. A particularly challenging, though somewhat speculative topic, is the field of molecular electronics in which molecules are used as electronic circuit elements in signal processing [490, 491]. This exciting idea is a logic consequence of theoretical and experimental limits in the miniaturization of conventional electronic devices. Thereby a tremendous storage capacity could become accessible, which exceeds that of the information carriers used up to now. In this context the spectroscopic investigation of an intramolecular through-bond electron transfer in solution would have its counterpart in the "conductance" of a single molecule in which electrons are transmitted through a single molecule from one conducting wire to another. The construction of a molecular switch would then be related to all chemical or physical processes by which the spacer group is changed in order to allow or not allow an electron transfer between the outer redox sites A [491–493]. Different structural features may be used for this purpose, e.g. 1) spin orientation within a molecule, 2) hydrogen bonding leading to a dipole moment which may be switched by electrical fields, 3) cis-trans isomerism that can be induced photochemically or by charging, and 4) donor-acceptor molecules which undergo charge separation upon excitation.

A major task is to find out how to use these well known processes in technical applications. Thereby it becomes important to address single molecules or clusters of molecules in the solid state. It has, indeed, been proposed that the system D-L-A can be considered as a zero-dimensional device which could rectify an electric current between two metal electrodes [494, 495]. If a good donor, such as TTF (120), and a good acceptor, such as TCNQ (121), are linked by a saturated spacer L, the energy of the ground state has to be much closer to the zwitterionic excited state $D^{+\bullet}\text{-L-A}^{-\bullet}$ than to the neutral state or even $D^{-\bullet}\text{-}L\text{-}A^{+\bullet}$ [496–498]. Accordingly, a current-voltage plot would be asymmetric. The actual detection of asymmetric current-voltage plots, preferably by scanning tunnelling microscopy [499], has been performed not actually for single molecules, but for experimental set-ups in which the active molecules are anchored to a substrate or organized in monolayers [500]. An interpretation of such plots, however, is by no means straightforward due to the presence of experimental artefacts.

Molecules of the general structure A-L-A can, in principle, adopt further electronic functions such as memory or amplification [492, 495]. This is best described by identifying a non-conducting conjugated π-chain as a pro-conductor since according to Sect. 3 it can be rendered conducting by an electron transfer. Consider an example of A-L-A in which linear π-chains are held in an orthogonal arrangement by, e.g., a spacer group (Scheme 15) and in which the ends of the π-chains are attached to electrodes. If an electron is removed from A-L-A, the charge can sit on either chain thus giving rise to a two-state system, which is actually one bit in a binary system since one state can be identified as

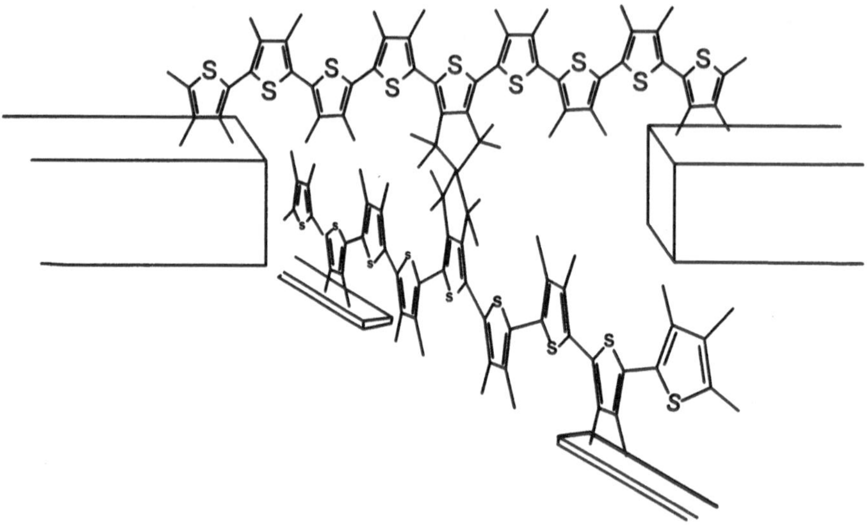

Scheme 15. Model for a molecular switch in which two identical π-chains are held in an orthogonal arrangement by the spacer **L**. The ends of the chains are connected to electrodes

"1" and the other state as "0" [490–495]. The information can, in principle, be read by measuring the conductivity of either chain. In the course of an electron transfer, however, the pro-conductor chain becomes conductive and the conductor chain becomes insulating. A possible way of switching is the application of an electric field perpendicular to both chains. This would correspond to chemical means of switching in which rapid processes such as protonation or photochemically induced conformational changes would affect the state of the bridging group **L**. As a consequence the initial state would be stable up to a threshold field and undergo a sudden change above the threshold.

One of the key features which qualifies molecules for use in information storage is the localization of charges on a molecule. The molecular chains connecting a bistable molecule which can be addressed, should be made of different conjugated polymers which can carry charged excitations such as polarons and bipolarons. It is thought that such electron-transfer storage can be brought about in molecules with an unsymmetric redox behavior, i.e. the uptake and release of electrons in a cyclic voltammogram occur at different potentials due to conformational changes as in COT (**11**) or zwitter-viologen **188** [491]. In **188** the charging with two electrons at negative potentials occurs in one step (-0.75 V), while the reoxidation is separated into two one electron transfer processes (-0.3 and -0.7V). Thus with a constant potential between the two separated redox steps (-0.5 V) an additional negative potential pulse would turn the zwitter-viologen into the doubly charged state which relaxes into the monocharged state, while a positive pulse turns it back into the neutral state [491, 501].

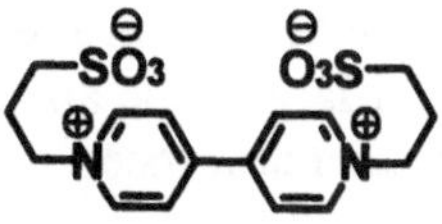

188

8.1.1 Electrochromic Devices

Besides the use of conjugated polymers as conductors their use in optical devices for technical application is most advanced [73, 502]. In electrochromic devices the application is due to the drastic color change observed upon doping to increase the conductivity. Electrochromism is defined as color change under application of an electric field or current [503]. Depending upon the conducting polymer chosen, the absorption of the doped state is dramatically red-shifted and the color can be modified using dopant ions which absorb in the visible. The effect can be obtained in solid polymer layers and in electrolyte solutions, where the film of the organic substrate covers an electrode [502, 504, 505]. The mechanism of electrochromism is assumed to occur upon injection of holes into the organic layer from the anode and anion migration from the electrolyte into the polymer to ensure electroneutrality (ion insertion materials). The advantage of the polymeric devices is their ease of preparation and the access of very thin film for use in displays [502, 506]. In a display device often an electrochemical cell is used in which an electrode (anode) is covered with the material and the cell is filled with the electrolyte medium.

Because the conducting polymers typically have very high absorption coefficients in the visible range, only very thin film coatings are required to provide display devices having high contrast and a very broad viewing angle. Extensively studied for suitable application have been polymers which can be directly polymerized on an electrode like polypyrrole (**25**), polythiophene (**26**) and polyaniline (**54**) [506–509]. During oxidation they cover a large spectral range, that expands for instance for **54** from yellow through green-blue and finally to brown. The transition time between the color changes lies below 100 msec even for relatively thick films of **54** (50 nm). The maximum number of cycles on the other hand is still limited to 10^5–10^6. Therefore a large scale production of electrochromic polymers for display application depends on the lengthening of the lifetimes up to 10^7 and the shortening of response times [510]. Research has been described using the electrochromic materials for mirrors and windows and control of solar heating [511].

8.1.2 Electroluminescence

The use of light emitting diodes (LED) continues to increase dramatically, for example in color displays. The inorganic semiconductors like GaN, ZnS, ZnSe, and SiC have low luminescence efficiency in the blue spectral range, and it is difficult to produce all colors for large-scale applications.

A major breakthrough has therefore been the discovery of electrolumine-scence from polymer materials which was first demonstrated to PPV **8** [6, 506]. This achievement opened a new branch in materials science, focusing on the optical properties of polymers after electrical excitation [512, 513]. The film of the polymer is usually deposited on a positive electrode like indium/tin-oxide and a negative electrode of aluminium or calcium is than evaporated on the upper surface. The emission is just detected by a photomultiplier tube. The molecular mechanism of the electroluminescence seems to be close to that in photoluminescence, and the emitted light in both kinds of experiments occurs in the same spectral region [6]. For electroluminescence a radiative recombination of the electrons and holes injected from opposite sides of the structure seems to be responsible. The color of the light produced by the polymer-based LED depends on the size of the energy gap and hence the electronic structure of the semiconducting polymer serving as the active light-emitting layer. In practice, this electronic structure is controlled through the design and synthesis of a variety of semiconducting polymers exhibiting a range of energy gaps between roughly 1.5 and 3 eV.

The first devices were based on **8** and emitted in the yellow-greenish (around 500–600 nm) spectral range through charge injection under a high applied field. This opened the search for other organic semiconductors to cover the visible spectral range [512–515]. The advantage of the polymeric materials over systems with low molecular weight is the tendency of the latter to recrystallize initiated by the heat produced in the device. Recently a poly(p-phenylene) device [514] has been made that emits in the blue spectral range, and block copolymers of PPV can be chemically tuned to emit light ranging from green-blue to orange-red [515]. Another derivative of PPV with much better processibility, namely MEH-PPV **189**, allows the preparation of flexible LED's on a polyethyleneter-ephthalate surface and using polyaniline **54** as an electrode [516].

189

Since the calcium or other metals used so far as electrodes must be rigorously protected from the atmosphere to prevent degradation further applications have to find stable electrode materials. Also, the attainable electric currents and the efficiency of the polymeric LED depends sensitively on the polymer-electrode interface.

8.2 Photoinduced Electron Transfer

The term photoinduced electron transfer describes the transfer of electrons from a photoexcited donor **D*** to a ground state acceptor **A**. This may occur in an intra- and intermolecular fashion. In the intramolecular electron transfer the time for charge separation and lifetime of the D^+/A^- state depends on the distance defined by the spacer **L** and the orientation of the photoactive centers (Scheme 14) [479, 485, 488, 516]. A great number of donor-acceptor molecules for photoinduced electron transfer has been synthesized, most of them for mimicking the charge separation and charge transfer in photosynthesis. The major concern is the distance and orientation of the photoactive components to enable a stabilization of the charge separation. These model systems have been extensively reviewed [11, 488]. No organic based system, however, capable of forward and reverse electron transfer, where a switch between the redox states could be assigned as logical unit like 0 and 1, have been prepared up to now.

The donor-acceptor structure **D-L-A** has already been mentioned for use in rectifiers, allowing charge transport only in one direction [490–497]. For use as a molecular switch in electronic devices, it is necessary to reverse the charge-separated state. This may be reached by a special setup of the **D-L-A** moiety, in which the spacer **L** can also be addressed. The spacer should then be a conjugated π-system, which supports polaron formation. This system has been called polaronic storage molecule [491]. A potential gradient within the extended molecule is necessary, with the HOMO of the donor should be higher in energy than the HOMO of the bridging π-chain (the valence band of the bridge) and the LUMO of the acceptor lower than the LUMO of the bridge. Upon donor excitation an electron transfer into the LUMO (conduction band) of the spacer is induced (anionic polaron formation) $D^{+\bullet}L^{-\bullet}A$. The polaron may move to the acceptor where it is trapped in the LUMO. The reverse process can be initiated by an excitation of the conjugated bridge, combined with hole tunnelling from the donor to the acceptor leading to charge recombination. Thereby a cationic polaron travels along the bridge. A number of molecules with similar features are available, which could be further tested for nanostructured devices [491].

For an effective charge-transfer process a large electronic interaction between the donor-acceptor groups is needed, which depends mainly on the spacer unit **L**. This interaction will generally decrease with an increase in the effective length of the bridging unit [493, 518]. For application as a molecular switch it is important that the bridge can change the interaction between the connected redox-groups drastically, under the influence of an external perturbation, inducing a jump from one state to another. For instance using bi- and oligoarylenes e.g. **2, 10, 19, 66** as a spacer, which serve as important chromophores and electrophores, the interaction between the redox centers depends on the angle of torsion between the arylenes (Sects. 1 and 3). If the aryl units are electronically decoupled, locally excited states may be formed [519, 520]. Symmetric biaryls sometimes exhibit solvent-induced charge separation in the excited state which has been called twisted intramolecular charge transfer (TICT) [521]. Experimentally the TICT state is established by a dual fluorescence. The prototype for molecules exhibiting TICT formation are aromatic amines like dimethylaminobenzonitril **190** [522]. In polar solvents, the photochemically excited state relaxes towards a twisted state with considerable degree of charge transfer (TICT) [521, 523, 524]. The planar excited state emits near 350 nm while the twisted state emits near 480 nm, and the electronic interaction between the donor and acceptor has vanished. Better candidates for application in switching devices should be more symmetrical molecules which can carry the same redox groups at the end. A standard example is 9,9'-bianthryl (**19**), where the anthracene units are arranged nearly orthogonal even in the ground state and the charge transfer step k_{CT} is very fast and determined by the dynamics of solvent relaxation [525]. This has been further established by comparison with symmetry disturbed mono-substituted bianthryls, where the charge separation process is even faster (by a factor 1.2–2.0) and the equilibrium between locally excited (LE) fluorescence and CT fluorescence is shifted towards the TICT state [526]. 9,10-Polyanthrylene and the corresponding oligomers **167** are promising candidates for a further increase and stabilization of the charge separated state [39, 219, 527]. All the oligomers of **167** studied so far exhibit dual fluorescence in the polar solvent ethanol. The compounds show long and mostly monoexponential fluorescence decay (11–13 ns) with short (425 nm) and long (500 nm) emission wavelength at room temperature, documenting the equilibrium of LE and TICT states. At lower temperatures a non exponential decay is envisaged, which is due to multistep charge separation processes [527]. Therefore the oligo-anthrylenes **167** are very effective for multiple charge separation.

190

In biaryls where the torsional angle between the aryl units is smaller one would expect that the formation of a TICT state will take more time. Comparing 3,3'-biperylenyl (**66**) (with 8,8',11,11'-tetra-*tert*-butyl side groups) with 9,9'-bianthryl (**19**), a solvent polarity induced redshift of the fluorescence for the former could be obtained [528]. The amount of charge transferred, however, is considerably lower than expected for a TICT state, and the CT emission is highly allowed with fluorescence quantum yields exceeding 50% in all solvents investigated. A model of two emissive states has been put forward to explain the observed excited-state dipole moments and the anomalous temperature dependence, exhibiting life-time lengthening on going from rigid matrix to a viscous solvent [528].

8.2.1 Photoconductors

While mostly the doping process is used to enhance the conductivity of a conjugated organic material, charge carriers for increasing the conductivity can also be generated photochemically. The charge carriers are created by charge separation, and they travel in the direction of an applied field [529]. Upon photogeneration, no counterion effects are present so that the charges should need lower activation energy for moving through the polymer sample than in the doped state. In photoconducting experiments, on the other hand, it seems to be more difficult to generate a large number of charge carriers and to prevent a fast charge recombination. The photoexcitation leads to an absorption spectrum which is very similar to the absorption spectra observed upon doping, indicating a common origin for both effects [3, 529].

The first commerical organic photoconductor was based on a charge transfer complex between polyvinylcarbazole **191** (donor) and trinitrofluorenone **192** (acceptor) in low concentrations or in 1:1 stoichiometry [530, 531]. These systems have been further optimized by use of dye-pigment systems, and the yields for charge-carrier production are in the order of 10%, which is equivalent or superior to their inorganic counterparts. The samples are mounted between aluminium electrodes produced by vapor deposition, thus, still being semi-transparent and allowing illumination. A short light pulse is then used to trigger a transient photocurrent that starts usually 10^{-2}–10^{-3}s after the original excitation, depending on applied voltage, sample thickness (ca. 10 μm), illumination intensity, and depth. The photoconductivity depends on the additional doping concentration (usually not higher than 5%), temperature, light intensity and in π-conjugated polymers on the degree of isomerization. The practically linear dependence of the photocurrent with light intensity and with applied field suggest that the photogeneration of radicals is a one-photon process [529]. While the response time is still too slow for fast switches, many organic materials are being tested for electrophotography and -printing (e.g. in xerographic devices). Thereby the above described setup has been changed into double layer systems, where one layer is used for high quantum efficiency of charge carrier generation and a second layer for high charge mobility. The holes migrate through the charge-transport layer and the electrons are neutralized at the front surface [3]. In cases of fast charge recombination photoluminescence is obtained. The luminescence often is red shifted compared to the absorption spectra

191   **192**

(Stokes shift). The bathochromic shift depends on the amount of lattice relaxation, e.g. change of geometry as in 12 and 13. Although the electron-hole pair generation is photoinduced (charge separation), the transport process through the sample is described as a ground state electron transfer mechanism [533].

The photoconductivity depends strongly on the conjugated π-system and the excitation wavelength. In PPV 8 and several phenyl and vinyl substituted derivatives, the influence of electron withdrawing and donating groups on the photoconductivity and excitation wavelength have been tested [50]. The comparison of PPV 8 and DPPPV 9 reveals a hypsochromic shift of the maximum of photoconductivity with excitation wavelength for the phenyl substitution, similar to the shift in optical absorption spectra. Upon introducing other substituents on the polymer chain the accessible excitation wavelength for high photoconductivities can even be tuned within the visible region. Furthermore the photoconductivity of PPV and its derivatives easily exceeds that of polyvinylcarbazole 191 [50], because optical generation of charge carriers appears to be more efficient.

Photoelectrical stabilities with 50–200 k cycles have already been reported for several organic photoreceptors, but the operating conditions and environment still limit the practical lifetime of the photoreceptor [3]. Therefore further stabilization of the charge transfer layers is important. The similarity in the fundamental steps between xerographic photoreceptors and organic energy conversion devices indicates that much higher efficiency for energy conversion becomes possible.

9 What About Synthesis?

The emphasis of this review clearly is on the electronic behavior of radical ions. The key ingredients of the approach taken are 1) the active physical function of the organic title systems and 2) the combination of solution and solid-state properties. Nevertheless, it should not be overlooked that some major problems of the materials science of organic radical ions can only be solved by way of synthesis. It is highly appropriate, therefore, to conclude this text by pointing out some important challenges for future work in organic and macromolecular synthesis.

The "inherent" effective conjugation length of radical ion sites on a π-chain can only be defined properly if there is no "false" confinement induced by chemical defects. Chemical defects such as sp^3-centers within a conjugated chain must also be avoided in view of the charge-carrier mobility. Problems associated with the synthesis of structurally homogeneous, defect-free π-systems should be even more serious when going from linear to two-dimensional, so-called ladder structures, which are known to have particularly attractive chemical and physical properties.

In the field of electrical conductors polyacetylene (6) and polypyrrole (25) are structurally different since in the formation of PAc one tries to create a structurally regular material with an array of parallel chains, while the macromolecules of PPy are "irregular" in that they possess various degrees of cross-linking and entangling of chains. It should be important in a reverse approach to obtain regular PPy and "irregular" PAc with conjugated crosslinks, where e.g. sp^3-centers are bridged via conjugated by-passes. The controlled production of sp^3-centers on an extended chain can also be a target of synthetic efforts as becomes clear from the tuning of electrochromic and electroluminescence properties (see Sect. 8). A closely related approach is the use of smaller oligomeric units where the synthesis can be much better controlled (see Sects. 2, 3 and 7). The design of many physical properties implies the organization of molecules into a regular solid-state structure, or, alternatively, the formation of amorphous and mechanically robust films. Consequently, synthetic efforts must be extended to include the aspects of processing and of supramolecular architecture. Phase formation from oligomeric and polymeric macromolecules e.g. by way of side-chain crystallization, is a typical example.

Properties such as conductivity of doped polymers and magnetism of high-spin structures are bound to the stability of radical and/or ion sites. Various attempts to stabilize doped materials on a molecular basis have been considered, e.g. via the introduction of polarizable sulfur substituents, or by methods of processing such as immobilization in films or isolation in matrices.

Problems associated with the one-dimensional character of the charge transport in conducting radical-ion salts and charge-transfer complexes, with stack-type structures suggest an increase in dimensionality by the creation of interstack interactions. This can be achieved by creating appropriate crystal structures, which is a matter of trial and error rather than of a "controlled engineering". Alternatively, one can try to anticipate an increase in dimensionality by the synthesis of extended donors with a higher chance for short interstack contacts in the crystal. The crystalline conductors, on the other hand, are brittle, non-processable materials; the synthesis of film-forming and yet conducting CT complexes is therefore an important research topic.

Even if unwanted chemical transformations of doped conjugated materials can be avoided by strict exclusion of air, the long-term stability is limited by a de-mixing of dopant and redox system. An alternative approach to persistent electrical conductors is therefore concerned with the synthesis of intrinsic conductors. Toward that end the lowering of the band gap is desirable which, on the other hand, is known to reduce the chemical stability. The synthetic approaches taken must therefore focus on a compromise between electronically attractive π-structure, stability and processability of the material. In conclusion, it is the flexible combination of different areas of research which provides new insights into the properties of monomeric, oligomeric and polymeric radical ions.

Acknowledgement. We wish to express our gratitude to our coworkers and collaborators who made this article possible by their skill and dedication. We are grateful to Prof. Dr. Bässler, Dr. S.

Schrader, Prof. Dr. J. Heinze, Priv.-Doz. Dr. V. Enkelmann and Prof. Dr. N. Tyutyulkov for careful reading of the manuscript, discussion and criticism. We have to thank our secretary Mrs. Pilger for helping us in preparing the manuscript and J. Pawlik for drawing the molecules. Financial support of our work by the Bundes Ministerium für Forschung und Technologie (BMFT), Deutsche Forschungsgemeinschaft (DFG), Fonds der Chemischischen Industrie, Stiftung Volkswagenwerk as well as by the BASF AG, Ludwigshafen, the Hoechst AG, Frankfurt, and the Wacker-Chemie GmbH is gratefully acknowledged.

10 References

1. Skotheim TA (ed) (1986) Handbook of conducting polymers. Marcel Dekker, New York
1b. Brédas JL, Silbey R (eds) (1991) Conjugated polymers. Kluwer, Dordrecht
2. Naarmann H (1990) In: Brédas JL, Chance RR (eds) Conjugated polymeric materials: Opportunities in electronics, optoelectronics, and molecular electronics. Kluwer, Dordrecht, p 11
3. Law KY (1993) Chem Rev 93: 449
4. Bohnen A, Räder HJ, Müllen K (1992) Synth Met 47: 37
5. Colaneri NF, Bradley DDC, Friend RH, Burn PL, Holmes AB, Spangler CW (1990) Phys Rev B42: 11670
6. Burroughes JH, Bradley DDC, Brown AR, Marks N, Mackay K, Friend RH, Burn PL, Holmes AB (1990) Nature 347: 539
7. Müllen K (1984) Chem Rev 84: 603
8. Müllen K (1987) Angew Chem 99: 192; Int Ed Engl 26: 204
9. Roth HD (1992) Topics Curr Chem 163: 131
10. Boche G (1988) Topics Curr Chem 146: 1
11. Fox MA, Cannon M (eds) (1988) Photoinduced electron transfer. Elsevier, Amsterdam
12. Metzger RM, Panetta CA (1989) J Mol Electronics 5: 1
13. Heilbronner E, Bock H (1978) (eds) Das HMO Modell und seine Anwendungen. Verlag Chemie Weinheim
14. Heinze J (1984) Angew Chem 96: 823, Int Ed Eng 23: 831
15. Gerson F (1967) Hochauflösende ESR Spektroskopie. Verlag-Chemie Weinheim
16. Kurreck H, Kirste B, Lubitz W (1988) ENDOR spectroscopy of radicals in solution. In: Marchand AP (ed) Methods in stereochemical analysis, vol 15. VCH, Weinheim
17. Enkelmann V, Göckelmann K, Wieners G, Monkenbusch M (1985) Mol Cryst Liq Cryst 120: 195
18. Kröhnke C, Enkelmann V, Wegner G (1980) Angew Chem 92: 941 Int Ed Engl 19: 912
19. Huber W, May A, Müllen K (1981) Chem Ber 114: 1318
20. Huber W, Müllen K (1986) Acc Chem Res 19: 300
21a. Gorman CB, Grubbs RH (1991) In: Brédas JL, Silbey R (eds) Conjugated polymers. Kluwer Dordrecht, p 1
21b. Hörhold HH (1972) Z Chem 12:41
22a. Meyer H (1992) Angew Chem 104: 1425; Int Ed Engl
22b. Mazzugato U, Momiccioli F (1991) Chem Rev 91: 1679
22c. Orlandi G, Zerbetto F, Zgierski MZ (1991) Chem Rev 91: 867
23. Bock H, Ruppert K, Fenske D (1989) Angew Chem 102: 548 Int Ed Engl 29: 525
24. Baumgarten M, Weitzel HP, Schulz A, Garay R, Müllen K (to be published)
25. Schenk R, Huber W, Schade P, Müllen K (1988) Chem Ber 121: 2201
26. Müllen K, Heinz W, Klärner FG, Roth WR, Kindermann, Adamczak O, Wette M, Lex J (1990) Chem Ber 123: 2349
27. Anet AL, Bourn AJR, Lin YS (1964) J Am Chem Soc 86: 3576
28a. Lhost O, Brèdas JL (1992) J Chem Phys 96: 5279
28b. Zerbi G, Gussoni M, Castiglioni C (1991) In: Brédas JL, Silbey R (eds) Conjugated Polymers. Kluwer, p 435

29a. Chien JCW (ed) (1984) Polyacetylene: Chemistry, Physics and Material Science. Academic, San Diego
29b. Ito T, Shirakawa H, Ikeda S (1974) J Polym Sci Polym Chem Ed 12: 11
29c. Gibson HW, Kaplan S, Mosher RA Prest WM, Weagley RJ (1986) J Am Chem Soc 108: 6843
 30. Pekker S, Janossy A (1986) In: Skotheim TA (ed) Handbook of Conducting Polymers. Marcel Dekker, New York, p 45
 31. Müllen K (1986) Pure & Appl Chem 58: 177
32a. Glasbeck M, van Voorst JDW, Hoijtink GJ (1966) J Chem Phys 45: 1852
32b. Krusic PJ, Wasserman E (1991) J Am Chem Soc 113: 2322
33a. Staley SW, Henry TJ (1969) J Am Chem Soc 91: 1239
33b. Schröder G (ed) (1965) Cyclooctatetraen. Verlag Chemie, Weinheim.
33c. Katz TJ, Garratt PJ (1964) J Am Chem Soc 86: 5194
 34. Müllen K (1974) Helv Chim Acta 57: 2399
 35. Oth JFM (1971) Pure & Appl Chem 25: 573
35b. Oth JFM, Anthoine G, Gilles JM (1968) Tetrahedron Lett 6265
 36. Müllen K, Meul T, Vogel E, Kürschner U, Schmickler H, Wennerström O (1985) Tetrahedron Lett. 26: 3091
 37. Gust D, Senkler GH, Mislow K (1972) J Chem Soc Chem Commun 1345
 38. Huber W (1985) Tetrahedron Lett 181
 39. Baumgarten M, Müller U, Bohnen A, Müllen K (1992) Angew Chem 104: 482, Int Ed Engl 31: 448
 40. Baumgarten M, Müller U (1993) Synth Met 57: 4751
 41. Rabinovitz M (1988) Top Curr Chem 146: 99
 42. Minsky A, Meyer AY, Rabinovitz M (1983) Angew Chem Int Ed Engl 22: 45
 43. Minsky A, Rabinovitz M (1984) J Am Chem Soc 106: 6755
 44. Gregorius H, Baumgarten M, Reuter R Tyutyulkov N, Müllen K (1992) Angew Chem 104: 1621 Int Ed Engl 31: 1653
 45. Brédas JL, Heeger AJ (1990) Macromolecules 23: 1150
 46. Baumgarten M, Huber W, Müllen K (1993) Adv Phys Org Chem 28: 1
47a. Heinze J (1990) Topics Curr Chem 152: 1
47b. Diaz AF, Bargon J (1986) In: Skotheim TA (ed) Handbook of conducting polymers. Marcel Dekker, New York, p 81
 48. Treloar LRG (1958) Physics and chemistry of rubbeer elasticity. Clarendon, Oxford
 49. Ward IM (1983) Mechanical properties of polymers. Wiley, Chichester
 50. Hörhold HH, Helbig M (1987) Makromol Chem Macromol Symp 12: 229
 51. Koßmehl G (1986) In: Skotheim TA Handbook of conducting polymers. M. Dekker, NY, p 351
 52. Heeger AJ, Smith P (1991) In: Brédas JL, Silbey R (eds) Conjugated polymers. Kluwer, Dordrecht, p 141
 53. Müllen K (1993) Pure & Appl Chem 65: 89
 54. Bohnen A, Heitz W, Mülleen K, Räder HJ, Schenk R (1991) Makromol Chem 192: 1679
 55. Rehahn M, Schlüter AD, Wegner G (1990) Makromol Chem 191: 1991
 56. Shi S, Wudl F (1990) Macromolecules 23: 2119
 57. Wegner G (1986) Makromol Chem Makromol Symp 1: 151
58a. Schäfer-Siebert D, Budrowski C, Kuzmany H, Roth S (1987) In: Kuzmany H, Mehring M, Roth S (eds) Electronic properties of conjugated polymers. Solid State Sciences 76: 38
58b. Kürti J, Kuzmany H (1987) In: Kuzmany H, Mehring M, Roth S (eds) Electronic properties of conjugated polymers, Solid State Sciences 76: 43
 59. Bradley DDC, Friend RH, Feast WJ (1987) Synth Met 17: 645
 60. Ringsdorf H, Schlarb B, Venzmer J (1988) Angew Chem 100: 117, Int Ed Engl 27: 113
61a. Ito T, Shirakawa H, Ikeda S (1974) J Polym Sci Polym Chem Ed 12: 11
61b. Naarmann H, Theophilou N (1987) Synth Met 22: 1
 62. Feast WJ (1986) In: Skotheim TA (ed) Handbook of conducting polymers. Marcel Dekker, New York, p 1
63a. Edwards JH, Fest EJ (1980) Polym Comm 21: 595
63b. Gagnon DR, Capistran JD, Korasz FE, Lenz RW (1984) Polym Bull 12: 93
 64. Wessling RA, Zimmermann RG (1968) US Patent 3: 401, 152
 65. Wessling RA (1986) J Polym Sci Polym Symp 72: 55
 66. Fahnenstich U, Koch KH, Müllen K (1989) Makromol Chem 10: 563
 67. Müller U, Baumgarten M, Müllen K (1994) submitted
 68. Weitzel HP, Müllen K (1990) Makromol Chem 191: 2837

69. Garay R, Baier U, Bubeck C, Müllen K (1993) Adv Mat 5: 568
70. Deronzier A, Moutet JC (1989) Acc Chem Res 22: 249
71a. Chiang CK, Park YW, Heeger AJ, Shirakawa H, Louis EJ, McDiarmid YW (1977) Phys Rev Lett 39: 1098
71b. Park YW, Heeger AJ, Druy MA, McDiarmid YW (1980) J Chem Phys 73: 946
72a. Maaxfield M, Mu SI, MacDiarmid AG (1985) J Electrochem Soc 132: 838
72b. Bittihn R, Ely G, Woffler F, Münstedt H, Naarmann H, Naegele D (1987) Makromol Chem Macromol Symp 8: 51
73. Baughman RH (1991) Makromol Chem Macromol Symp 51: 193
74. Lee BI (1992) Polymer Engineering and Science 32: 36
75. Tyutyulkov N, Karabunarliev S, Müllen K, Baumgarten M (1993) Synth Met 53: 205
76. Tanaka K, Koike T, Ueda K, Ohzeki K, Yamabe T (1985) Synth Met 11: 61
77. Lee YS, Kertesz M, Elsenbaumer RL (1990) Chem Mater 2: 526
78. Baumgarten M, Karabunarliev S, Koch KH, Müllen K, Tyutyulkov N (1992) Synth Met 47: 21
79. Scherf U, Müllen K (1992) Synthesis 1/2: 23
80. Scherf U Müllen K (1992) Polym Commun 33: 2443
81. Hong SY, Kertesz M, Lee YS, Kim OK (1992) Macromolecules 25: 5424
82. Rughooputh SDDV, Hotta S, Heeger AJ, Wudl F (1987) J Polym Sci Polym Phys 25: 1071
83. Schimmel T, Gläser S, Schwoerer M, Naarmann H (1991) In: Brédas JL, Silbey R (eds) Conjugated polymers. Kluwer Academic Publishers, Netherlands, p 49
84. Rost A (1978) "Messung dielektrischer Stoffeigenschaften", Akademie-Verlag, Berlin
85. Montgomery HC (1971) J Appl Phys 42: 2971
86. Wegner G (1981) Angew Chem 93: 352 Int Ed Engl 20:361
87. Alcazar L (1980) "The physics and chemistry of low dimensional solids", Reidel, Dordrecht
88. Menke K, Roth S (1986) ChIUZ 20: 1, 33
89. Chance RR, Boudreaux DS, Brédas JL, Silbey R (1986) In: Skotheim TA (ed) Handbook of conducting polymers. Marcel Dekker, New York, p 825
90. Heeger AJ, Kivelson S, Schrieffer JR, Su WP (1988) Rev Mod Phys 60: 781
91. Bernier P (1986) In: Skotheim TA (ed) Handbook of conducting polymers. Marcel Dekker, New York, p 1099
92. Thomann H, Dalton LR (1986) Skotheim TA (ed) Handbook of conducting polymers, Marcel Dekker, New York, p 1157
93. Mehring M, Grupp A, Hofer P, Kass H (1989) Synth Met 28: D399
94. Nechtschein M, Devreux F, Genoud F, Gugliani M, Holcer K (1983) Phys Rev B 27: 61
95. Brédas JL, Street GB (1985) Acc Chem Res 18: 309
96. Brédas JL (1986) In: Skotheim TA (ed) Handbook of conducting polymers, Marcel Dekker, New York, p 859
97. Vogel P, Campbell DK (1990) Phys Rev B 41: 12797
98. Ofer D, Crooks RM, Wrighton MS (1990) J Am Chem Soc 112: 7869
99. Schenk R, Gregorius H, Meerholz K, Heinze J, Müllen K (1991) J Am Chem Soc 113: 2634
100. Schenk R, Ehrenfreund M, Huber W, Müllen K (1990) J Chem Soc Chem Commun 23: 1673
101. Schenk R, Gregorius H, Müllen K (1991) Adv Mat 3: 492
102. Heinze J, Mortensen J, Müllen K, Schenk R (1987) Chem Commun 701
103. Meerholz K, Schenk R, Müllen K, Heinze J (1994) to be published
104. Tian B, Zerbi G, Schenk R, Müllen K (1991) J Chem Phys 95: 3191
105. Tian B, Zerbi G, Müllen K (1991) J Chem Phys 95: 3198
106. Woo HS, Lhost O, Graham SC, Bradley DDC, Friend RH, Quattrocchi C, Brédas JL, Schenk R, Müllen K (1993) Synth Met 59: 13–28
107. Brendel P, Grupp A, Mehring M, Schenk R, Müllen K, Huber W (1991) Synth Met 45: 49
108. Plato M, Biehl R, Möbius K, Dinse KP (1976) Z Naturforsch A 31: 169
109. Huber W, May A, Müllen K (1981) Chem Ber 114: 1318
110. Rabinovitz M, Willner L, Minsky A (1983) Acc Chem Res 16: 298
111. Chang R, Markgraf JH (1972) Chem Phys Lett 13: 575
112. Schenk R, Hucker J, Hopf H, Räder HJ, Müllen K (1989) Angew Chem 101: 942-944; Int Ed Engl 28: 904
113. Thulin B, Wennerström O (1976) Acta Scand B B30: 369
114. Huber W, Müllen K, Wennerström O (1980) Angew Chem 92: 636; Int Ed Engl 19: 624
115. Müllen K, Unterberg H, Huber W, Wennerström O, Norinder U, Tanner D (1984) J Am Chem Soc 106: 7514
116. Schenk R, Huber W, Schade P, Müllen K (1988) Chem Ber 121: 2201

117. Schenk (1991) PhD thesis Mainz
118. Spangler CW, Hall TJ, Sypochak LS, Liu PK (1989) Polymer 30: 1166
119. Hogen-Esch TE (1977) Adv Phys Org Chem 15:153
120a. Koch KH, Fahnenstich U, Baumgarten M, Müllen K (1991) Synth Met 41: 1619
120b. Koch KH, Müllen K (1992) Chem Ber 124: 2091
121a. Baumagarten M, Anton U, Gherghel L, Müllen K (1993) Synth Met 57: 4801
121b. Baumgarten M, Koch KH, Müllen K (1993) J Am Chem Soc (submitted)
122. Viruela-Martin R, Viruela-Martin PM, Orti E (1992) J Chem Phys 97: 8470
123. Bakhshi AK, Ladik J (1989) Synth Met 30: 115
124. Böhm A, Mauermann H, Gherghel L, Baumgarten M, Müllen K (1993) to be published
125. Böhm A, Adam M, Mauermann H, Stein S, Müllen K (1992) Tetrahedron Lett 33: 2795
126. Böhm A (1991) PhD thesis
127. Karabunarliev S, Baumgarten M, Gregorius H, Müllen K, Tyuytulkov N (1994) to be published
128. Heitz W, Ullrich R (1966) Makromol Chem 98: 29
129. Gregorius H, Heitz H, Müllen K (1993) Adv Mat 5: 279
130. Dawe EA, Land EJ (1975) J Chem Soc Faraday Trans I 72: 2162
131. Lafferty J, Roach A, Sinclair RS, Truscott TG, Land EJ (1977) J Chem Soc Faraday Trans I 73: 416
132. Knoll K, Schrock RR (1989) J Am Chem Soc 111: 7989
133. Bally T, Roth K, Tang W, Schrock RR, Knoll K, Park LY (1992) J Am Chem Soc 114: 2440
134. Kiehl A, Eberhard A, Adam M, Enkelmann V, Müllen K (1992) Angew Chem 104: 1623 Int Ed Engl
135. Kiehl A (1993) PhD thesis Mainz
136. Kiehl A, Eberhard A, Müllen K, Gherghel L, Baumgarten M (1993) (to be published)
137. Tolbert LM, Ogle ME (1990) J Am Chem Soc 112: 9512
138. Deussen M, Bässler H (1992) Chem Phys 164: 247
139. Bässler H, Deussen M, Heun S, Lemmer U, Mahrt RF (1993) Z Phys Chem N F (in press)
140. Kohler BE, Spangler CW, Westerrfield C (1988) J Chem Phys 89: 3422
141. Frank J, Grimme W, Lex J (1978) Angew Chem 90:1002; Int Ed Engl 17: 943
142. Huber W, Müllen K, Busch R, Grimme W, Heinze J (1982) Angew Chem 94: 294; Int Ed Engl 21: 301
143. Heinze J, Dietrich M, Hinkelmann K, Meerholz K, Rashwan (1989) Dechema Monogr 112: 61; Chem Abstr 110: 221 359r
144. Heinz W, Langensee P, Müllen K (1986) J Chem Soc Chem Commun 947
145. Heinze J (private communication)
146. Huber W (1985) Tetrahedron Lett 181
147. It should be noted in the characterization of radical anions by absorption spectroscopy that even traces of water give rise to rapid protonations of the charged species originally obtained. This quenching process is particularly critical for the sparingly soluble higher oligomers. The dihydro products thus formed still possess shorter phenylenevinylene electrophores, which undergo further reduction and thus give rise to false signal assignment.
148. Fichou D, Xu B, Horrowitz G, Garnier F (1991) Synth Met 41: 463
149. Horrowitz G, Peng X-Z, Fichou D, Garnier F (1991) J Molec Electr 7: 85
150. Havinga EE, Rotte I, Meijer EW, Hoeve WT, Wynberg H (1991) Synth Met 41: 473
151. Yassar A, Delabouglise D, Hmyene M, Nessak B, Horrowitz G, Garnier F (1992) Adv Mater 4: 490
152. Bäuerle P, Adv Mater (1992), 4, 102
153. Bäuerle P, Segelbacher U, Gaudl GU, Huttenlocher D, Mehring M (1993) Angew Chem, 105: 125.
154. Segelbacher U, Sariciftci NS, Grupp A, Bäuerle P, Mehring M (1993) Synth Met 55-57: 4728
155. Ehrendorfer C, Neugebauer H, Neckel A, Bäuerle P (1993) Synth Met 55-57: 493
156. Bäuerle P, Götz G, Segelbacher U, Huttenlocher D, Mehring M (1993) Synth Met 55-57: 4768
157. Bäuerle P, Gaudl KU (1991) Synth Met 43: 3037
158. Martina A, Enkelmann V, Schlüter AD, Wegner G (1992) Synth Met 51: 299
159. Zotti G, Martina S, Wegner G, Schlüter AD (1992) Adv Mater 4: 798
160. Martina S, Enkelmann V, Wegner G, Schlüter AD, Zotti G, Zerbi G (1993) Synth Met (in press)
161. Wegner G (1984) In: Van der berg EJ (ed) Contemporary topics in polymer science. Plenum, New York p 281

162. Soos ZG, Hayden GW (1991) In: Skotheim TA (ed) Electroresponsive molecular and polymeric systems. Marcel Dekker, New York, p 197
163a. Roth S (1986) In: Pollak M, Shklovskii BI (eds) Hopping transport in solids. Elsevier, Amsterdam p 378
163b. Meerholz K, Heinze J (1993) Synth Met 55–57: 5040
164. Kuzmany H, Mehring M, Roth S (eds) (1989) Electronic properties of conjugated polymers III, part 1. Springer, Berlin Heidelberg New York
165. Sheng P (1980) Phys Rev Lett 45: 60
166. Karl N (1974) Adv Sol State Phys 14: 261
167. Garnier F (1993) Lecture hold at the Kern-Symposium, Mainz
168. Becker B, Bohnen A, Ehrenfreund M, Wohlfarth W, Sakata Y, Huber W, Müllen K (1991) J Am Chem Soc 113: 1121
169. Böttger H, Bryksin VV (1985) (eds) Hopping conduction in solids. VCH, Weinheim
170. Fritsche H, Pollak M (1990) (eds) Hopping and related phenomena. World Scientific Publishers, Singapore
171. Pollak M, Shklovskii BI (1991) (eds) Hopping transport in solids Elsevier. Amsterdam
172. Mott NF, Davis EA (eds) (1979) Electronic processes in non-crystalline materials. Clarendon, Oxford
173. Gorham-Bergeron E, Emin D (1977) Phys Rev B 15: 3667
174. Schimmel TH, Denninger G, Riess W, Voit J, Schwoerer M, Schoeppe W, Naarmann H (1989) Synth Met 28: D11
175. Halim J, Enkelmann V, Fischer H, Wegner G, Albouy PA (1991) Macromol Chem, Rap Commun 12: 301
176. Kaiser AB (1991) Synth Met 41: 183
177. Tritthart W, Leising G (1993) Synth Met 55-57: 4878
178. Enkelmann V, Halim J, Fischer H, Wegner G, Albouy PA (1992) Synth Met 51: 1
179. Baughman RH, Brédas JL, Chance RR, Elsenbaumer RL, Schacklette LW (1982) Chèm Rev 82: 209.
180. Roth S, Menke K (1983) Kunststoffe 73: 520.
181. Nowak RJ, Schultz FA, Umana M, Lam R, Murray RW (1980) Anal Chem 52: 315.
182. Murray RW, Bard AJ (eds) (1984) Electroanalytical chemistry. Marcel Dekker, New York, 13: 191.
183. Shacklette LW, Chance RR, Ivory DM, Miller GG, Baughman RH (1979) Synth Met 1: 307.
184. Diaz AF, Kanazawa KK, Gardini GP (1979) J Chem Soc Commun 635.
185. Tourillon G, Garnier F (1982) J Electroanal Chem 135: 173
186. Diaz AF, Logan JA (1980) J Electroanal Chem 111: 111
187. Heinze J, Mortensen J, Störzbach (1987) In: Kuzmany H, Mehring M, Roth S (eds) Electronic properties of conjugated polymers. Springer, Berlin Heidelberg New York p 385
188. Bartz T, Klapper M, Müllen K, Schulz RC (1993) Polym Int 31: 153
189. Bender D, Przybylski M, Müllen K (1989) Makromol Chem 190: 2071
190. Smith TW, Kuder JE, Wychick D (1976) J Polym Sci 14: 2433
191. Flanagan JB, Margel S, Bard AJ, Anson FC (1978) J Am Chem Soc 100: 4248
192. Saji T, Pasch NF, Webber SE, Bard AJ (1978) J Phys Chem 82: 1101
193. Hörhold HH, Opfermann J, Atrat P, Tauer KD, Drefahl G (1976) Polycondensation Processes. 5th Int Symp Polycondensation, Varna, Publication of the Bulgarian academy of Science, Bulgaria, Sofia p 171
194. Alexander J, Ehrenfreund M, Fiedler J, Huber W, Räder HJ, Müllen K (1989) Angew Chem 101: 1530; Int Ed Engl 28: 1531
195. Kovacic P, Kyriakis A (1963) J Am Chem Soc 85: 454
196. Ohlemacher A, Schenk R, Weitzel HP, Tyutyulkov N, Tasseva M, Müllen K (1992) Makromol 193: 81
197. Huber W, Irmen W, Lex J, Müllen K (1982)Tetrahedron Lett 23: 3889
198. Mortensen J, Heinze J, Herbst H, Müllen K (1992) J Electroanal Chem 324: 201
199. Hörhold HH, Helbig M, Raabe D, Opfermann J, Scherf U, Stockmann R, Weiß D (1987) Z Chem 27: 126
200. Genies EM, Pernaut JM (1985) J Electroanal Chem 191: 111
201. Meerholz K, Heinze J (1990) Angew Chem 102: 695 Int Ed Engl 29: 692
202. Genies EM, Pernault JM (1984) Synth Met 10: 117
203. Zhou QX, Kolaskie CJ, Miller LL (1987) J Electroanal Chem 223: 283
204. Daum P, Murray RW (1981) J Phys Chem 85: 389

205. Heinze J, Störzbach M, Mortensen J (1987) Ber Bunsenges Phys Chem 91: 960
206. Diaz AF, Crowley J, Bargon J, Gardini GP, Torrance JB (1981) J Electroanal Chem 121: 355
207. Nigrey PJ, MacDiarmid AG, Heeger AJ (1982) Mol Cryst Liq Cryst 83: 309
208. Tourillon G, Garnier F (1984) J Electroanal Chem 161: 55
209. Heinze J, Dietrich M, Mortensen J (1987) Makromol Chem Macromol Symp 8: 73
210. Feldberg SW (1984) J Am Chem Soc 106: 4671
211. Bohnen A, Koch KH, Lüttke W, Müllen K (1990) Angew Chem 102: 548; Int Ed Eng 29: 525
212. Bohnen A (1992) PhD thesis, Mainz, FRG
213. Schulz A, Koch KH, Müllen K, Heinze J (1994) to be published
214. Heinze J, Dietrich M (1989) Mater Sci For 42: 63
215. Heinze J, Bilger R, Meerholz K (1988) Ber Bunsenges Phys Chem 92: 1266
216. Hutton Rs, Kalaji M, Peter LM (1989) J Electroanal Chem 270: 429
217a. Gamba A, Rusconi E, Simonetta (1970) Tetrahedron 26: 871
217b. Wilson KE, Pincock PE (1977) Can J Chem 55: 889
218. Dietrich M, Heinze J (1990) J Am Chem Soc 112: 5142
219. Subaric-Leitis a, Monte C, Roggan A, Rettig W, Zimmermann P, Heinze J (1990) J Chem Phys 93: 4543
220. Mataga N, Yao H, Okada T, Rettig W (1989) J Phys Chem 93: 3383
221. Dietrich M, Mortensen J, Heinze J (1985) Angew Chem 97: 502 Int Ed Engl 24: 508
222. Müllen K, Baumgarten M, Karabunarliev S, Tyutyulkov N (1991) Synth Met 40: 127
223. Baumgarten M, Anton U, Gherghel L, Müllen K (1993) Synth Met 55-57: 4801
224. Weitzel HP, Bohnen A, Müllen K (1990) Makromol Chem 191: 2815
225. Heunt S, Mahrt RF, Greiner A, Lemmer U, Bässler H, Halliday DA, Bradley DDC, Burn PL, Holmes AB (1993) J Phys Condens Matter 5: 247
226. Anton U, Bohnen A, Koch KH, Naarmann H, Räder HJ, Müllen K (1992) Adv Mater 4: 91
227a. Oth JFM, Gilles JM, Woo EP, Sondheimer F (1972) J Chem Soc Perkin II
227b. Müllen K, Huber W, Nakagawa M, Iyoda M (1982) J Am Chem Soc 104: 5403
228. Becker BC, Huber W, Schnieders C, Müllen K (1983) Chem Ber 116: 1573
229. Müllen K, Huber W, Meul T, Nakagawa M, Iyoda M (1983) Tetrahedron 39: 1575
230. Becker BC, Huber W, Müllen K (1980) J Am Chem Soc 102: 7803
231. Müllen K, Oth JFM, Engels HW, Vogel E (1979) Angew Chem 91: 251; Int Ed Engl 18: 229
232. Auchter-Krummel P, Müllen K (1991) Angew Chem 103: 996, Int Ed Engl 30: 1003
233. Fry AJ, Hutchins CS (1975) J Am Chem Soc 97: 591
234. Allendoerfer RD, Rieger (1965) J Am Chem Soc 87: 2336
235. Staley SW, Dustman CK, Facchine KL, Linkowsky GE (1985) J Am Chem Soc 107: 4003
236. Tourillon G, Garnier F (1983) J Phys Chem 87: 2289
237. Salmon M, Diaz AF, Logan AJ, Krounbi M, Bargon J (1982) Mol Cryst Liq Cryst 83: 1297
238. Hotta S, Hosaka T, Shimotsuma W (1983) Synth Met 6: 69
239. Chung TC, Kaufman JH, Heeger AJ, Wudl F (1984) Phys Rev 30B: 702
240. Shaklette LW, Elsenbaumer RL, Baughman RH (1983) J Phys Coll 44C3: 559
241. Dietrich M, Mortensen J, Heinze J (1986) J Chem Soc Chem Commun: 1131
241b. Bittihn R (1985) In: Kuzmany H, Mehring M, Roth S (eds) Electronic properties of conducting polymers I. Springer, Berlin Heidelberg New York, p 206
242. Shacklette LW, Maxfield M, Gould S, Wolf JF, Jow TR, Baughman RH (1987) Synth Met 18: 611
243. Anton U, Müllen K (1993) Makromol Chem 14, 223–229 (1993)
244. Beck F (1974) Elektroorganische Chemie. VCH Weinheim
245. Baizer MM, Lund H (1983) Organic electrochemistry. Marcel Dekker, New York.
246. Breitenbach M, Heckner KH (1973) J Electroanal Chem 43: 267.
247. Kobayashi T, Yaneyama H, Tamura H (1984) J Electroanal Chem 161: 419
248. Huang WS, Humphrey BD, MacDiarmid AG (1986) J Chem Soc Faraday Trans I, 82: 2385
249. Chiang JC, Mac Diarmid AG (1986) Synth Met 13: 193
250. Boudreaux DS, Chance RR, Wolf JF, Shacklette LW, Brédas JL, Thémans B, André JM, Silbey R (1986) J Chem Phys 85: 4584
251. Genies EM, Lapkowski M (1987) J Electroanal Chem 220: 67
252. Genies EM, Lapkowski M (1987) Synth Met 21: 117
253. Kovacic P, Koch FW (1965) J Org Chem 30: 3177
254. Kovacic P, (1981) J Polym Sci Polym Lett 19: 359
255. Heinze J (1981) Angew Chem 93: 183; Int Ed Engl 20: 202
256. Giese B (1985) Nachr Chem Tech Lab 33: 298

257. Giese B (1989) In: Regitz M, Giese B (eds) C-Radikale. Houben-Weyl Methoden der organis- chen Chemie Band E19a/Teil 1, p 1
258. Dodd JW (1971) J Chem Soc (B) Phys Org 2427
259. Kira A, lmamura M (1979) J Phys Chem 83: 2267
260. Huber W, Unterberg H, Müllen K (1983) Angew Chem 96: 800; Int Ed Engl 22: 242
261. Heinze J, Serafimov O, Zimmermann HW (1974) Ber Bunsenges 78: 652
262. Roth HD (1987) Acc Chem Res 20: 343
263. Parker VD (1983) Adv Phys Org Chem 19: 131
264. Torrance JB (1979) Acc Chem Res 12: 79
265. Freeman GR, Patai S, Zabicky J (1970) The Chemistry of Functional Groups: The chemistry of the carbonyl group, Wiley, London, vol 2, p 343
266. Metzger JO (1989) In: Regitz M, Giese B (eds) C-Radikale. Houben-Weyl Methoden der organischen chemie Band E19a/Teil 1, p 109 a 192
267. Oda M, to be published
268. Hirabayashi T, Naoi K, Osaka T (1987) J Electrochem Soc 134: 758
269. Edlund U, Eliasson B (1982) J Chem Soc Chem Commun 950
270. Baizer MM (1964) J Electrochem Soc 111: 215
271. Beck F (1965) Chem-lng Techn 37: 607
272. Szwarc M (1956) Nature 178: 1168
273. Schäfer H (1970) Chem Ing Tech 42, 164
274. Ziegler K (1936) Angew Chem 49: 499
275. Morton M (1983) Anionic polymerization: principles and practice. Academic, New York, p 103
276. Morton M, Fetters LJ (1975) Rubber Chem Technol 48: 359
277. Uranek CA (1971) J Polym Sci Polym Chem Ed 9: 2273
278. Schue F, Worsfold DJ, Bywater S (1970) Macromolecules 3: 509
279. Vracken A, Smid J, Szwarc M (1962) J Chem Soc Farad Trans 55: 2036
280. McCormick HW (1957) Pol Sci 25: 488
281. Bargon J, Mohammed S, Waltmann RJ (1983) IBM J Res Develop 27: 330
282. Daub J (1987) Chimia 41: 52
283. Street GB (1986) In: Skotheim TA (ed) Handbook of conducting polymers. Marcel Dekker, New York, p 265
284. Genies EM, Bidan G, Diza AF (1983) J Electroanal Chem 149: 101
285. Inoue T, Yamase T (1983) Bull Chem Soc Jpn 56: 985
286. Kossmehl G, Chatzitheodorou G (1982) Makromol Chem Rapid Commun 2: 551
287. Diaz AF, Castillo JI, Logan JA, Lee WY (1982) J Electroanal Chem 129: 115
288. Waltman RJ, Diaz AF, Bargon J (1984) J Phys Chem 88: 4343
289. Waltman RJ, Bargon J (1984) Can J Chem 64: 76
290. Street GB, Lindsy SE, Nazzal AlWynne KJ (1985) Mol Cryst Liq Cryst 118: 137
291. Aalstad B, Ronlan A, Parker VD (1981) Acta Chem Scand B35: 649
292. Shirota Y, Noma N, Shimizu Y, Kanega H, Jeon IR, Nawa K, Kakuta T, Yasui H, Namba K (1991) Synth Met 41-43: 3031
293. Koch W, Heitz W (1983) Makromol Chem 184: 779
294. Bunnett JF (1978) Acc Chem Res 11, 413
295. Bellville DJ, Wirth DD, Bauld NL (1981) J Am Chem Soc 103: 718
296. Harirchian B, Bauld NL (1987) Tetrahedron Lett 927
297. Mlcoch J, Steckhahn E (1987) Tetrahedron Lett 1081
298. Müllen K, Huber W (1978) Helv Chim Achta 61: 1310
299. Böhm A, Müllen K (1992) Tetrahedron Lett 33: 611
300. Böhm A, Meerholz K, Heinze J, Müllen K (1992) J Am Chem Soc 114: 688
301. Laarhoven WH, Duppen THJHM (1972) J Chem Soc Perkin Trans 1: 2074
302. ophetVield PHG, Laarhoven WH (1977) J Chem Soc Perkin Trans 2: 268
303. Laarhoven WH, Cuppe TJHM, Niverd RJF (1970) Tetrahedron 26: 1069
304. Bartz T, Böhm A, Klapper M, Müllen K, Weitzel HP (1992) Makromol Chem 54-55: 495
305. Böhm A, Fiesser G, Mauermann H, Stein S, Müllen K (1994) in preparation
306. Schlüter AD (1991) Adv Mater 3: 282
307. Wegener S, Müllen K (1993) Macromolecules 26: 3037
308. Anton U, Müllen K (1993) Macromolecules 26: 1248
309. Hörhold HH (1993) Proc. Macromol. 70/71: "34th IUPAC Sympos on Macromolecules, 13- 18th July 1992, Prague"

310. Enkelmann V (1988) In: Ebert B (ed) Polynuclear aromatic compounds. Chap 11, Adv Chem Ser 217: 177
311. Chiang TC, Reddoch AH, Williams J (1971) J Chem Phys 54: 2051
312. Jost W, Adam M, Enkelmann V, Müllen K (1992) Angew Chem 104: 883
313. Keller HJ, Nöthe D, Pritzkow H, Wehe D, Werner M, Koch P, Schweitzer D (1981) Mol Cryst Liq Cryst 62: 181
314. Endres H, Keller HJ, Müller B, Schweitzer D (1985) Acta Cryst C41: 607
315. Eichele H, Schwoerer M, Kröhnke C, Wegner G (1981) Chem Phys Lett 77: 311
316. Sigg J, Prisner T, Dinse KP, Brunner H, Schweitzer D, Hausser KH (1983) Phys Rev B 27: 5366
317. Denninger G, Stöcklein W, Dormann E, Schwoerer M (1984) Chem Phys Lett 107: 222
318. Maresch GG, Mehring M, vonSchütz JU, Werner HU, Gökelmann K, Enkelmann V, Müllen K, Klabunde KU (1989) J Chem Phys 91: 4543
319. Alexander J, Ehrenfreund M, Fiedler J, Huber W, Räder HJ, Müllen K (1989) Angew Chem 101: 1530; Int Ed Engl 28: 1531
320. Irmen W, Huber W, Lex J, Müllen K (1984) Angew Chem 96: 800; Int Ed Engl 23: 818
321. Bock H, Herrmann HF, Fenske D, Goesmann H (1988) Angew Chem 100: 1125; Int Ed Engl 27: 1067
322. Jost W, Adam M, Enkelmann V, Müllen K (1992) Angew Chem 104: 883
323. Zinke A, Linner F, Wolfbauer O (1925) Ber Dtsch Chem Ges 58: 323
324. Akamatu H, Inokuchi H, Matsunaga Y (1954) Nature 173: 168
325. Coffen DL, Garret PE (1969) Tetrahedron Lett 2043
326. Ferraris J, Cowan DO, Bloch AN (1974) J Chem Soc Chem Commun 937
327. Garito AF, Heeger AJ (1974) Acc Chem Res 7: 232
328. Almen G, Bauer T, Hünig S, Kupcik V, Langohr U, Metzenheim T, Meyer K, Rieder H, vonSchütz JU, Tillmanns E, Wolf HC (1991) Angew Chem 103: 608; Int Ed Engl 30: 561
329. Torrance JB, Tomkiewicz Y (1976) Bull Am Phys Soc 21: 313
330. Torrance JB, Silverman BD (1977) Phys Rev B 15: 788
331. Wudl F (1982) Pure & Appl Chem 54: 1051
332. Epstein AJ, Lipari NO, Sandman DJ, Nielson (1976) Phys Rev B 13: 1569
333. Hubbard J (1978) Phys Rev B 17: 494
334. Comés R (1977) In: Keller HJ (ed) Chemistry and physics of one dimensional metals. NATO Adv Study Institutes Series, B-Physics, Plenum, NY, 25: 315
335. Pauling L (1960) The nature of chemical bond. Ithaca, NY, p 511
336. Klots CE, Compton RN, Raaen VF (1974) J Chem Phys 60: 1177
337. Sandman DJ (1979) Mol Cryst Liq Cryst 50: 235
338. Metzger RM (1975) J Chem Phys 63: 5090
339. Kistenmacher TJ, Phillips TE, Cowan DO (1974) Acta Cryst B30: 763
340. Phillips TE, Kistenmacher TJ, Bloch AN, Ferraris JP, Cowan DO (1977) Acta cryst B33: 422
341. Aumüller A, Hünig S (1984) Angew chem 96:437; Int Ed Engl 23: 447
342. Hünig S, Aumüller A, Erk P, Meixner H, von Schütz JU, Gross HJ, Langohr U, Werner HP, Wolf HC, Burschka C, Klebe G, Peters K, von Schnering HG (1988) Synth Met 27: B181
343. Enkelmann V (1991) Angew Chem 103: 1142
344. Aumüller A, Erk P, Klebe G, Hünig S, vonSchütz JU, Werner HP (1986) Angew Chem 98: 759; Int Ed Engl 25: 740
345. Bechgaard K, Jakobsen CS, Mortensen K, Pederson MJ, Thorup N (1980) Solid State Commun 33: 1119
346. Bechgaard K, Carneiro K, Rasmussen FG, Olsen K, Rindorf G, Jakobsen CS, Pederson HJ, Scott JE (1981) J Am Chem Soc 103: 2440
347. Mizuno M, Garito AF, Cava MP (1978) J Chem Soc Chem Commun 18
348. Williams JM, Ferraro JR, Thorn RJ, Carlson KD, Geiser U, Wang HH, Kini AM, Whangbo MH (eds) Organic Superconductors, Synthesis, Structure, Properties, and Theory. Prentice Hall, Englewood Cliffs NJ
349. Urayama H, Yamochi H, Saito G, Nozawa K, Sugano T, Kinoshita M, Sato S, Oshima K, Kawamoto A, Tanaka J (1988) Chem Lett 55
350. Kini AM, Geiser U, Wang HH, Carlson KD, Williams JM, Kwok WK, Vandervoort KG, Thompson JE, Stupka DL, Jung D, Whangbo MH (1990) Inorg Chem 29: 2555
351. Pitman CU, Narita, Liang YF (1976) J Org Chem 41: 2855
352. Bryce MR, Moore AJ (1990) Pure & Appl Chem 62: 473

353. Sugimoto T, Awaji H, sugimoto I, Misaki Y, Kawase T, Yoneda S, Yoshida Z, Kobayashi T, Anzai H (1989) Chem Mater 1: 535
354. Sugiomoto T, Awaji H, Misaki Y, Yoshida Z, Kai Y, Nakagawa H, Kasai N (1985) J Am Chem Soc 107: 5792
355. Adam M, Wolf P, Räder HJ, Müllen K (1990) J Chem Soc Chem Commun 1624
356. Adam M, Müllen K (1993) Adv Mater submitted
357a. Iwamura H (1990) Adv Pys Org. Chem 26: 179
357b. Iwamura H, Miller JS (1993) Proceedings of the Symposium on the "CHEMISTRY AND PHYSICS OF MOLECULAR BASED MAGNETIC MATERIALS". Mol Cryst Liqu Cryst 232/233: 1–360/1–366
358. Dougherty DA (1991) Acc Chem Res 24: 88
359. Miller JS, Epstein AJ, Reiff WM (1988) Acc Chem Res 21: 114
360. Miller JS, Dougherty DA (eds) (1989) Proceedings of the Symposium of Ferromagnetic and high spin molecular based materials. Mol Cryst Liq Cryst 176: 1
361. Heisenberg W (1928) Z Phys 49: 619
362. Bencini A, Gatteschi D (1990) EPR of exchange coupled system. Springer Verlag, Heidelberg.
363. Anderson PW (1963) In: Rado GT, Suhl H (eds) Magnetism. Academic, New York, vol 1 p 25
364. Hay PJ, Thibeault JC, Hoffmann RH (1975) J Am Chem Soc 97: 4884
365. Kramers HA (1934) Physica 1: 182
366. Anderson PW (1959) Phys Rev 115: 2
367. Tyutyulkov N, Karabunarliev S, Müllen K, Baumgarten M (1992) Synth Met 52: 71
368. Buchachenko A (1989) Mol Cryst Liq Cryst 176: 307
369. LePage TJ, Breslow R (1987) J Am Chem Soc 109: 6412
370. Miller JS, Epstein AS (1993) Angew Chem in press
371. McConnell HM (1963) J Chem Phys 1963 39, 1910
372. Mataga N, Theor Chim Acta 1968, 10, 372
373. Ovchinnikov AA (1978) Theor Chim Acta 47: 297
374. Aoki Y, Imamura A (1992) Theor Chim Acta 84: 155
375. McConnell (1967) Proc RA Welsh Found Conf Chem Res 11: 144
376. Breslow R, Juan B, Klutz RQ, Xia CZ (1982) Tetrahedron 38: 863
377. Miller JS, Epstein AJ (1989) Mol Cryst Liq Cryst 176: 347
378. Miller JS, Epstein AJ (1993) in press
379. Wassermann E, Murray RW, Yager WA, Trozzolo AM, Smolinsky G (1967) J Am Chem Soc 89: 5076
380. Itoh K (1967) Chem Phys Lett 1: 235
381. Schlenk W, Brauns M (1915) Ber Dtsch Chem Ges 48: 661: 669
382. Iwamura H (1987) Pure & Appl Chem 59: 1595 ibid 65: 57
383. Tyutyulkov N, Karabunarliev S (1986) Int J Quantum Chem 29: 1325
384. Veciana J, Vidal J, Jullian N (1989) Mol Cryst Liq Cryst 176: 443
385. Breslow R (1989) Mol Cryst Liq Cryst 176: 199
386. Coulson CA, Longuet-Higgins HC (1947) Proc R Soc Ser A 191: 39 and 192: 16
387. Dietz F, Müllen K, Baumgarten M, Tyutyulkov N (1993) Chem Phys in press
388. Even in: Buckminsterfullerene the highest spin state available is $S = 1$, although the neutral molecules is assumed to possess a threefold degenracy of the LUMO
    Baumgarten M, Gügel a, Gherghel L (1993) Adv Mater in press
389. Berson JA (1978) Acc Chem Res 11: 446
390. Berson JA (1987) Pure & Appl Chem 59: 1571
391. Tyutyulkov N, Karabunarliev S, Ivanov K (1989) Mol Cryst Liq Cryst 176: 139
392. Veciana (1989) Mol Cryst Liq Cryst 176: 75
393a. Kirste B, Kurreck H, Lubitz W, Schubert K (1978) J Am Chem Soc 100: 2292
393b. Kurreck H (1993) Angew Chem 105: 1472
394. Grimm M, Kirste B, Kurreck H (1986) Angew Chem 98: 1095 Int Ed Engl 25: 1097
395. Horn T, Baumgarten M, Gherghel L, Müllen K (1993) Tetrahedron Lett, 34: 5889
396. Longuet-Higgins HC (1950) J Chem Phys 18: 265
397. Klein DJ (1982) J Chem Phys 77: 3098
398. Itoh K (1978) Pure & Appl Chem 50: 1251
399. Dowd P, Chang W, Paik YH (1986) J Am Chem Soc 108: 7416
400. Du P, Borden WT (1987) J Am Chem Soc 109: 5284
401. Nash JJ, Dowd P, Jordan KD (1992) J Am Chem Soc 114: 10071
402. Nachtigall P, Jordan KD (1993) J Am Chem Soc 115: 270

403. Schmauss G, Baumgarte H, Zimmermann H (1965) Angew Chem Int Ed Engl 4: 596
404. Brickmann J, Kothe G (1973) J Chem Phys 59: 2807
405. Novak C, Kothe G, Zimmermann H (1974) Ber Bunsen-Ges 78: 265
406. Rajca A (1990) J Am Chem Soc 112: 5890
407. Nakamura N, Inoue K, Iwamura H, Fujioka, Sawaki Y (1992) J Am Chem Soc 114: 1484
408. Higuchi J (1963) J Chem Phys 38: 1237; ibid 39: 1847
409. Iwamura H (1986) Pure & Appl Chem 58: 187
410. Murata S, Iwamura H (1991) J Am Chem Soc 113: 5547
411. Baumgarten M, Müllen K, Tyutyulkov N, Madjarova G (1993) Chem Phys 169: 81
412. Tyutyulkov N, Ivanov Cl, Schopov I, Polansky OE, Olbrich G (1988) Int J Quantum Chem 34: 361
413. Yoshizawa K, Hatanaka M, Ito A, Tanaka K, Yamabe T (1993) Chem Phys Lett 202: 483
414. Ishida T, Iwamura H (1991) J Am Chem Soc 113, 4238–4241
415. Tanaka K, Yoshizawa K, Takata A, Yamabe T, Yamauchi J (1991) Synth Met 41-43: 3297
416. Mukai K, Nagai H, Ishizu K (1975) Bull Chem Soc Jpn 48: 2381
417. Calder A, Forrester AR, James PG, Luckhurst GR (1969) J Am Chem Soc 91: 3724
418. Yoshizawa K, Chano A, Ito A, Tanaka K, Yamabe T, Fujita H, Yamauchi J, Shiro M (1992) J Am Chem Soc 114: 5994
419. Yoshizawa J, Tanaka K, Yamabe T, Yamauchi J (1992) J Chem Phys 96: 5516
420. Torrance JB, Oostra S, Nazal A (1987) Synth Met 19: 809
421. Baumgarten M, Wehrmeister T, Karabunarliev S, Tyutyulkov N, Müllen K (1994) to be published
422. Wassermann E, Schueller K, Yager WA (1968) Chem Phys Lett 2: 259
423. Tukada H, Mutai K, Iwamura H (1987) J Chem Soc Chem Commun 1159
424. Matsumoto T, Ishida T, Ishida T, Koga N, Iwamura H (1992) J Am Chem Soc 114: 9952
425. Ling C, Minato M, Lahti PM, vanWilligen H (1992) J Am Chem Soc 114: 9959
426. Dowd P (1966) J Am Chem Soc 88: 2587
427. Dowd P, Paik YH (1986) J Am Chem Soc 108: 2788
428. Jain R, Snyder GJ, Dougherty DA (1984) J Am Chem Soc 106: 7294
429. Novak JA, Jain R, Dougherty DA (1989) J Am Chem Soc 111: 7618
430. Dougherty DA, Jacobs SJ, Silverman SK, Murray MM, Shultz DA, West AP, Clites JA (1993) Mol cryst Liq Cryst in press
431. Fukutome H, Takahashi A, Ozaki Ma (1987) Chem Phys Lett 133: 34
432. Tukada H (1993) J Am Chem Soc in press
433. Müller U, Baumgarten M (1994) J Am Chem Soc submitted
434. Baumgarten M, Gherghel L, Gregorius H, Karabunarliev K, Müllen K (1993) to be published
435. Breslow R, Chang HW, Hill R, Wasserman E (1967) J Am Chem Soc 89: 1112
436. Wassermann E, Hutton RS, Kuck VJ, Chandross EA (1974) J Am Chem Soc 96: 1965
437. Saunders M, Berger R, Jaffe A, McBride JM, O'Neill J, Breslow R, Hoffman JM, Perchonock C, Wassermann E, Hutton RS, Kuck VJ (1973) J Am Chem Soc 95: 3017
438. Breslow R, Maslak P, Thomaides J (1984) J Am Chem Soc 106: 6453
439. Breslow R, Hill R, Wassermann E (1964) J Am Chem Soc 86: 5349
440. Fukunaga T (1976) J Am Chem Soc 98: 610
441. Veciana J, Rovira C, Ventosa N, Crespo MI, Palacio F (1993) J Am Chem Soc 115: 57
442. Hoshino M, Kimura K, Imamura M (1973) Chem Phys Lett 20: 193
443. Tukada H (1991) J Am Chem Soc 113: 8991
444. Tukada H, Mutai K (1993) Tetrahedron Lett in press
445. Rajca A, Utamapanya (1992) J Org Chem 47: 1760
446. Dougherty DA, Kaisaki DA (1990) Mol Cryst Liq Cryst 183: 71
447a. Nakamura N, Inoue K, Iwamura H (1992) J Am Chem Soc 114: 1484
447b. Nakamura N, Inoue K, Iwamura H (1993) Angew Chem 105: 900
448. Rajca A, Utamapanya S, Thayumanavan S (1992) J Am Chem Soc 114: 1884
449. Miller JS (1992) Adv Mater 4: 298; ibid 435
450. Winter H, Gotoschy B, Dormann E, Naarmann H (1990) Synth Met 341–352
451. Cosmo R, Dormann E, Gotschy B, Naarmann H, Winter H (1991) Synth Met 41-43: 369
452. Cosmo R, Naarmann H (1990) Mol Cryst Liq Cryst 185: 89
453. Yoshioka N, Nishide H, Tsuchida E (1990) Mol Cryst Liq Cryst 190: 45
453b. Nishide H, Yoshioka N, Kaneko T, Tsuchida E (1990) Macromolecules 23: 4487
454. Iwamura H, McKelvey RD (1988) Makromolecules 21: 3386
455. Fujii A, Ishida T, Koga N, Iwamura H (1991) Macromolecules 24: 1077

456. Iwamura H, Murata S (1989) Mol Cryst Liq Cryst 176: 33
457. Wegner G (1977) Pure & Appl Chem 49: 443
458. Wegner G (1979) in: Hartfield EW (ed) Molecular metals. Plenum Press, NY, p 209
459. Korshak YV, Madvedeva TV, Ovchinnikov AA, Spector VN (1987) Nature 326: 370
460. Zhang JH, Epstein AJ, Miller JS, O'Connor CJ (1989) Mol Cryst Liq Cryst 176: 271
461. Cao Y, Wang P, Hu Z, Li S, Zhang L, Zhao J (1988) Solid State Commun 68: 817
462. Wiley DW, Calabrese JC, Miller JS (1989) Mol Cryst Liq Cryst 176: 277
463. Inoue K, Koga N, Iwamura H (1991) J Am Chem Soc 113: 9803
464. Inoue K, Iwamura H (1992) Adv Mater 4: 801
465. Awaga K, Sugano T, Kinoshita M (1986) J Chem Phys 85: 2211
466. Harrer W, Kurreck H, Reusch J, Gierke W (1975) Tetrahedron 31: 625
467. Shiomi D, Tamura M, Sawa H, Kato R, Kinoshita M (1993) Synth Met 1993 in press
468. Awaga K, Maruyama Y (1989) Chem Phys Lett 158: 556
469. Turek P, Nozawa K, Shiomi D, Awaga D, Awaga K, Inabe T, Maruyama Y, Kinoshita M (1991) Chem Phys Lett 180: 327
470. Kinoshita M, Turek P, Tamura M, Nozawa K, Shiomi D, Nakazawa Y, Ishikawa M, Takahashi M, Awaga K, Inabe T, Maruyama Y (1991) Chem Lett 1225
471. Chiang LY, Goshorn DP (1989) Mol Cryst Liq Cryst 176: 229
472. Miller JS, Epstein AJ (1989) Mol Cryst Liq Cryst 176: 347
473. Miller JS, Calabrese JC, Harlow RL, Dixon DA, Zhang JH, Reiff WM, Chittipaldi S, Selover MA, Epstein AJ (1990) J Am Chem Soc 112: 5496
474. Epstein AJ, Miller JS (1993) Proceedings of Electrical and related properties or organic solids. Capri I, (1992) Mol Cryst Liq Cryst 228: 99
475. Cannon RD (1980) "Electron Transfer Reactions", Butterworth, London
476. Paddon-Row MN, Jordan KD (1988) In: Liebman JF, Greenberg A (eds) Modern models of bonding and delocalization, chap 3. VCH Publishers, New York
477. Marcus RA (1956) J Chem Phys 24: 979
478. Marcus RA (1965) J Chem Phys 43: 679
479. Closs G, Miller JR (1988) Science 240, 440
480. Santamaria J (1988) In: Fox MA, Cannon M (eds) Photoinduced electron transfer. Elsevier, Amsterdam
481. Grampp G, Jaenicke W (1984) Ber Bunsenges Phys Chem 88: 325; ibid 88: 335
482. Gerson F, Kowert B, Peake PM (1974) J Am Chem Soc 96: 118
483. Kuznetsov AM, Ulstrup J, Vorotyntsev MA (1988) Solvent effects in charge transfer processes. In: Dogonadze RR, Kalman E, Kornyshev AA, Ulstrup J (eds) The chemical physics of solvation. Studies in physical and theoretical chemistry, vol 38, chap 3, Elsevier, Amsterdam
484. Gerson F, Wellauer T, Oliver AM, Paddon-Row MN (1990) Helv Chim Acta 73: 1586
485. Liang N, Miller JR, Closs GL (1990) J Am Chem soc 112: 5353
486. Jordan KD, Paddon-Row MN (1992) Chem Rev 92: 395
487. Kochi JK (1988) Angew Chem 100: 1331; Int Ed Engl 27: 1227
488. Wasielewski MR (1992) Chem Rev 92: 435
489. Baumgarten M, Müllen K (1992) AIP Conf Proc; St. Thomas, Virgin Islands 262: 68–76
490. Hopfield J, Onuchic JN, Beratan DN (1988) Science 214: 817
491. Mehring M (1989) In: Kuzmany H, Mehring M, Roth S (eds) Electronic properties of conjugated polymers III. Springer, Berlin Heidelberg New York, p 242
492. Aviram A (1988) J Am Chem Soc 110: 5687
493. Joachim C, Launay JP (1990) J Mol Electronics 6: 37
494. Aviram A, Ratner MA (1974) Chem Phys Lett 29: 277
495. Aviram A, Seiden PE, Ratner MA (1982) In: Carter FL (ed) Molecular electronic devices. Marcel Dekker, New York, p 5
496. Metzger RM, Panetta CA, Miura Y, Torres (1987) Synth Met 18: 787
497. Torres E, Panetta CA, Metzger RM (1987) J Org Chem 52: 2944 (1987)
498. Metzger RM, Panetta CA (1991) New J Chem 15: 209
499. Binnig G, Rohrer H, Gerber Ch, Weibl E (1982) Phys Rev Lett 49: 57
500. Roberts GG (ed) (1990) "Langmuir Blodgett Films" Plenum Press, New York
501. Sariciftci NS, Neugebauer H, Mehring M (1991) Synth Met 41-43: 2971
502. Burroughes JH, Friend RH (1991) In: Brédas JL, Silbey R (eds), Conjugated polymers. Kluwer, Dordrecht, p 555
503. Kmetz AR, Willisen FK (1976) Non-emissive electrooptic displays. Plenum, New York
504. Inganas O, Lundström I (1987) Synth Met 21: 13

505. Inganas O, Lundström I, Skotheim TA (1986) In: Skotheim TA (ed) Handbook of conducting polymers. Marcel Dekker, New York, p 524
506. Friend RH (1992) Synth Met 51: 357
507. Yoneyama H, Wakamato K, Tamura H (1985) J Electrochem Soc 132: 2414
508. Yosino K, Kaneto K, Inuishi Y (1983) Jpn J Appl Phys 22: L157
509. Kobayashi T, Yoneyama H, Tamura H (1984) J Elektroanal Chem 177: 281
510. LaCroix JC, Kanazawa KK, Diaz AF (1989) J Electrochem Soc 136: 1308
511. Wolf JF, Miller GG, Shacklette, Elsenbaumer RL, Baughman RH US Patent 4, 893, 908
512. Braun D, Heeger AJ (1991) Appl Phys Lett 58: 1982
513. Adachi C, Tsutsui T, Saito S (1993) Appl Phys
514. Grem G, Leditzky G, Ulrich B, Leising G (1992) Adv Mater 4: 36
515. Holmes AB (1992) Nature356: 47
516. Heeger AJ (1992) Nature 357: 477
517. Reimers JR, Hush NS (1990) Inorg Chem 29: 3886
518. Joachim C, Launay JP, Woitellier S (1990) Chem Phys 147: 131
519. Heine B, Sigmund E, Maier S, Port H, Wolff HC, Effenberger F, Schlosser H (1990) J Molec Electronics 6: 51
520. Yao H, Okada T, Mataga N (1989) J Phys Chem 93: 7388
521. Rettig W (1986) Angew. Chemie 98: 969; Int. Ed. Engl 125: 971
522. Lippert E, Lüder W, Boos H (1962) In: Mangini A (ed) Adv in molecular spectroscopy. Pergamon, Oxford, p 443
523. Bonacic-Koutecky V, Koutecky J, Michl J (1987) Angew Chem 99: 216, Int. Ed. Engl 26: 218
524. Rettig W (1991) Nachr Chem Tech Lab 39: 398406
525. Barbara PF, Jarzeba W (1988) Acc Chem Res 21: 195
526. Mataga N, Yao H, Okada T, Rettig W (1989) J Phys Chem 93: 3383–86
527. Fritz R, Rettig W, Müller U, Müllen K (1994) (to be published)
528. Dobkowski J, Rettig W, Paeplow B, Koch KH, Müllen K, Lapouyade R, Grabowski ZR (1994) J Am Chem Soc in press
529. Haarer D, Blumen A (1988) Angew Chem 100: 1252; Int Ed Engl 27: 1210
530. Hoegl H, Sus O, Neugebauer W (1962) US Pat 3 037 861
531. Shattuk DM, Vahtra U (1969) US Pat 3 484 237
532. Abkowitz M, Bässler H, Stolka M (1991) Phil Mag B 63: 201

Photoinduced Charge Transfer Processes at Semiconductor Electrodes and Particles

Rüdiger Memming

FB Physik der Carl-von Ossietzky-Universität Oldenburg and Institut für Solarenergieforschung (ISFH), Sokelantstr. 5, 30165 Hannover 1, FRG

Table of Contents

Topics in Current Chemistry, Vol. 169
© Springer-Verlag Berlin Heidelberg 1994

In contrast to reactions at metal electrodes, charge transfer processes at semiconductor electrodes can be controlled by light excitation if minority carriers are involved. Since electron and hole transfer always occur via one of the energy bands, valuable information on the energy parameters determining the reaction rates can be obtained. In the present paper, models of the charge transfer processes at semiconductor electrodes are presented, and the kinetics of various reactions in the dark and under illumination are discussed in detail. During the last decade it has become very popular to investigate photoinduced charge transfer reactions at small semiconductor particles, i.e. at microoheterogeneous systems, because of the large surface area. Here, various fundamental processes in suspensions or colloidal solutions are compared with corresponding reactions at extended electrodes. Finally, several applications of photoinduced reactions at semiconductor electrodes and particles are briefly described.

1 Introduction

The modern work on semiconductor electrochemistry dates back to around 1960, when well-defined single crystals of germanium and silicon became available. Several fundamental effects, such as the potential distribution across the semiconductor-electrolyte interface, the mechanism of the anodic decomposition of germanium and silicon and few redox processes were studied at an early stage of this research. Later on, other semiconductors, e.g. various II-VI and III-V-compounds, some stable oxide semiconductors and transition metal chalcogenides (layer compounds) also became available. These compounds were of special interest because one could discriminate more easily between electron and hole transfer processes via the conduction and valence band, respectively, because most of these compounds exhibit a larger bandgap. Within a rather short period of time, a qualitative understanding of many effects was reached. During this period, theoretical models concerning the kinetics of charge transfer reactions were developed by Gerischer [1], Marcus [2], Dogonadze [3] and Levich [4]. These models are essential for the quantitative understanding of processes at semiconductor electrodes. Especially the theories developed by Gerischer and Marcus are widely applied in semiconductor electrochemistry. Models and many results have been summarized in several books and review articles [5–12].

The whole field received a new impetus after the first oil crisis, when Fujishima and Honda reported on the photoelectrolysis of water at TiO_2-electrodes [13]. Whereas, before the oil crisis, most basic models and results had been published only by 3–4 research groups in the world, many other scientists entered the field after this crisis and studied solar applications, and hundreds of papers were published. Since then, many processes at semiconductor electrodes have been studied more quantitatively by using not only standard electrochemical methods, but also new techniques, such as spectroscopic surface analysis (see e.g. [12]). Naturally, photoeffects played a dominant role in these investigations. These were not only restricted to reactions induced by light excitation within the semiconductor electrode [11], but were also extended to the excitation of adsorbed dye molecules [14,15].

During the last decade, investigations of microheterogeneous systems, such as suspensions and colloidal solutions of semiconductor particles have also been started. Many researchers were attracted to study light-induced reactions with particles, because they are easily prepared and their surface area is very large. Most interesting here is the observation that the bandgap of the semiconductor particles increases with decreasing size below a diameter of some nanometers. Some of these results have also been summarized in few review articles [16–18].

In the present article, various fundamental photoelectrochemical effects are quantitatively described and discussed, with the main emphasis on the kinetics of charge transfer processes. Although in principle the same reaction mechanisms are valid for extended semiconductor electrodes and particles, different factors govern the reaction rate, as will be discussed in detail. Finally, a brief overview of various applications will be given.

2 Some Fundamentals of Semiconductors

2.1 Energy Levels in Solids and Carrier Densities

The quantum theory of solids presents a complete and rigorous description of energy levels in a semiconductor, of the nature of charge carriers and of laws governing their motion. The energy spectrum of electrons in an ideal crystal consists of filled and empty energy states, i.e. of the valence (E_v) and conduction bands (E_c), respectively. The energy bands which can have a considerable width are separated by the bandgap E_G, as illustrated in Fig. 1. In an intrinsic semiconductor, the generation of electrons and holes occurs in the dark as a result of thermal excitation of electrons from the valence into the conduction

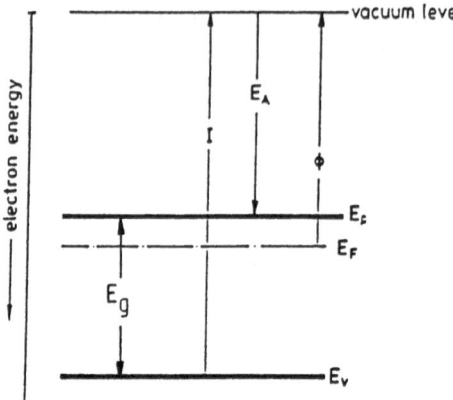

Fig. 1. Position of energy bands and Fermi level with the vacuum level as a reference; E_A: electron affinity, I: ionization energy, ϕ: work function

band. In the case of large bandgaps, the density of carriers generated by thermal excitation is very small. In a semiconductor doped by a corresponding electron donor or acceptor, the thermal energy required for electron excitation is much smaller (usually not more than 0.1–0.2 eV), so that a considerable electron- or hole density exists at room temperature.

The density of energy states within the energy bands increases with the square root of energy above the conduction band or below the valence band (see e.g. Refs. [19, 20]). Near the band edges, i.e. within 1 kT, the density of energy states can be approximated by

$$N_c = \frac{2\pi(4/3m_e kT)^{3/2}}{h^3}\left(\frac{m_e^*}{m_e}\right)^{3/2} \tag{1}$$

for the conduction band and

$$N_v = \frac{2\pi(4/3m_e kT)^{3/2}}{h^3}\left(\frac{m_h^*}{m_e}\right)^{3/2} \tag{2}$$

for the valence band,

in which h is the Planck constant, m_e the electron mass in free space, m_e^* and m_h^* the effective mass of electrons and holes, respectively. The probability that an energy state is occupied by an electron is described by the Fermi-Dirac function. The electron- and hole densities in the conduction- and valence band, respectively, are related to the Fermi level by

$$n = N_c \exp\left(-\frac{E_c - E_{F,n}}{kT}\right) \tag{3a}$$

$$p = N_v \exp\left(-\frac{E_v - E_{F,p}}{kT}\right) \tag{3b}$$

in which N_c and N_v are given by Eq. (1). At equilibrium, the Fermi- levels of electrons and holes are identical, i.e. $E_{F,n} = E_{F,p} = E_F$, and $n = n_0$ and $p = p_0$. Inserting these values into Eqs. (3a) and (3b) and multiplying these two equations, one obtains at equilibrium:

$$n_0 p_0 = N_c N_v \exp\left(-\frac{E_c - E_v}{kT}\right) = n_i^2 = \text{const} \tag{4}$$

The relative position of the Fermi level E_F depends of course on the electron and hole concentration, i.e. on the doping of the semiconductor crystal. One possible position is indicated in Fig. 1. From Eqs. (3a) and (3b), the density of free carriers in the conduction- and valence band can be calculated. Assuming, for instance, that the effective mass of electrons is equal to that in free space, then the density of energy states at the lower edge of the conduction band, as given by Eq. (1a), amounts to about $N_c = 5 \times 10^{19} \text{ cm}^{-3}$. Assuming further that the energy distance between E_F and E_c is 0.1 eV, then the electron density is about $n_0 = 1.5 \times 10^{18} \text{ cm}^{-3}$, i.e. only 1/30 of the energy states within an energy interval of 1 kT at the lower edge of the conduction band are occupied by electrons at equilibrium.

In an absolute energy scale, the position of the conduction band with respect to the vacuum level is given by the electron affinity E_A (Fig. 1). The distance of the valence band with respect to the vacuum level is given by the ionization potential I, and that of the Fermi level is the work function Φ. Values of these quantities are usually determined by surface spectroscopic methods [21].

2.2 Light Absorption by Semiconductors

Various electronic transitions upon light absorption are possible. The most effective absorption is the band to band transition, leading to an equal additional density of electrons and holes, as illustrated in Fig. 2. The absorption starts at photon energies $E_{ph} = h\nu = E_g$, rises, usually strongly, with increasing photon energies. The absorption coefficient reaches values for direct bandgaps (see below) of up to 10^5–10^6 cm^{-1}. Electrons excited into higher energy states

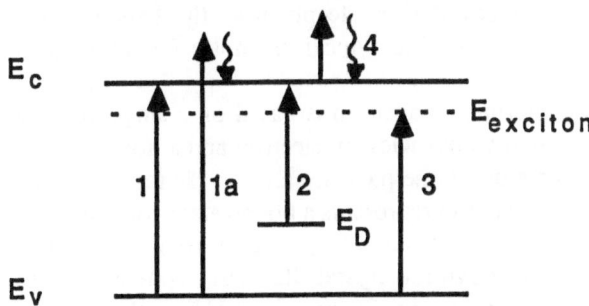

Fig. 2. Optical transitions in a semiconductor

(transition 1a in Fig. 2) are thermalized to the lower edge of the conduction band within about 10^{-12}–10^{-13} s. Besides this basic band to band transition, an excitation of an electron in a donor state or in an impurity level into the conduction is also possible (transition 2). However, since the concentration of impurities is very small, the absorption cross section and therefore the corresponding absorption coefficient α will be smaller by many orders of magnitude than that for an inter-band transition. At lower photon energies, i.e. at $E_{ph} < E_g$, frequently an absorption increase with decreasing E_{ph} has been observed. This absorption has been related to an intraband transition (transition 4 in Fig. 2) and is approximately described by the Drude theory [23]. The absorption coefficient increases with the free carrier density. It is small for carrier densities below about 10^{18} cm^{-3}.

Regarding the fundamental interband transition and the corresponding photogeneration of electron-hole pairs, the interband transitions have to be divided into direct and indirect transitions. The meaning of these terms is as follows:

Within each band the different electron states are characterized not only by their energy E, but also by their momentum p. The electron energy E is a function of the momentum p, which is specific to each crystal and its structure, and to each of its energy bands. In the simplest case, the minimum of $E_c(p)$ and the maximum of $E_v(p)$ occur at $p = 0$. An optical transition from the valence- into the conduction band can occur for $E_{ph} = E_g$ without changing the momentum of the electron (direct transition). The momentum of the photon is in the order of hv/c and is negligible compared to the momentum of an electron. There are, however, many semiconductors for which the maxima and minima of the bands do not coincide. The law of conservation of momentum excludes here the possibility of the absorption of a photon of an energy close to the bandgap. A photon absorption becomes possible, however, if a phonon supplies the missing momentum to the electrons (indirect transition). Such a transition requires a "3-body"-collision (photon, electron, phonon), which occurs less frequently than a "2-body"-collision, i.e. the absorption coefficient will be considerably smaller for a semiconductor with an indirect bandgap. This becomes obvious by measuring the absorption spectra of a semiconductor, a selection is given in Fig. 3. For instance, GaAs and CuInSe$_2$ are examples for a direct bandgap, i.e. the absorption coefficient rises steeply near the bandgap and reaches very high values. Si and GaP are typical examples for an indirect transition, i.e. α is low.

The interpretation of the interband transition is based on a single particle model, although in the final state two particles, an electron and a hole, exist. In some semiconductors, however, a quasi one-particle state, an exciton, is formed upon excitation [23, 24]. Such an exciton represents a bound state, formed by an electron and a hole, as a result of their Coulomb attraction, i.e. it is a neutral quasi-particle, which can move through the crystal. Its energy state is close to the conduction band (transition 3 in Fig. 2), and it can be split into an independent electron and a hole by thermal excitation. Therefore, usually

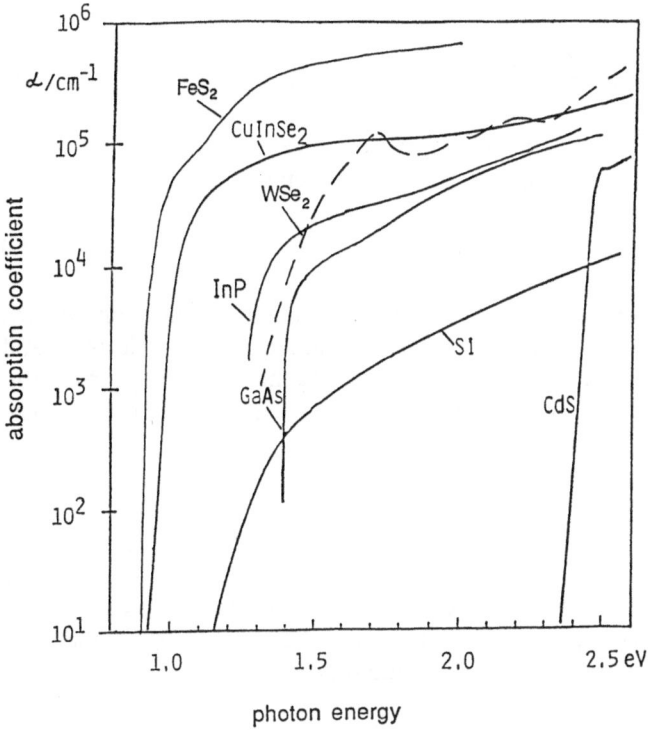

Fig. 3. Absorption spectra of various semiconductors

a sharp absorption peak just below the bandgap energy can only be observed at low temperature, whereas at room temperatures only the usual interband transition is visible in the absorption spectrum. The situation is different in organic crystals [25] and also for small inorganic semiconductor particles (see Sect. 2.4.).

2.3 Recombination of Electrons and Holes

After excitation, the electrons and holes must recombine again. Such a recombination can be either a direct band-band transition or can occur – as has frequently been observed – via impurity centers within the gap. The corresponding transitions can be a radiative or a non radiative process. The first one can be measured as fluorescence, and in this case, it is rather easy to detect whether an impurity state is involved or not. Frequently, a strong recombination at the surface of a crystal has been detected, which has been explained by a high density of surface states.

Upon excitation, the densities of electrons and holes are increased above their equilibrium concentrations n_0 and p_0. We have then

$$n = n_0 + \Delta n; \; p = p_0 + \Delta p \tag{5}$$

111

The decay of excess carriers is given by

$$\frac{d(\Delta n)}{dt} = g_n - r_n; \quad \frac{d(\Delta p)}{dt} = g_p - r_p \tag{6}$$

in which g_n and g_p represent the excess generation rate due to light excitation. Phenomenologically the recombination rates are defined by

$$r_n = \frac{n - n_0}{\tau_n}; \quad r_p = \frac{p - p_0}{\tau_p} \tag{7}$$

The lifetime of electrons and holes, τ_n and τ_p, does not depend alone on the elementary recombination process itself, but also on the density of electrons and holes. The latter result can be proven by considering the rather simple recombination process via a band-band transition. In this case, $r_n = r_p = r$, and the recombination rate is given by

$$r = C_d(np - n_0 p_0) \tag{8}$$

in which C_d is the recombination cross section. Since also $\Delta n = \Delta p$, the lifetimes of electrons and holes must be equal ($\tau_n = \tau_p = \tau$). Considering now an n-type semiconductor and assuming that the number of carriers produced by light excitation is smaller than the equilibrium electron density ($\Delta n \ll n_0$), then one obtains from Eq. (7) by inserting Eqs. (5) and (8):

$$\tau = \frac{1}{C_d n_0} \tag{9}$$

According to this equation, the lifetime of excited carriers decreases with increasing majority carrier density and consequently with doping. A similar result is obtained if the recombination process occurs via impurity centers (Shockley-Read equation [20]), which will not be shown here. The recombination rate also influences the stationary density of electrons and holes produced by light excitation. One obtains from Eqs. (6) and (7):

$$\Delta n(\text{stat.}) = g\tau \tag{10}$$

Finally it should be emphasized that under stationary illumination conditions, which means non-equilibrium, always

$$np \gg n_0 p_0 \tag{11}$$

Since the relation between carrier density and Fermi level still holds (Eqs. (3a) and (3b)), the Fermi-levels for electrons and holes are now different.

2.4 Energy States of Semiconductor Particles

In principle, the optical properties of semiconductor particles, prepared as suspensions in a solution or as colloidal solutions, are identical to that of extended crystals. Differences in the absorption spectra of colloidal and macro-crystalline semiconductors occur, however, if the dimensions of colloidal particles become smaller than about 10 nm, as it was first recognized for CdS [26] and AgBr [27] around 25 years ago. Systematic investigations of this effect were started about 10 years ago, mainly by Henglein [28, 16, 17] and Brus [29] [30] and their co-workers. Various methods have been developed for preparing colloidal particles of a rather narrow size distribution. Usually it has been observed that – upon decreasing the particle size – the absorption rises more slowly with increasing photon energy, then the edge is blue-shifted, and finally at very small sizes the absorption spectrum becomes structured, i.e. various absorption peaks occur. Accordingly, the colour of the material changes considerably. For instance, in the case of Cd_3P_2, which is a low bandgap semiconductor ($E_G = 0.5$ eV), the colour changes from black to colourless when decreasing the particle size to 2 or 3 nm [31], as shown in Fig. 4.

This size effect is generally described by the quantum mechanics of a "particle in a box". The electron and hole are limited by potential walls of small dimensions, which leads to a quantization of the energy levels. Therefore, particles of this small size are usually called Q-semiconductors (Q stands for quantized). The effect described above occurs when the size of the small particles comes in the order of the De Broglie wavelength of charge carriers, which is given by

$$\lambda_B \sim \frac{h}{(2\pi k T m_e^*)} \tag{12}$$

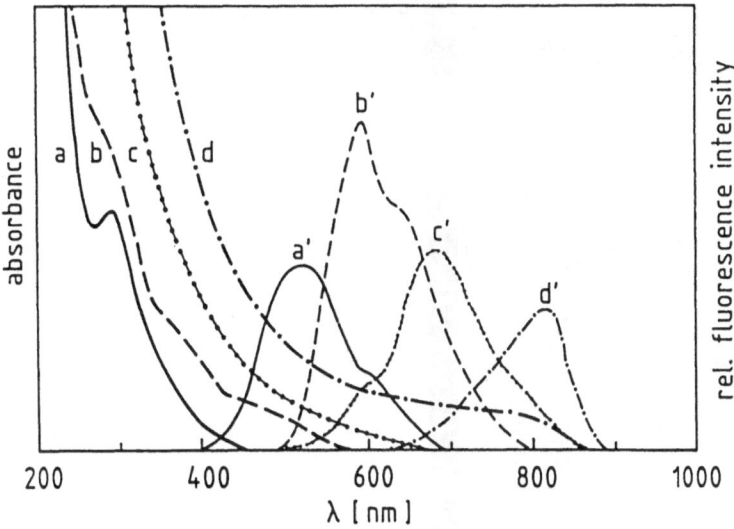

Fig. 4. Absorption (a, b, c, d) and fluorescence (a', b', c', d') spectra of colloidal solutions of Cd_3P_2 of different particle size (size increasing from $a \rightarrow d$) [31]

Although λ_B of a free electron is in the order of 0.1 nm, the De Broglie wavelength of an electron in a crystal can be much larger, because the effective mass m_e^* is usually considerably smaller than the free electron mass. Therefore, quantization effects can be observed with bigger particles. Various theories, all based on the model of hard potential walls for the box, have been developed [32–35]. Figure 5 illustrates the change of bands and the occurance of discrete energy states for small particle sizes [36], as derived on the basis of a qualitative molecular orbital picture [37].

It should be mentioned that the fluorescence spectra also change if the particles become smaller. The evaluation of the spectra is sometimes not easy, because surfaces become more dominant at very small sizes.

Quantization effects have not only been observed with colloidal solutions, but also with nano crystals deposited on Pt or SnO_2 [38, 39] and with semiconductor clusters (CdS, PbS) produced in zeolites [40, 41]. The same effect has been found with thin single crystalline semiconductor layers (super lattices) (see e.g. Refs. [42, 43]. In this case the quantization occurs only in one dimension, and the shift of energy states is much less, i.e, in the order of 0.1 eV, whereas the shift in a colloidal particle (quantization in 3 dimensions) amounts to several eV.

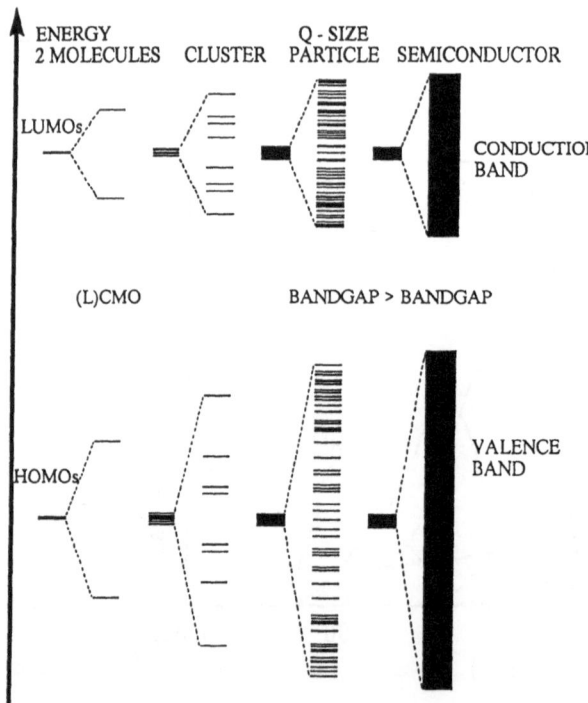

Fig. 5. MO model for different particle sizes [36]

3 Energy Levels at the Semiconductor-Electrolyte Interface

3.1 Energy Levels in the Electrolyte

Taking a simple one-step redox couple as an example, the electrochemical potential of electrons is given by

$$\bar{\mu}_{e,\,redox} = \mu^0_{redox} + kT \ln\left(\frac{c_{ox}}{c_{red}}\right) \tag{13}$$

where c_{ox} and c_{red} are the concentrations of the oxidized and reduced species of the redox system, respectively. Usually, the corresponding redox potentials are given on the conventional scale, using the Normal Hydrogen Electrode [NHE] or Saturated Calomel Electrode (SCE) as a reference electrode. Using theoretically the absolute scale with a vacuum level as a reference point, then the electrochemical potential of electrons in a redox system is equivalent to the Fermi level $E_{F,\,redox}$ [44], i.e.

$$E_{F,\,redox} = \bar{\mu}_{e,\,redox} \tag{14}$$

on the absolute scale. Accordingly, the electrochemical potential of electrons is defined in solids and in electrolytes in the same way.

As mentioned above, the electrochemical potential of a redox couple is usually given with respect to the Normal Hydrogen Electrode (NHE). Using an absolute energy scale with the vacuum level as a reference, the energy of a redox couple is given by

$$E_{F,\,redox} = E_{ref} - eU_{redox} \tag{15}$$

in which U_{redox} is the redox potential vs NHE, and E_{ref} the energy of the reference electrode versus the vacuum level. The determination of E_{ref} has been subject of various calculations, as indicated in Ref. [10]. The data derived by various authors scatter from 4.3 to 4.7 eV. Usually, an average value of $E_{ref} = 4.5$ eV for NHE is used, so that Eq. (15) yields

$$E_{F,\,redox} = -4.5\,eV - eU_{redox} \tag{16}$$

with respect to the vaccum level.

Information about further energy levels in a redox system can be derived from the theory of electron transfer between a redox system and an electrode, which has been derived by various authors [2–5]. In all these theories it is assumed that the vibration of redox molecules and their surrounding solvation shell is slow, compared to the actual electron transfer, i.e. it is assumed that the Frank-Condon principle is valid. As shown by Gerischer [44], this assumption leads to the consequence that the energy levels involved in the charge transfer differ from the thermodynamic value $E_{F,\,redox}$. This model leads to a distribution of empty and occupied states versus electron energy, as illustrated in Fig. 6. These electron states are not discrete energy levels, but are distributed over

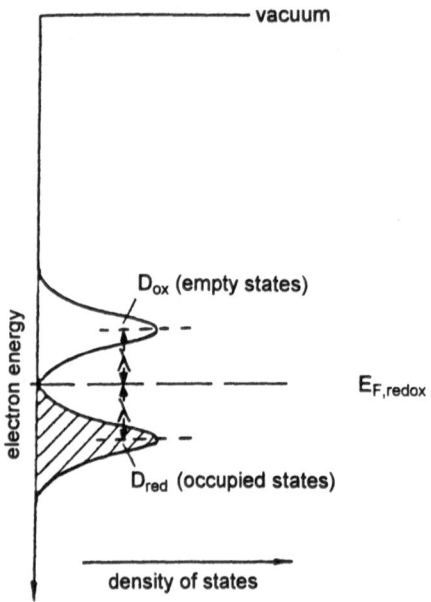

Fig. 6. Distribution of energy states in a redox system

a certain energy range, due to fluctuations in the solvation shell surrounding the redox molecules. The exact derivations of this model can be found in other review articles [5, 10]. The distribution curves, D_{ox} and D_{red}, are described by a Gaussian type of function, which can be derived quantitatively assuming a harmonic oscillation for the solvation shell. One obtains

$$D_{red} = D_{red}^0 \exp\left[-\frac{(E - E_{F,\,redox} - \lambda)^2}{4\,kT\lambda} \right] \tag{17a}$$

$$D_{ox} = D_{ox}^0 \exp\left[-\frac{(E - E_{F,\,redox} + \lambda)^2}{4\,kT\lambda} \right] \tag{17b}$$

in which λ is the reorientation energy which is in the range of $0.5-2\,eV$, depending on the interaction of the molecule with the solvent [6, 45, 46].

3.2 Potential Distribution at the Semiconductor-Electrolyte Interface

Combining a semiconductor with an electrolyte containing a redox system, equilibrium is achieved if the electrochemical potential is constant throughout the whole system, i.e. the Fermi-levels of the semiconductor and the redox system must be equal on both sides of the interface:

$$E_F = E_{F,\,redox} \tag{18}$$

To achieve equilibrium, electrons cross the interface until a corresponding potential difference occurs. The latter depends on the difference of the two Fermi levels before the contact is made. Neglecting any specific adsorption of H^+- and OH^--ions or the formation of a dipole layer on the surface, the potential difference is given by

$$\bar{\mu}_{sem} - \bar{\mu}_{redox} = eU_{sc} \qquad (19)$$

It can be calculated, provided that the electron affinity E_A (see Sect. 2.1.) and $E_c - E_F$ are known . Examples are given in Figs. 7a and b for redox systems and n- and p-type semiconductors.

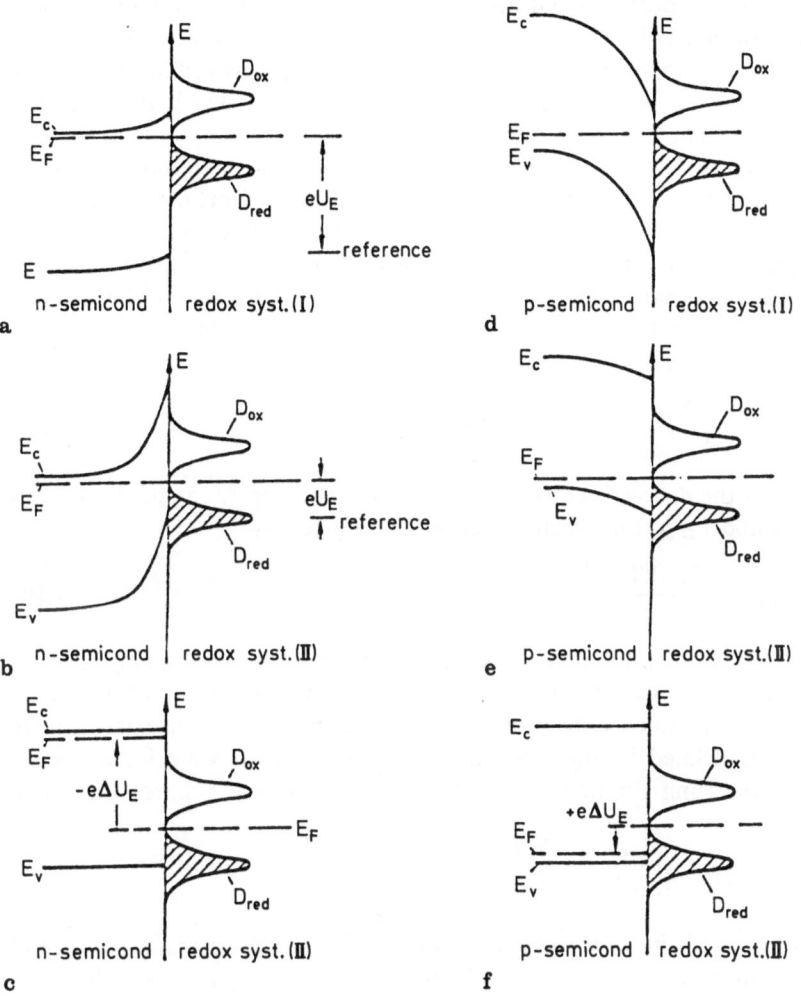

Fig. 7. Relative position of energy levels at the interface semiconductor-redox system for two different redox systems: redox system I (a, d); redox system II (b, c, e, f);
   $a)$ $b)$ $d)$ $e)$ at equilibrium
   $c)$ $f)$ at flatband potential

It can be clearly seen from this figure that the potential difference developed for achieving equilibrium occurs within the semiconductor below its surface. The magnitude of band bending and its sign depends on the redox potential (compare Figs. 7a and b) and on the doping of the semiconductor. It should be emphasized that the potential occurs only across the space charge layer, provided that the ion concentration in the electrolyte is sufficiently large (the Gouy-layer is neglected).

In principle, the energy diagram is also valid for "real" interfaces, at which H^+ or OH^- or other ions are adsorbed or bonded to the semiconductor, which leads to the formation of a Helmholtz double layer. In this case, however, the position of the conduction band with respect to the vacuum level is not anymore entirely determined by the electron affinity E_A, and the potential difference cannot be simply calculated from Eq. (19). The distance between the conduction band and vacuum level is now given by $E_A + eU_H$, in which the last term represents the potential across the Helmholtz layer.

This result makes it impossible to predict theoretically the position of energy bands without further experiments, since U_H is unknown. Fortunately, however, the position of energy levels can be obtained experimentally by capacity measurements. The differential capacity of the space charge layer below the semiconductor surface can be derived quantitatively by solving the Poisson equation (see e.g. Ref. [6]). For doped semiconductors one obtains the so-called Mott-Schottky equation:

$$1/C_{sc}^2 = \left(\frac{2L_{D, \text{eff}}}{\varepsilon\varepsilon_0}\right)^2 \left(\frac{e\phi_{sc}}{kT} - 1\right) \tag{20}$$

in which ε is the dielectric constant of the semiconductor, ε_0 the permittivity of free space and $L_{D, \text{eff}}$ the effective Debye length given by

$$L_{D, \text{eff}} = \left(\frac{\varepsilon\varepsilon_0 kT}{2n_0 e^2}\right)^{1/2} \tag{20a}$$

Equation (20) is an approximation being valid only in the depletion layer, where the majority carrier density at the surface (n_s for n-type; p_s for p-type) is smaller than the corresponding bulk concentration. The thickness of the space charge layer can be defined by the relation $d_{sc} = \varepsilon\varepsilon_0/C$ which is valid for a normal capacitor. Inserting Eq. (20), for the thickness of the space charge layer, one obtains:

$$d_{sc} = 2L_{D, \text{eff}}\left[\frac{e\phi_{sc}}{kT} - 1\right]^{1/2} \tag{21}$$

The thickness of the space charge layer increases with increasing potential Φ_{sc} across this layer, i.e. $d_{sc} \approx 9L_{D, \text{eff}}$ for $\Phi_{sc} = 0.5$ V. Taking a semiconductor of a typical doping, for instance $n_0 = 10^{17}\,cm^{-3}$, one obtains $L_{D, \text{eff}} = 10^{-6}\,cm$ ($\varepsilon = 10$).

Since the space charge capacity C_{sc} is usually much smaller than the capacity C_H of the Helmholtz double layer, it can easily be determined experimentally.

One example of $1/C_{sc}^2$ vs electrode potential U_E is given in Fig. 8 [47]. Since the slope of these straight Mott-Schottky curves are identical to the theoretical value given by Eq. (20), it must be concluded that the potential across the Helmholtz layer remains constant, i.e.

$$\Delta U_E = \Delta\phi_{sc} \tag{22}$$

An extrapolation of the Mott-Schottky plot to $1/C_{sc}^2$ yields the electrode potential at which the potential across the space charge layer becomes zero ($\Phi_{sc} \rightarrow 0$). Accordingly, we have

$$1/C_{sc}^2 \rightarrow \phi_{sc} \approx 0 \rightarrow U_E = U_{fb} \tag{23}$$

U_{fb} is usually called the flatband potential. The result that the flatband potential differ considerably for n- and p-type electrodes becomes clear by considering the energy scheme in Fig. 7c. In order to reach flatband, an n-type electrode has to be polarized negatively and a p-type positively with respect to equilibrium. In the case of rather strongly doped semiconductors, the two flatband potentials should differ by E_g/e, as actually found for GaP and also for other semiconductor electrodes. The latter conclusion is only valid, however, if the positions of the energy bands are identical for the n- and p-type electrodes.

According to this result, the energy bands are pinned, and their position is independent of the doping. This result has been confirmed for most other semiconductor electrodes in aqueous electrolytes. In many cases, a shift of energy bands has been found upon variation of the pH of the electrolyte due to a corresponding change of the potential across the Helmholtz layer. Especially in the case of oxide semiconductors, the pH-dependence has been measured

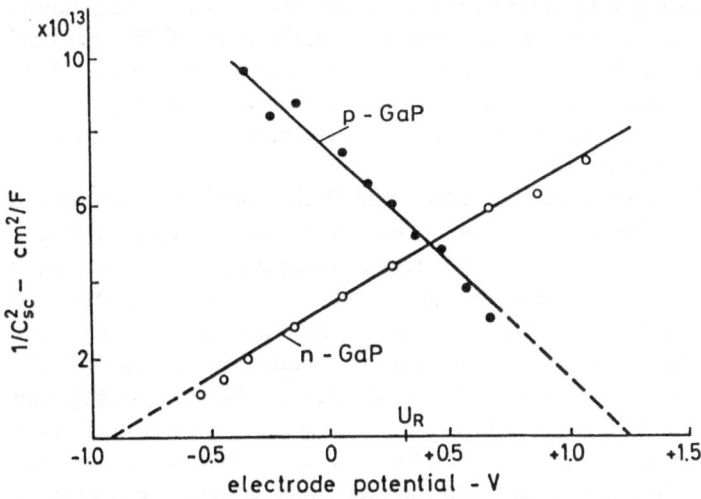

Fig. 8. Mott-Schottky plot of the space charge capacity vs electrode potential at n- and p-type GaP in 0.1 M H_2SO_4 [47]

quantitatively. Such experiments have yielded a shift of 0.059 V/pH, as shown by various authors [45, 48–50]. The results indicate that the surface bonding of hydroxyl groups determines the behavior of the Helmholtz layer. A pH-dependence has also been observed for some III-V compounds in a certain pH range [51, 52]. In some cases, such as Ge and GaAs, the flatband potential depends on the prepolarization of the electrode [53, 54]. This phenomenon has been interpreted as the formation of a hydroxyl surface during anodic polarization and a change into a hydride surface after cathodic treatment.

According to these results, the position of energy bands of semiconductors in aqueous electrolytes is entirely determined by the strong interaction of the solvent with the semiconductor surface. A selection of band position of various semiconductors is given in Fig. 9. In some cases, the flatband potentials reported in the literature vary over a considerable range. This is mostly due to contaminations of the semiconductor surface. This became quite obvious for CdS, for which the flatband potentials as measured by various authors, differ over a range of about 1 V. This can be explained by the adsorption of sulfur on the surface, formed by etching or by anodic polarization [58]. A really clean surface can only be formed by cathodic polarization in the presence of oxygen [57]. In this case, a rather negative flatband potential of -1.8 V was found [57]. Interestingly, the same flatband potential was found in solutions of Na_2S, which dissolves the S^0 from the CdS (polysulfide formation); i.e. the surface can also be cleaned chemically [57].

In the literature, sometimes $1/C_{sc}^2 - U_E$ measurements are presented, yielding not straight but curved Mott-Schottky curves, which are frequently interpreted by two types of donor- or acceptor states within the space charge region of the semiconductor. In other cases, corresponding measurements are only performed within a very small potential range. Careful investigations with few materials, such as e.g. CdS, have shown, however, that a bent Mott-Schottky curve can also be due to surface contaminations. In the case of CdS it has been proved that excellent straight lines over a large potential range (about 6 V!) can be obtained after having cleaned the surface as described above [57]. Another example is TiO_2. Here, the origin of a bent curve is due to inhomogeneous doping after H_2-treatment [59].

The interaction between semiconductor and liquid is sufficiently strong that the energy bands at the surface remain unchanged if a redox couple is added to the electrolyte. Since the energy bands remain completely pinned, any band bending can be obtained at equilibrium by using a proper redox couple, as shown in Fig. 7. Some semiconductors, such as layer compounds, however, show a weaker interaction. This becomes obvious, for instance, in the case of WSe_2, with which no shift of bands upon a change of pH was observed [60, 61]. This has been explained by the layer structure, i.e. the metal atom which could form a hydroxide is shielded from the solution by Se-atoms. Only if many steps from one layer to the other are present, can the metal then contact the electrolyte. Accordingly, a pH-dependence of the flatband potential occurs if steps are created by scratching the surface [60]. On the other hand, the interaction with

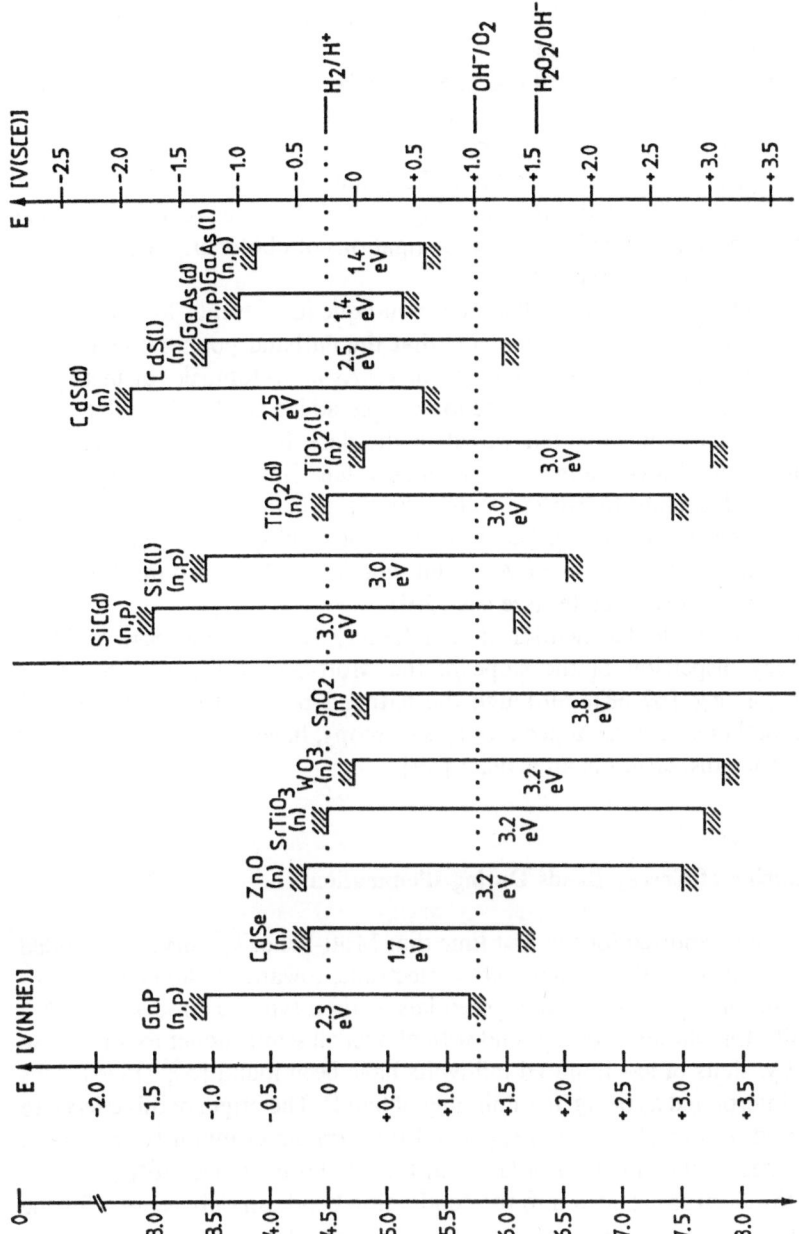

Fig. 9. Position of energy bands at the surface of various semiconductors at pH 0; (*d*): dark (*I*): illuminated

the solvent is still sufficiently strong in order to prevent any shift of energy bands upon addition of a redox system.

Bard et al. have emphasized that there may be exceptions in so far as Fermi level-pinning by surface states may occur similarly as at semiconductor-metal junctions [62]. Such an effect would lead to an unpinning of bands. There are some examples in the literature, such as FeS_2 in aqueous solutions [63, 64] and Si in CH_3OH [65], for which an unpinning of bands has been reported. In some cases, the interpretation is based on investigations of photocurrents, which will be discussed in the next section.

The situation can be quite different in non-aqueous solvents. Recently, it has been shown by capacity measurements that the flatband potential and consequently the position of energy bands of the GaAs in acetonitrile- or methanol-solutions depends strongly on the redox couple added to the electrolyte [66]. Actually, the flatband potential varied in the dark by more than 1 V, if the standard potential was changed by the same value. This result was interpreted as Fermi level pinning by surface states, the latter being located about 0.3 to 0.4 eV above the valence band. According to these results, the GaAs/acetonitrile interface behaves similarly as GaAs/metal junctions, which also exhibit Fermi level pinning by surface states (see e.g. [20]).

Finally, it should be mentioned that frequently, as in the case of TiO_2, a frequency dispersion of the slope of the Mott-Schottky curves has been observed (see e.g. [67, 68]), although the flatband potential was not affected. Modern methods, such as impedance spectroscopy, have shown, however, that this frequency dispersion is an artifact [59].

3.3 Unpinning of Energy Bands During Illumination

In 1980, it was reported for the first time that Mott-Schottky curves are shifted upon illumination of the semiconductor electrode, towards cathodic potentials with p-type and toward anodic potentials with n-type electrodes [69, 70]. Meanwhile, this effect has been found with almost all semiconductors studied so far. Mostly, shifts of few hundred millivolts have been found [57, 71–76]. This shift was interpreted as being an unpinning of bands. The origin of this effect can be multifold. In most cases, it is explained by trapping of minority carriers in surface states, as illustrated for p-GaAs in Fig. 10. Using an electrolyte such as H_2SO_4, the electrons excited into the conduction band can only be used for the reduction of protons. These electrons, however, can also be trapped in surface states, as illustrated in the center part of Fig. 10. Since the hole density at the surface is small, the recombination rate of the trapped electrons with holes is low. Provided that a sufficient number of surface states is available, a considerable charge can be stored, leading finally to a change of the potential distribution. The stored charge ΔQ_s can be calculated from

$$\Delta Q_s = C_H \Delta U_H (h\nu) \qquad (24)$$

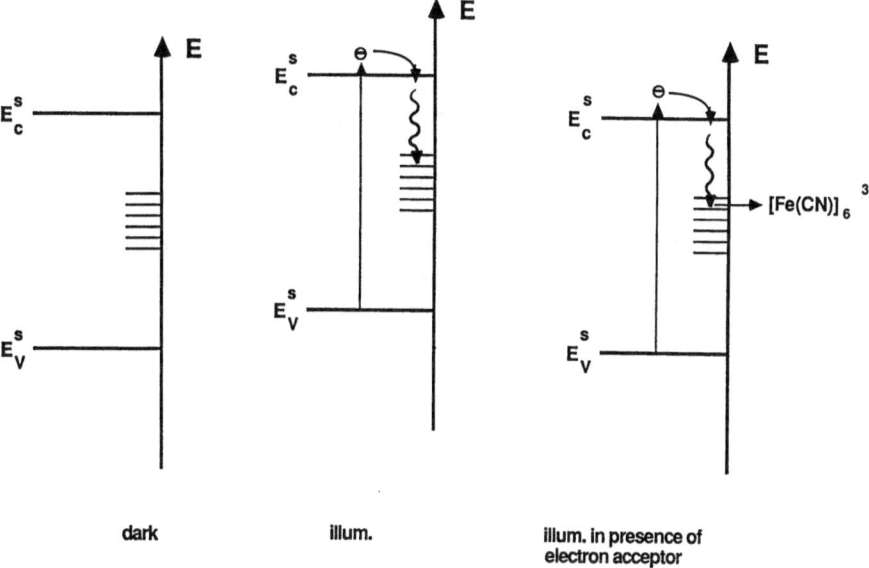

Fig. 10. Position of energy bands at the surface of p-GaAs (pH 1) in the dark and under illumination (data from [70])

in which C_H is the Helmholtz capacity and $\Delta U_H(h\nu)$ the shift of the Mott-Schottky curve. The density of surface states $N_t = \Delta Q/e$ is in the order of about $10^{13}\,cm^{-2}$ for a shift of $U_{fb} = 0.2\,V$, and assuming $C_H = 10^{-5}Fcm^{-2}$. It is interesting to note that $\Delta U_H(h\nu)$ occurs mostly at very low light intensities, and it saturates at higher intensities if all surface states are filled [70]. The density of surface states estimated above are derived under the latter conditions.

The accumulation of minority carriers and the corresponding shift of energy bands can be avoided if a suitable redox system is added to the electrolyte. For instance in the case of p-GaAs, the energy bands are shifted downward back to its original dark value upon addition of $[Fe(CN)_6]^{3-}$ as an electron acceptor. It is believed here that the electrons captured by surface states, are transferred from these states to the electron acceptor in the solution, as illustrated on the right side of Fig. 10 [70].

As already mentioned, this phenomenon has been found with many semiconductor electrodes. However, the origin of band edge shift by illumination must not necessarily be due to trapping of minority carriers by surface states. For instance, an accumulation of minority carriers can occur at the surface if no electron acceptor or donor is available in the solution. One example is n-WSe$_2$ or MoSe$_2$, with which a shift of the Mott-Schottky curves of about 0.6 V has been observed, using an electrolyte without any redox system [61]. This large shift has been interpreted by a very slow kinetics of the anodic decomposition of the material, so that the holes (minority carriers) are accumulated in the valence band at the surface [78]. Adding a redox couple such as

$[Fe(phen)_3]^{2+}$ to the solution, the standard potential of which is located very close to the valence band of WSe_2, the energy bands remain pinned to their dark values [78], i.e. holes created by light excitation are efficiently transferred from the valence band to the hole acceptor in the solution. In the absence of a redox system, typical photocurrent transients have been observed in the range $U_{fb(dark)}$ and $U_{fb(hv)}$. Details of the charge transfer mechanism will be given in Sect. 4.3.

It should be mentioned here that the shift of energy bands is not limited to the formation of minority carriers by light excitation. The same effect occurs

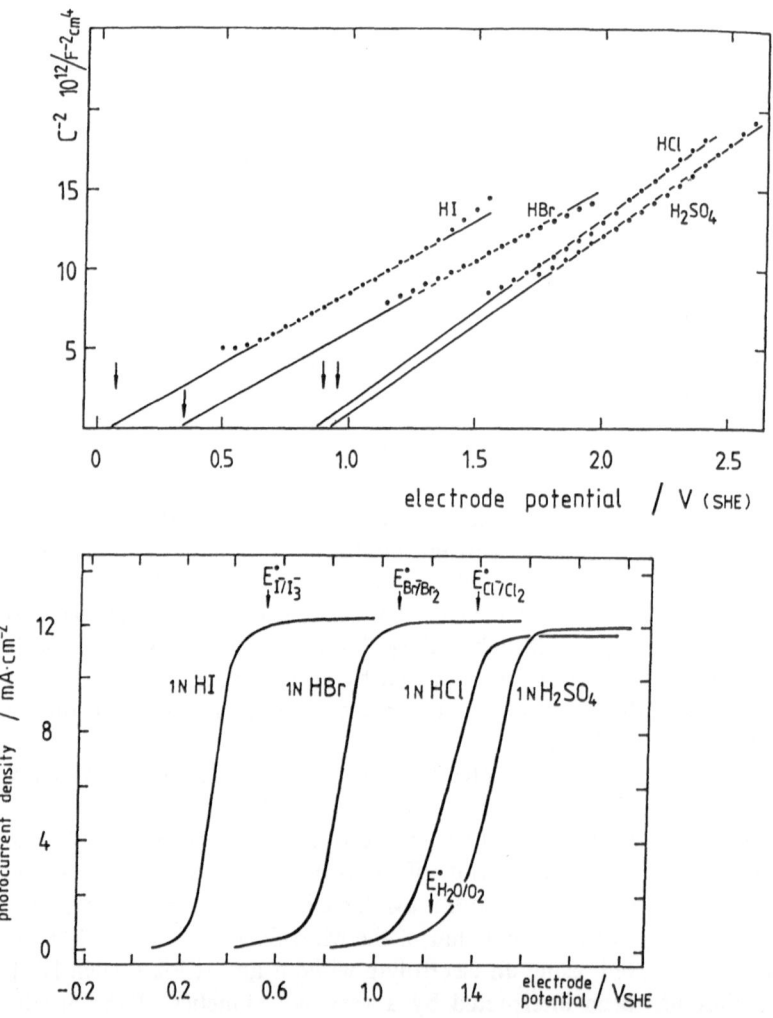

Fig. 11. Mott-Schottky plot (*upper figure*) and photocurrent vs. electrode potential (*lower figure*) for n-RuS$_2$ [79]

124

also if minority carriers are created in the semiconductor by injection from the electrolyte. Examples are the injection of holes into the valence band of n-GaAs [56] and n-WSe$_2$ in the dark [61]. In both cases, exactly the same shift of Mott-Schottky curves has been found as with illumination. According to investigations with Ce^{4+} at n-WSe$_2$ and n-MoSe$_2$, Scholz et al. have concluded, however, that the shift of bands is not due to hole injection, but is caused by adsorption of Ce^{4+} [77].

A rather extreme shift of bands of up to 2.0 V has been found with RuS$_2$ [79]. This shift is even considerably larger than the bandgap [$E_G = 1.25$ eV). It cannot be interpreted on the basis of a simple surface state model, because an estimation of corresponding density of surface states by using Eq. (24) would lead to $N_T > 10^{14}$ cm^{-2}, which is an unreasonable number, because it would correspond to more than 10% of the surface atoms. More information can be obtained by comparing the photocurrent-potential dependence and the Mott-Schottky curves measured by the same authors [79] (Fig. 11). The photocurrent onset in IM H$_2$SO$_4$ occurs around 1.4 V [NHE], and the flatband potential during illumination at $U_{fb}(h\nu) = +1.0$ V, whereas in the dark a flatband of $U_{fb} = -1$ V (SCE) was found. This result is reasonable because only at $U_E > U_{fb}$ there is an electrical field across the space charge layer, which drives the holes created by light excitation towards the surface. Since RuS$_2$ is stable against corrosion, the anodic process corresponds to O$_2$ − formation. Interestingly, the photocurrent onset occurs at less anodic potential in the presence of

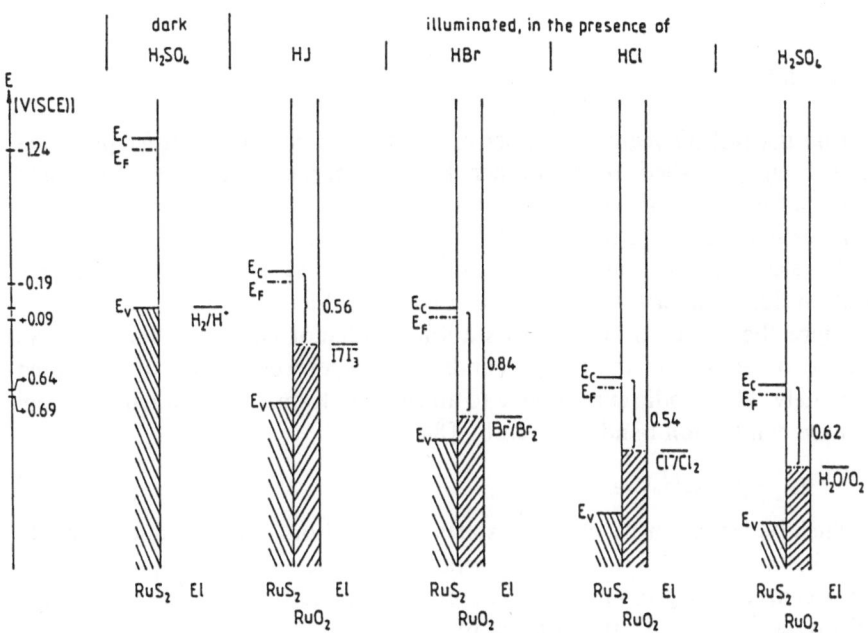

Fig. 12. Position of energy bands at the surface of n-RuS$_2$ [80]

halide ions, such as J^-, Br^-, or Cl^- in the electrolyte (Fig. 11). Simultaneously, the shift of the Mott-Schottky curves upon illumination is also smaller (upper part of Fig. 11). In all cases, the flatband potential occurs close to the onset of the photocurrent. These results can be interpreted by a model as follows [80]:

According to a surface spectroscopic analysis (XPS), one or two monolayers of RuO_2 are formed during anodic polarization [12]. This result indicates that the formation of the oxide prevents RuS_2 from any anodic dissolution. Since RuO_2 is a metal-like conductor, a kind of semiconductor-metal Schottky-junction exists on the surface, as illustrated in Fig. 12. Assuming a low overvoltage for the oxidation of halides at the RuO_2-layer, the Fermi level in RuO_2 should be pinned close to the redox potential of the corresponding halide. A comparison of the redox potential with the flatband potentials during illumination yields finally the energy diagram as given in Fig. 12. This diagram leads to a more or less constant barrier at the RuS_2/RuO_2-interface, as it is typical for semiconductor-metal Schottky junctions. According to this model, the origin of the flatband shift upon illumination is a complete pinning of the quasi-Fermi level of holes to that in the RuO_2-layer.

The same type of Fermi level pinning occurs also when the surface of InP is reduced by cathodic polarization leading to a thin In-layer, or when another metal, such as e.g. Pt, is deposited [83].

4 Charge Transfer Kinetics

4.1 Models

During the last 30 years, a number of theories on electron transfer processes have been published by Gerischer [1, 5], Marcus [2], Levich [4] and Dogonadze [3, 82]. Especially the models and theories developed by Marcus and Gerischer are applied for electrochemical reactions at metal and semiconductor electrodes by many other scientists. On the basis of these theories, the electron flux or interfacial currents can be derived as follows:

Since the charge transfer across a semiconductor electrolyte-interface can only occur via one of the energy bands, the two processes have to be treated separately. The anodic current, due to an electron transfer from a redox system into the conduction band, is given by [84]:

$$j_c^+ = e\,k_{c,ox}\,N_c\,c_{red} \qquad (25)$$

in which N_c is the density of energy states at the bottom of the conduction band, given in units of cm^{-3}, c_{red} is the concentration of the reduced component of the redox couple, also given in cm^{-3}. Accordingly, the rate constant $k_{c,ox}$ occurs in units of $cm^4\,s^{-1}$. The latter is given by

$$k_{c,ox} = k_{c,ox}^{max}\,D_{red}(E_c) \qquad (26a)$$

Inserting D_{red} (E_c) by using eq. (17a), one obtains

$$k_{c,ox} = k_{c,ox}^{max} \exp\left[-\frac{(E_c - E_{F,redox} - \lambda)^2}{4kT\lambda} \right] \tag{26b}$$

Since the conduction band remains fixed with respect to $E_{F,redox}$ during polarization (compare e.g. Fig. 7b with c and Fig. 7e with f), the rate constant and j_c^+ are independent of the potential. The cathodic current is given by

$$j_c^- = e k_{c,red} n_s c_{ox} \tag{27}$$

in which n_s is the electron density at the surface as defined by eq. (3), c_{ox} the concentration of the oxidized species of the redox system, and $k_{c,red}$ is defined by

$$k_{c,red} = k_{c,red}^{max} D_{ox}(E_c) \tag{28a}$$

With eq. (17b) we have

$$k_{c,red} = k_{c,red}^{max} \exp\left[-\frac{(E_c - E_{F,redox} + \lambda)^2}{4kT\lambda} \right] \tag{28b}$$

Similar expressions are obtained for valence band processes, i.e. for an anodic current we have

$$j_v^+ = e k_{v,ox} p_s c_{red} \tag{29}$$

in which p_s is the surface hole density and

$$k_{v,ox} = k_{v,ox}^{max} D_{red}(E_v) = k_{v,ox}^{max} \exp\left[-\frac{(E_v - E_{F,redox} - \lambda)^2}{4kT\lambda} \right] \tag{30}$$

The cathodic current is given by

$$j_v^- = e k_{v,red} N_v c_{ox} \tag{31}$$

with

$$k_{v,red} = k_{v,red}^{max} D_{ox}(E_v) = k_{v,red}^{max} \exp\left[-\frac{(E_v - E_{F,redox} + \lambda)^2}{4kT\lambda} \right] \tag{32}$$

In the case of a valence band process, the cathodic current is independent of the electrode potential.

These equations can be derived from Gerischer's [5] as well as from Marcus' [2] theories. The main reason of using either theory is that the exponential terms in Eqs. (26b), (28b), (30) and (32) originate from the fluctuation of the solvation shell around the redox molecules or ions, and in both theories the fluctuation is described by a harmonic oscillation [6].

Charge transfer between energy states in semiconductor and in the redox system only occurs when there is a sufficient overlap of occupied and empty states. The actual currents across the interface can be derived for conduction or valence band processes by taking into account the equilibrium conditions, where $j_c^+ = j_c^- = j_c^0$ and $j_v^+ = j_v^- = j_v^0$, and also $n_s = n_s^0$ and $p_s = p_s^0$. In the case

127

of a conduction band process, by using Eqs. (25) and (27), one obtains:

$$j_c = - j_c^0 \left[\frac{n_s}{n_s^0} - 1 \right] \quad \text{(conduction band process)} \tag{33}$$

in which n_s^0 is the surface electron density at equilibrium and

$$j_c^0 = e\, k_{c,ox} N_c c_{red} \tag{33a}$$

A similar expression can be derived for the charge transfer via the valence band. One obtains

$$j_v = j_v^0 \left[\frac{p_s}{p_s^0} - 1 \right] \quad \text{(valence band process)} \tag{34}$$

in which p_s^0 is the surface hole density at equilibrium and

$$j_v^0 = e\, k_{v,red} N_v c_{ox} \tag{34a}$$

To obtain relations between carrier density at the interface and at the inner edge of the depletion layer (the thickness of the space charge layer d_{sc} is defined by Eq. (22)), we assume Boltzmann equilibrium for the carriers across the space charge layer. Using Eqs. (3a) and (3b), we have

$$n_s = n_w \exp\left(- \frac{e\phi_{sc}}{kT} \right) \tag{35a}$$

$$p_s = p_w \exp\left(\frac{e\phi_{sc}}{kT} \right) \tag{35b}$$

in which n_w and p_w are the carrier densities at the inner edge of the space charge region. The densities n_w and p_w are not identical to the bulk densities n_0 and p_0 if minority carriers are involved in the reactions.

The carrier densities and the actual currents also depend on the doping of the semiconductor (n- or p-type doping) and on the generation of charge carriers by light excitation. Minority carriers, generated by light excitation in the bulk $(x > d_{sc})$ can only diffuse towards the surface, whereas those which are generated within the space charge region are driven towards the interface by the electric field. Restricting ourselves to an n-type semiconductor and following Reichman's derivation [85], the latter is given by

$$j_{SPR} = e I_0 [1 - \exp(- \alpha d_{sc})] \tag{36}$$

in which I_0 is the light intensity and α the absorption coefficient. The diffusion current, due to generation of minority carriers in the bulk, can be obtained by solving the corresponding diffusion equation. The resulting diffusion current is then given by

$$j_{diff} = - j_0 \left[\frac{p_w}{p_0} - 1 \right] + \frac{e I_0 \alpha L_p}{1 + \alpha L_p} \exp(- \alpha d_{sc}) \tag{37}$$

in which

$$j_0 = \frac{e n_i^2 D_p}{N_D L_p} \tag{37a}$$

In Eqs. (37) and (37a), p_0 is the equilibrium bulk hole density, N_D the donor density, n_i the intrinsic carrier density (see Eq. (4)), D_p and L_p are the diffusion constant and diffusion length of holes respectively. The two components of the hole current must be equal to the total hole current, i.e.

$$j_{SPR} + j_{diff} = j_v \tag{38}$$

Inserting Eqs. (34), (36) and (37) into (38), one obtains

$$j_v = \frac{j_G - j_0 \exp\left(\frac{-e\eta}{kT}\right)}{1 + \left(\frac{j_0}{j_v^0}\right) \exp\left(\frac{-e\eta}{kT}\right)} \tag{39}$$

in which η is the overvoltage with respect to equilibrium, i.e.

$$\eta = U_E - U_{redox} \tag{40}$$

and the generation current is given by

$$j_G = j_0 + e I_0 \left[1 - \frac{\exp(-\alpha d_{sc})}{1 + \alpha L_p} \right] \tag{41}$$

The second term in Eq. (41) is actually the photocurrent, so that

$$j_G = j_0 + j_{ph} \tag{41a}$$

It should be mentioned here that Eq. (41) had already been derived by Gärtner [86] for the photocurrent of a semiconductor-metal junction in reverse bias, assuming $P_w = 0$. In the derivation of Reichman [85], however, P_w was obtained in a manner consistent with the interface boundary conditions. Although the derivation of Reichman is much more general, most scientists only applied the Gärtner-equation (see e.g. [87]). Later on, Wilson extended the Gärtner model by including recombination via surface states [88].

So far, only the minority carrier process has been considered. Since the reorientation energy is usually rather large, i.e. in the order of 1 eV, charge transfer can occur via both energy bands if the standard redox potential is near the middle of the energy gap, and if the gap is rather small (compared Figs. 13a and b). Accordingly, a majority carrier process is also possible. In the case of an n-type semiconductor, the majority carrier current is given by

$$j_c = -j_c^0 \left[\exp\left(-\frac{e\eta}{kT}\right) - 1 \right] \tag{42}$$

Equations (39) and (42) describe the complete current-potential behaviour under illumination and in the dark. Without any light excitation (i.e. $I_0 = 0$), one

129

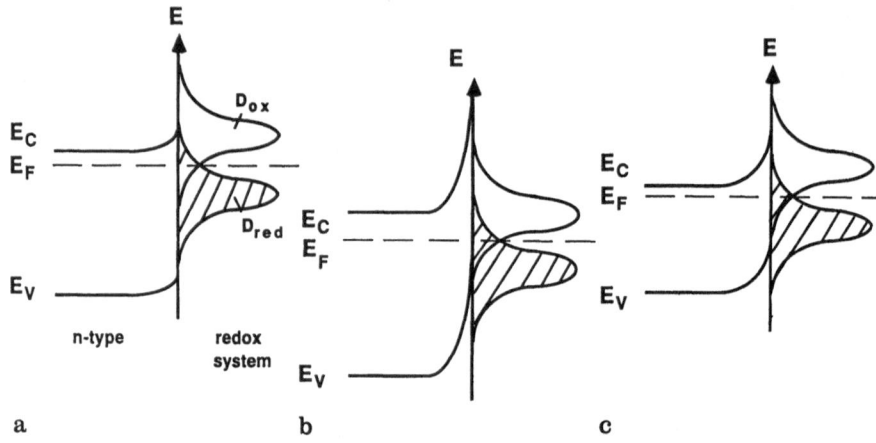

Fig. 13. Electrode potential and energy bands of a semiconductor being in contact with a redox system; **a** redox couple with a relatively negative standard potential; **b** with positive standard potential; **c** the Fermi-level of the redox system occurs around the middle of the energy gap

obtains from Eq. (39):

$$j_v(\text{dark}) = -\frac{j_0\left[\exp\left(-\frac{e\eta}{kT}\right) - 1\right]}{1 + \left(\dfrac{j_0}{j_v^0}\right)\exp\left(-\dfrac{e\eta}{kT}\right)} \tag{43}$$

Using Eqs. (31) and (43), current-voltage curves can be calculated as displayed qualitatively for a pure valance band process at an n-type electrode in Fig. 14. At anodic polarization (positive η-values), the dark current will become $j_v(\text{dark}) \to j_0$, and the total current under illumination $j_v \to j_G$. The cathodic current-potential behaviour depends on the ratio of j_0/j_v^0. Referring to Eqs. (34a) and (37a), these two currents represent the generation/recombination rate of minority carriers in the bulk of the semiconductor (j_0) and the rate of hole transfer at the interface (j_v^0). The first current depends entirely on the properties of the semiconductor, whereas j_v^0 is mainly controlled by the overlap of occupied and empty states on both sides of the interface and by the concentration of the redox system. Accordingly, the ratio j_0/j_v^0 actually controls whether the generation/recombination process or the surface kinetics are rate determining. Assuming that the hole injection is relatively slow ($j_v^0 \ll j_0$), then one obtains from Eq. (43) for large negative η-values

$$j_v(\text{dark}) \to -j_v^0 \tag{44}$$

as indicated by the dotted curve in Fig. 14. This kinetically controlled current should increase linearly with c_{ox}. On the other hand, if the recombination

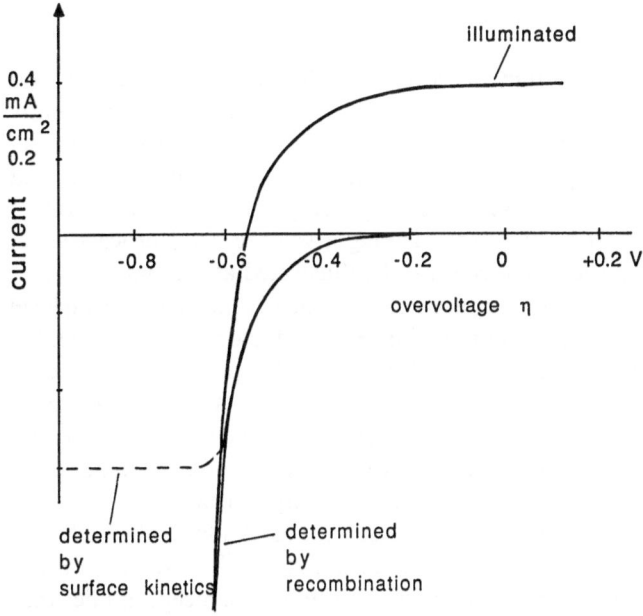

Fig. 14. Theoretical current-potential curve for a n-type semiconductor being in contact with a redox system, assuming a valence band process

controls the current $j_v^0 \gg j_0$, then one obtains:

$$j_v \rightarrow -j_0\left[\exp\left(-\frac{e\eta}{kT}-1\right)\right]+j_{ph} \tag{45}$$

This relation is identical to that derived for a pn-junction (see e.g. [89]) (solid curve in Fig. 14). It also looks similar to the current-voltage relation derived for a majority carrier transfer, as given by Eq. (42), both relations differ only by the pre-exponential factor. The first case, i.e. limitation by surface kinetics (Eq. (44)), is difficult to realize, because the majority carrier transfer becomes dominant for redox systems, the standard potential of which is located in the middle of the gap.

In the model presented here, only recombination in the bulk of the semiconductor has been considered. It is well known from solid state junction (minority carrier devices) that also recombination within the space charge layer can taken place. In this case, a quality factor n, ranging between 1 and 2, is introduced. Then Eq. (45) has to be replaced by

$$j_v = -j_0\left[\exp\left(-\frac{e\eta}{nkT}\right)-1\right]+j_{ph} \tag{45a}$$

Experimental results will be discussed in Sect. 4.4.

4.2 The Quasi-Fermi Level Concept

In the theory of non-equilibrium processes at solid state junction and also semiconductor-liquid interfaces, as developed in the previous section, frequently quasi-Fermi levels have been used for the description of minority carrier reactions [90, 91]. A concept for a quantitative analysis for reactions at n- and p-type electrodes has been derived [92, 93], using the usual definition of a quasi-Fermi level (Eqs. (3a) and (3b)). Taking a valence band process as an example, the quasi-Fermi level concept can be illustrated as follows:

If the density of holes p_s at the surface – or equivalently the quasi-Fermi level $E_{F,p}$ – are equal at the surface of an n- and p-semiconductor electrode, then the same reaction with identical rates, i.e. equal currents, takes place at both types of electrodes (Fig. 15). Since holes are majority carriers in a p-type semiconductor, the position of the quasi-Fermi level $E_{F,p}^s$ is identical to the electrode potential (see right side of Fig. 15), and therefore – with respect to the reference electrode – directly measurable. The density of p_s can easily be calculated, provided that the positions of the energy bands at the surface are known. The measurements of a current-potential curve also yields automatically the relationship between current and quasi-Fermi level of holes. The basic concept implies that the position of the quasi-Fermi level $E_{F,p}^s$ at the surface of an n-type semiconductor and the corresponding hole density p_s can be derived for a given photocurrent, since the same relationship between current and the quasi-Fermi level of holes holds.

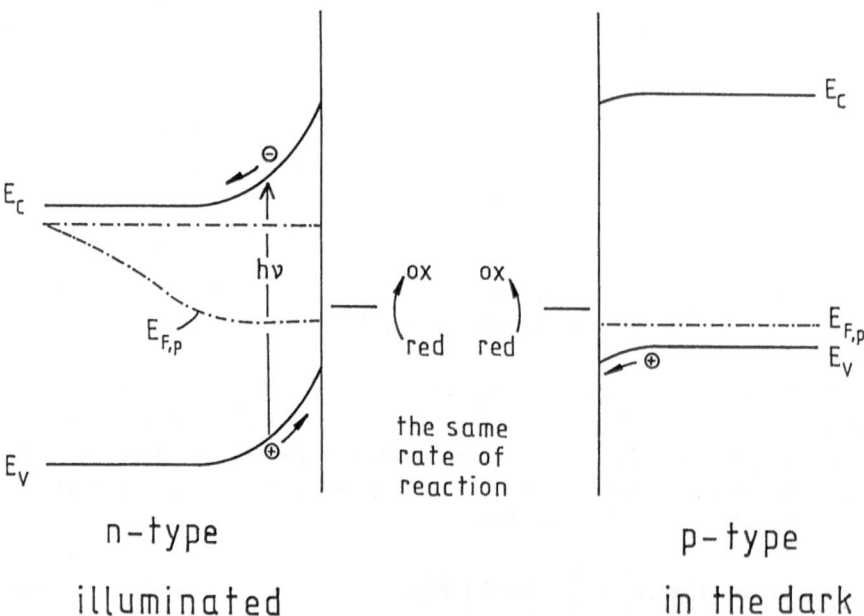

n–type
illuminated

p–type
in the dark

Fig. 15. Principle of comparibility of reactions at n- type p-type electrodes

This model is only applicable if three conditions are fulfilled:

(i) At equilibrium, the conduction and valence band edges at the surface of the n- and p-type electrode have the same position.
(ii) All reactions at the electrode can be described as a function of the surface hole density.
(iii) The holes at the surface of the p-type electrode are nearly in equilibrium with those in the bulk, i.e. the Fermi level of the majority carriers is constant within the electrode.

This model has been proved experimentally by studying the competition of the anodic decomposition reaction and the oxidation of Cu^{1+} at p-GaAs in the dark and at n-GaAs under illumination [93]. This is a suitable redox system, because reduction and oxidation occur via the valence band, and because the anodic oxidation of Cu^{1+} proceeds independently from the corrosion. Accordingly, the total current is given by

$$j = j_{corr} + j_{ox} \tag{46}$$

i.e. the ratio of j_{ox}/j is independent of the total current at p-GaAs. The same result has been obtained with n-GaAs under illumination, and an identical ratio of j_{ox}/j was found. Other authors also have observed that reactions at n- and p-type electrodes are comparable [94–96]. The advantage of the model presented here is that the quasi-Fermi level of the majority carriers (holes in p-type) can be determined because it is identical to the electrode potential. Accordingly, by measuring the current-potential curve at a p-type electrode, the relationship between current and quasi-Fermi level $E_{F,p}$ can be determined. Since the same relationship holds for processes at the n-type electrode, the position of the quasi-Fermi level of holes (minorities) at the surface of n-type can directly be related to the hole current (photocurrent).

It should be emphasized here that an anodic current (valence band process) at a p-type as well as at an n-type electrode is only possible if $E_{F,p}$ occurs in both cases below the redox potential, as illustrated in Fig. 15. This is valid in the dark and during light excitation. Polarizing an n-type electrode anodically, then usually a very small, but constant anodic current is observed in the dark. During this polarization, the band bending is increased, and the quasi-Fermi level of electrons (majority carriers) is moved downwards. The quasi-Fermi level of holes, however, remains at the same position, because the current remains constant.

More insight into the quasi-Fermi level concept and its application has been obtained by the following example [93]. In Fig. 16, curve a) represents the current-potential curve as obtained with p-GaAs in an electrolyte without any redox system. The anodic current corresponds to the decomposition of GaAs. After addition of Cu^{2+} as an oxidizing agent, a corresponding cathodic current is visible (curve b)). At the mixed potential $U_{M,1} = 0.34$ V ($j = 0$), the two currents are equal. The position of the quasi-Fermi level $E_{F,p}$ in the p-electrode

R. Memming

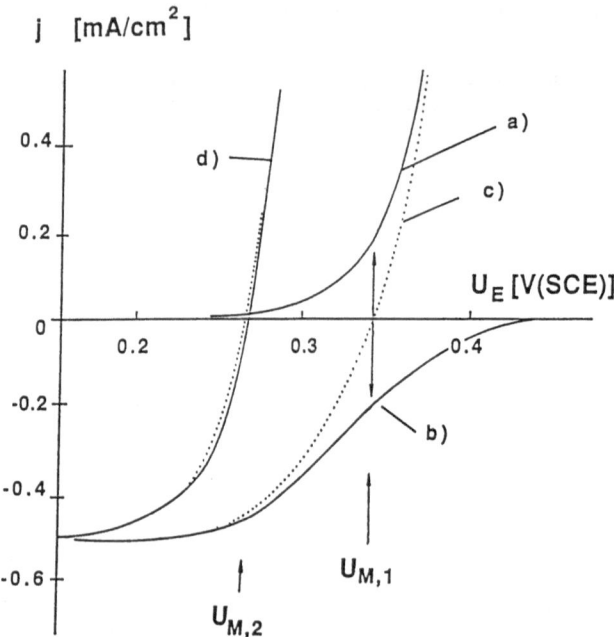

Fig. 16. Current-potential curve for a rotating (1000 rpm) p-GaAs electrode in 6 M HCl [93] *a*) anodic decomposition current; *b*) partial current of Cu^{2+}-reduction (0.7 mM); *c*) total current (dotted line); *d*) total current upon addition of Cu^{1+}-ions (50 mM)

at the potential is illustrated in Fig. 17a. The same experiment has also been performed with *n*-GaAs. Here, the total current is nearly zero over a large potential range (> 0.1 V), as shown in Fig. 18, i.e. the quasi-Fermi level of holes should also occur at 0.34 V, as illustrated for two potentials in Fig. 17b and c. In the range where the total current is zero, the latter is composed of two partial currents (dashed curved in Fig. 18), as determined by using a rotating ring-disc electrode. The partial currents are identical to those obtained with the p-electrode at $U_{M,1}$ (j = 0). After addition of Cu^{1+}-ions, the mixed potential $U_{M,2}$ was found at 0.26 V (curve d in Fig. 16), i.e. in this case $E_{F,p}$ occurs also at 0.26 V at the p-type electrode as illustrated in Fig. 16. The partial currents, anodic corrosion and cathodic reduction of Cu^{2+}, are considerably smaller than in the first case. The same observation was made with the n-type electrode. The position of the quasi-Fermi levels at the surface of n-type is shown in Figs. 17b and c for two different potentials.

These examples show very nicely how well the quasi-Fermi concept operates. There are various results given in the literature which can be interpreted on the basis of this concept. Especially the investigations of reactions of $[Fe(CN)_6]^{3-/4-}$ at GaAs and the etching behaviour confirm this model [97, 98]. The quasi-Fermi level concept also explains multiple step redox reactions.

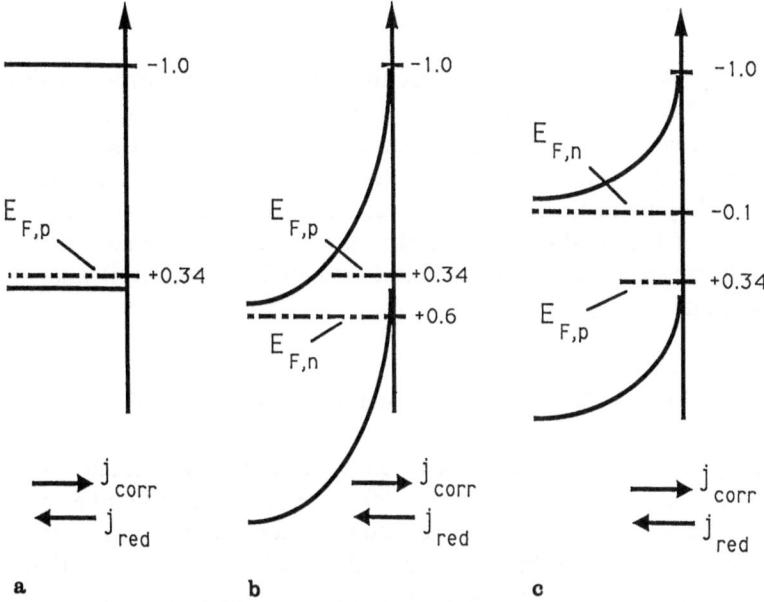

Fig. 17. Position of the quasi-Fermi levels in the presence of 0.7 mM Cu^{2+}, according to data given in Figs. 16 and 18 [93] **a** p-GaAs at the potential $U_{M,1} = +0.34$ V (SCE); **b** n-GaAs at $U_E = +0.6$ V; **c** n-GaAs at $U_E = -0.1$ V

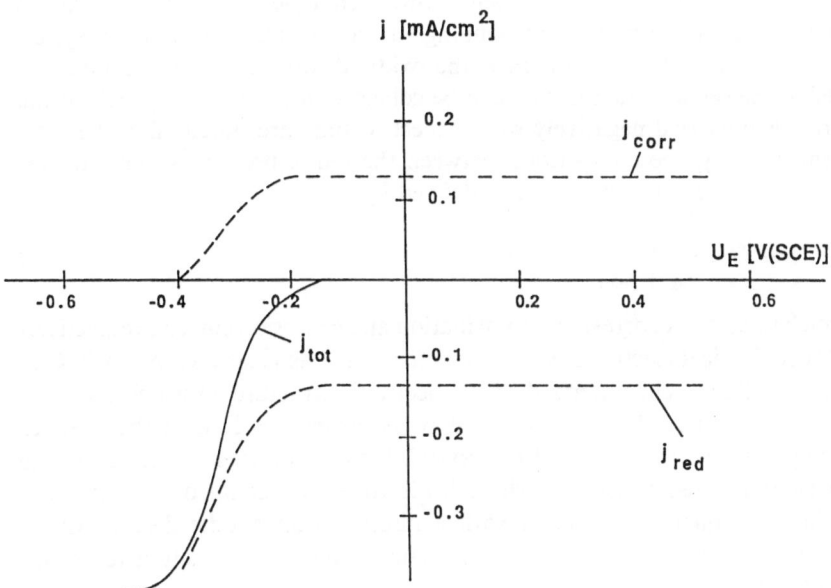

Fig. 18. Current-potential curve for a rotating (1000 rpm) n-GaAs-electrode in the dark in 6 M HCl with 0.76 mM Cu^{2+}. Dashed curves are the partial currents of anodic decomposition (j_{corr}) and of Cu^{2+}-reduction (j_{red}), as determined by a rotating ring-disk electrode [93]

Examples are H_2O_2/H_2O [99], I_3^-/I^- [72] at GaAs and Br^-/Br_2 at InP [100, 101].

Finally it should be noted that the relative position of the corresponding quasi-Fermi level with respect to the redox potential yields actually the thermodynamic force which drives an electrochemical reaction. This has sometimes been overlooked, and instead it has been assumed that the driving force is simply provided by holes at the upper edge of the valence band or electrons at the lower edge of the conduction band. In view of thermodynamics, this is simply the difference of an electrochemical potential (Fermi level) and an electrical potential.

4.3 Competition Between Redox Reactions and Anodic Decomposition

Studying charge transfer reactions between semiconductor electrodes and redox systems in aqueous electrolytes, frequently difficulties arise, because other processes, such as proton reduction or anodic decomposition occur simultaneously. Especially the latter process has been the subject of several investigations. Information on the competition between a redox process and the anodic decomposition cannot be obtained from current-potential curves alone, because in many cases they do not look much different upon addition of a redox system, especially if the current is controlled by the light intensity. Therefore, a rotating ring-disc electrode assembly (RRDE) consisting of a semiconductor disc and a Pt-ring is mostly applied, i.e. a well-known technique which makes it possible to determine the current corresponding to the oxidation of a redox system, separately. [102–104]. In this case, the oxidized form of a redox couple produced at the semiconductor disc, can be collected at the Pt-ring, provided that the ring is polarized negatively with respect to standard potential of the redox system. Usually, the competition between the redox process is quantitatively described by the stability factor, as defined by

$$s = \frac{j_{ox}}{j_{tot}} = \frac{j_{ox}}{j_{ox} + j_{corr}} \tag{47}$$

in which j_{ox} and j_{corr} correspond to oxidation and corrosion current, respectively.

Most classical electronic semiconductors, such as Ge, Si, GaAs, InP, GaP, CdSe or CdS, are characterized by an electronic structure, in which the upper state of the valence band is occupied by electrons involved in the chemical bonding between the atoms of the crystal. Holes occurring at the surface are equivalent to missing bonds, which leads to a weakening of the structural stability and finally, in contact with a liquid, to an anodic decomposition. Taking a binary semiconductor as an example, the decomposition reaction is given for instance by

$$AB + 2zp^+ + zH_2O \rightarrow A^{z+} + B_3^{z-} + 2zH^+ \tag{48}$$

Although this is an irreversible reaction, it can be treated as a reversible process, and a thermodynamic redox or decomposition potential can be defined and calculated from the thermodynamic quantities of all species involved [105–107]. Two cases are illustrated in Fig. 19. In one case (A), the decomposition potential occurs below the valence band, in the other somewhere within the bandgap (B). In the first case, the semiconductor would be thermodynamically stable, because the energy of holes would be too low for an anodic decomposition reaction. In the presence of a suitable redox system, the standard potential of which occurs, for instance, slightly above the valence band, the holes could then only be transferred to the redox system. The situation is completely different in Fig. 19B. Here, the holes have sufficient energy for the corrosion process. Using the same redox couple as in A, then the corrosion reaction is thermodynamically still much more favourable. Many experimental investigations have shown, however, that such a mechanism is far too simple.

However, the competition between anodic decomposition and hole transfer to a redox system upon light excitation cannot be described only by this simple thermodynamic model, as proved by many experimental data. One reason is that the anodic decomposition is a very complicated reaction, which involves a number of charge transfer steps (e.g. 6 holes in the case of III-V-compounds). It is reasonable to assume that the first reaction-step is rate-determining, and the free energy of holes required to cause the first reaction step may be considerably different from that of the overall reaction given above in Eq. (48), as discussed even in the very early stage of the research on semiconductor electrochemistry (see e.g. [108, 109]). In addition, the corrosion reaction and redox processes are influenced by kinetic factors, which may even play a dominant role. The kinetics may also be controlled by the surface chemistry, surface structure and crystal orientation. Accordingly, it is very difficult to predict whether corrosion or redox process dominates under given circumstances.

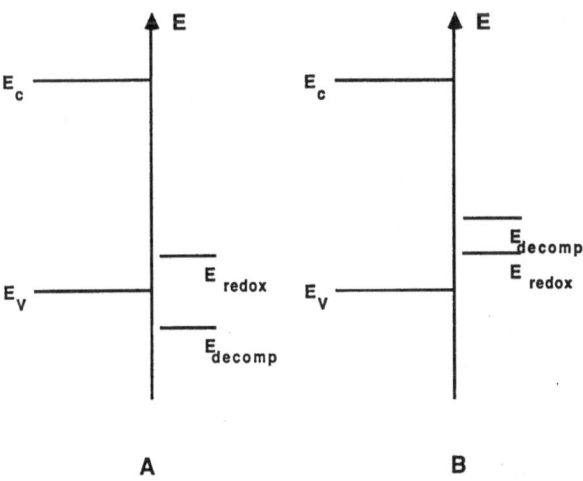

Fig. 19. Decomposition and redox potentials with respect to the position of energy bands

During the last decade, various mechanisms have been suggested [110–113]. The principles can be summarized as follows:

$$S + h^+ \rightarrow S^{\bullet +} \tag{49}$$

$$S^{\bullet +} + (m - 1)h^+ + mX^- \rightarrow SX_m \tag{50}$$

$$Red + h^+ \rightarrow Ox \tag{51}$$

$$S^{\bullet +} + Red \rightarrow S + Ox \tag{52}$$

in which S represents a semiconductor surface molecule, $S^{\bullet +}$ a surface radical, X^- a species in the solution (e.g. OH^-), and m the number of holes required for the complete dissolution of one surface molecule. Reaction (52) was first introduced by Frese et al. [110], because they had observed a decrease of the stability factor s with increasing light intensity (see below). These reactions are illustrated in an atomistic picture in Fig. 20b. The energy level of the radical is not located in the valence band, but somewhere within the gap, as shown in Fig. 20a, i.e. it corresponds energetically to a surface state. Reaction (51) corresponds then to a direct hole transfer from the valence band to the redox system, and reaction (52) to a transfer via the surface state. Since the reaction scheme (49)–(52) was

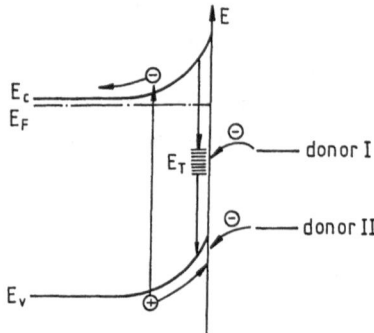

a

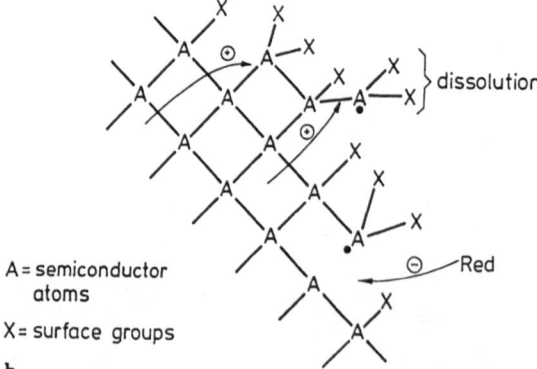

A = semiconductor
   atoms

X = surface groups

b

Fig. 20. Stabilization by surface state reactions

still too simple, various detailed models have been derived [111, 113], some of them being rather complicated. The essential aspects have been summarized in a reaction scheme [112], as given in Fig. 21, in which the reaction rates v_i depend on the corresponding rate constants and concentrations. In the first part of the complete reaction scheme (Eqs. (53) and (54), the generation (g) of holes (h^+) in the valence band and of the surface radical $S^{\bullet +}$ are described. The holes can also be consumed by recombination with electrons (rate $V_{n,1}$) or by direct hole transfer to the redox system ($V_{r,1}$). The surface radical $S^{\bullet +}$ can react with an electron from the conduction band ($V_{n,2}$) or with the redox system ($V_{r,2}$), processes by which the radical disappears. Accordingly, the original bond is repaired by the latter reaction, as illustrated in Fig. 20.

The lower part of the reaction scheme describes three possible dissolution pathways (Eqs. (55)–(56)). Only the first step in each sequence is essential. It is important to note that in one path (case A) the first step is a pure chemical reaction, characterized by the rate v_x, whereas in the others (case B) a hole transfer is involved. According to the kinetics of this reaction scheme, an intensity dependence of the stability factor s (defined in Eq. (47)) or of the ratio j_{ox}/j_{corr} occurs only for case B, as shown in a previous review article [114]. There are various examples published in the literature which demonstrate this effect

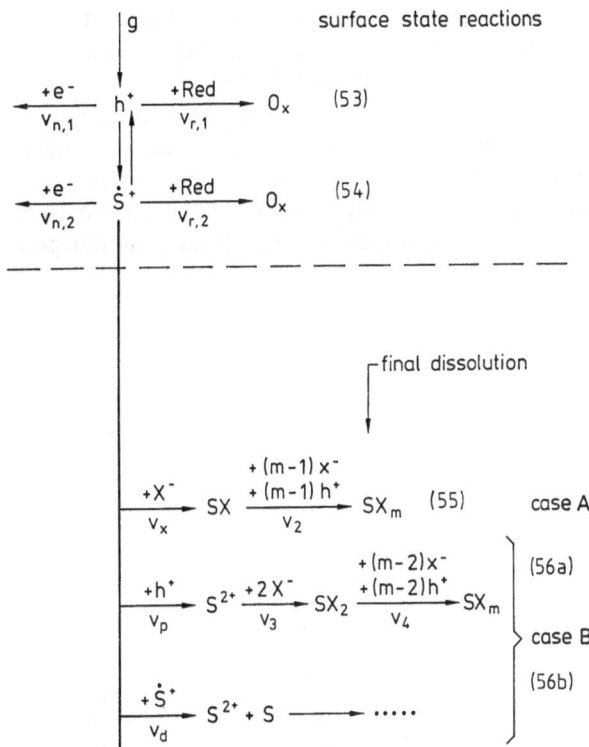

Fig. 21. General scheme of dissolution reactions

[110, 113]. Especially Gomes and coworkers have analyzed the competition between redox reaction (oxidation of Fe^{2+} and of $Fe(EDTA)^{2+}$) and corrosion reaction at GaAs and GaP [113, 116, 117]. They evaluated their data quantitatively in terms of the stability factor as a function of photocurrent. Such a quantitative analysis is not always unambiguous because of layer formation by the products formed in these reactions, as found for the Fe^{2+}-oxidation at GaAs [93]. Such an effect is not really visible by studying only the photocurrent and the stability factor at n-type semiconductors, because the light intensity determines the photocurrent. Since according to the quasi-Fermi level concept the same reactions occur at the corresponding p-type electrode, it is advantageous to check the reaction at this electrode. Here the valence band process takes place in the dark, and the current increases with increasing anodic polarization. Any layer formation at the surface can be recognized by evaluating the slope of the $\log j - U_E$-curve, especially if the latter changes upon addition of a redox system, as it happens with p-GaAs with Fe^{2+} in the electrolyte [93].

In several cases it has been found that the oxidation of the redox system occurs entirely via hole transfer directly from the valence band to the reduced form of the couple. Then both processes, oxidation of the redox system and corrosion, proceed independently. This is usually not visible from measurements with an n-type electrode, because the photocurrent is entirely determined by the light intensity. As already mentioned above, p-type electrodes are more suitable, because the current is determined by majority carrier transfer (reaction rate $v_{r,1}$ in Fig. 21). From the thermodynamic point of view, the oxidation of Cu^{1+} at GaAs is an interesting case. The corresponding current-potential curves are given in Fig. 22 [93]. The corrosion current is not changed upon addition of Cu^{1+}, i.e. corrosion and redox process are completely independent. In this case, the kinetics of the direct hole transfer is obviously very fast, i.e. the redox current is considerably larger than the corrosion current. Both processes occur indepen-

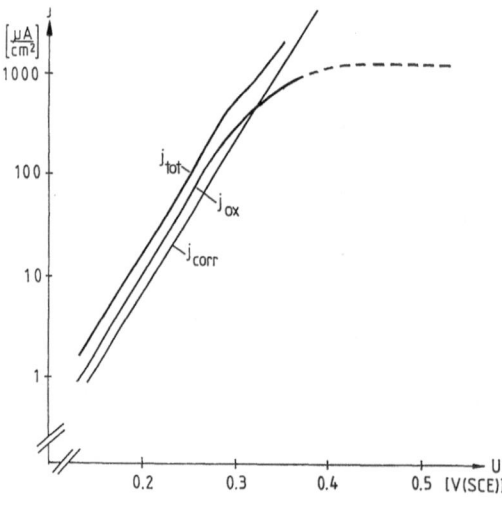

Fig. 22. Current vs potential at a rotating p-GaAs-electrode (6 M HCl; 2.5 nM CuCl; 1000 rpm); j_{corr}: anodic dissolution current; j_{ox}: partial current of Cu^{1+}-oxidation [93]

dently, i.e. the stability factor is here independent of the light intensity, a result which has also been found with n-GaAs by applying the RRED-technique. Although from the thermodynamic point of view this process is rather unfavourable with respect to the dissolution potential, the redox reaction dominates over the anodic decomposition even at rather low concentrations. Similar observations have made for Eu^{2+} at GaAs [188, 119].

As already mentioned in Sect. 3.3, frequently a shift of the Mott-Schottky curves has been observed upon illumination, which has been interpreted as a corresponding shift of the flatband potential. In a recent study, Allongue et al. [119] have correlated this effect with the formation of surface radicals by using the same model proposed in Ref. [112, 114]. The shift of flatband is then given by

$$\Delta U_{fb} = \frac{c_{s^{\cdot+}}}{C_H/e} \tag{57}$$

in which $c_{s^{\cdot+}}$ is the concentration of the surface radicals and C_H the capacity of the Helmholtz layer.

From the reactions scheme in Fig. 21 one can derive the oxidation and corrosion currents j_{ox} and j_{corr}, respectively, in terms of $S^{\cdot+}$ as given by

$$j_{corr} = \frac{mk_p(k_{r,2}c_{s+} + k_{1-})c_{s+}}{k_1 - k_p c_{s+}}$$

$$\xrightarrow{k_{r,2} \to 0}$$

$$\frac{mk_p k_{1-}c_{s+}}{k_1 - k_p c_s^+} \tag{58}$$

$$j_{ox} = \left[\frac{k_{r,1}(k_{r,2}c_R + k_{1-}) + k_{r,2}(k_1 - k_p c_{s+})}{k_1 - k_p c_{s+}} \right] c_R c_{s+}$$

$$\xrightarrow{k_{r,2} \to 0}$$

$$\left[\frac{k_{r,1}, k_{r-}}{k_1 - k_p c_{s+}} \right] c_R c_{s+} \tag{59}$$

whereas the total current (photocurrent at n-type) is given by

$$j_{tot} = j_{corr} + j_{ox} \tag{60}$$

One example, oxidation of iodide at n-GaAs, is illustrated in terms of photocurrent vs shift of flatband potential in Fig. 23. These authors found a very high stability factor of about $s = 0.97$. On the basis of Eqs. (57) and (59), they were able to simulate the experimental curves very well if $k_{r,2} = 0$, i.e. the reaction rate between $S^{\cdot+}$ and Red is negligible (not shown). For details, one must refer to their paper [119]. Similar results have been obtained with polysulfide.

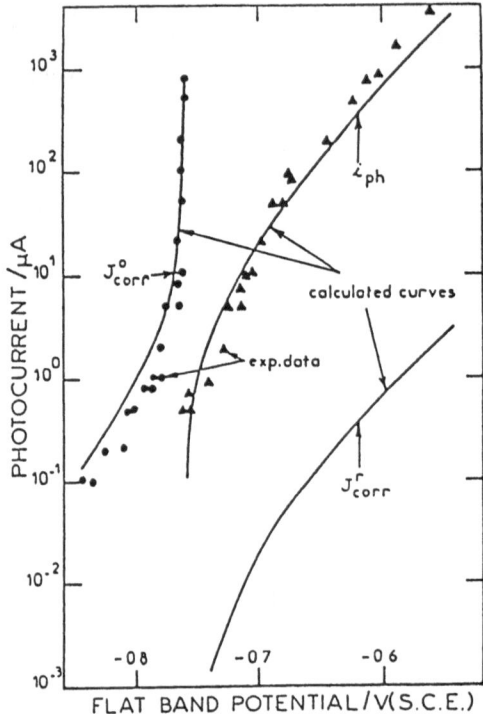

Fig. 23. Photocurrent vs flatband potential at n-GaAs in H_2SO_4 with 7 M NaJ; *dots*: experimental values; *solid lines*: theoretical curves [119]

Interestingly, no charge transfer between $S^{\cdot+}$ and Red occurs, although a considerable concentration of $S^{\cdot+}$ has been found under stationary illumination. It should be mentioned that similar results have been obtained with p-type electrodes in the dark, which the same authors evaluated on the basis of the quasi-Fermi level concept.

In other cases, such as the oxidation of Cu^{1+} [93] and Eu^{2+} [119] at n-GaAs, no shift of the flatband potential was observed. This result indicates that the kinetics of hole transfer is very fast.

A particular class of electrode materials are the transition metal chalcogenides, such as n-WSe_2, n-$MoSe_2$ and others, which form layer crystals. Tributsch originally expected a high stability of these materials against corrosion, because the electronic states of the valence band are formed by non-bonding d-electron states of the metal [120–122, 12]. Investigations of these electrodes, the basal planar surfaces (perpendicular to the c-axis) of which were contacting the electrolyte, it has been found that the flatband potential is independent of pH [60]. This result, observed with surfaces of low density of steps, indicates only a weak interaction between the metal (W in WSe_2) and the solvent. The corrosion exhibits a very anisotropic behaviour with respect to the crystal plane [123]. Nevertheless, the anodic photocurrent is entirely due to anodic decomposition, leading to selenates as corrosion products [121]. However, the photocurrent onset occurs with a very large overvoltage (≥ 0.6 V) with respect to the flatband potential in the dark. According to capacity measurements, this over-

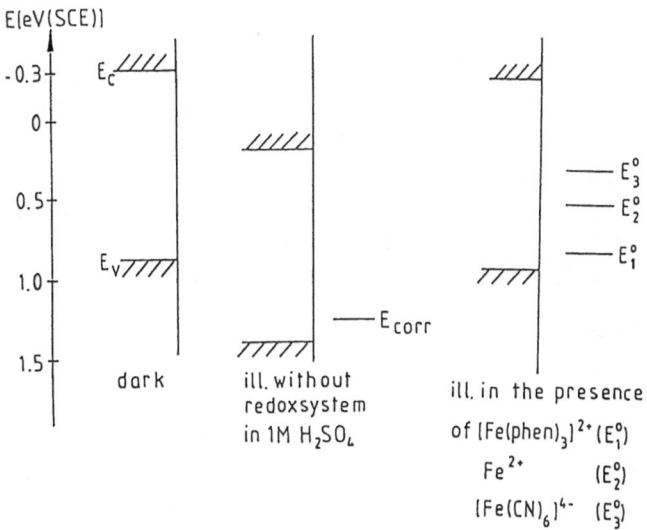

Fig. 24. Position of energy bands of n-WSe₂ in the dark and under illumination

voltage originates in a downward shift of the energy bands at the interface, as illustrated in the left and center part of Fig. 24). This shift band is caused by an accumulation of holes, because the decomposition is inhibited, although it is not completely clear whether it is kinetically or thermodynamically limited. This effect must be attributed to the layer structure, since the metal d-state does not interact with the solvent. It has been suggested by Gerischer that H_2O is oxidized to some radical intermediate which in turn reacts with selenium atoms of the layer [107]. This mechanism makes sense insofar as the reaction takes place only if the valence band is shifted downward to relatively high energies (Ev → + 1.5 V). Upon addition of redox systems, the standard potential of which is located at or above the valence band, the energy bands remain pinned (right side of Fig. 24)), even under illumination, and the photocurrent occurs already at much more cathodic potentials [61, 78]. Detailed studies with $Fephen_3^{2+/3+}$ as a redox system with a standard potential being very close to the valence have shown that the anodic photocurrent is entirely due to the oxidation of the $Fephen_3^{2+}$, i.e. the anodic decomposition could be completely avoided, even in the potential range where dissolution occurred in solutions free from any redox system [78]. On the same basis, the excellent stability of several layer compounds reported in the literature [124–126] can be explained. It should be mentioned here that in the case of n-WSe₂ and n-MoSe₂ the shift of bands was originally explained by trapping of holes in surface states [60]. Since, however, the shift could be avoided also by using a redox system, the standard potential of which is located at the valence band of WSe₂ ($Febipy_3^{2+/3+}$), the shift of energy bands in the absence of a redox system can simply be interpreted by the slow kinetics of the anodic decomposition reaction, leading to an accumulation of photoexcited holes in the valence band at the surface [78].

4.4 Evaluation of Current-Potential Curves and Determination of Rate Constants

As shown in the previous section, rather complex reaction mechanisms occur in aqueous electrolytes because of the competition of redox processes with other reactions, such as e.g. anodic decomposition. Therefore, only sufficiently fast redox reactions, for which other side reactions can be neglected, will be considered in this section. Before analyzing few examples, the maximum possible rates have to be derived for semiconductor electrodes.

In the case of metal electrodes, the current is usually given by

$$j \text{ (metal)} = ek_m c_{redox} \tag{61}$$

or the exchange current by

$$j_0 \text{ (metal)} = ek_M^0 c_{redox} \tag{61a}$$

The electron density is included in k_M or k_M^0 because it is constant. This rate constant has the dimension of a velocity $[cm \, s^{-1}]$. It is given by

$$k_M^0 = k_M^{max} D_{ox} \quad \text{or} \quad k_M^{max} D_{red} \tag{62}$$

in which D_{ox} and D_{red} are given by Eqs. (17a) and (17b), i.e. at equilibrium we have [6]

$$k_M^0 = k_M^{max} \exp\left(-\frac{\lambda}{4kT}\right) \tag{63}$$

The maximum possible value for k_M^{max} (frequently denoted by Z in the literature [127] has been approximated as the gas-phase collision frequency of molecules with a plane, i.e. $10^5 \, cm \, s^{-1}$. Using this value, one obtains e.g. for a reorientation energy of $\lambda = 0.5 \, eV$, $k_M^0 = 7 \times 10^{-3} \, cm \, s^{-1}$ or $j_0 \text{ (metal)} = 0.7 \, A \, cm^{-2}$, i.e. a typical value which has been reported in the literature [128].

In the case of semiconductor electrodes, the situation is different insofar as the carrier density is usually much smaller and varies with the electrode potential. Accordingly, it is useful to define the rate in terms of the product of a potential-independent rate constant and of carrier density at the semiconductor surface, as it has already been introduced in Eqs. (25)–(32). The maximum second order rate constant k_c^{max}, k_v^{max} etc. given in units of $cm^4 \, s^{-1}$ have recently been estimated by Lewis [129] on the basis of various models [130, 131, 7]. He has obtained $k^{max} = 10^{-17} - 10^{-16} \, cm^4 \, s^{-1}$. Applying this value also to metal electrodes, one has

$$k_M^{max} = k^{max} n_M \tag{64}$$

in which n_M is the free electron density in the metal. Inserting $k_M^{max} = 10^5 \, cm \, s^{-1}$ and the above value of k^{max}, then an electron concentration of $n_M = 10^{21} - 10^{22} \, cm^{-3}$ is obtained, which is reasonable for metals.

At a semiconductor electrode, the actual rate constants depend strongly on the overlap of energy states at both sides of the interface. There may only be an overlap with the conduction bands, or only with the valence band, or with both energy bands if the energy gap is rather small, as it has already been illustrated in Fig. 13 a–c. The corresponding second order rate constants for a charge transfer process via the conduction and the valence band are given by Eqs. (26b), (28b) and (30), (32), respectively. In contradiction to reactions at a metal electrode, at which the electron transfer occurs close to the Fermi level (see Fig. 25), the rate constant at a semiconductor electrode may be lower or also much higher, depending on the position of the maximum of the distribution curve D_{ox} and D_{red}. Accordingly, if the maximum occurs at the edge of one of the bands, k_v or k_c may reach k^{max}, as indicated in Fig. 25.

The best way of obtaining experimental data are measurements of current-potential curves under forward bias in the dark, i.e. under conditions that only majority carriers are transferred. Experimental data, however, are rather rare. One reason is that other reactions, such as anodic decomposition, interfere with the redox reaction. Another reason is the result that the slope of the log j vs. U_E-curves is mostly considerably larger than 60 mV/decade. One example where these problems do not occur is the reduction of various redox systems at n-ZnO (Fig. 26) [132]. Taking the reduction of $Fe(CN)_6^{3-}$ at pH 3.8 as an example, one obtains $k_{c, red} = 10^{-17} cm^4 s^{-1}$, i.e. a value close to the maximum value predicted theoretically. With $E_c = - 0.25$ eV (SCE) and a standard redox potential of $U^0 = + 0.2$ V (SCE), such a high value is only possible for $\lambda \approx 0.4 - 0.5$ eV (see Eq. (28b)). In alkaline solutions (pH 12), one obtains $k_{c, red} = 2 \times 10^{-19} cm^4 s^{-1}$. Here the energy bands occur at higher energies because of the pH-dependence of the Helmholtz double layer, i.e. $E_c = - 0.55$ eV. Using $\lambda = 0.4$ eV, then the theoretical value of $k_{c, red} = 4 \times 10^{-2} k^{max}$ is obtained from Eq. (28b). This value seems to be reasonable, since the same λ-value has been

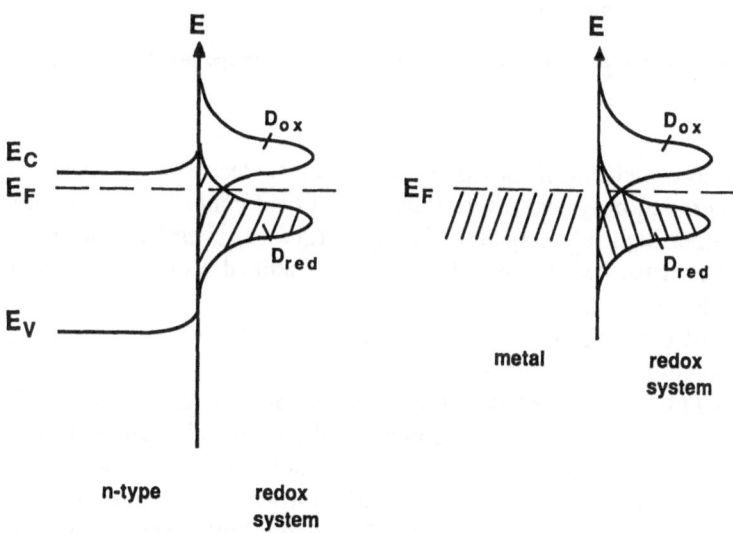

Fig. 25. Energy states at the interface of a semiconductor redox system and a metal-redox system

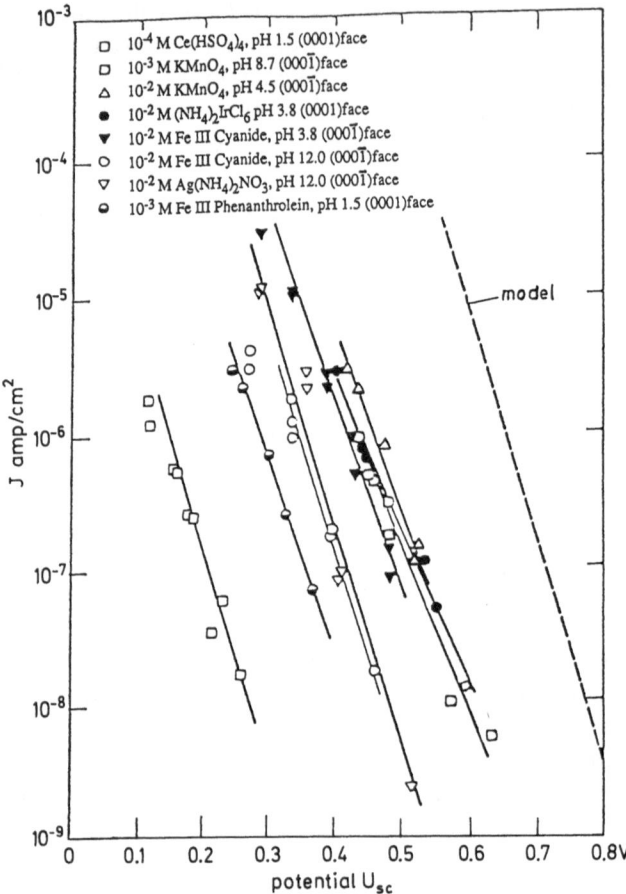

Fig. 26. Variation of cathodic current with potential across the space charge layer for n-ZnO [132]

obtained from electrochemical measurements at SnO_2-electrodes [6, 46]. Another outersphere redox system investigated at ZnO is $Fephen_3^{3+}$ (see Fig. 26).
Here, also a rather high rate constant of $k_{c,red} = 1.5 \times 10^{-17}\ cm^4\ s^{-1}$ has been
obtained. In this case, the same value is only obtained from Eq. (28b) if
$\lambda = 0.7$–0.8 eV is assumed.

Since in most cases the reorientation energy is not known very well, the
theoretical predictions are rather uncertain. Lewis recently made some calculations assuming much higher λ-values for the same redox couples (e.g. $= 1.6$ eV
for $[Fe(CN)_6]^{3-/4-}$) [129]. His calculation is based on a model recently derived
by Marcus [133], according to which the reorientation energy of molecules at
the interface of two dielectric media, such as a semiconductor/liquid junction,
can differ considerably from the corresponding metal-liquid value, even by
a factor of 2. Since this purely electrostatic model yields too high values, the

problem of correct λ-values remains open. In addition, it has to be remembered that the reorientation energy λ in a heterogenous reaction is only one half of that in a homogenous reaction [134]. Since electrochemical studies with SnO_2-electrodes have yielded much lower values [46] than those given by Lewis [129], the electrostatic model leads to an overestimate, at least for doped semiconductors.

Another interesting redox system is Cu^{1+}/Cu^{2+}, studied extensively at GaAs-electrodes [135]. In this case, the standard potential ($U_{redox}^0 = +0.33$ eV (SCE)) is close to the valence band of GaAs ($E_v = +0.43$ eV). Polarizing a p-type electrode anodically, one obtains a very nice straight line of log j vs U_E with an ideal slope of 60 mV/decade (Fig. 22) [93]. A detailed investigation of this reaction has shown that it is a diffusion-controlled process (reversible reaction), and the 60 mV-slope is entirely determined by Nernst-law with respect to the concentration of the reduced and oxidized species at the surface, i.e. the 60 mV-slope is not due to the variation of the surface electron density. From impedance spectroscopy-studies, a transfer rate of $k_{v,ox}p_s$ of approximately 2×10^{-2} cm s^{-1} has been obtained at the standard redox potential [136]. This rate is by about one order of magnitude higher than that controlled by diffusion at a rotation speed of 1000 rpm. With $p_s = 2 \times 10^{16}$ cm^{-3} at U_{redox}^0, we have then $k_{v,ox} = 1 \times 10^{-18}$ cm^4 s^{-1}, i.e. again a relatively high value which can be explained theoretically with $\lambda \approx 0.4$–0.5 eV.

These few examples yield rate constants in the right order of magnitude. There are, however, other systems with which completely different results have been obtained. One example is the oxidation of Eu^{2+} at p-GaAs. Since the standard redox potential occurs at -0.7 V (SCE), a value rather close to the conduction band (-0.9 V at pH 1), one would expect on the first sight that both, reduction and oxidation, would be a conduction process. Experimentally it has been found, however, that only the reduction occurs via the conduction, whereas the oxidation is a pure valence band process, as proved by measurements with p- and n-GaAs (Fig. 27) [118, 137]. The latter result, oxidation of Eu^{3+} via hole transfer, can only be explained assuming a rather high λ-value, as illustrated in Fig. 28. Evaluating these data, one obtains $k_{v,ox} \approx 10^{-13}$ cm^4 s^{-1}, i.e. a value which is by 3 or 4 orders of magnitude higher than the maximum theoretical value derived above. Recently, Allongue et al. reported an even higher value ($k_{v,ox} > 10^{-11}$ cm^4 s^{-1}) [119, 138]. In the case of the oxidation of Eu^{3+} at p-GaAs, the resulting currents reach even values which are also obtained by using the thermionic emission model usually applied for semiconductor-metal junctions [84], as given by

$$j = AT^2 \exp\left[-\frac{(E_v - E_{F,redox})}{kT}\right] \exp\left[\frac{e(U_E - U_{F,redox}^0)}{kT}\right] \quad (65)$$

with $A = 120$ A cm^{-2} (compare with Fig. 29).

Similar high rate constants have been reported for the oxidation of ferrocence at p-GaAs in acetonitrile solutions [139]. This oxidation is a conduction band process, i.e. it occurs only upon light excitation at p-GaAs. The rate

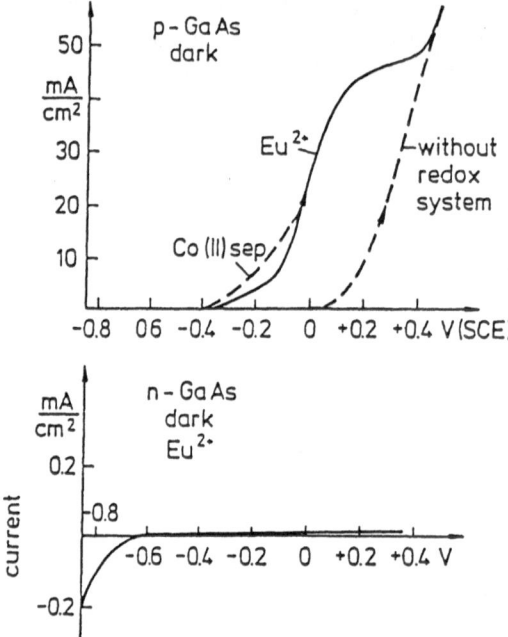

Fig. 27. Oxidation current vs. potential for *n*- and *p*-GaAs in the presence of Eu²⁺ in H₂SO₄ [118]

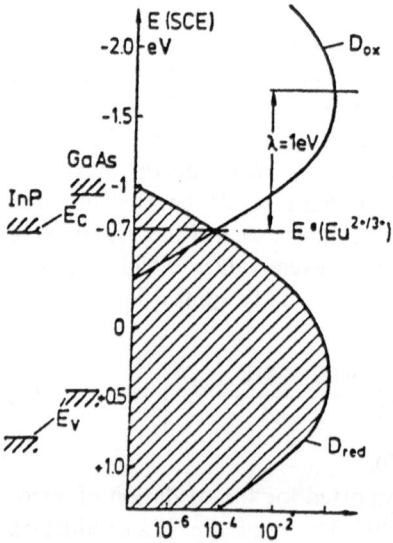

Fig. 28. Distribution of electron states of Eu²⁺/Eu³⁺ with respect to energy bands at the surface of GaAs and InP [132]

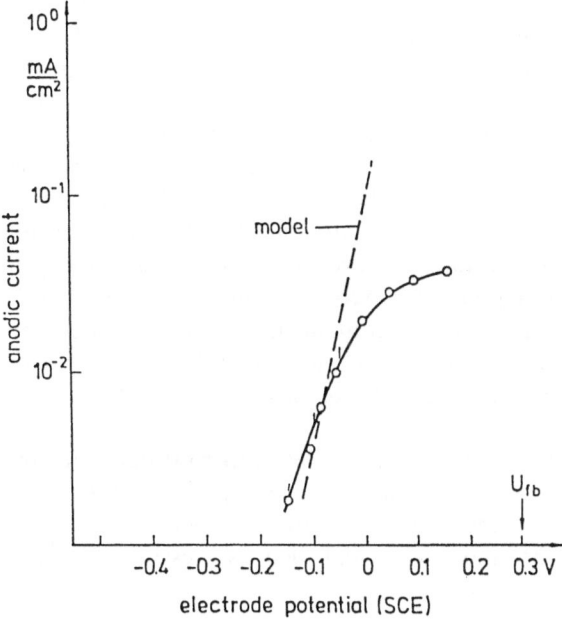

Fig. 29. Anodic current vs electrode potential for p-GaAs in solutions of 5×10^{-2} M Eu^{2+} (pH 1); dashed line calculated according to eq. (65)

constant was derived from fluorescence decay measurements, and a value of $k_{c,ox} = 2.3 \times 10^{-12}$ cm^4 s^{-1} has been reported.

The question remains, how to explain these extremely high rate constants. Gerischer has suggested that high rates can be possible for reactions via surface states, if the charge transfer from one of the bands into the surface state is rate-limiting [131]. However, this would require an even faster rate for the transfer from the surface state to the redox system, for which the same maximum rate should be valid. Accordingly, this cannot really explain the unusually high rates. Interestingly, a high rate has also been found for the reduction of H_2O_2 at n-GaAs [140, 141]. In this case, surface, radicals are formed by pure chemical etching, according to the reaction:

$$- Ga \cdot \cdot As - + H_2O_2 \longrightarrow \overset{HO^- \cdot OH}{- Ga_+ \cdot As -} \tag{66}$$

i.e. a surface state is formed which acts as an effective electron trap, i.e.

$$\overset{HO^- \quad \overset{\bullet}{O}H}{- Ga_+ \cdot As -} + e^- \longrightarrow \overset{\cdot OH}{- Ga \cdot \cdot AS - OH^-} \tag{67}$$

The $\overset{\bullet}{O}H$-radical is further reduced by hole injection into the valence band

$$\overset{\bullet}{O}H \rightarrow OH^- + p^+ \tag{68}$$

149

Since the electron transfer from the conduction band into the surface state (Eq. 67) can be very fast and the corresponding rate may be determined by the thermal velocity of electrons toward the surface, it has to be assumed that the initial chemical etching reaction (66) is even faster. However, it is not clear whether this assumption is correct. Very recently it has been found that also the reduction of protons (H_2-formation) at n-GaAs is a very fast reaction. The current potential dependence can actually be described by the thermionic emission model (see Eq. (65)) [142]. This result indicates that the electron transfer can occur at much higher rates if the electron acceptor is adsorbed on the surface. This assumption is supported by recent results reported by Nozik [143]. He repeated his fluorescence decay measurements by using nitrobenzene as an electron acceptor and found a much lower rate than for ferrocene. Nozik assumed that the high rate constant for ferrocene may also be due to adsorption.

In conclusion, it must be stated that rather few experimental data are available, and further careful measurements with simple redox systems at various concentrations at well-defined semiconductor electrodes are required before it can be decided whether the models and theories presented here are really applicable.

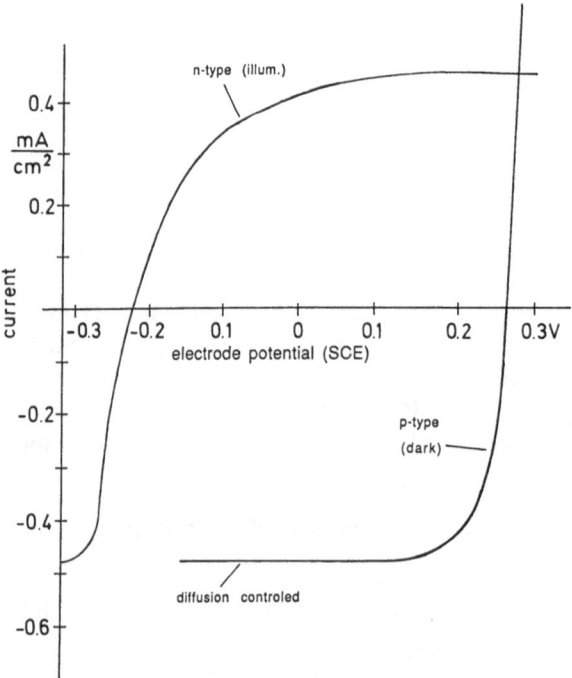

Fig. 30. Current-potential curves for n- and p-GaAs in 6 M HCl (0.7 mM Cu^{2+}; 50 mM Cu^{1+}); n-type electrode illuminated

So far, only majority carrier processes have been considered. In principle, the same mechanisms occur if minority carriers are involved. Considering for instance a valence band reaction, such as the oxidation of Cu^{1+} at GaAs, then an anodic photocurrent occurs at the n-type electrode, as illustrated in Fig. 30. The anodic photocurrent is determined by the light intensity. According to the quasi-Fermi level concept, however, the same reaction of the same rate as at the p-type electrode when the anodic photocurrent at n-type and the dark current are equal (see Sect. 4.2). In the whole potential range where the photocurrent is constant, the quasi-Fermi level of holes at the surface of n-GaAs is identical to that in p-GaAs at an electrode potential at which the same current (dark) occurs, as illustrated in terms of quasi-Fermi levels and overpotentials in Fig 31. This is also valid if the reaction is partly or completely diffusion controlled. In the latter case, a corresponding diffusion overvoltage ($e\eta_{diff}$) occurs in the energy scheme [135]. According to the quasi-Fermi level concept, the photocurrent-potential dependence can be predicted quantitatively from a j-U_E-curve measured in the dark.

In the case of a cathodic reaction via the valence band, minority carriers are injected into an n-type electrode in the dark. The injected holes diffuse into the bulk of the semiconductor until they recombine with the electrons (Fig. 32), and the current is then determined by the difference of the two quasi-Fermi levels $E_{F,n}^b$ and $E_{F,p}^s$, i.e.

$$j_{rec} = -j_0 \exp\left(\frac{E_{F,n}^b - E_{F,p}^s}{nkT}\right) \tag{69}$$

in which j_0 is given by Eqs. (37a) or (37b), depending on whether the recombination occurs in the bulk or within the space charge region. Frequently it has been observed that the two reaction steps occur via different energy bands. Since the quasi-Fermi level of electrons $E_{F,n}^b$ is identical to the electrode potential $U_E(n)$ and the quasi-Fermi levels of holes at the surface $E_{F,p}^s$ identical to the electrode potential $U_E(p)$ of the p-type electrode, Eq. (69) can be transformed into

$$U_E(n) = U_E(p) - \frac{nkT}{e} \ln\left(\frac{-j_{tot}}{j_0}\right) \tag{70}$$

in which $j_{tot} = j_{rec} + j_0$ [135, 137].

According to this equation, the current-potential curve for the n-type electrode can be calculated from an experimental curve measured with the corresponding p-type electrode, using j_0 and n as free parameters. Since $j_v^0 \ll j_0$ (see Eq. (43)) it can be neglected here. This has been proved with a Cu^{1+}/Cu^{2+} redox system at GaAs. For further details of the analysis it must be referred to the literature [135, 137].

In conclusion it should be emphasized that – although the same surface kinetics are valid at the surface of n- and p-type electrodes – the current, due to injection of minority carriers, is entirely governed by the recombination. A detailed analysis of injection and recombination process at Si-electrodes has been given by Lewis and coworkers [144–147] and by Kobayashi et al. [148, 149].

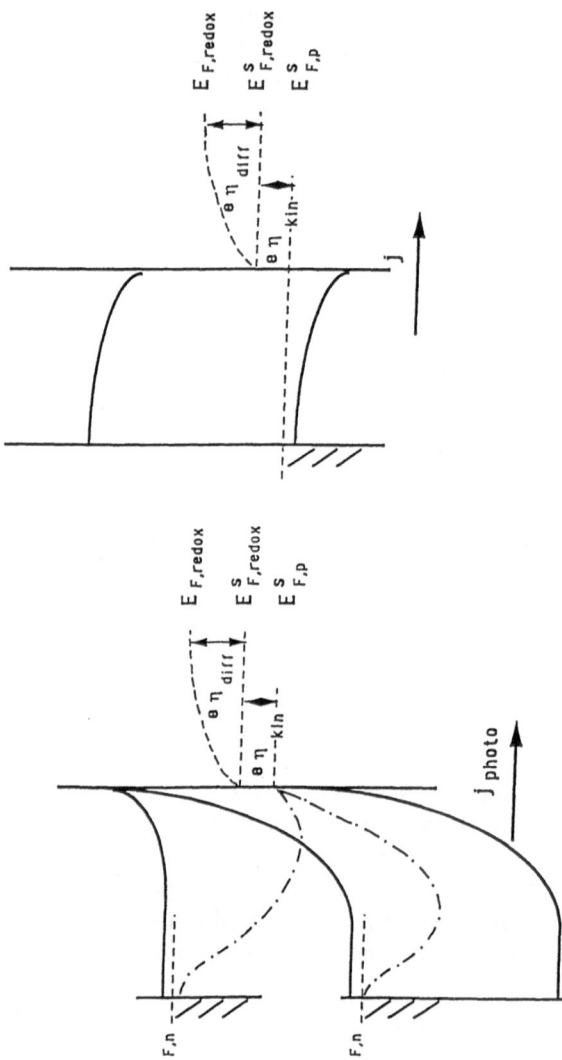

Fig. 31. Concept of overpotential for minority carrier reactions at n-type electrode (*left*) and for majority carrier reactions at p-type electrode (*right*)

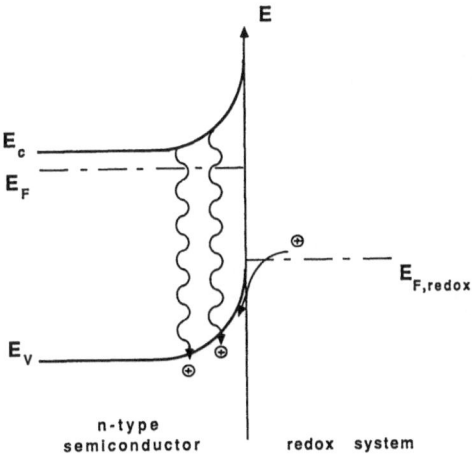

Fig. 32. Injection and recombination of minority carriers for a n-type semiconductor electrode

4.5 Two Step Redox Processes

Two step redox processes are understood as reactions in which two electrons are transferred at the electrode, until a stable state of the redox system is reached. This occurs mainly in the oxidation and reduction of organic molecules, and also in the reduction of H_2O_2 and $S_2O_8^{2-}$. Frequently, it has been observed that the two reaction steps occur via different energy bands. In the first step, usually a very reactive radical is formed. Taking the reduction of H_2O_2 as an example, it has been proved at first at n- and p-GaP that the first reduction step occurs via an electron transfer from the conduction band to H_2O_2, leading to the formation of an $OH^{\cdot}$-radical, and the latter was reduced to OH^- by hole injection into the valence band, according to the reaction [150, 10].

$$H_2O_2 + e^- \rightarrow \dot{O}H + OH^- \tag{71}$$

$$\dot{O}H \qquad \rightarrow OH^- + p^+ \tag{72}$$

This leads to the so-called "current-doubling"-effect, as shown for p-GaP under illumination in Fig. 33. The origin of this effect is based on the result that only for the first step an excitation by light is required. At n-type electrodes, the complete reaction occurs already in the dark. The hole injection was proved by luminescence measurements in n-GaP [151, 15]. The same result has been obtained with $S_2O_8^{2-}$ [150] and for the reduction of quinones [152].

Since the reduction of H_2O_2 consists of two consecutive steps, it is reasonable to describe its redox properties by two standard potentials, given by

$$E_1 = eU_1 = eU_1^0 + \frac{kT}{e} \ln \frac{C_{H_2O_2}}{C_{OH_\bullet} C_{OH^-}} \tag{73a}$$

$$E_2 = eU_2 = eU_2^0 + \frac{kT}{e} \ln \frac{C_{OH_\bullet}}{C_{OH_\bullet} C_{OH^-}} \tag{73b}$$

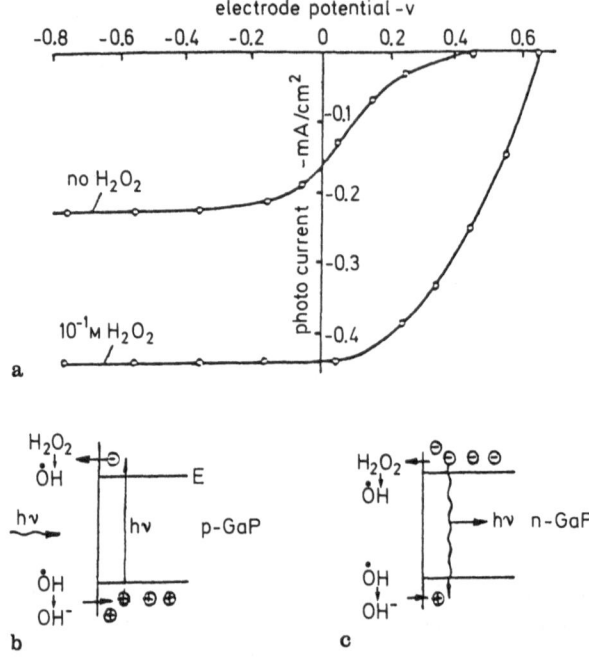

Fig. 33. Current-doubling process; **a** cathodic photocurrent at p-GaP with 10^{-1} M H_2O_2 at pH 1 [47]; **b** two-step electron transfer at p-GaP upon light excitation; **c** the same as **b** for n-type in the dark

in which U_1^0 and U_2^0 are the standard potentials of each step, whereas the standard potential U_0 of the whole system is an average value given by

$$E_0 = eU^0 = \frac{eU_1^0 + eU_2^0}{2} \tag{74}$$

At equilibrium, the concentration of OH-radicals is very low, i.e.

$$E = E_1 = E_2 = E_{F,redox} \tag{75}$$

As far as the energetics of the charge transfer is concerned, the standard potential of each step is of importance. According to the experimental results obtained with GaP, the energy eU_1^0 must be located near the conduction band, and eU_2^0 far below the upper edge of the valence band, as indicated in Fig. 33. As mentioned before, very similar results have been obtained with $S_2O_8^{2-}$ and quinones. In the latter case, the current-doubling has also been analyzed quantitatively with other semiconductors, such as Si and Ge [152]. It is interesting to see that this current-doubling at GaP occurs only at pH < 5.5 for benzoquinone and at pH < − 0.5 for duroquinone, indicating that the protonated radical accepts an electron from the conduction band.

The current-doubling effect for solutions containing H_2O_2 has also been studied with GaAs-electrodes [153, 154]. In connection with etching experiments, Kelly et al. have shown that the first reduction step is not a direct transfer from the conduction band to H_2O_2, but that H_2O_2 reacts chemically with the GaAs-surface, leading to the formation of a surface radical and $OH^\bullet$, as given by Eq. (66) (see previous section). Accordingly, a surface state is formed. During cathodic polarization of illuminated GaAs, an electron is transferred from the conduction band into the surface state, i.e. the original bond is repaired (Eq. (67)). From the $OH^\bullet$-radical, a hole is injected into the valence band (Eq. (68)) [140, 141]. The same model [155] has also been applied for interpreting the current-doubling effect observed with Br_2 [141], CIO^- [154] and BrO_3^- [141]. We shall return to this model in Sect. 6.4.

An analogous description can be given for the oxidation of various organic compounds. For instance, Morrison and co-workers [156, 157] have investigated quantitatively the oxidation of formic acid at n-ZnO and found a doubling of the anodic photocurrent upon addition of HCOOH. They interpreted this result in terms of the reaction

$$HCOO^- + p^+ \rightarrow HCOO^\bullet \tag{76}$$

$$HCOO^\bullet \rightarrow CO_2 + H^+ + e^- \tag{77}$$

In this case, a hole created by light excitation is transferred in the first reaction step. Similar observations have been made for the oxidation of alcohols at CdS-[158], ZnO [161] and TiO_2-electrodes [159], [160]. Interestingly, it has been observed with the latter electrode material that the current-doubling effect increased with increasing doping, with respect to the oxidation of CH_3OH in

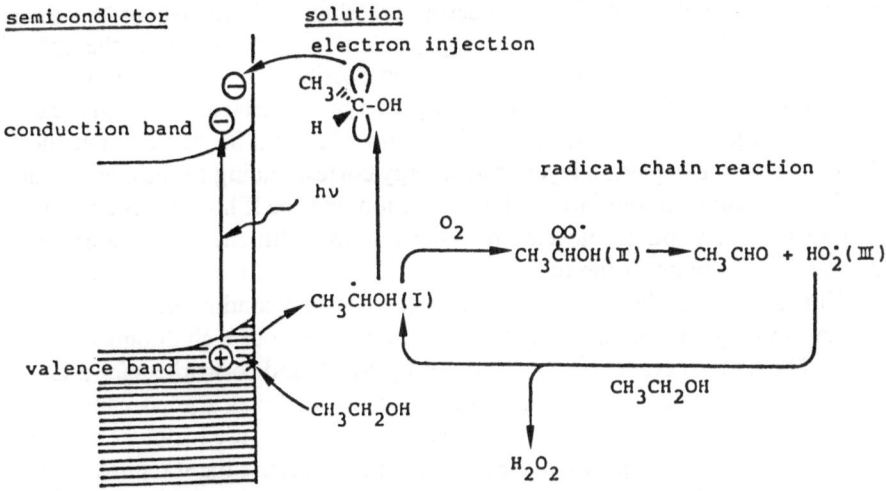

Fig. 34. Reaction mechanism of ethanol oxidation at illuminated TiO_2 in the presence of O_2 [159]

acetonitrile [160]. The authors related this effect to tunneling of electrons from the radical being formed in the first step into the conduction band. This implies that the standard potential of the $CH_2OH^\bullet/CH_2O$-couple occurs below the conduction band in acetonitrile. This is rather surprising, because a true current-doubling has been found for CH_3OH in aqueous solutions at CdS [158] and ZnO [159], i.e. at semiconductors the energy bands of which occur at much higher energies. In addition, the result obtained with TiO_2 is in contradiction to pulse radiolysis and polarographic investigations, according to which the standard potential of $CH_2OH^\bullet/CH_2O$ is much more negative than the conduction band of TiO_2 [162].

Various authors have also observed that the current-doubling is decreased by purging oxygen through the solution [159, 160]. This observation has been interpreted by a reaction between the radical and oxygen, as illustrated in Fig. 34. Other authors have shown that interesting additional informations on the current-doubling effect can be obtained by temperature change-measurements [163], which cannot be discussed here.

4.6 Hot Carrier Transfer Processes

In semiconductors, photogenerated electrons and holes created by photons with energies greater than the bandgap initially have excess kinetic energy. The question arises whether these hot carriers can be transferred across a semiconductor/liquid interface, or whether they thermalize to the bandgap before they are transferred [164]. This problem is of great interest, because a theoretical thermodynamic analysis of hot carrier processes for solar energy quantum conversion has shown that the conversion efficiency can be nearly doubled, compared to a fully thermalized quantum process [165]. Hot carrier processes occur when the thermalization processes are slow with respect to the time required for the photogenerated charge carrier to travel through the space charge layer. Figure 35 illustrates two different possible hot carrier-processes for an n-type semiconductor electrode [166]. In type I, electrons are fully thermalized to the bottom of the conduction band while still in the bulk, but then driven across the depletion layer at an energy corresponding to the edge of the conduction band without further thermalization. In type II hot carries are never fully thermalized and are injected into the solution with energies greater than the conduction band in the bulk.

Hot carrier transfer is most likely expected for semiconductors having high carrier mobility, low minority carrier effective mass and of high doping density [167]. The first experiments were reported by Nozik and co-workers for p-GaP and p-InP liquid junctions [168, 166]. Especially InP was a good candidate, because of its high electron mobility. The authors used p-nitrobenzene ($U_{redox}^0 = -0.86$ V (SCE)) as an electron acceptor, because the standard potential occurs 0.44 eV above the conduction band at the interface, as shown in Fig. 36. A photocurrent was observed at potentials negative of $U_E = +0.15$ V.

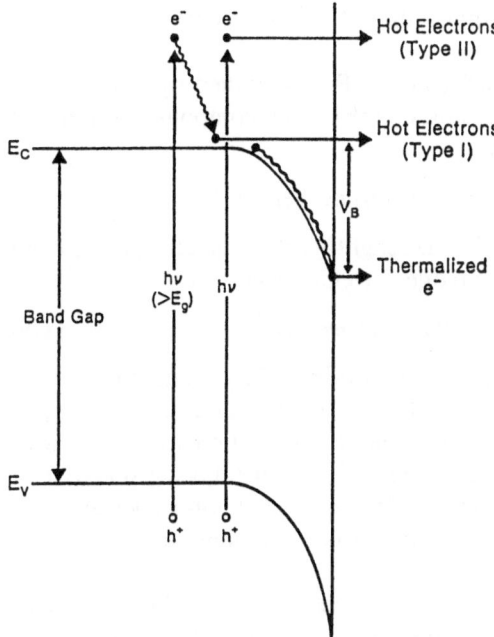

Fig. 35. Concept of generation and transfer of hot electrons at a p-type electrode [166]

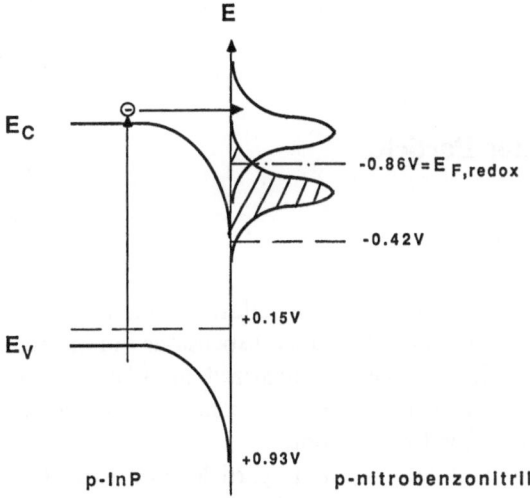

Fig. 36. Energy configuration at the interface p-InP and acetonitrile (redox couple: p-nitrobenzotrile) at the potential ($+0.15$ V) at which the cathodic current is detectable (data from [166])

Since at this potential the conduction band in the bulk occurred already at -1.17 V, the authors interpreted the photocurrent as hot electron ejection to p-nitrobenzene. This interpretation relied, however, on the assumption that the energy bands remain pinned upon illumination. They checked this by Mott-

Schottky measurements and found that a shift of bands occurred only at much more negative potentials.

More recently, hot electron transfer at p-InP was studied by Koval et al., who used a copper (*trans*-diene) complex [169]. This compound is reduced according to the following reactions:

$$[Cu(II)(trans\text{-diene})]^{2+} + e^- \rightleftarrows [Cu(I)(trans\text{-diene})]^{1+} \tag{78}$$

The reversible wave occurs at -0.97 V, i.e. slightly below the conduction band of InP. The Cu(I)-complex is further reduced to Cu-metal:

$$[Cu(I)(trans\text{-diene})]^{1+} e^- \rightarrow Cu(O) + trans\text{-diene} \tag{79}$$

at -2.3 V at glassy carbon, and at -2.2 V at n-InP. They found a photocurrent at potentials negative of -0.5 V for low-doped as well as for high-doped p-InP. Interestingly, these authors found Cu(O) after illumination only with high-doped, and not with low-doped electrodes, the Cu-metal being analyzed by anodic stripping technique. This is an excellent proof for hot-carrier ejection, indicating that an electron transfer is possible without thermalization, when the depletion layer is sufficiently thin [169].

Another type of hot electron transfer from a metal-covered n-Si electrode to a redox system ($[Fe(CN)_6]^{4-/3-}$) was studied by Chen and Freese [170, 171]. The hot electrons were actually produced in the metal by transfer from the conduction band of n-Si during cathodic polarization. In sufficiently thin layers (< 500 Å) they were transported to the metal/liquid interface without essential thermalization.

5 Reactions at Semiconductor Particles

5.1 Basic Processes

During the last decade, many investigations have been performed with semiconductor particles, either dissolved as colloids or used as suspensions in aqueous solutions. Essential results have already been summarized in other review articles [16, 18, 172]. The main purpose of this section is to compare processes at particles with those obtained with extended electrodes.

In principle, the same reactions should occur at particles and extended electrodes. One essential advantage of using particles is the large surface. The photogenerated charge carriers can easily reach the surface before they recombine, so that rather high quantum yields can be expected. However, one difficulty arises insofar as always two reactions, an oxidation and a reduction must occur simultaneously, as indicated in Fig. 37. Otherwise, the particles would be charged up, which would lead to a complete stop of the total reaction. Accordingly, the slowest process determines the rate of the total reaction.

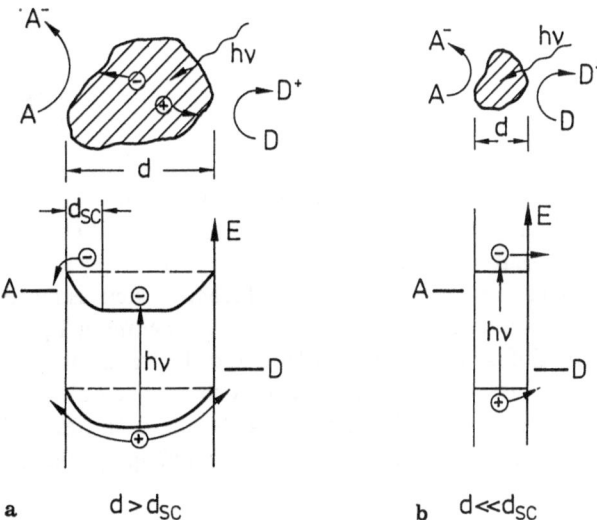

Fig. 37. Electron and hole transfer at large **a** and small **b** semiconductor particles to an electron acceptor A and donor D

A particle is actually a micro-electrode, always kept under open circuit potential, at which the anodic and cathodic currents are equal. At extended electrodes, these currents are mostly rather small, because the majority carrier density at the surface is rather small because of the depletion layer below the semiconductor surface, as illustrated in Fig. 37a. The thickness of such a space charge layer depends on the doping and on the potential across the space charge layer. It is given by Eq. (21) (Sect. 3.2). Taking typical values such as $\varepsilon = 10$; $n_0 = 10^{17}$ cm^{-3} and $\phi_{sc} = 1$ V, the thickness of the space charge layer is about 10^{-5} cm. In the case of much smaller particles of a diameter $d \ll d_{sc}$, of course, no space layer exists (Fig. 37b). On the first sight one may expect that the electron transfer at the interface of bigger particles would be limited to a very low rate because of the large upward band bending. This is not true, because upon light excitation few holes may be transferred to an electron acceptor in the solution, leading to a negative charging of the particle which reduces the positive space charge. The latter effect causes a flattening of energy bands (see dashed line in Fig. 37a) – which is equivalent to a negative shift of the rest potential of an extended electrode upon light excitation – and electrons can be transferred to an acceptor. Since most investigations have been made with undoped semiconductor particles, the diameter of particles was probably much smaller than d_{sc}.

Producing electron-hole pairs by light excitation in a small particle ($d \ll d_{sc}$), electrons and holes can easily be transferred to an electron acceptor and donor, respectively, provided that the energetic requirements are fulfilled. The quantum efficiency of the reaction depends on the transfer rate at the interface, on the recombination rate within the particle and on the transit time, the latter being given by [173].

$$\tau = \frac{R^2}{\pi^2 D} \tag{80}$$

where R is the radius of the particle and D the diffusion coefficient of the excited charge carriers. Taking a typical value of $D \approx 0.1 \text{ cm}^2\text{s}^{-1}$ and a radius of 100 Å, the average transit time is only about 1 ps. This value is much smaller than that for recombination, which is usually greater than 10 ns.

In the case of larger particles $(d \gg d_{sc})$, the sequence of reaction steps is more complex because of the space charge layer below the surface. Exciting a 1 μm-particle, the minority carriers diffuse toward the surface, and the transient time is then about 10 ns, i.e. recombination effects may be important. The holes are easily transferred because of the upward band bending. In the first stage of illumination this leads to a negative charging until the original space charge given by [10, 114]

$$Q_{sc} = (2\varepsilon\varepsilon_0 n_0 e)^{1/2} \phi_{sc}^{1/2} \tag{81}$$

is compensated. Using again $\varepsilon = 10$; $n_0 = 10^{17} \text{ cm}^{-3}$ and $\phi_{sc} = 1 \text{ V}$ and $R = 0.5 \mu\text{m}$, one obtains $Q_{sc} = 5 \times 10^{-15}$ As per particle, and the number of charges compensated are 3×10^4 per particle. Taking a suspension (10^8 particles of 1 μm in 1 cm^3), so that the incident light is just completely absorbed, then it takes in the average about 1 ns between the absorption incidents of 2 photons in one particle for a photon flux of $10^{17} \text{ cm}^{-2}\text{s}^{-1}$. Since at least 3×10^4 photons are required to compensate the space charge, it should take around 30 ms before the majority carriers are transferred. This also leads – compared to extended electrodes – to extremely low current densities at a particle. In terms of a current-potential curve one operates with a particle very close to the open circuit potential.

As discussed in the previous sections, charge transfer reactions at extended electrodes have been mainly studied by using simple one-step redox systems which are reversible. There is no point in studying corresponding reactions at particles, because a redox system being oxidized by a hole would be immediately reduced by an electron transfer from the conduction band. Therefore, only irreversible reactions of organic compounds have been investigated. Since small particles tend to conglomerate, stabilizers such as SiO_2 or polymers have been used. This may lead to problems in the case of a polymer, because it can be oxidized by holes. This problem can be avoided by using bare particles in solutions of a low ion concentration. Since the particles are charged, the counter charge extends relatively far into the electrolyte (Gouy Chapman layer), and conglomeration does not take place due to the repulsive forces between equally charged particles. When organic molecules, such as e.g. alcohols, are oxidized by hole transfer, usually oxygen in the solution acts as an electron acceptor. A whole sequence of reaction steps can occur which are frequently difficult to analyze, because also cross reactions may be possible and a new product is formed. One example is the formation of 2-phenylindazole from azobenzene at illuminated at TiO_2-particles in methanol solutions. Minoura et al. proposed a mechanism as follows [174]:

Azobenzene acts as an electron acceptor, i.e. it is reduced according to the reaction:

$$CH_3O-\!\!\langle\!\!\bigcirc\!\!\rangle\!\!-N{=}N-\!\!\langle\!\!\bigcirc\!\!\rangle + e^- \longrightarrow \left[CH_3O-\!\!\langle\!\!\bigcirc\!\!\rangle\!\!-N{=}N-\!\!\langle\!\!\bigcirc\!\!\rangle\right]^{\cdot-} \qquad (82)$$

whereas methanol is oxidized by hole transfer:

$$CH_3OH + h^+ \rightarrow CH_2OH^{\cdot} + H^+ \qquad (83)$$

The reaction of the two radicals formed at the same TiO_2-particle leads to:

$$\left[CH_3O-\!\!\langle\!\!\bigcirc\!\!\rangle\!\!-N{=}N-\!\!\langle\!\!\bigcirc\!\!\rangle\right]^{\cdot-} + H^+ + \cdot CH_2OH$$

$$\longrightarrow \underset{CH_3O}{\overset{}{\bigcirc\!\!\bigcirc}}\!\!\!\!\!\!\underset{N}{\overset{N}{\diagdown}}\!\!-N-\!\!\langle\!\!\bigcirc\!\!\rangle + H_2 + H_2O \qquad (84)$$

They checked this mechanism by using TiO_2-electrodes instead of particles and found that the 2-phenylindazole was only formed in a small potential range around the photocurrent onset, i.e. at potentials at which also a comparable cathodic current occurs. At more anodic potentials, the hydroxymethyl radical is further oxidized [175]. It should be mentioned here that the oxidation of CH_3OH (Eq. (83)) or other organic compounds must not necessarily occur via direct hole transfer. There are strong indications that at first an OH -radical is formed at the surface of TiO_2-particles by hole transfer, and that this radical oxidizes the organic molecule in a second step, as proved in the case of acetate oxidation [176].

As mentioned above, the slowest reaction step determines the rate of the total reaction. For instance, if no electron acceptor is present in the solution, then the photoexcited electrons may be trapped in surface sites via

$$Ti^{IV} + e^- \rightarrow Ti^{III} \qquad (85)$$

These trapped electrons exhibit a characteristic absorption spectrum, as found with colloidal TiO_2-particles [177] (Fig. 38). Because of the large surface area, this transient spectrum can easily be studied by laser flash spectroscopy. According to these investigations, the trapping occurs instantaneously as the electrons reach the surface, i.e. it is a very fast process [178]. Also the trapping of holes has been observed when electrons are efficiently scavenged. Corresponding spectra for the absorption of trapped holes have been found with TiO_2^- [177] and CdS-particles [179]. Such a light-induced trapping of electrons or holes should lead to a shift of energy bands in the same way as found with electrodes (Sect.

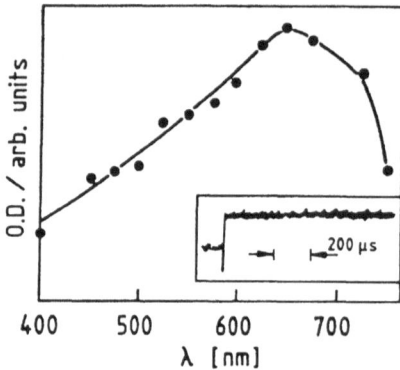

Fig. 38. Absorption spectrum of a laser-flashed solution of 6.3×10^{-3} M TiO_2 particles containing 5×10^{-3} M PVA [177]. Insert: time profile of absorption signal

3.3.). It is interesting to recognize that trapping of charges at the semiconductor surface upon illumination can be determined in particles by spectroscopic methods and at extended electrodes by capacity measurements.

The position of energy bands at the surface of particles cannot be determined exactly, because capacity measurements are not possible. Their position can only be estimated by checking which reaction is possible. Frequently, methyl viologen (MV^{2+}) has been used as an electron acceptor which can accept an electron from the conduction band upon illumination, provided the conduction band is above the reduction potential of MV^{2+}. The radical ($MV^{\bullet+}$) formed in this reaction is usually spectroscopically [16] or electrochemically [180] analyzed. These methods, however, give a very rough estimate, because usually it is not known whether surface states are involved in the charge transfer process.

As already discussed in Sect. 2.2., the bandgap of semiconductor particles increases considerably when their size becomes smaller than about 100 Å (Figs. 4 and 5). Accordingly, the position of energy bands is shifted, and it is expected that certain reactions should become possible with quantized particles which do not occur with bulk materials. This has been demonstrated for H_2-evolution in 50 Å PbSe- and HgSe-colloids, which has not been observed with large particles [181, 182]. An extreme negative shift of the conduction band by about 1.2 eV has been found with 50 Å-CdTe-colloids due to their low effective mass. Since CO_2-reduction to formic acid was observed with photoexcited CdTe-colloids, the conduction band must be at ≤ -1.9 eV, compared to the flatband potential of n-CdTe electrodes of -0.6 V [181].

Frequently, reactions have been investigated at particles loaded with a catalyst such as a metal (Pt) or a metal oxide (RuO_2). Such catalysts enhance the reaction rate, and sometimes other products are formed. This has been discussed in detail in various papers [114, 183]. Although it is generally assume that for instance noble metals (Pt) catalyzes a reduction process such as H_2-formation, it is impossible to get information whether the reduction or the oxidation occurs via the catalyst. Few years ago, a semiconductor monograin membrane technique was developed by which the particles are fixed, as illustrated in Fig. 39

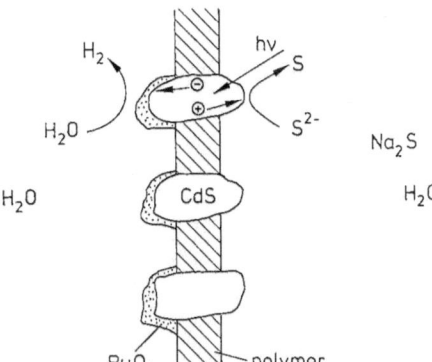

Fig. 39. Principle scheme of a semiconductor monogram membrane and photolysis of H_2S [184]

[184]. This technique makes it possible to load one side of the membrane with a catalyst. Using a two-compartment cell in which both are separated by the membrane, it is possible to determine whether a product is formed on the free or on the catalyst-loaded side. So far this technique has only been applied successfully to CdS and SiC.

5.2 The Role of Surface Chemistry

There are many indications in the literature that surface chemistry plays an important role in photoelectrochemical reactions at extended electrodes and at particles. There are, however, only a few quantitative investigations on this problem (see e.g. [12, 57, 58]), probably due to the lack of sufficiently sensitive methods. In the case of metal oxide particles, the adsorption of H_2O plays already an important role. Due to the amphoteric behaviour of most metal hydroxides, two surface equilibria have to be considered [18]:

$$-M-OH + H^+ \rightleftarrows -M-OH_2^+ \qquad (pK_1) \qquad (86)$$

$$-M-OH \qquad \rightleftarrows -M-O^- + H^+ \quad (pK_2) \qquad (87)$$

The zero-point-of-charge (pH_{zpc}) of metal oxides is defined as the pH where the concentrations of protonated and deprotonated surface groups are equal:

$$pH_{zpc} = 1/2(pK_1 + pK_2) \qquad (88)$$

According to the equilibria given in Eqs. (86) and (87), the surface is predominantly positively charged below pH_{zpc} and negatively charged above this value. The influence of the surface charge on the photocatalytic activity is demonstrated in Fig. 40, according to which the oxidation of an anion such as trichloroacetate at TiO_2-particles was only observed at low pH and of cation (chloroethylammonium) at high pH-values [185]. This interpretation is entirely based on an electrostatic model, according to which the reaction rate is reduced

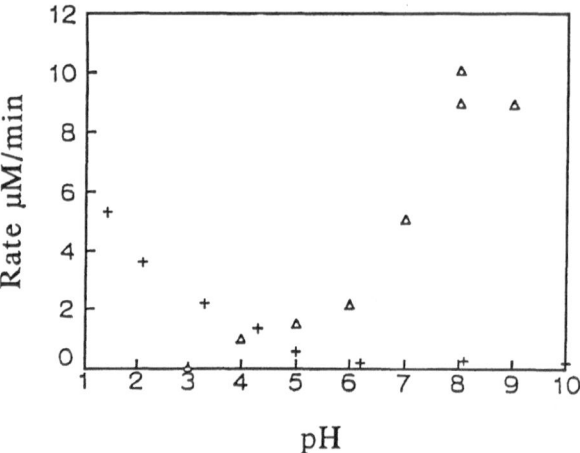

Fig. 40. pH-dependence of the rate of degradation of trichloroacetate (+) and chloroethylam-monium (Δ) -ions is illuminated suspensions of TiO_2 (0.5 g/l) [185a]

to a very low value if the surface charge and that of the ions involved in the reaction have equal signs. In such a case, the photoexcited charge carriers either recombine, or the holes are used for oxidation of water, leading to the formation of O_2. Since oxygen is reduced again by the photoexcited electrons, no overall reaction occurs, i.e. we have a recombination via the solution. In principle, the same process should occur at extended electrodes in dilute solutions. At an anodically polarized electrode, however, the photoexcited holes are forced across the interface of an n-type electrode, i.e. an external recombination should not occur. Assuming that here also the oxidation of anions or cations is pH-dependent, as described above, then the relative yield of H_2O-oxidation should vary correspondingly with pH. Such an effect has not been reported yet.

Another aspect of surface chemistry is the composition of the surface. In the case of small particles, even their preparation has a strong influence on the surface composition, the formation of surface states [186] and consequently on the reaction mechanisms, as it has been investigated particularly with ZnS [187]. Various authors have shown that ZnS-particles prepared by precipitation from solutions containing an excess of SH^--ions does not show any fluorescence [186, 188]. Synthesizing the particles in solutions with an excess of Zn^{2+}-ions, however, fluorescence occurs peaking around 430 nm. Since this fluorescence could be quenched upon addition of MV^{2+}, it has been concluded that the fluorescence is due to a transition via surface states [186], as illustrated in the energy scheme of Fig. 41. This variation in the synthesis of ZnS-particles of nm-size has a strong influence also on the photoreactions. For instance, illuminating a colloidal solution of ZnS containing an excess of 17 mol % SH^- in the presence of ethanol, butanediol has been found besides acetoaldehyde as oxida-

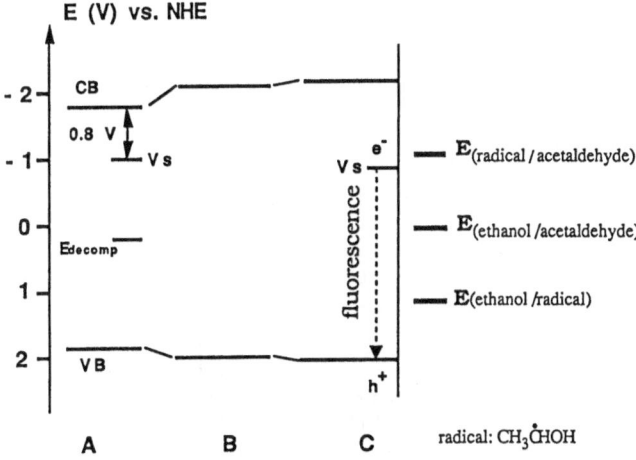

Fig. 41. Energy scheme of ZnS-particles (3 nm) prepared with an excess of Zn^{2+} (4 mol %)

tion products (Fig. 42a) in a ratio of $1:2.5$, according to the reaction scheme [187]

$$CH_3CH_2OH + h^+ \rightarrow CH_3CHOH^\bullet + H^+ \tag{89}$$

$$2CH_3\dot{C}HOH \begin{cases} \xrightarrow{\text{disprop.}} CH_3CH_2OH & (90a) \\ \\ \xrightarrow{\text{dimer.}} H_3C\text{--}CH(OH)\text{--}CH(OH)\text{--}CH_3 & (90b) \\ \qquad\qquad \text{(butanediol)} \end{cases}$$

In the first reaction step (Eq. (89)), a radical was formed in a one-hole process. Radicals formed at different ZnS-particles react further via disproportionation, leading to aldehyde or via dimerization, which yields butanediol (reaction (90)). Parallel to the oxidation of the alcohol, protons are reduced by electron transfer from the conduction band, i.e.

$$2H^+ + 2e^- \rightarrow H_2 \tag{91}$$

Performing the same experiment with ZnS-colloids with an excess of 4 mol % Zn^{2+}, no butanediol was formed (Fig. 42b) [187]. This result has been interpreted by assuming an electron transfer from the radical into the surface states, which is energetically possible (see Fig. 41). The corresponding reaction is given by

$$CH_3CHOH^\bullet \rightarrow CH_3CH{=}O + H^+ + e_{ss}^- \tag{92}$$

The electrons stored in the surface states (e_{ss}^-) are then used for the H_2-formation [187]. Surprisingly, butanediol is formed again with nm-ZnS, prepared with an excess of 45 mol % Zn^{2+} (Fig. 42c). It is interesting to note that aldehyde

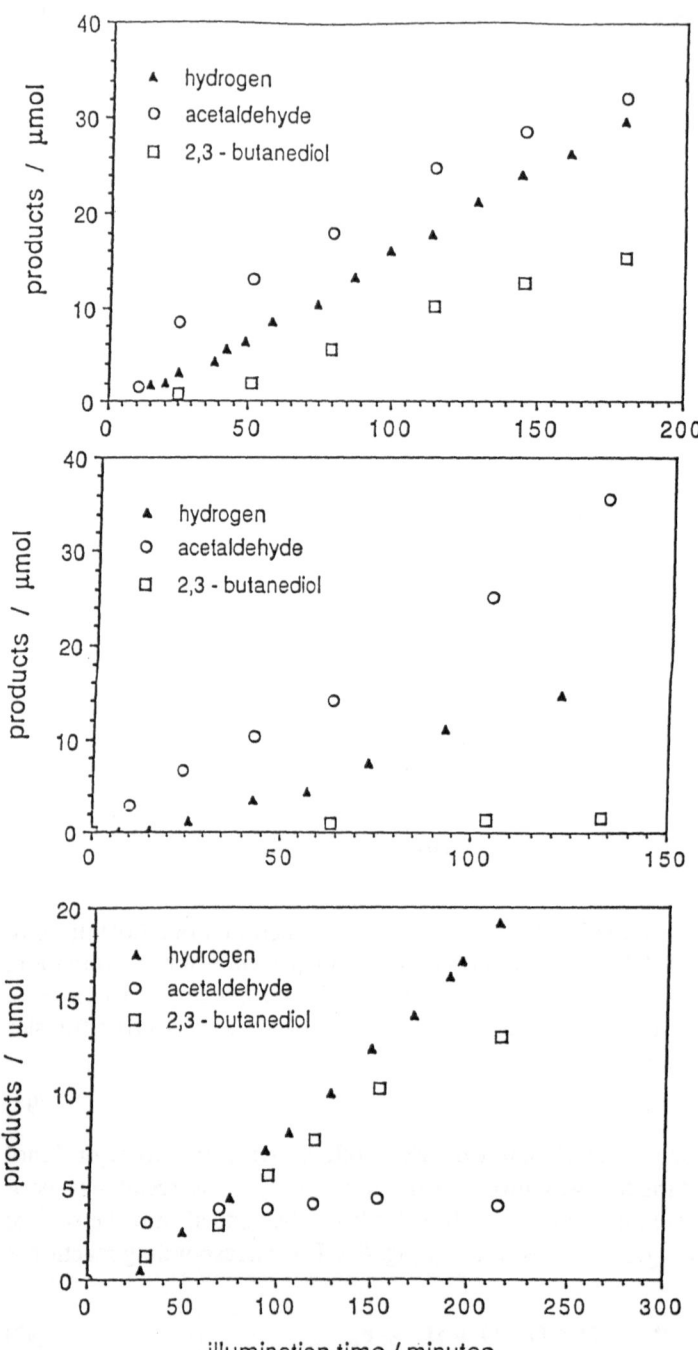

Fig. 42. Concentration of products vs illumination time for ZnS-particles (3 nm) in ethanol/water solutions (ratio 1 : 5) prepared with an excess of **a** 17 mol % SH^-; **b** 4 mol % Zn^{2+}; **c** 45 mol % Zn^{2+}

formation rises only within the first period of illumination and levels off at longer illumination. In addition, a considerable amount of Zn^0 is deposited. The latter result has been interpreted by the following mechanism [187, 189]: The formation of aldehyde occurs in the same way as in the case of the ZnS-particles with 4 mol % Zn^{2+} excess. Since the aldehyde concentration remained constant, it has been assumed that the aldehyde is reduced at the Zn^0-clusters, leading to the formation of a radical as an intermediate. These radicals, here produced in a reduction step, can dimerize according to reaction (90b).

Henglein et al. have also found dimerization-products at ZnS-colloids, using 2-propanol as an electron donor and CO_2 as an electron acceptor [190]. The dimerization produced was pinacol, and its formation has been interpreted by the same mechanism as given for the ethanol oxidation. Inoue et al. performed the same experiment, but did not find pinacol as a dimerization-product [191]. According to Müller et al., these differences may be due to slight differences in the preparation of the ZnS-colloids, which seems to be extremely critical [187].

5.3 Particle Size Effect

As already discussed, reactions at extended electrodes and particles differ only insofar as, at particles, both an oxidation and a reduction process always occur simultaneously. There is, however, one further aspect which may be of importance for using big or small particles. Taking two solutions containing semiconductor particles of different sizes, i.e. for instance of 3 nm and 4 μm, then many more particles are present in the solution containing the 3 nm-particles than in that of 4 μm-particles, provided that the concentration of the semiconductor material is identical in both solutions. As it can easily be calculated, a time interval of 5.4 ms exists between the absorption of two photons in one individual 3 nm-particle for a photon-flux of 4×10^{17} cm^{-2} s^{-1}, assuming that all photons are absorbed in the colloidal solution [114]. In the case of the 4 μm-particles, the time interval is about 20 ps for the same photon flux, i.e. it is shorter by a factor of 10^8, compared to the time interval estimated for the 3 nm-particles. This can be important for reactions where two or more electrons are involved, typical in many oxidation- and reduction reactions with organic molecules [114].

This phenomenon has recently been analyzed by studying the oxidation of ethanol at ZnS-particles [187, 189]. As already discussed above (Sect. 5.2.), the oxidation of ethanol occurs in two steps, i.e. in the first step, a radical is formed by hole transfer at the surface of an individual particle. Since it takes in the average several milliseconds before another hole is generated by photon absorption in the same individual 3 nm-particle, the radical can diffuse into the solution. There, the radicals can disproportionate and dimerize according to Eqs. (90a) and (90b) leading to the formation of acetoaldehyde and butanediol, respectively, as proved with 3 nm ZnS-particles. Performing the same experiment with much larger particles (4 μm), no butanediol was found [187, 189]. Since the time interval between the absorption of two photons in one particle is only 20 ps, the radical can be oxidized to acetaldehyde by a further hole transfer at the same particle [187].

It should be emphasized that the experiments have been performed with ZnS-particles synthesized with an excess of S^{2-} in order to avoid complications by surface states (see previous section). As shown above, ZnS was a very suitable semiconductor for these investigations because of its high energy position. The particle size-effect did not occur for instance with CdS, the energy bands of which occur at relatively low energies. In this case, current doubling occurs, i.e. the radical formed by a photoexcited hole can be further oxidized at the same particle by electron injection into the conduction band of CdS (189].

6 Applications

6.1 Photoelectrochemical Solar Energy Conversion

During the last 15 years, many photoelectrochemical systems have been studied. with respect to their applicability for solar energy conversion. Since most of these results have been summarized in various review articles [114, 107, 191–197], only some more recent developments will be considered here.

In principle, photoelectrochemical cells can be used for the conversion of solar energy into electrical energy or for the production of a storable fuel. The first type (regenerative cells) consists of a semiconductor and inert counter electrode and a redox system in the electrolyte. The current-voltage behaviour is described by the diode equation, which is also valid for pure solid state devices (pn-junction, Schottky diode) i.e.

$$j = j_0 \left[\exp\left(\frac{eU}{kT}\right) - 1 \right] - j_{ph} \tag{93}$$

in which U is the externally applied voltage. As a matter of fact, this equation is identical to Eq. (45) if one replaces the overvoltage η by U (see Sect. 4.1.). The photovoltage is obtained for j = 0, i.e. $U = U_{ph}$. We then have:

$$U_{ph} = \frac{kT}{e} \ln\left(\frac{j_{ph}}{j_0} + 1\right) \tag{94}$$

It is evident from Eq. (94) that the maximum photovoltage depends critically on the exchange current j_0. In the case of pn-junctions, j_0 is determined by the injection and recombination (minority carrier device), Whereas in Schottky-type of cells j_0 can be derived from the thermionic emission model (majority carrier device). The analysis of solid state systems has shown that j_0 is always smaller for minority carrier devices [20, 21]. Using semiconductor-liquid junctions, both types of cells can be realized. If in both processes, oxidation and reduction, minority carrier devices are involved, then j_0 is given by Eq. (37a), similarly as for pn-junctions. There are however, many examples in the literature where the forward current is determined by a transfer of majority carriers from the semiconductor to the redox system [194]. In such a case, however, j_0 is not determined by the thermionic emission, but by the surface kinetics [197], (Eqs. (27) and (29)), as discussed in detail in Sect. 4.4. This leads usually to much lower exchange currents j_0 than in solid state Schottky cells, i.e. in majority carrier devices, consisting of semiconductor liquid-junction, higher photovoltages can be obtained. Many systems have been investigated, and in some cases conversion efficiencies of up to 16% have been obtained, as summarized e.g. in Ref. [194]. There are some examples in which the efficiency is entirely determined by the quality of the semiconductor, similarly as in pn-junctions. The corrosion problem is also reduced to a very low level (see also Sect. 4.3).

During the last couple of years, several attempts have also been made to develop new materials, such as layer compounds and metal chalcogenides. For instance, Tributsch and coworkers have concentrated their research on pyrite (FeS_2). This material is not only of interest for semiconductor-liquid cells, but also for pure solid state-devices, because FeS_2 has a direct bandgap at $E = 1.03$ eV and a very large absorption coefficient ($\alpha = 6 \times 10^5 \, cm^{-1}$), which is of importance for thin layer solar cells. Rather stable photoelectrochemical cells, using I^-/I_3^- as a redox couple, have been produced. Although the quantum efficiency for the photocurrent is large, the photovoltage did not exceed 0.2 V [63, 198]. This is certainly due to the properties of FeS_2 itself, i.e. its electronic quality has to be considerably improved. A further development of this material is of great interest, since it would be favourable for applications because of ecological reasons. The present status of research on FeS_2 has recently been summarized in Ref. [199].

Rather recently, sensitizers have been used in regenerative solar cells. The sensitization-effect is based on the excitation of a dye-molecule adsorbed on the surface of a semiconductor electrode, followed by an electron injection into the conduction band of the electrode, as illustrated in Fig. 43. In the presence of a suitable electron donor, such as e.g. I^-/I_3^-, the oxidized dye-molecule is reduced again. The current in the solution is carried by the redox system, which is then reduced at the metal counter electrode (Fig. 43). Several research groups have studied these sensitization-processes, and the mechanisms are well known (compare with [14, 15]). At a rather early stage, Tributsch and Calvin suggested an application in solar cells [200]. Although the quantum yield of this process can reach more than 90%, the photocurrent efficiency with respect to the incident light was less than 1%, due to the fact that a dye-monolayer absorbs less than 1% of the light. A few attempts have been made to increase the surface area by using sintered electrodes, however, without much success [201, 202]. A breakthrough became possible after Stalder et al. succeeded in preparing highly porous TiO_2-electrodes [203]. Using these electrodes, high sensitization current-efficiencies have been obtained [204]. Further improvement of the electrode preparation has led to a cell—using a ruthenium complex as a dye—for which a solar conversion efficiency of more than 7% has been

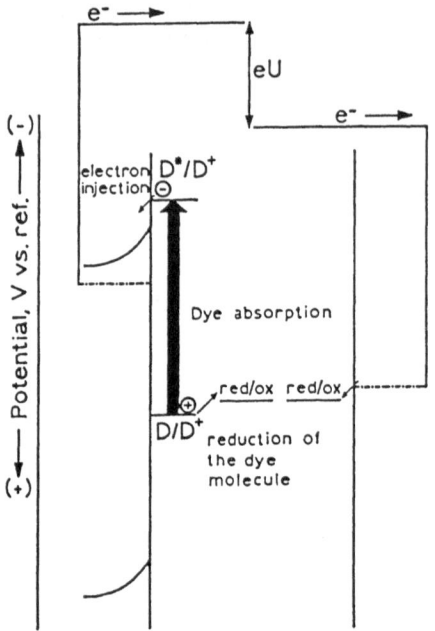

Fig. 43. Energy diagram of a sensitized photoelectrochemical cell

reported [205]. The same authors stated that the cell is stable when using
acetonitrile as an electrolyte. This is surprising, because the diffusion of I^-
within the small channels of the TiO_2 should be hindered, especially at currents
in the order of several mA cm^{-2}. The transport of the reducing agent is essential,
because otherwise the oxidized dye may undergo side reactions which would
lead to a destruction of the dye.

Recently, these very porous TiO_2-electrodes have also been applied by using
an inorganic sentitizer, such as CdS. In this case, small CdS-clusters with a size
of some nanometers were deposited. Action spectra of the photocurrent, measured in the presence of $[Fe(CN)_6]^{4-/3-}$ or polysulfide, proved the excitation
within CdS and a subsequent electron transfer into TiO_2 [210]. Similar results
have been obtained by depositing 15 nm FeS_2-particles on porous TiO_2 [211].
It is interesting to note that here a photovoltage of up to 0.6 V has been
obtained, whereas with FeS_2-electrodes only 0.2 V have been measured (see
above).

The application of semiconductor-liquid junctions is of special interest for
the direct production of a chemical fuel. Especially the production of hydrogen
by photoelectrolysis of H_2O has been studied by many research groups (compare with [114, 194]. It has been demonstrated by many authors that
H_2-formation is rather easy at semiconductor electrodes. The crucial point is the
simultaneous oxidation of H_2O. So far, photoelectrolysis was only achieved
with $SrTiO_3$, a semiconductor of a large bandgap ($E_g = 3.1$ eV) [206]. Very
recently, photocleavage of H_2O was also found with some niobates under open

circuit conditions [212, 213]. In other cases, such as TiO_2 [207], Fe_2O_3 [208] and RuS_2 [209], the direct photoelectrolysis of H_2O was only achieved by applying an additional voltage to the cell. Other semiconductors, including SiC [214], are not stable. The question remains whether one can modify the semiconductor surfaces to improve the oxidation of H_2O and to avoid anodic decomposition of the semiconductor. One strategy is the development of compounds which induce an interfacial coordination electrochemistry for improving the photocatalysis of multi-electron-reactions [195]. Another approach is the deposition of a metal or a metal oxide as a catalyst on the semiconductor surface [215]. However, just covering the surface by a complete metal layer would lead to a metal-semiconductor Schottky junction which usually exhibits high loss currents. The disadvantage can be avoided by depositing small metal clusters. In this type of surface modification, the favourable quality of the semiconductor-liquid junction is kept, as shown for redox reactions by Nakato et al. (see Ref. [216] and literature cited there). Whether this approach will be successful for H_2O-oxidation remains to be proved.

6.2 Photo-Detoxification of Waste Water

Suspensions of semiconductor particles have been applied for the degradation of inorganic and organic pollutants. Similarly as described in Sect. 5, organic molecules are oxidized by hole transfer, the holes being excited within the particle. Usually, oxygen is used as an electron acceptor. Investigations with model substances, such as e.g. chloroform, have shown that the organic molecule is quantitatively oxidized to CO_2. The overall reaction is given by [217]

$$2CHCl_3 + O_2 + 2H_2O \xrightarrow[TiO_2]{h\nu} CO_2 + 6H^+ + 6Cl^- \tag{95}$$

This method of photocatalytic detoxification is especially suitable for elimination of small traces of pollutants in H_2O because of the large surface area of TiO_2-suspensions. The reaction mechanism and product distributions have been studied by various research groups (compare with [234, 235]). During the last couple of years, some groups have concentrated their research on the quantitative analysis of the reactions and on the improvement of the technology. It has been shown, for instance, that the reaction rate does not necessarily increase linearly with light intensity. Typically, the rate increases linearly with the square root of intensity, as proved in the case of chloroform, which may be explained by the formation of $O\dot{H}$-radicals as intermediates [218]. These radicals can either oxidize $CHCl_3$ or recombine (H_2O_2-formation). An important question for this application is the nature of the rate limiting step. A key role here is played by the 2-electron reduction of oxygen, a process which is considerably slower than the oxidation of organic molecules. According to theoretical studies, the rate depends on the presence of surface electron traps and can be influenced by the size of particles [219, 220].

171

The application of TiO_2-catalysts limits the use of solar energy, because the efficiency is below 5%, due to the large bandgap of this material. Therefore, various attempts have been made to find stable semiconductors of a lower gap. One example are mixed oxides of titanium and iron. The absorption edge shifts considerably with increasing iron content. The decomposition rate, however, passes a maximum for 2.5% Fe, as shown for the oxidation of dichloroacetate [221].

The first large scale applications of solar waste water detoxification are being tested in the US and in Spain. Researchers of the Sandia National Laboratory are testing TiO_2-suspensions in a plant in New Mexico, and various European groups are using the test center in Almeria in Spain.

A. Heller and his group has suggested using these photocatalytic methods for treatment of oil slicks, which is a very interesting and economical application [222]. In order to keep the photocatalytic material on the surface of the sea, these researchers have deposited the TiO_2 on hollow microbeads of aluminosilicate. The properties of these microsystems are still under investigation.

6.3 Light-Induced Metal Deposition

Selective metal deposition is of interest in several applications, such as the formation of conducting patterns for integrated circuits and semiconductor devices. Instead of depositing a complete metal film and producing the pattern by selective etching makes it attractive to form the pattern directly by photodeposition. The basic concept of the procedure was developed as long as 20 years ago [223–225]. The results, however, were not well reproducible. The situation has recently improved because of a better understanding of the primary reaction steps. The principles of the photodeposition are as follows:

Illuminating an n-type semiconductor which is in contact with a metal ions-containing electrolyte, then two equal partial currents occur under open circuit conditions (Fig. 44a). The anodic photocurrent is due to O_2-formation, whereas the cathodic partial current corresponds to the reduction of the metal ions. Since the holes cannot diffuse very far, most of them are collected at the illuminated interface. In the case of a doped semiconductor, sufficient electrons are available everywhere, so that metal deposition should occur at illuminated as well as at dark surfaces (Fig. 44b), a conclusion which is unfavourable for selective metal deposition. According to experimental results, however, a selective metal deposition has been observed, e.g. at CdS at illuminated surfaces [226] and at TiO_2 at the dark surfaces [227]. In the case of CdS, this phenomenon has been interpreted by a downward shift of the energy bands at the illuminated surface, which makes the electron transfer there more favorable [226]. The result obtained with TiO_2 has been explained by strong internal and external recombination [227, 228]. Since TiO_2 is more suitable for application, much effort was laid on influencing the area where deposition occurs. Many attempts have been made applying different surface treatments or using other

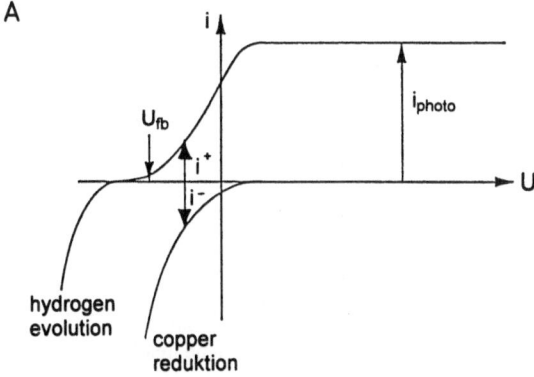

A

B

electrolyte   semiconductor   electrolyte

Fig. 44. a Current-voltage curve (theor.) of an illuminated TiO_2-electrode in the presence of metal-ions in the electrolyte; **b** energy model

hole acceptors in the solution. For instance, the result described above was only obtained with very well polished and etched TiO_2-surfaces, whereas with rough surfaces Cu-deposition was found everywhere on the crystal. A real improvement was finally obtained by using another hole acceptor, such as methanol or formic acid. In this case, the metal deposition was found almost entirely on the illuminated side, and the deposition rate is increased (Fig. 45). This result has been explained as follows:

In solutions without Cu^{2+}, the photocurrent is enhanced upon addition of CH_3OH, due to the current-doubling effect (see Sect. 4.5, Fig. 34), i.e.

$$H_3COH + h^+ \rightarrow H_2C^{\bullet}OH + H^+ \tag{96a}$$

$$H_2C^{\bullet}OH \rightarrow H_2CO + H^+ + e^- \tag{96b}$$

173

R. Memming

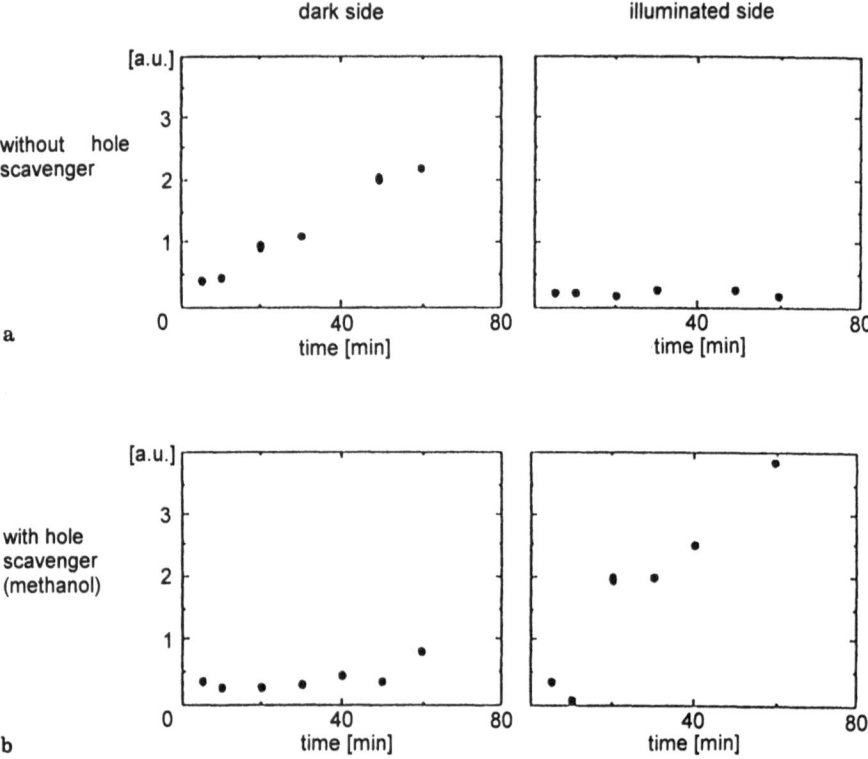

Fig. 45. Concentration (arbit units) of deposited Cu vs illumination time in solutions with and without methanol [228]

Upon addition of Cu^{2+}-ions, the current-doubling effect disappears. Accordingly, reaction (95b) did not take place. It has been concluded that the radical must be capable of reducing Cu^{2+}, i.e.

$$2H_2C^{\bullet}OH + Cu^{2+} \rightarrow 2H_2CO + 2H^+ + Cu^0 \tag{97}$$

Since the radicals are formed by hole transfer only at the illuminated side, Cu^0 must also be deposited there [228].

6.4 Photoetching of Semiconductors

Etching of semiconductors in the dark and under illumination plays an important role in device fabrication. Since mostly Si and GaAs are used in devices, the research on etching processes is concentrated on these materials. Taking GaAs as an example, the dissolution at low pH can be described by [229]:

$$GaAs + 3H_2O + 6h^+ \rightarrow Ga^{3+} + H_3AsO_3 + 3H^+ \tag{98}$$

174

When etching is carried out at open circuit potential, these holes must be supplied by an oxidizing agent of a rather positive standard potential, such as Ce^{4+} [230].

Photoetching is of special interest, because material can be locally etched away by focussing a light beam on a certain spot (or by using a laser beam). In order to achieve enhanced etching under illumination at open circuit potential, it is necessary to involve both, majority and minority carriers, in the dissolution mechanism. The principle of photoetching is illustrated in Fig. 46. During light excitation, the anodic decomposition occurs via hole consumption (valence band process), whereas electrons are transferred to the Ox-form of the redox system (conduction band process). This principle can be applied to n- as well as to p-type semiconductors. In the case of a p-type electrode, the anodic dissolution can occur in the dark, but the electron transfer reaction requires light excitation. With n-type, the reduction of the redox system occurs in the dark, but holes have to be excited for the dissolution process. Accordingly, corresponding anodic and cathodic partial currents are required under open circuit conditions, similarly as for the light-induced metal deposition (compare Fig. 44a). Suitable redox couples should be those the standard potential of which occurs somewhat below the conduction band. Investigations with $Eu^{2+/3+}$, $V^{3+/2+}$ or $Cr^{3+/2+}$ have shown, however, that no etching takes place at GaAs [231]. This result is due to extremely small exchange currents at the open circuit potential. Especially the small rate for the electron transfer from the conduction band to the redox system is responsible for this effect.

Interestingly, certain more complex redox systems are much more effective with respect to photoetching. A typical example for GaAs is H_2O_2. As already discussed in Sect. 4.4., the reduction of H_2O_2 occurs at a very high rate at n-type as well as at p-type GaAs. A key role plays obviously a surface radical, being formed by chemical etching (see Eq. (66)) [231]. Other redox systems, such as for instance Br_2 and BrO_3^{2-} behave similarly. Kelly and coworkers have studied the corresponding reaction mechanisms in detail. They proposed a unified model which they applied for all three redox systems [140, 141, 232].

Finally it should be mentioned that this technique can also be applied for etching small and deep grooves without having severe diffusion problems. The photoetching operates well with n-GaAs, because the reduction of the redox

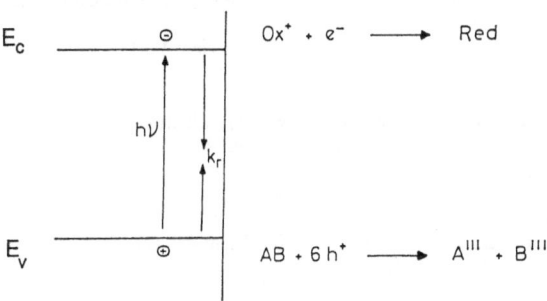

Fig. 46. Model for photoetching at open circuit potential [231].

system must not necessarily occur deep inside the groove, but anywhere in the dark areas of the surface [233].

Acknowledgement. The author is indebted to Dr. H. Cachet (CNRS, Paris), Dr. A.G. Nozik (NREL, Golden) and Prof. H. Tributsch (HMI, Berlin) foɪ making available manuscripts prior to publication. Thanks are also due to Dr. D. Bahnemann, Dr. D. Meissner and Dr. R. Reineke for many valuable discussions.

7 References

1. Gerischer H (1960) (Z Phys Chem N F) 26: 223; 26: 325: (1961) 27: 48
2. Marcus RA (1964) Ann Rev Phys Chem 15: 155
3. Dogonadze RR, Kuznetsov AM, Chizmadzhev (1964) Russ I Phys Chem 38: 652
4. Levich VG (1966) In: Delahay P, Tobias CW (eds) Advances in electrochemistry and electrochemical engineering. Wiley Interscience, New York, vol 4, p 249
5. Gerischer H (1970) In: Eyring M. Jost W, Henderson D, (eds) Physical chemistry, Academic, New York, vol 4 A, p 463
6. Memming R (1979) In: Bard AJ (ed) Electroanalytical chemistry. Marcel Dekker, New York vol 11, p 1
7. Morrison SR (1980) Electrochemistry at semiconductor and oxidized metal electrodes. Plenum, New York
8. Gomes WP, Cardon F (1982), Progr Surf Sci 12: 155
9. Pleskov Yu V (1980) In: Conway BE, Bockris IO'M, Yeager E (eds) Comprehensive treatise of electrochemistry. Plenum, New York, vol 1, p 291
10. Memming R (1983) In: Conway BE, Bockris IOM, Yeager E (eds) Comprehensive treatise of electrochemistry. Plenum, New York, vol 7, p 529
11. Pleskov Yu V; Gurevick YU (1986) Semiconductor Photoelectrochemistry, Consultant Bureau, New York
12. Jaegermann W, Tributsch H (1988) Prog Surf Sci 29: 1
13. Fujishima A, Honda K (1972) Nature 238: 37
14. Gerischer H, Willig F (1976) In: Topics in current chemistry. Springer, Berlin Heidelberg New York, vol 61, p 33
15. Memming R (1984) Progr Surf Sci 17: 7
16. Henglein A (1988) Topics of curr chem 143: 113
17. Henglein A (1989) Chem Rev 89: 1861
18. Bahnemann DW (1991) In: Pelizzetti E, Schiavello M (eds) Photochemical conversion and storage of solar energy, Kluwer, The Netherlands, p 251
19. Kittel Ch (1976) Introduction to solid state physics. John Wiley, New York
20. Sze S (1981) Physics of semiconductor devices, 2nd edn. John Wiley, New York
21. Brillson LI (1982) Surf Sci Rep 2: 123
22. Zimann IM (1972) Principles of theory of solids. Cambridge University Press, Cambridge, UK
23. Pankove IL (1971) Optical Processes in semiconductors. Prentice Hall, Englewood Cliffs, UK
24. Knox RS, Dexter KL (1965) Excitons, New York
25. Gutmann F, Lyons LE (1967) Organic semiconductors. John Wiley, New York
26. Berry CR (1967) Phys Rev 161: 611
27. Meehan EI, Miller IK (1968) J Phys Chem 72: 1523
28. Fojtik A, Weller H, Koch U, Henglein A (1984) Ber Bunsenges Phys Chem 88: 969
29. Rossetti R, Ellison IL, Gibson IM, Brus LE (1984) J Chem Phys 80: 4464
30. Chestnoy N, Hall R, Brus LE (1986) J Chem Phys 85: 2237
31. Weller H, Fojtik A, Henglein A (1985) Chem Phys Lett 117: 485
32. Brus LE (1986) J Phys Chem 90: 2555
33. Bawendi MG, Steigerwald ML, Brus LE (1990) Annu Rev Phys Chem 41: 477
34. Schmidt HM, Weller H (1986) Chem Phys Lett 129: 615

35. Weller H, Schmidt HM, Koch U, Fojtik A, Baral S, Henglein A, Kunath W, Weiss K, Dieman E (1986) Chem Phys Lett 124: 557
36. Bahnemann DW, Kormann C, Hoffmann MR (1987) J Phys Chem 91: 3789
37. Atkins PW (1978) Physical chemistry. Oxford University Press, Oxford, UK
38. Hodes G, Albu-Yaron A, Decker F, Motisuke P (1987) Phys Rev B 36: 4215
39. Hodes G, Howell IDJ, Peter LM (1992) J Electrochem Soc 139: 3136
40. Ekimor AI, Onushchenko AA (1984) Pisma Zh Eksp Teor Fis 40: 337
41. Wang Y, Herron N (1987) J Phys Chem 91: 257
42. Dingle R (1975) In: Queisser HI (ed) Advances in solid state physics. Pergamon Viehweg, Braunschweig, vol 53, p 136
44. Gerischer H (1960) Z Phys Chem NF 26: 232 and 325; (1961): 27: 48
45. Möllers F, Memming R (1976) Ber Bunsenges Phys Chem 76: 469
46. Memming R, Möllers F (1976) Ber Bunsenges Phys Chem 76: 475
47. Memming R (1969) J Electrochem Soc 116: 785
48. Lohmann F (1966) Ber Bunsenges Phys Chem 70: 428
49. Bolts IM, Wrighton MS (1976) J Phys Chem 80: 2641
50. Pleskov Yu V (1973) Prog Surf Membr Sci 7: 57
51. Memming R, Schwandt G (1968) Electrochem Acta 13: 1299
52. Van den Berghe RAL, Cardon F, Gomes WP (1973) Surf Sci 39: 368
53. Gerischer H, Mindt W (1966) Surface Sci 4: 440
54. Gerischer H, Hoffmann-Perez M, Mindt W (1965) Ber Bunsenges Phys Chem 69: 130
55. Memming R, Neumann G (1969) J Electroanal Chem Interfac Electrochem 21: 295
56. Schröder K, Memming R (1985) Ber Bunsenges Phys Chem 89: 385
57. Meissner D, Memming R, Kastening B (1988) J Phys Chem 92: 3476
58. Meissner D, Benndorf C, Memming R (1987) Appl Surf Sci 27: 423
59. Rimmasch J (1992) Die Untersuchung von Ladungstransferprozessen an der Phasengrenze TiO₂/Electrolyt. Thesis, University of Hamburg, Germany
60. Lewerenz HJ, Gerischer H, Lübke M (1984) J Electrochem Soc 131: 100
61. Mc Evoy AI, Etman M, Memming R (1985) Electroanal Chem 190: 225
62. Bard AI, Bocarsly AB, Fan F, Walton EW, Wrighton MS (1980) J Am Chem Soc 102: 3671
63. Ennaoui A, Tributsch H (1986) J Electrochem Soc 204: 185
64. Mishra KK, Osseo-Asare K (1992) J Electrochem Soc 139: 749
65. Fantini MCA, Shen WM, Tonkiewicz M, Gambino JP (1989) J Appl Phys 65: 4884
66. Ba B, Fotouhi BB, Gabouze N, Gorochov O, Cachet H (1993) J Electroanal Chem, 334, 263
67. Dutoit EC, Cardon F, Gomes WP (1976) Ber Bunsenges 80: 475
68. Kiwiet NJ, Fox MA (1990) J Electrochem Soc 137: 1240
69. Memming R, Kelly JJ (1980) Connolly (ed), Photochemical conversion and storage of solar energy. Academic Press, New York, p 243
70. Kelly JJ, Memming R (1982) J Electrochem Soc 129: 730
71. Mc Intyre R, Gerischer H (1985) Ber Bunsenges Phys Chem 88: 963
72. Van den Meerakker JEAM (1985) Electrochim Acta 30: 435
73. Allongue P, Cachet H (1985) Sol State Electron 55: 44
74. Lincot D, Vedel J (1987) J Electroanal Chem 220: 179
75. Van den Meerakker JEAM, Kelly JJ, Notten PHL (1985) J Electrochem Soc 132: 638
76. Meissner D, Lauermann I, Memming R, Kastening B (1988) J Phys Chem 92: 3484
77. Scholz GA, Gerischer H (1992) J Electrochem Soc 139: 165
78. Sinn Ch, Meissner D, Memming R (1990) J Electrochem Soc 137: 168
79. Kühne HM, Tributsch H (1986) J Electroanal Chem 201: 263
80. Meissner D, Memming R (1988) In: Grassi G, Hall DO (eds) Photocatalytic production of energy-rich compounds. Elsevier, London, p 138
81. Jägermann W, Kühne HM (1986) Appl Surf Sci 26: 1
82. Dogonadze RR, Kutzenov (1983) In: Conway BE, Bockris JOM, Yeager E, Khan SUM, White RE (eds) comprehensive treatise of electrochemistry, vol 7
83. Schefold J, Kühne HM (1991) J Electroanal Chem 300: 211
84. Memming R (1987) Ber Bunsenges Phys Chem 91: 353
85. Reichman J (1980) Appl Phys Lett 36: 574
86. Gärtner WW (1959) Phys Rev 116: 84
87. Butler MA (1977) J Appl Phys 48: 1914
88. Wilson RH (1977) J Appl Phys 48: 4292

177

89. Memming R (1991) In: Pelizzetti E, Schiavello M (eds) Photochemical conversion and storage of solar energy. Kluwer, The Netherlands, p 193
90. Shockley W (1950) Electrons and Holes in Semiconductors, Van Norstrand, New York
91. Williams F, Nozik AI (1984) Nature 311: 5989
92. Reineke R, Memming R (1989) In: Hall DO, Grassi G (eds) Photo conversion pro-cesses for energy and chemicals. Elsevier, Essex, UK, p 129
93. Reineke R, Memming R (1992) J Phys Chem 96: 1310
94. Vanmaekelbergh D, Gomes WP, Cardon F (1982) J Electrochem Soc 129: 564
95. Lu Shou Yun, Vanmaekelbergh D, Gomes WP (1987) Ber Bunsenges Phys Chem 91: 390
96. Vanmaekelbergh D, Lu Shou Yun, Gomes WP (1987) J Electroanal Chem 221: 187
97. Notten PHL (1987) Electrochim Acta 32: 575
98. Notten PHL, Kelly JJ (1987) J Electrochem Soc 134: 444
99. Kelly JJ, Reynders AC (1987) Appl Surf Sci 29: 149
100. Notten PHL, Damen AAJM (1987) Appl Surf Sci 28: 331
101. Notten PHL (1984) J Electrochem Soc 131: 2641
102. Inoue T, Watanabe T, Fujishima A, Honda, K, Kohayakawa (1977) J Electrochem Soc 124: 719
103. Memming R (1977) Ber Bunsenges Phys Chem 81: 732
104. Memming R (1978) J Electrochem Soc 125: 117
105. Gerischer H (1977) J Electrochanal Chem 82: 133
106. Bard AJ, Wrighton MS (1977) In: Heller A (ed) Semiconductor-liquid junctions. Electrochemical Society, Princeton NY, Proc. vol 77-3, p 195
107. Gerischer H (1979) In: Seraphin BO (ed) Topics in appl physics. Springer, Berlin Heidelberg New York, p 115
108. Gerischer H, Mindt W (1968) Electrochim Acta 13: 1329
109. Memming R, Schwandt G (1966) Surf Sci 5: 97
110. Frese K, Madou MJ, Morrison SR (1980) J Phys Chem 84: 3172; (1981) J Electrochem Soc 128: 1528
111. Gerischer H, Lübke M (1983) Ber Bunsenges Phys Chem 87: 123
112. Memming R (1985) In: Schiavello (ed) Photoelectrochemistry, photocatalysis and photoreactors. D. Reidel, Dordrecht, p 107
113. Cardon F, Gomes WP, Kerchove F, Vanmaekelbergh D, v. Overmeire F (1980) Faraday Discussions 70: 153
114. Memming R (1988) In: Topics in current chemistry. Springer Verlag, Berlin, vol 143, p 79
115. Menezes S, Miller B (1983) J Electrochem Soc 130: 517
116. Gomes WP, v. Overmeire F, Vanmaekelbergh D, v. d. Kerchove F, Cardon F (1981) In: Nozik AJ (ed) Photophysics at semiconductor-elektrolyte interfaces. ACS-Series No. 146, p 120
117. Vanmaekelbergh D, Gomes WP (1990) J Phys Chem 94: 1571
118. Meissner D, Sinn Ch, Memming R, Notten PHL, Kelly JJ (1986) In: Pelizzetti E, Serpone N (eds) Homogeneous and heterogeneous photocatalysis. D Reidel, The Netherlands, p 343
119. Allongue P, Blonkowsky S, Lincot D (1991) J Electroanal Chem 300: 261
120. Tributsch H (1977) Ber Bunsenges Phys Chem 81: 361
121. Tributsch H, Bennett JC (1977) J Electroanal Chem 81: 97
122. Tributsch H (1982) In: Structure and bonding. Springer, Berlin Heidelberg, New York, vol 49, p 129
123. Kautek W, Gerischer H (1982) Surf Sci 119: 46
124. Kautek W, Gerischer H (1981) Electrochim Acta 26: 1771
125. Kautek W, Willig F (1981) Elektrochim Acta 26: 1709
126. Tributsch H (1979) Sol Energy Mat 1, 257
127. Hale JM (1971) In: Hush NS (ed) Reactions of molecules at electrodes. Wiley, New York, p 229
128. Vetter K (1961) Elektrochemische Kinetik, Springer, Berlin Heidelberg New York
129. Lewis NS (1991) Ann Rev Phys Chem 42: 543
130. Marcus RA (1990) J Phys Chem 94: 4152 and 94: 7742
131. Gerischer H (1991) J Phys Chem 95: 1356
132. Morrison SR (1969) Surf Sci 15: 363
133. Marcus RA (1990) J Phys Chem 94: 1050
134. Gurevich Yu Ya, Pleskov Yu V (1982) Elektrokhimiya 18: 1477
135. Reineke R, Memming R (1992) J Phys Chem 96: 1317
136. Uhlendorf I, Rimmasch J, Reineke R, Meissner D (1991), presented at the 42nd Meeting of the Internat Soc of Electrochem, Montreux

137. Meissner D, Memming R (1992) Electrochim Acta 37: 799
138. Allongue P, Blonkowski S, Souteyrand E (1992) Electrochim Acta 37: 781
139. Rosenwaks Y, Thacker BR, Ahrenkiel RK, Nozik AJ (1993) Nature, in press
140. Minks BP, Oskam G, Vanmaekelbergh D, Kelly JJ (1989) J Electroanal Chem 273: 119
141. Minks BP (1991) Photoetching of GaAs: The Cathodic Reaction. Thesis, University of Utrecht (Holland)
142. Uhlendorf I, Reineke R, Memming R (publication in preparation)
143. Nozik AJ (1993) Second Internat Conference on solar energy storage and appl photochemistry (cairo), Proc (in press)
144. Rosenbluth ML, Lieber CM, Lewis NS (1984) Appl Phys Lett 45: 423
145. Rosenbluth ML, Lewis NS (1986) J Am Chem Soc 108: 4689
146. Lewis NS (1990) Acc Chem Res 23: 176
147. Kumar A, Lewis NS (1991) J Phys Chem 95: 7021
148. Kobayashi H, Tsubomura H (1989) J Electroanal Chem 272: 37
149. Kobayashi H, Chigami A, Takeda N, Tsubomura H (1990) J Electroanal Chem 287: 239
150. Memming R (1969) J Electrochem Soc 116: 785
151. Beckmann KH, Memming R (1969) J Electrochem Soc 116: 368
152. Memming R, Möllers F (1972) Ber Bunsenges Phys Chem 76: 609
153. Gerischer H, Müller N, Haas O (1981) J Electroanal Chem 119: 41
154. Notten PHL (1987) J Electroanal Chem 224: 211
155. Kelly JJ, Minks BP, Verhaegh AM, Stumper J, Peter LM (1992) Electrochim Acta 37: 909
156. Morrison SR, Freund T (1968) Electrochim Acta 13: 1343
157. Morrison SR, Freund T (1967) J Chem Phys 47: 1543
158. Micka K, Gerischer H (1972) J Electroanal Chem 38: 397
159. Yamagata S, Nakabayashi S, Sancier KM, Fujishima A (1988) Bull Chem Soc Jap 61: 3429
160. Hykaway N, Sears WM, Morisaki H, Morrison SR (1986) J Phys Chem 90: 6663
161. Lee J, Kato T, Fujishima A, Honda K (1984) Bull Chem Soc Jap 57: 1179
162. Lilie J, Beck G, Henglein A (1971) Ber Bunsenges Phys Chem 75: 458
163. Maeda Y, Fujishima A, Honda K (1981) J Electrochem Soc 128: 1731
164. Williams F, Nozik AJ (1978) Nature 271: 137
165. Ross RT, Nozik AJ (1982) J Appl Phys 53: 3813
166. Cooper G, Turner JA, Parkinson BA, Nozik AJ (1983) J Appl Phys 54: 6463
167. Boudreaux DS, Williams F, Nozik AJ (1980) J Appl Phys 51: 2158
168. Turner JA, Nozik AJ (1982) Appl Phys Lett 41: 101
169. Koval CA, Segar PR (1989) J Am Chem Soc 111: 2004
170. Frese KW, Chen C (1992) J Electrochem Soc 139: 3234
171. Chen C, Frese KW (1992) J Electrochem Soc 139: 3243
172. Fox MA (1987) In: Topics in current chemistry. Springer Verlag, vol 143: 42
173. Grätzel M, Frank AJ (1982) J Phys Chem 86: 2964
174. Minoura H, Katoh Y, Sugiura T, Ueno Y, Matsui M, Shibata K (1990) Chem Phys Lett 173: 220
175. Minoura H, Inayoshi N, Ueno Y, Matsui M, Shibata K (1992) J Electroanal Chem 332: 279
176. Wolf K, Bahnemann DW (publication in preparation)
177. Bahnemann DW, Henglein A, Spanhel L (1984) Faraday Discuss Chem Soc 78: 151
178. Rothenburger G, Moser I, Grätzel M, Serpone N, Sharma DK (1985) J Am Chem Soc 107: 8054
179. Baral S, Fojtik A, Weller H, Henglein A (1986) J Am Chem Soc 108: 375
180. Chen G, Zen IM, Fan F, Bard AI (1991) J Phys Chem 95: 3682
181. Nedeljkovic IM, Nenadovic MT, Micic OI, Nozik AJ (1986) J Phys Chem 90: 12
182. Micic OI, Rajh T, Comor MV (1992) In: Mackay RA, Texter J (eds) Electrochemistry in colloids and dispersions. VCH, New York (USA), p 457
183. Kraeutler B, Bard AJ (1977) J Am Chem Soc 99: 7729; (1978) 100: 5985
184. Meissner D, Memming R, Kastening B (1983) Chem Phys Lett 96: 34
185. Kormann C, Bahnemann DW, Hoffmann MR (1990) Langmuir 6: 555
185a. Kormann C, Bahnemann DW, Hoffmann MR (1991) Environ Sci Techn 25: 494
186. Henglein A, Gutiérrez M (1983) Ber Bunsenges Phys Chem 87: 852
187. Müller B, Majoni S, Meissner D (publication in preparation)
188. Yanagida S, Ishimaru Y, Miyake Y, Shiragami T, Pac C, Hashimoto K, Sakata T (1989) J Phys 93: 2576

189. Müller B (1993) Photoeletrochem Untersuchungen an Halbleiterteilchen unterschiedlicher Größe Thesis, University of Hamburg (Germany)
190. Henglein A, Gutiérrez M, Fischer CH (1984) Ber Bunsenges Phys Chem 88: 170
191. Inoue H, Torimoto T, Sakata T, Mori H, Yoneyama H (1990) Chem Lett 1483
191a. Tributsch H (1982) In: Solar energy materials, structure and bonding. Springer, Berlin Heidelberg New York, vol 49, p 127
192. Lewis NS (1984) Ann Rev Mater Sci 14: 95
193. Hodes G (1985) In: Grätzel M (ed) Energy resources through photochemistry and catalysis. Academic, New York, p 521
194. Memming R (1990) In: Rabek JF (ed) Photochemistry and photophysics. CRC Boca Raton (USA), vol II, p 143
195. Tributsch H (1988) In: Schiavello (ed) New trends and applications of photocatalysis and photochemistry for environmental problems. D. Reidel. Dordrecht (Holland), p 297
196. Honda K, Fujishima A, Watanabe T (1982) In: Ohta T (ed) Solar hydrogen energy systems. Pergamon Press, Oxford, UK, p 137
197. Memming R (1991) In: Pelizzetti E, Schiavello M (eds) Photochemical conversion and storage of solar energy. Kluwer (Holland), p 193
198. Ennaoui A, Tributsch H (1986) Sol Energy Mat 14: 461
199. Ennaoui A, Fiechter S, Pettenhofer Ch, Allonso-Vante N, Büker K, Bronold M, Höpfner C, Tributsch H (1993) In: Solar energy materials and solar cells, in press
200. Tributsch H, Calvin M (1971) Photochem Photobiol 14: 95
201. Matsumura N, Nomura Y, Tsubomura H (1977) Bull Chem Soc Jap 50: 2533
202. Alonso N, Beley VM, Chartier P, Erns N (1981) Rev Phys Appl 16: 5
203. Stalder C, Augustynsky J (1979) J Electrochem Soc 126: 2007
204. Desilvestro I, Grätzel M, Kavau L, Moser I, Augustynsky J (1985) J Am Chem Soc 107: 2988
205. O'Regan B, Grätzel M (1991) Nature 353: 737
206. Mavroides JG, Dafalos JA, Kolesar DF (1976) Appl Phys Lett 28: 241
207. Nozik AJ (1975) Nature 257: 5527
208. Yeh LSR, Hackermann N (1977) J Electrochem Soc 124: 833
209. Piazza S, Kühne HM, Tributsch H (1985) J Electroanal Chem 196: 53
210. Vogel R Pohl K, Weller H (1990) Chem Phys Lett 174: 241
211. Ennaoui A, Fiechter S, Tributsch H, Giersig M, Vogel R, Weller H (1992) J Electrochem Soc 139: 2514
212. Sayama K, Tanaka A, Domen K, Maruya K, Onishi T (1991) J Phys Chem 95: 1345
213. Kudo A, Sayama K, Tunaka A, Asakura K, Domen K, Maruya K, Onishi T (1989) J Catalysis 120: 337
214. Lauermann I, Meissner D, publication in preparation
215. Meissner D (1992) In: Veziroglu TN, Derive C, Pottier I (eds), Hydrogen energy progress IX, Proc 9th World Hydrogen Energy Conf Paris, vol 1, p 517
216. Nakato Y, Ueda K, Yano H, Tsubomura H (1988) J Phys Chem 92: 2316
217. Kormann C, Bahnemann DW, Hoffmann MR (1989) J Photochem Photobiol, A: Chemistry 48: 161
218. Bahnemann DW, Bockelmann D, Goslich R (1991) Solar Energy Mat 24: 564
219. Gerischer H, Heller A (1991) J Phys Chem 95: 5261
220. Gerischer H, Heller A (1992) J Electrochem Soc 139: 113
221. Bahnemann DW, Bockelmann D, Hilgendorff M, Weichgrebe D, Goslich R (1993) In: Ollis D, Al-Ekabi H (eds) TiO$_2$ photocatalytic purification and treatment of water and air. Elsevier Science Publishers Amsterdam, in press
222. Jackson NB, Wang CM, Luo Z, Schwitzgebel J, Eckerdt JG, Brock JR, Heller A (1991) J Electrochem Soc 138: 3660
223. Möllers F, Tolle HJ, Memming R (1974) J Electrochem Soc 121: 1160
224. Jacobs JWM (1986) J Phys Chem 90: 6507
225. Jacobs JWM (1988) Laser initiated metal deposition on semiconductors from aqueous solutions. Thesis, University of Eindhoven (Holland)
226. Lauermann I, Meissner D, Memming R (1988) In: Ginley DS et al (eds) Proc. of the symposium of photoelectrochemistry and electrosynthesis on semiconductor materials. Proc vol 88-14, the Electrochemical Society, Pennington, NY, p 190
227. Kobayashi T, Taniguchi Y, Yoneyama H, Tamura H (1983) J Phys Chem 87: 768
228. Richter W, Rimmasch J, Kastening B, Memming R, Meissner D (1991) In: Datta M, Sheppard

K, Snyder D (eds) Electrochemical Microfabrication, Proc. vol 2–3, The Electrochemical Society, Pennigton, NY, p 149
229. Ostermayer FW, Kohl PA, Burton RH (1983) Appl Phys Lett 43: 642
230. Kelly JJ, Notten PHL (1984) Electrochim Acta 29: 589
231. Kelly JJ, van den Meerakker JEAM, Notten PHL (ed) (1986) In: Behrens H (ed) Grundlagen von elektrodenreaktionen, dechema monographien. Verlag Chemie, vol 102, p 453
232. Minks BP, Vanmaekelbergh D, Kelly JJ (1989) J Electroanal Chem 273: 133
233. van de Ven J, Nabben HJP (1990) J Appl Phys 67: 7572
234. Ollis DF (1985) Environ Sci Technol 19: 486
235. Serpone N (1989) In: Norris JR, Meisel D (eds) Photochemical energy conversion. Elsevier, New York, p 297

Umpolung of Ketones via Enol Radical Cations

Michael Schmittel

Institut für Organische Chemie der Universität Würzburg, Am Hubland, 97074 Würzburg, FRG

Table of Contents

Topics in Current Chemistry, Vol. 169
© Springer-Verlag Berlin Heidelberg 1994

For several tautomeric systems (ketones/enols, imines/enamines and others) a distinct reversal of the stability order is observed when going from the neutral compounds to the radical cations, the first use of which in a new preparative α-Umpolung reaction has been documented for keto/enol systems. The present review provides a critical evaluation of the chemistry of enol radical cations in solution with a special emphasis on the Umpolung reaction and the intermediates thereof. Other enol type of radical cations are discussed with respect to their potential to provide α-carbonyl radical and α-carbonyl cation intermediates. Hence, this article does not constitute a comprehensive summary on all enol type of radical cation reactions. All potentials in this review are referenced versus SCE, unless noted otherwise. Potentials measured against the ferrocene/ferrocenium couple were converted to SCE by adding 0.334 V.

1 Introduction

Electron transfer concepts [1, 2] play an important role in inorganic chemistry and detailed knowledge has been accumulated about stabilities, reactivity patterns and structural particularities of compounds at various oxidation levels [3–5]. In contrast, thinking in terms of electron transfer has not yet received wide recognition within the organic community, although from a conceptual point of view electron transfer constitutes the simplest elementary transformation in chemistry [6] – besides photochemical excitation [7]. Thus, for most organic molecules, there is only scattered knowledge about the chemistry of one-electron oxidized or reduced species [8], unless viewed in the context of electrochemical transformations [9–12], radiation chemistry [13–15] and photoinduced electron transfer (PET) reactions [16, 17]. As a result the potential of electron transfer chemistry for synthetic purposes still needs to be uncovered. The high reactivity and sometimes low selectivity of charged odd-electron species and the long established preference for even-electron mechanisms in organic chemistry and the simplicity of mechanistic formulations thereof have long delayed a thorough and careful analysis of the chances of electron transfer activation in organic chemistry.

$$RH^{\cdot-} \xleftarrow{\;+e^-\;} RH \xrightarrow{\;-e^-\;} RH^{\cdot+}$$

While Nature successfully employs electron transfer activation for important synthetic transformations [18–21], only few organic reactions are based on one-electron oxidation or reduction processes [8]. For a rational approach to the design of new reactions using electron transfer concepts an even better understanding of fundamental principles is needed, although a first insight has started to emerge. For example, it has been demonstrated that electron transfer can lead to significant changes in terms of structure, stability and reactivity. As shown in Table 1 carbon-carbon bond dissociation energies (BDE_{C-C}) are severely affected by simply removing one electron. The change in the BDE_{C-C} when going from ethane to *neo*-pentane radical cation is most illustrative, especially when compared to the BDE_{C-C} of the neutral compounds. But electron removal not

Table 1. Carbon–carbon bond dissociation energies (BDE_{C-C}) in neutral compounds and their radical cations [kJ mol^{-1}]. Data are gas-phase numbers from thermochemical cycle calculations by Dinnocenzo [22]

RH	$BDE_{C-C}(RH)$	$BDE_{C-C}(RH^{+\bullet})$
H_3C-CH_3	377	213
$Et-CH_3$	360	109
$tBu-CH_3$	351	13
H_2N-CH_3	356[a]	431[a]
 △	251	59
 ⟂	146	0
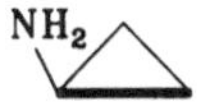 NH_2 ⟋△	180	-117

[a] C–N bond dissociation energy

necessarily entails weakening of a bond. Depending on the structural situation the bond in the radical cation can be even strengthened (e.g. CH_3NH_2).

C–C bond dissociation energies are not the only molecular properties affected by electron transfer. In addition, radical cations exhibit a number of unique characteristics: flexible structures [23], a low sensitivity towards steric effects [24], low activation barriers for inter- and intramolecular reactions [25–28], high acidities [29–32] and the inversion of the thermochemical stability order for certain tautomeric systems. Examples in the recent literature demonstrate that it is worthwhile thinking about how the changed molecular properties can be used for the design of new reactions that complement the thermal and photochemical reactivity patterns [26, 33–35].

For the generation of radical cations in solution several methods are feasible, the most important of which are: (1) use of chemical oxidants [8, 36], (2) electrochemical oxidation [9–12] and (3) photoinduced electron transfer [16, 17]. All these methods have been described in detail. It should be mentioned that several formal one-electron oxidants (Mn(III), Ce(IV), Co(III), radical cation salts) may react via inner-sphere electron transfer thus not involving free radical cation intermediates [8, 37, 38] and that only in few reports a rigorous mechanistic analysis according to the Chanon/Tobe diagnostic criteria was undertaken [39]. Hence, in the following, special emphasis was given to reports employing well-defined outer-sphere one-electron oxidants: anode, Fe(phen)$_3^{3+}$ (**FePHEN**), Ru(phen)$_3^{3+}$ (**RuPHEN**), IrCl$_6^{2-}$ (**IrCl**) and heteropoly anions [40].

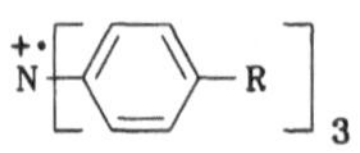

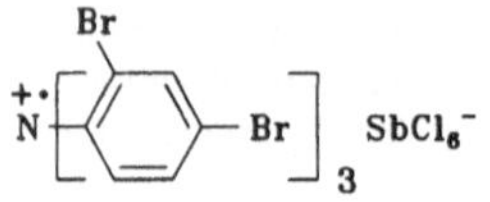

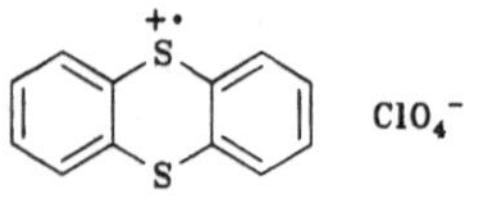

Michael Schmittel

Br

$\overset{+\cdot}{N}$ —R |₃

TBPA⁺· (R: Br)
TTA⁺· (R: CH₃)
TMPA⁺·(R: OCH₃)

TDBPA ⁺·   SbCl₆⁻

TH ⁺·   ClO₄⁻

Fe(phen)₃(PF₆)₃   Ru(phen)₃(PF₆)₃   Na₂IrCl₆   Ce(NH₄)₂(NO₃)₆

FePHEN   RuPHEN   IrCl   CAN

2 Inversion of Stability Order of Tautomers upon One-Electron Oxidation

An important, but not widely recognized effect in electron transfer chemistry is the phenomenon that the thermochemical stability order of several tautomeric systems can be inverted upon one-electron oxidation. Hitherto, most of the data stem from gas-phase measurements (photoelectron and mass spectrometry data) or from calculations at several levels of theory (Table 2). While the explicit numbers still differ for each method they all agree on the inversion of the stability order for keto/enol, alkine/allene, imine/enamine, nitrile/isonitrile and aldimine/aminocarbene pairs. One-electron reduction, on the other hand, does not necessarily lead to a thermochemical stability inversion, as demonstrated in the case of acetaldehyde⁻·/ethenol⁻· [41].

In spite of this interesting effect and its potential for synthetic purposes, only the one-electron oxidation chemistry of the keto/enol pair has been studied in

Table 2. Inversion of the thermochemical stability order of several tautomeric systems upon one-electron oxidation

tautomers (A/B)	$\Delta\Delta H_f^0(B-A)/$ kJ mol⁻¹	$\Delta\Delta H_f^0(B^{+\cdot}-A^{+\cdot})/$ kJ mol⁻¹	method	Ref.
acetaldehyde/ethenol	+ 32	− 138	MNDO	41
acetaldehyde/ethenol	+ 55	− 63	AE[a]	42–44
acetone/propen-2-ol	+ 58	− 59	AE[a]	42–44
propanal/propen-1-ol	+ 21[b]	− 105	AE[a]	42–44
propine/allene	+ 5[c]	− 56, − 63[c]	6-31G*	45, 46
acetimine/vinylamine	+ 28	≈ − 138	PE[d]	47
HCN/HNC	+ 40[c]	− 128[c]	4-31G*	48
H₂CNH/HCNH₂	+ 163[c]	− 26[e)]	6–31G**	49

[a] Appearance energy measurements, [b] see Ref. [50, 51], [c] calculated relative energies, [d] photoelectron spectroscopy.

186

some detail. A rationalization of the stability order inversion for the keto/enol pair was recently advanced: Neutral ketones are more stable than enols because of the stronger bond energies involved (i.e. C–H, C–C and C = O vs C = C, C–O and O–H). On the other hand, a lower destabilization of the core/core repulsion and a stronger stabilization of the electronic energy term (it takes much less energy to remove an electron from the π-orbital of the enol than from the n-orbital of the ketone) thermodynamically favor the enol over the keto radical cation [41]. After removal of one electron from the enol tautomer the C–O bond shortens and C–C bond elongates while the preferred conformation switches from *syn* to *anti* in the cation. The structural changes when going from ketone → ketone$^{+\cdot}$ are much less pronounced, because the electron is removed from the oxygen lone pair.

Similarly, the reversal of the thermochemical stability order upon one-electron oxidation has been demonstrated theoretically and experimentally for several heteroatom substituted carbonyl/enol pairs, e.g. esters [52, 53] and acids [54, 55]. A recent detailed evaluation of the substituent effect by Heinrich, Frenking and Schwarz using ab initio molecular orbital calculations [56] is summarized in Table 3. Both σ- and π-donors X stabilize the two cationic tautomeric forms, but with π-donating groups (X: F, OH, NH$_2$) the enol radical cations are much more stable than the corresponding keto ions. On the other hand, with σ-donor/π-withdrawing substituents this thermochemical preference is less pronounced and in the case X: BeH the order of relative stabilities of ionic keto/enol pairs is even reverted.

Despite the importance of keto/enol tautomers [57] only a small amount of work has been devoted to the study of enol radical cations in condensed phase. This is directly related to the fact that simple enols as the thermodynamically less stable tautomers [58] are usually not isolable, since the kinetic barrier for ketonization is rather low [59, 60]. Much more is known about the chemistry of enol and keto radical cations in the gas-phase [61]. For details the reader is referred to recent comprehensive reviews [62]. The only available data on the thermodynamics of enol/keto radical cations in solution stem from a recent study [63]. Using stable dimesityl substituted enols the relative stabilities were determined by a thermochemical cycle approach.

Table 3. Calculated (HF/6-31G*//3-21G) energy differences $\Delta E[\text{kJ mol}^{-1}]$ of neutral and cationic keto/enol pairs (CH$_3$COX → CH$_2$ = C(OH)X) [56]

keto/enol X	ΔE(enol → ketone)	ΔE(enol$^{+\cdot}$ → ketone$^{+\cdot}$)
F	132	− 59
OH(anti)	150	− 48
NH$_2$(planar)	139	− 88
CH$_3$	70	− 20
BH$_2$(planar)	23	− 21
BeH	93	19
H	68	− 33

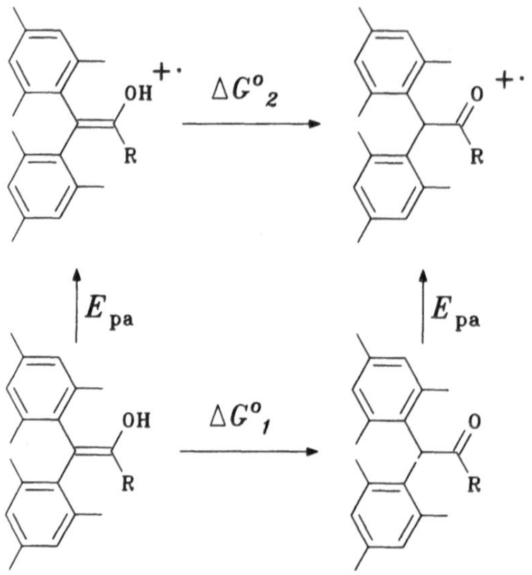

Fig 1. Thermochemical cycle used for the evaluation of the relative stabilities for enol and keto radical cations in solution [63]

1 (R: H)   2 (R: H)
3 (R: tBu)   4 (R: tBu)

Table 4. Thermochemistry of keto and enol radical cations in CH_3CN [63]

enol(E)/ ketone(K)	$E_{pa}(E)/V$	$E_{pa}(K)/V$	$\Delta G_1^0(E \rightarrow K)/$ $kJ\,mol^{-1}$	$\Delta G_2^0(E^{+\cdot} \rightarrow K^{+\cdot})/$ $kJ\,mol^{-1}$
1/2	1.00	1.98	+ 13	+ 108
3/4	0.97	1.85	− 11	+ 74

Since the resulting radical cations proved to be highly reactive with lifetimes much below 10^{-5} s in acetonitrile (results from fast scan cyclic voltammetry [64]), only irreversible oxidation potentials of the enols and ketones were obtained. Therefore, the data can only be viewed as a good estimate (Table 4). Nevertheless, in agreement with gas-phase results, it is evident that, in solution as well, enol radical cations are more stable than the corresponding ketone ions. Unfortunately, no solution data are so far available for simple aliphatic systems.

How can one use this thermochemical effect for a preparative route to enol radical cation intermediates in solution? Since one usually has to start with the ketone tautomer, two possibilities are conceivable at first from Fig. 2: (1) direct oxidation of the ketone to the ketone radical cation followed by a 1,3-hydrogen migration to provide the enol radical cation, or (2) selective one-electron oxidation of the enol that is present in the equilibrium situation under fast enolization conditions.

From several early calculations [65, 66], it soon became evident that the rearrangement of ketone to enol radical cations is a slow process with activation

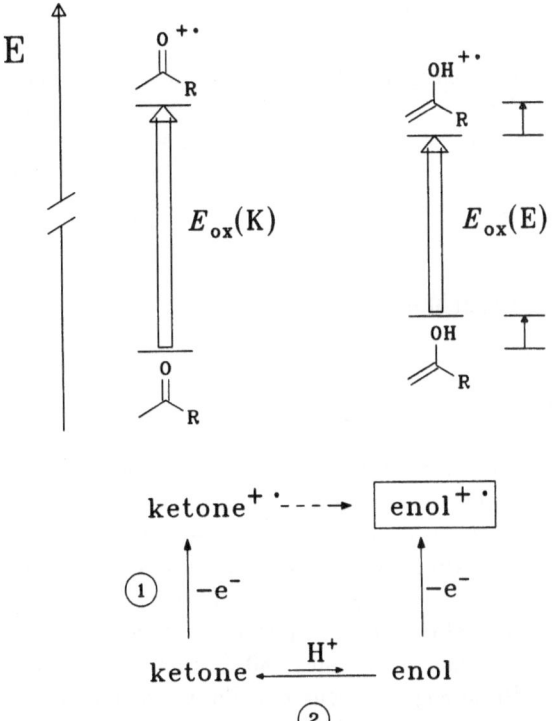

Fig. 2. Relative stabilities of neutral and ionized keto/enol pairs in solution

energies not amenable to unstable intermediates. Later ab initio calculations by Hoppilliard and Bouchoux [67] as well as by Apeloig and Schwarz [68] confirmed high barriers for rearrangement of aldehyde and keto radical cations. For example, isomerization of $CH_3CHO^{+\cdot}$ to $CH_2 = CHOH^{+\cdot}$ via direct 1,3-hydrogen migration or alternatively two successive 1,2-hydrogen shifts should proceed with a barrier in the order of $250 \, kJ \, mol^{-1}$. Similarly, using RRKM-QET calculations on an ab initio potential energy surface a barrier of $163 \, kJ \, mol^{-1}$ was obtained for the rearrangement of $CH_3COCH_3^{+\cdot}$ to $CH_2 = C(OH)CH_3^{+\cdot}$ by Lifshitz and Schwarz [69]. These high barriers definitely preclude any isomerization reactions of ketone radical cations in solution, since low energy processes like deprotonation, hydrogen abstraction and bond dissociation will prevail. On the other hand, the rearrangement of *o*-methylbenzophenone and *o*-methylacetophenone radical cations to the corresponding enol ions via a 1,5-hydrogen migration was observed in matrix at low temperatures (30–93 K) by UV and IR [70, 71]. A kinetic analysis of the enolization indicated a significant contribution of quantum-mechanical hydrogen atom tunnelling.

All together, route 2 (Fig. 2) seems to be much more promising for offering access to enol radical cations in solution. As a consequence of the inversion of the stability order in keto/enol systems upon one-electron oxidation, the oxidation potential (or the ionization potential in the gas-phase) of the enol tautomer is lower than that of the ketone, often by more than 1 V. If an appropriate one-electron oxidant is chosen, selective oxidation of the enol tautomer, even in the presence of a large excess of the ketone (as given at the equilibrium situation), is possible. However, to transform quantitatively any ketone via the enol to the enol radical cation enolization has to be faster than slow endergonic oxidation of the ketone.

3 Reactions of Enol Radical Cations

3.1 One-Electron Oxidation of Stable Enols

Whereas little is known about ketone and enol radical cations in solution, the related one-electron oxidation of phenols has been extensively studied [72]. Nowadays, anodic oxidation of phenols constitutes a valuable synthetic access to phenoxenium ions [73] which are important intermediates for carbon–carbon bond formation processes [74–76] and to various natural products [77]. In light of the biological relevance of phenol oxidation [78, 79] redox potentials of phenols [72, 80] and phenolates [80–85] as well as pK_a values of phenol radical cations [80, 86, 87] are documented in various solvents. Some of the data will be quoted later in comparison with enol systems.

In general, ketone radical cations are difficult to prepare in solution because of their high oxidation potentials [88, 89]. Thus, only few appropriate oxidizing systems are available, e.g. strong metal oxidants or anodic oxidation. Similar to photoexcited carbonyl compounds, ketone radical cations may undergo γ-hydrogen abstraction [90–92], which has some potential for remote functionalization reactions, and α-bond cleavage [93–95]. ESR-spectra of several aliphatic ketone radical cations obtained by γ-irradiation in CCl_3F matrices [96–98] showed unequivocally that the unpaired spin mainly resides in the formally non-bonding n-orbital of the $C = O$ group. Internal hydrogen transfer [96] was observed for several of the longer chain species, while for cyclobutanone radical cation the ESR results are in agreement with ring opening [99]. From thermochemical cycle calculations the acidities of carbonyl radical cations were derived, being in the order of $pK_a = -26$ to -31 for aliphatic ketones [89]. On the other hand, ketones containing aryl or other readily oxidizable groups are usually not oxidized at the carbonyl functionality. In this context, several examples of carbon–carbon bond cleavage reactions of radical cations containing the carbonyl group have been reported [100–105].

Enols are much more easily oxidized than the corresponding ketone tautomers, however, their tendency to ketonize rapidly has precluded extensive electrochemical and other one-electron oxidation studies so far. Hence, oxidation potentials were only determined of few relatively stable enols of β-dicarbonyl compounds or α-cyano ketones, where the enol content may be as

Table 5. Oxidation potentials of selected enols [63, 64, 106–108] and ketones [89, 94]. The potentials are referenced to SCE unless noted otherwise

enol	E_{ox} [V]	ketone	E_{ox} [V]
		acetone	3.06
	1.45	3-pentanone	2.96
		$(CH_3)_2CCOCH_3$	2.64
		$(CH_3)_3CCHO$	2.81
		$(CH_3)_3CCOC(CH_3)_3$	2.51
			2.71

R^1, R^2:

				2.86
C_6H_5, H	1.32 vs Ag/Ag$^+$			
C_6H_5, CH$_3$	1.28 vs Ag/Ag$^+$			
p–CH$_3$OC$_6$H$_4$, H	0.99 vs Ag/Ag$^+$			
p–CH$_3$OC$_6$H$_4$, CH$_3$	0.95 vs Ag/Ag$^+$			2.64

	R^1	R^2				2.33
1	Mes	H	1.00			
3	Mes	tBu	0.97			
5	Mes	CH$_3$	1.01			
6	CH$_3$	Mes	1.26	X: CN	3.21	
7	Mes	C$_6$H$_5$	0.94	X: H	2.82	
8	Mes	Mes	1.09	X: OMe	2.09	
9	Mes	p–C$_6$H$_4$CH$_3$	0.90			
10	Mes	p–C$_6$H$_4$OCH$_3$	0.84			

high as 100% [58] and of the aforementioned diarylsubstituted stable enols [63, 64].

Fortunately, the low enol content in simple ketone systems does not necessarily impose an obstacle to generating the corresponding enol radical cations in solution. As outlined in Sect. 2 the selective oxidation of the enol tautomer even in the presence of a vast excess of the ketone opens up an indirect, but quantitative access to enol radical cation intermediates for all systems, if an appropriate oxidant has been chosen. The first, albeit *indirect* evidence for this selective oxidation step stems from kinetic studies by Henry [109] and Littler [110–112] and will be discussed in more detail in Sect. 3.3. *Direct* evidence for a specific oxidation of enols was provided by Orliac-Le Moing and Simonet [108]. Using voltammetry at a rotating disc electrode they were able to establish a linear correlation between the anodic current and the enol content for various α-cyano ketones **11**. In electrolysis experiments the corresponding 1,4-diketones **13** were obtained in high current yield (ca. 90%).

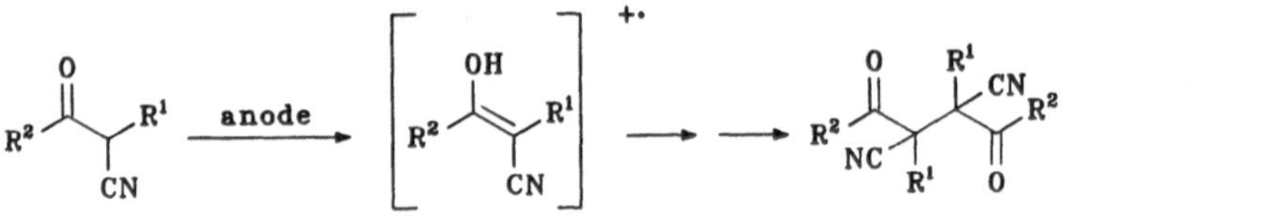

R^1, R^2: C_6H_5, H

         C_6H_5, CH_3

         $p-CH_3OC_6H_4$, H

         $p-CH_3OC_6H_4$, CH_3

         $p-CH_3OC_6H_4$, $CH_2C_6H_5$

A scheme was presented to explain formation of the diketones. Accordingly, two reactions of enol radical cations are plausible: (1) deprotonation and (2) reaction with a nucleophile. The role of water as base or nucleophile in the proposed ECC process was tested, but no decisive mechanistic results were obtained to rigorously differentiate between the alternative pathways (Scheme 1). The potential involvement of α-carbonyl cations (path 3) was not considered.

That deprotonation of enol radical cations is a plausible reaction can be deduced from Yoshida's work [107, 113] on the electrooxidative addition of 1,3-dicarbonyl compounds 14 in acetonitrile to substituted olefins yielding dihydro-furans 18 as [3 + 2] cycloaddition products in good to excellent yields. Al-

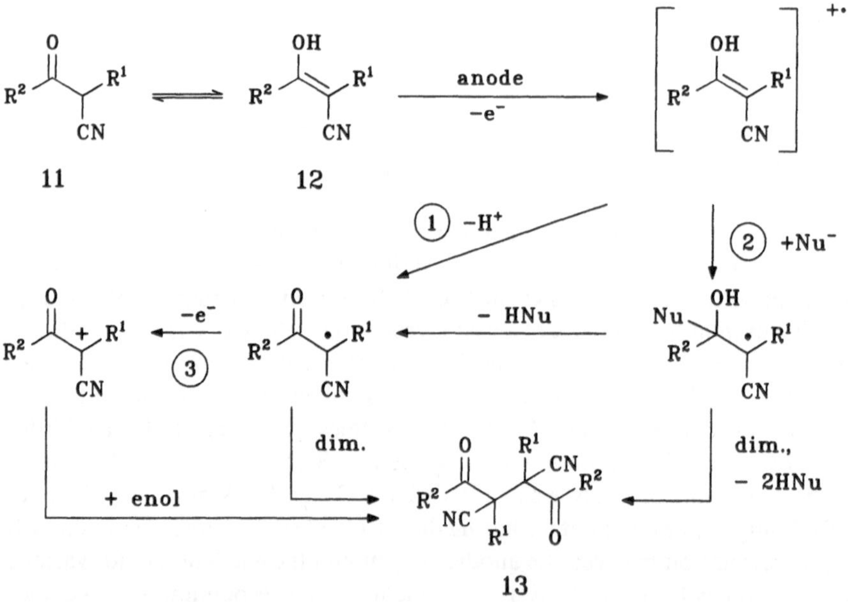

Scheme 1.

though the authors did not specifically probe the mechanism, it is likely that only the enol tautomer was oxidized. After deprotonation of the enol radical cation the resulting α-carbonyl radical 16 adds in a highly regioselective way to the olefinic component, thus providing after a further one-electron oxidation step and cyclization the observed products. The overall reaction is closely related to anodic oxidation reactions of enolates [114] and several oxidative radical reactions mediated by metal salts such as Co(III) [37] and Mn(III) [38] in the presence of olefins.

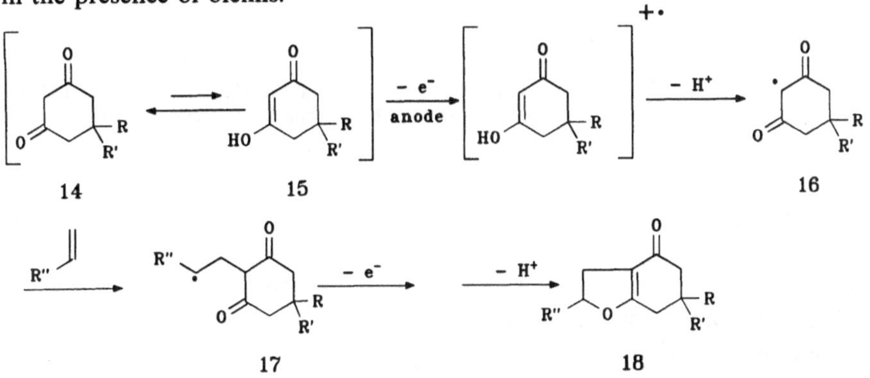

R,R': H,H; CH₃, CH₃; and related diketones
R'': Ph, CH₂SiMe₃, CO₂Et, and other olefins

R,R': H,H; CH_3, CH_3; and related diketones
R'': Ph, CH_2SiMe_3, CO_2Et, and other olefins

As a general problem, early studies were always obscured in terms of interpretation by the simultaneous presence of both enol and keto tautomers. Only recently, a direct study of the chemistry of enol radical cations was undertaken using stable simple enols of the Fuson type [63, 64]. Extensive work originally by Fuson [115–117] and later by Rappoport [118] showed for a series of sterically hindered, mesityl-substituted systems that both the kinetically stabilized enols and the corresponding ketones can be prepared in tautomerically pure form.

When the enols 1, 3, 5–10 were reacted with 2 equivalents of well known one-electron oxidants, e.g. triarylaminium salts, iron(III)phenanthroline (FePHEN)

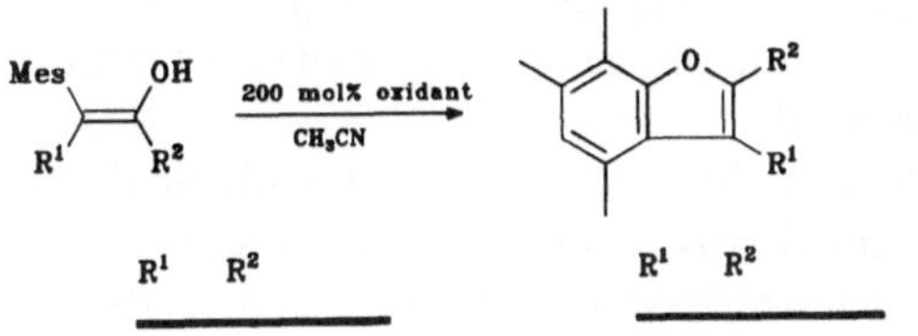

	R¹	R²		R¹	R²
1	Mes	H	19	Mes	H
3	Mes	tBu	20	Mes	tBu
5	Mes	CH₃	21	Mes	CH₃
6	CH₃	Mes	22	CH₃	Mes
7	Mes	C₆H₅	23	Mes	C₆H₅
8	Mes	Mes	24	Mes	Mes
9	Mes	p–CH₃C₆H₄	25	Mes	p–CH₃C₆H₄
10	Mes	p–CH₃OC₆H₄	26	Mes	p–CH₃OC₆H₄

Michael Schmittel

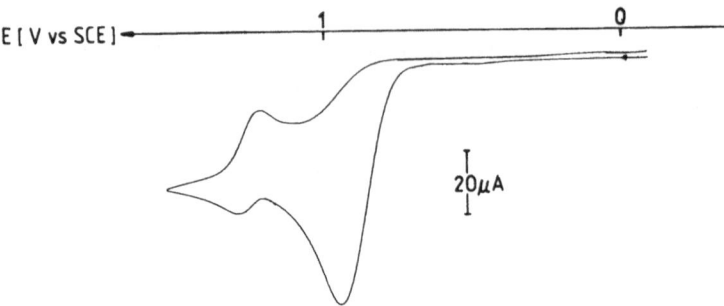

Fig. 3. Cyclic voltammogram of **3** in acetonitrile using a scan rate of 100 mV s^{-1} [63]. The first anodic wave corresponds to the irreversible oxidation of **3**, the second to the reversible oxidation of benzofuran **20**. *Copyright* 1990 VCH

or thianthrenium perchlorate (**TH$^{+\cdot}$**), within seconds the color of the oxidant was discharged and the benzofurans **19–26** were obtained in good yields [64, 119]. Significantly, the same products were formed in cyclic voltammetry studies (Fig. 3) and under preparative electrooxidation conditions, thus proposing enol radical cations as plausible intermediates in this reaction [119].

From a mechanistic point of view two possible hypotheses were discussed (Scheme 2). Since radical cations are intrinsically very acidic [29–32] one would expect enol radical cation **27$^{+\cdot}$** to deprotonate efficiently and rapidly thus providing an α-carbonyl radical (mechanism 1). In a further one-electron oxidation step the α-carbonyl cation is formed, that cyclizes to an intermediate cyclohexadienyl cation **28**. After a 1,2-methyl shift and deprotonation the benzofuran **29** is obtained. The mechanistic proposal is in line with benzofuran formation from α-carbonyl cations as demonstrated by Okamoto [120]. Interestingly, the above mechanism was first proposed by Bailey to explain the formation of 3% of benzofuran **24** in the ozonization of enol **8** [121]. Years later, however, the mechanistic hypothesis was proven to be untenable [122] under ozonization conditions. A priori, it cannot be excluded that intramolecular cyclization of the enol radical cation **27$^{+\cdot}$** is faster than deprotonation (mechanism 2). The distonic radical cation formed is expected to lose a proton readily and after a second one-electron oxidation the same cyclohexadienyl cation intermediate as in mechanism 1 is formed.

Even in the presence of nucleophiles like methanol, these sterically hindered enols only afforded benzofurans under oxidative conditions. Importantly, when stable enols of the Fuson type, which are less sterically hindered in the β-position, were oxidized with one-electron oxidants α-substituted carbonyl compounds **31** and **34** were obtained with water and various alcohols [64]. With acetonitrile, formation of the oxazole **32** was observed. A related example of this chemistry can be found in the context of Fuson's oxidation studies using Pb(OAc)$_4$ [123] that provided the α-acetoxy aldehyde **31c** from **30**. To explain the products **31** and **34**, three different mechanistic hypotheses were advanced (Scheme 3) [64] similar to the ones rationalizing the benzofuran formation.

194

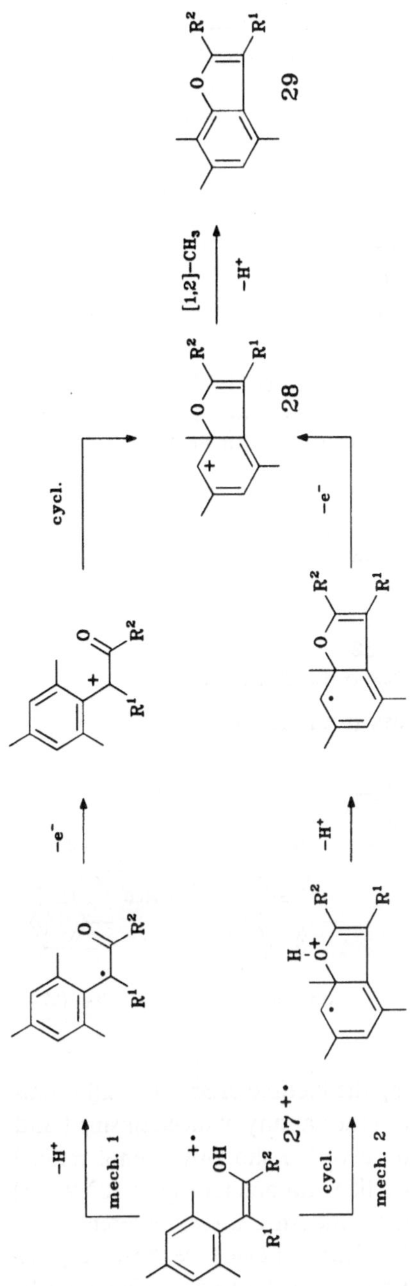

Scheme 2.

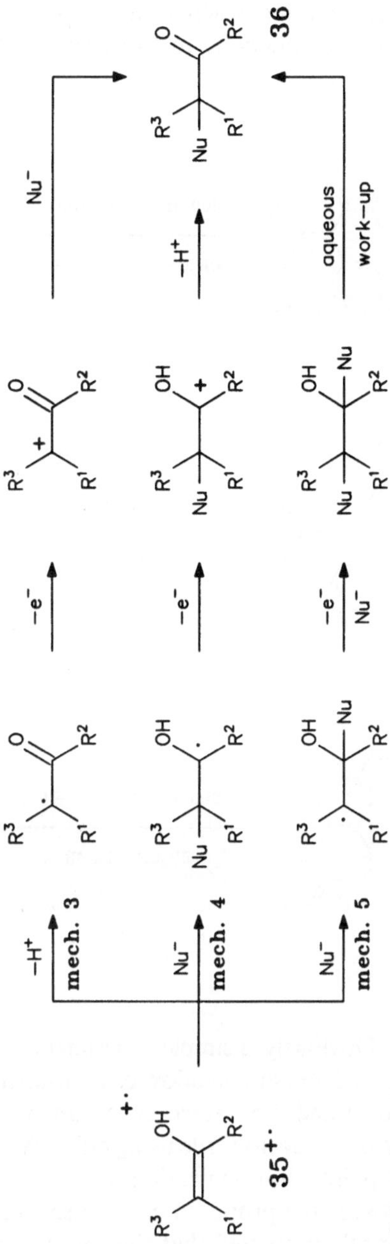

Scheme 3.

195

Thus, according to mechanism 3 the primary reaction of an enol radical cation is deprotonation, whereas mechanisms 4 and 5 postulate addition of a nucleophile at the double bond in either the β-or α-position.

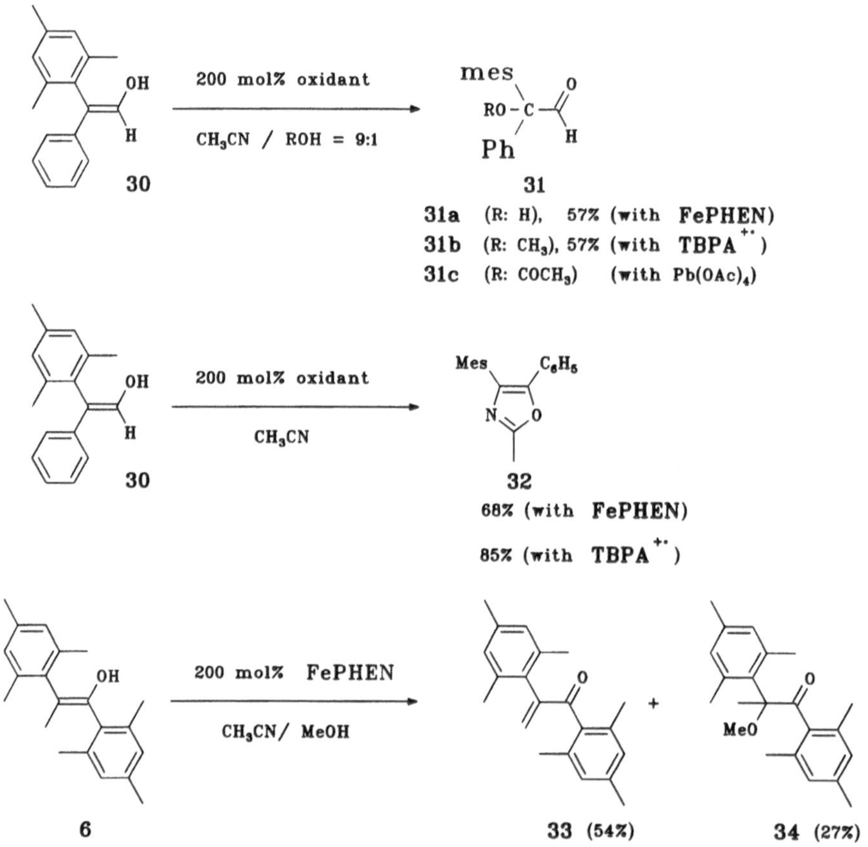
OH
H
30
200 mol% oxidant
CH₃CN / ROH = 9:1
mes
RO-C
O
H
Ph
31
31a (R: H), 57% (with FePHEN)
31b (R: CH₃), 57% (with TBPA⁺˙)
31c (R: COCH₃) (with Pb(OAc)₄)
OH
H
30
200 mol% oxidant
CH₃CN
Mes
C₆H₅
N
O
32
68% (with FePHEN)
85% (with TBPA⁺˙)
OH
6
200 mol% FePHEN
CH₃CN/ MeOH
O
+
MeO
O
33 (54%)
34 (27%)

Obviously, a simple experiment comparing the lifetime of $\mathbf{35}^{+\bullet}$ (O–D) versus $\mathbf{35}^{+\bullet}$ (O–H) should allow us to discern between the validity of mechanisms 1 and 2 or 3 and 4/5. However, all attempts to observe the short-lived enol radical cations in acetonitrile using either fast scan cyclic voltammetry (up to 10^5 V s^{-1}) or pulse photoionization have failed so far, rendering *direct* differentiation between the proposed mechanisms impossible. Indirect evidence, however, was recently presented that rigorously excluded pathways 4 and 5 as viable mechanisms [64]. Whereas enol radical cations $\mathbf{30}^{+\bullet}$ and $\mathbf{6}^{+\bullet}$ are not stable even on the μs time scale, the corresponding enol ether radical cations $\mathbf{37}^{+\bullet}$ and $\mathbf{38}^{+\bullet}$ exhibit lifetimes in the order of seconds. Thus, replacement of OH versus OMe prolongs the lifetime of the radical cations by at least five orders of magnitude, suggesting that the primary reaction of enol radical cations is loss of one proton.

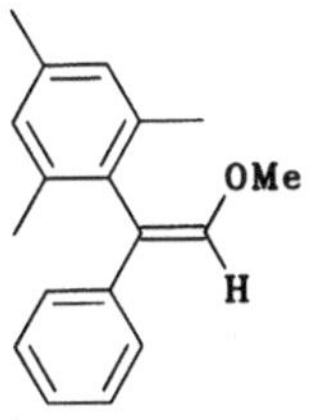

37 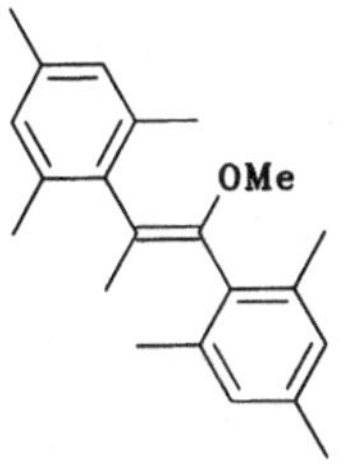   38

In a kinetic study, it was demonstrated that enol ether radical cations react very sluggishly with nucleophiles like acetonitrile ($k = 7\,M^{-1}s^{-1}$ (37$^{+\cdot}$), $k < 0.1\,s^{-1}$ (38$^{+\cdot}$)) or methanol ($k < 700\,M^{-1}s^{-1}$ (37$^{+\cdot}$), $k < 300\,M^{-1}s^{-1}$ (38)$^{+\cdot}$)). If we accept the plausible assumption that enol ether radical cations are good models for enol radical cations in their reaction with nucleophiles, the short lifetime of the enol radical cations can only be rationalized by assuming a fast deprotonation step.

The proposed proton loss from enol radical cations is in line with the high acidity that was estimated via a thermochemical cycle in acetonitrile [64]. Since these pK_a values depend significantly on the oxidation potential of the enols [32] comparison to phenols should be restricted to the ones with similar potentials. According to Bordwell [80] the pK_a of the hydroquinone radical cation is -5.5 and that of the 3-aminophenol radical cation 3.0, both in DMSO.

The following general mechanism for benzofuran formation and α-substitution was proposed for the mesityl substituted enols [64]. Indeed, this mechanistic scheme parallels the one for phenol oxidation, where phenoxyl radicals and

Table 6. pK_a data of various enol radical cations in acetonitrile [64]

enol$^{+\cdot}$	pK_a (in CH$_3$CN)
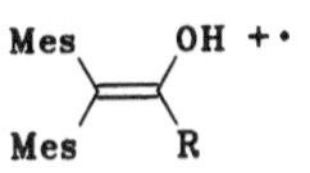 Mes — OH +· / Mes — R	6.2 (R : tBu) 5.7 (R : H) 5.6 (R : CH$_3$)
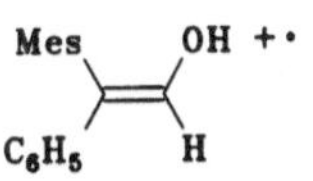 Mes — OH +· / C$_6$H$_5$ — H	4.4
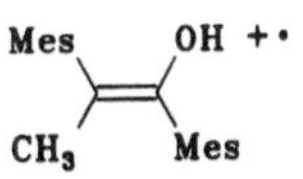 Mes — OH +· / CH$_3$ — Mes	1.3 Mes:

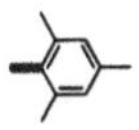

Michael Schmittel

phenoxenium ions are invoked [124]. Analogously, intramolecular cyclizations [74] and reactions with nucleophiles [73, 77, 125] have been observed during phenol oxidation reactions.

Since Scheme 4 implies formation of α-carbonyl radicals after deprotonation of enol radical cations, the same oxidation chemistry should potentially be accessible from various enol derivatives as enolates, silyl enol ethers and enol esters (Scheme 5). On the other hand, enol ether radical cations do not fit in this systematization since they are attacked by nucleophiles at the double bond faster than providing α-carbonyl radical intermediates through O–C bond cleavage (Sect. 4.3).

A rational approach to either α-carbonyl radical or cation chemistry from enol oxidation thus necessitates explicit knowledge of the oxidation potentials of

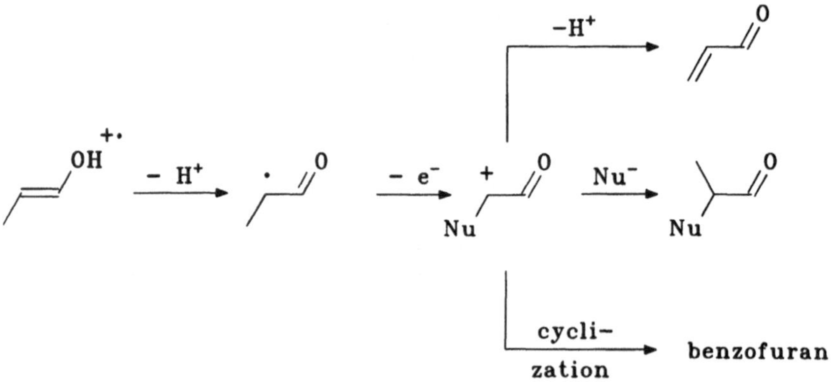

Scheme 4.

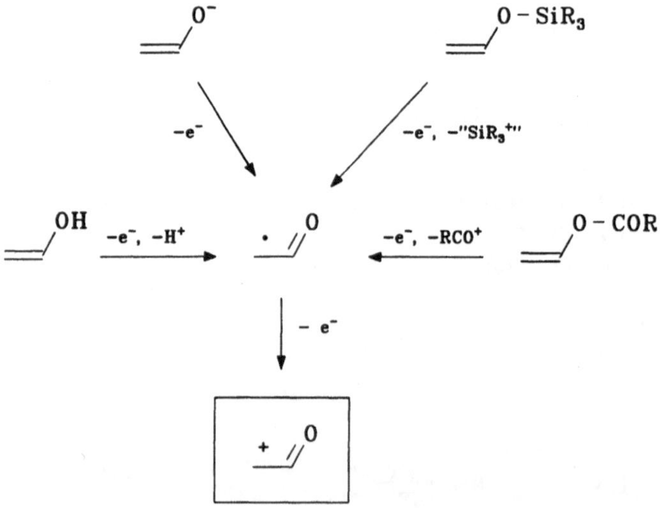

Scheme 5.

198

the enol precursors and the intermediate radicals. Under strongly oxidative conditions, when one-electron oxidation of α-carbonyl radicals is fast, it should be possible to suppress other radical reactions to a great extent and to trigger carbocation chemistry. According to Scheme 5 this approach is not restricted to enols and hence, in the following, reactions of other enol systems will be analyzed under oxidative conditions.

A novel pathway to enol radical cations in solution through protonation of β-hydroxy vinyl radicals **39** has recently been postulated by Gilbert [126]. Using ESR techniques it was demonstrated that β-hydroxyl vinyl radicals rearrange to α-carbonyl radicals **41** when the pH was lowered, and enol radical cations were proposed as intermediates.

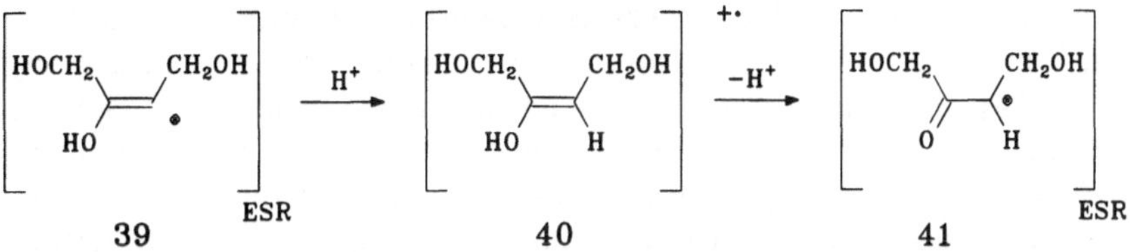

Enol radical cation intermediates have recently been invoked in the ribonucleotide reductase process. According to a hypothesis by Stubbe [127], they are formed through water loss from the 3'-ribonucleotide radical. They are supposed to react subsequently with "H$^-$" (or alternatively via a very unlikely two-electron reduction followed by protonation) to an intermediate 3'-hydroxy radical that is finally transformed to the deoxyribonucleotide. The above mechanistic evidence on simple enol radical cation chemistry, however, argues against this mechanistic model, since deprotonation should be much faster than nucleophilic attack even under physiological conditions.

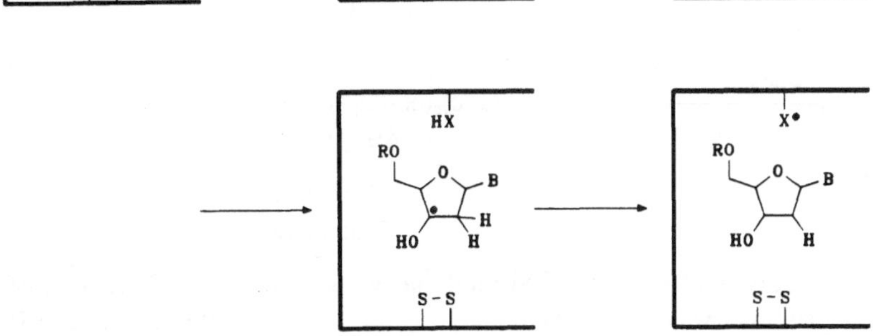

3.2 Oxidation of Enolates to α-Carbonyl Radicals and α-Carbonyl Cations

Surprisingly little preparative work has been done on the one-electron oxidation of enolates [9], although the resulting α-carbonyl radicals are important intermediates in synthetic organic chemistry [128–130] and biological systems [131, 132]. This is certainly related to the fact that enolate oxidation under anodic conditions leads to synthetically less interesting symmetrical dimers [133–135]. In addition, α-carbonyl radicals are readily obtained from various precursors, such as α-haloketones [128, 136], by radical addition to α,β-unsaturated ketones [129] and by oxidative radical formation from 1,3-dicarbonyl compounds [137–141], ketones and aldehydes using Mn(III) [38]. The preparative value of enolate oxidation, however, could be greatly improved, if they not only served as stable precursors to α-carbonyl radicals but also to α-carbonyl cations (after transfer of two electrons), a point already emphasized by Schäfer [114]. In his pioneering studies on the anodic oxidation of enolates [114, 142, 143] in the presence of unsaturated compounds the isolated products such as **42** and **43** are clearly derived from intermediate α-carbonyl radicals **45**. No evidence was obtained that intermediate α-carbonyl cations were formed under anodic oxidation conditions from sodium dimethyl malonate or sodium acetylacetonate. Products, like **47–49** most likely originate from oxidation of the intermediate alkyl radical **46**, which depending on the nature of Y (alkyl, OEt, vinyl or aryl) has a much lower oxidation potential than the enolates (E_{ox} between 0.6 and 1.4 V) themselves.

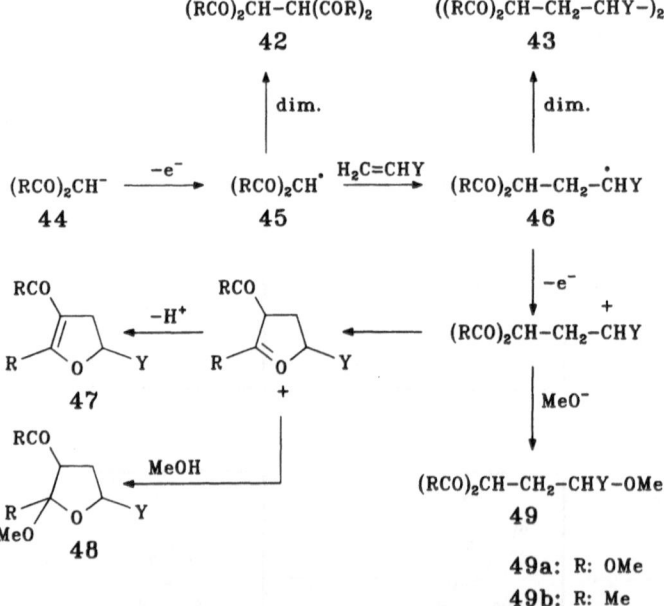

In the electrolysis of sodium ethyl methylacetoacetate (**50**) in DMF one of three main products was 4-ethoxycarbonyl-3,4-dimethylcyclohex-2-enone (**53**)

[144]. It arises from **54** by cyclization and retro-Claisen loss of an ethoxycarbonyl group. Brettle suggested that **54** was formed from radical disproportionation followed by Michael addition of the carbanion to the resulting ethyl methyleneacetoacetate. Alternatively, **52** could likewise be derived from α-carbonyl cation intermediates formed by one-electron oxidation of **51**.

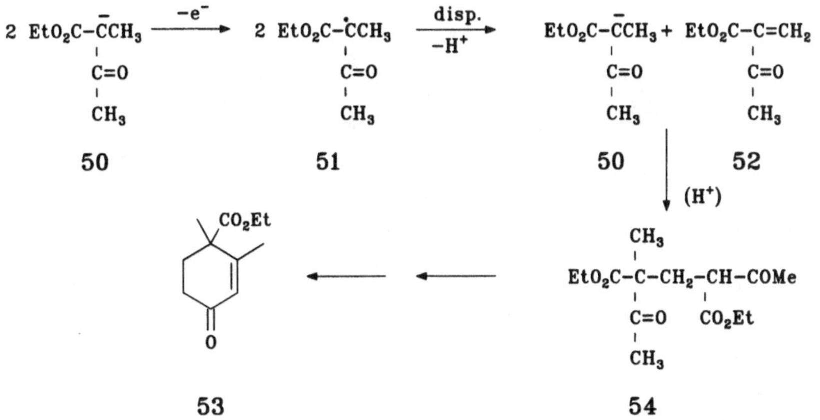

A useful preparative application of enolate oxidation was presented by Torii in the context of a facile synthesis of 4-hydroxyindole **59** [145]. Similar to Schäfer's work [114], the anion of 1,3-cyclohexadione was added anodically to ethyl vinyl ether providing products **56** and **57** in 65%. The mixture of both can be transformed by reaction with $(NH_4)_2CO_3$ in methanol into **58** that is finally converted to **59**.

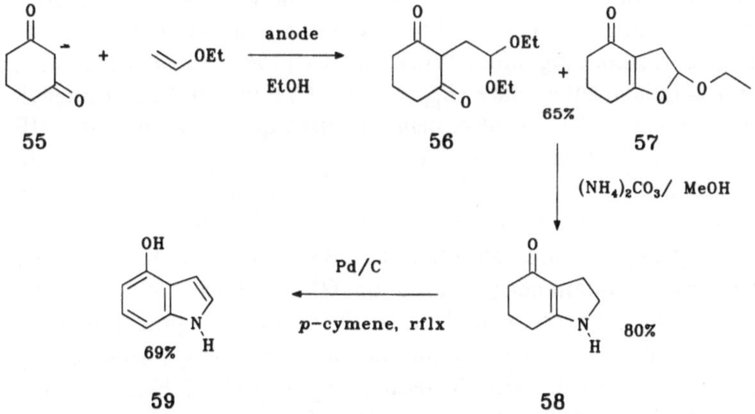

Pandey has tentatively proposed the involvement of enolates in the photo-induced electron transfer oxidation of arylacetones by the system 1,4-dicyanonaphthalene (8 mol%)/acetonitrile/aqueous NaOH [146]. The corresponding benzofurans were obtained in 50–60% yield, but the mechanism is speculative since no oxidant in stoichiometric amounts was present. It was assumed that cyclization takes place on the stage of the α-carbonyl radical.

Recent work has now provided a comprehensive view on enolate oxidation since the controlled generation of either α-carbonyl radicals or α-carbonyl cations was possible depending on the conditions [147]. In line with the prediction made in Scheme 5 enolates **60–63** could be oxidized by 2 eq. of FePHEN ($E_{1/2} = 1.09$ V) to the corresponding benzofurans **19–21, 24**.

$$\text{Mes} \diagdown \overset{O^-}{\underset{R^1 \quad R^2}{\diagup}} \quad \xrightarrow[\text{CH}_3\text{CN}]{200 \text{ mol\% FePHEN}} \quad \text{benzofuran}$$

	R^1	R^2			R^1	R^2
60	Mes	H		19	Mes	H
61	Mes	tBu		20	Mes	tBu
62	Mes	CH$_3$		21	Mes	CH$_3$
63	Mes	Mes		24	Mes	Mes

Cyclic voltammetry investigations revealed that the enolates were oxidized reversibly, and the resulting α-carbonyl radicals proved to be stable at least for hours. Using a weaker one-electron oxidant, the **TMPA**$^{+\cdot}$ SbCl$_6$ ($E_{1/2}$ = 0.52 V), the radicals were formed quantitatively and characterized by magnetic susceptibility measurements, ESR and ENDOR [148].

Despite the dearth of knowledge on one-electron oxidation chemistry of enolates, extensive compilations of data on their oxidation potentials are available from the work of Bordwell [89, 149] and Federlin [150–153], wherein a correlation was established between the oxidation potentials of enolates and the pK_a's of the corresponding enols. Work by Fox [154] provided oxidation potentials of 1,2-diphenylethenolate (E_{pa} = -0.40 V) and of 1,2,2-triphenyl-ethenolate ($E_{1/2}$ = -0.49 V), the latter being oxidized quasi-reversibly in THF at 500 mV s^{-1} at ambient temperature. These oxidation potentials have significant mechanistic meaning since single electron transfer has been invoked in various reactions of enolates, e.g. aldol condensation [155, 156].

Thermodynamically relevant oxidation potentials of enolates were recently obtained from cyclic voltammetry studies on **60–63**. Since the α-carbonyl radicals proved to be sufficiently stable, also their oxidation potentials were determined. They are much higher than the ones from the corresponding enolates and agree qualitatively with the reduction potentials of three related α-carbonyl cations as determined by Okamoto [157, 158]. Thus, depending on the oxidation power of the used oxidant either α-carbonyl radical or α-carbonyl cation chemistry can be triggered from enolates as demonstrated above.

Certainly, it is not unexpected to see the oxidation potentials of the enolates to be much lower than the ones of the parent enols. This behavior has been observed for numerous RH/R$^-$ pairs [8, 9]. However, it is important to stress that the enolate oxidation potentials are lower than the ones of the correspond-

Table 7. Irreversible oxidation potentials of enolates [89, 151]

enolate	E_{ox} [V]	enolate	E_{ox} [V]
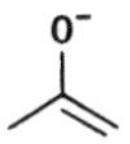 (structure)	−0.16	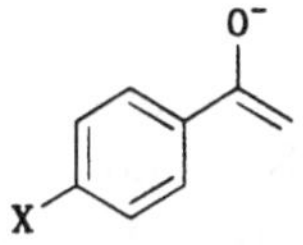 (structure)	
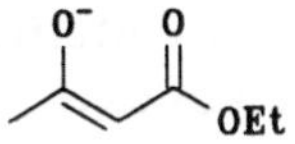 (structure, OEt)	+0.71	X: CN	+0.07
		X: H	−0.10
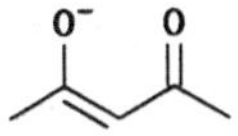 (structure)	+0.82	X: OMe	−0.15
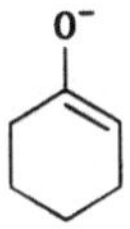 (structure)	−0.43	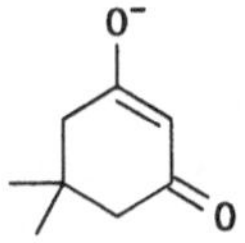 (structure)	+0.47
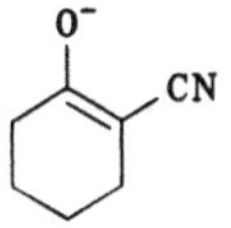 (structure, CN)	+0.10	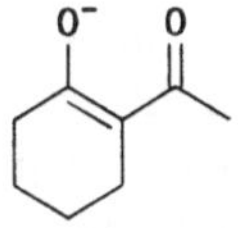 (structure)	+0.36

Table 8. Oxidation potentials of enolates, α-carbonyl radicals, enols and ketones in comparison [147]

R	$E_{1/2}$/V $Mes_2C = C(O^-)R$	E_{pa}/V $Mes_2C = C(O\,)R$	E_{pa}/V $Mes_2C = C(OH)R$	E_{pa}/V $Mes_2CH-C(=O)R$
Mes	− 0.42	+ 0.59	+ 1.09	nk[a]
t-Bu	− 0.68	+ 0.48	+ 0.97	+ 1.85
Me	− 0.57	+ 0.54	+ 1.01	nk[a]
H	− 0.40	+ 0.69	+ 1.00	+ 1.98

[a] not known

ing ketones by more than 2 V [89, 147]. Thus, in basic media the formal oxidation of ketones will most likely proceed by selective oxidation of the enolate.

The above oxidation potential order for enolates and α-carbonyl radicals parallels that of related phenolates and phenoxyl radicals [124]. Recently, reversible oxidation potentials for phenolates have been determined in either organic or aqueous media [80–85], whereas oxidation potentials of phenoxyl radicals are as scarce as for α-carbonyl radicals [124].

3.3 α-Umpolung of Ketones via Enol Radical Cations

In spite of the fact that numerous oxidation reactions are known, that lead to α-functionalization of ketones [159, 160], in most cases enol radical cations are not involved in these transformations, and rigorous evidence for their formation through selective oxidation of the enol tautomer (Fig. 2, path 2) has only been obtained in a few cases. For example, it could be inferred from kinetic studies that in many cases enols are not intermediates in aqueous oxidation reactions with V(V), Co(III), Ce(IV) and Mn(III) [161–163], whereas in acetic acid Mn(III) was postulated to attack the enol form of ketones [164, 165], but not by electron transfer [166]. On the other hand, oxidants as Cr(VI), Tl(III), Hg(II) and Mn(VII) [167] as well as Pb(IV) [168] definitely react with the enol form, but since with these inner-sphere oxidants electron transfer is assumed to occur in a bonded fashion, radical cation intermediates are most likely not implicated.

Hence, the first clearcut evidence for the involvement of enol radical cations in ketone oxidation reactions was provided by Henry [109] and Littler [110, 112]. From kinetic results and product studies it was concluded that in the oxidation of cyclohexanone using the outer-sphere one-electron oxidants, tris-substituted 2,2'-bipyridyl or 1,10-phenanthroline complexes of iron(III) and ruthenium(III) or sodium hexachloroiridate(IV) (**IrCl**), the cyclohexenol radical cation ($65^{+\cdot}$) is formed, which rapidly deprotonates to the α-carbonyl radical **66**. An upper limit for the deuterium isotope effect in the oxidation step ($k_H/k_D < 2$) suggests that electron transfer from the enol to the metal complex occurs prior to the loss of the proton [109]. In the reaction with the ruthenium(III) salt, four main products were formed: 2-hydroxycyclohexanone (**67**), cyclohexenone, cyclopentanecarboxylic acid and 1,2-cyclohexanedione, whereas oxidation with **IrCl** afforded 2-chlorocyclohexanone in almost quantitative yield. Similarly, enol radical cations can be invoked in the oxidation reactions of aliphatic ketones with the substitution inert dodecatungstocobaltate(III), $CoW_{12}O_{40}^{5-}$ complex [169]. Unfortunately, these results have never been linked to the general concept of inversion of stability order of enol/ketone systems (Sect. 2) and thus have never received wide attention.

64   65   65$^{+\cdot}$ 66   67
+ other products

In 1990, two papers from two independent groups were published, addressing the issue of the α-Umpolung of ketones via enol radical cations. In a mechanistic study the α-methoxylation of anisylsubstituted ketones was reported in the presence of typical one-electron oxidants [170]. Other alcohols [171] and acetonitrile can equally well be used as nucleophiles. For quantitative conversion two equivalents of oxidant are needed and the oxidation can easily be followed by the color discharge of the blue aminium salts.

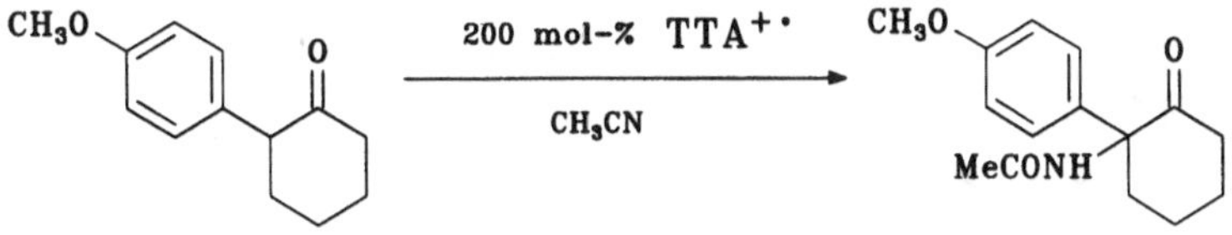

Table 9. Yields of the reaction of **72** with various nucleophiles ROH in the presence of 2 equivalents of one-electron oxidants TBPA$^{+ \cdot}$ or TTA$^{+ \cdot}$ [170, 171]

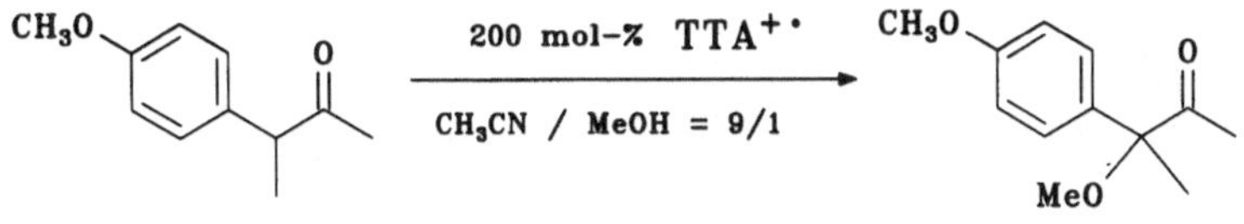

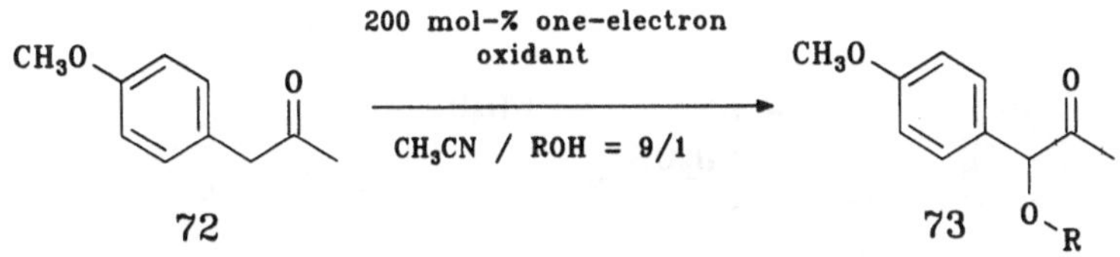

R	yield of 73 (%) with	
	TBPA$^{+ \cdot}$	TTA$^{+ \cdot}$
Me	60%	77%
Et	62%	76%
i-Pr	57%	68%
sec-Bu	n.d.[a]	63%
t-Bu	15%	n.d.[a]
neo-Pent	78%	52%
COMe	35%	n.d.[a]
Me[b]	8%[b]	n.d.[a]

[a] not determined. [b] In the presence of 2 equivalents of di-tert-butyl pyridine

The stoichiometry indicates that two protons are liberated per product molecule, and thus highly acidic conditions are finally produced during the conversion. Evidence was presented that mechanism 1 is operative in the α-methoxylation reaction and mechanism 2 (Scheme 6) was ruled out on the basis of a base criterion [170].

205

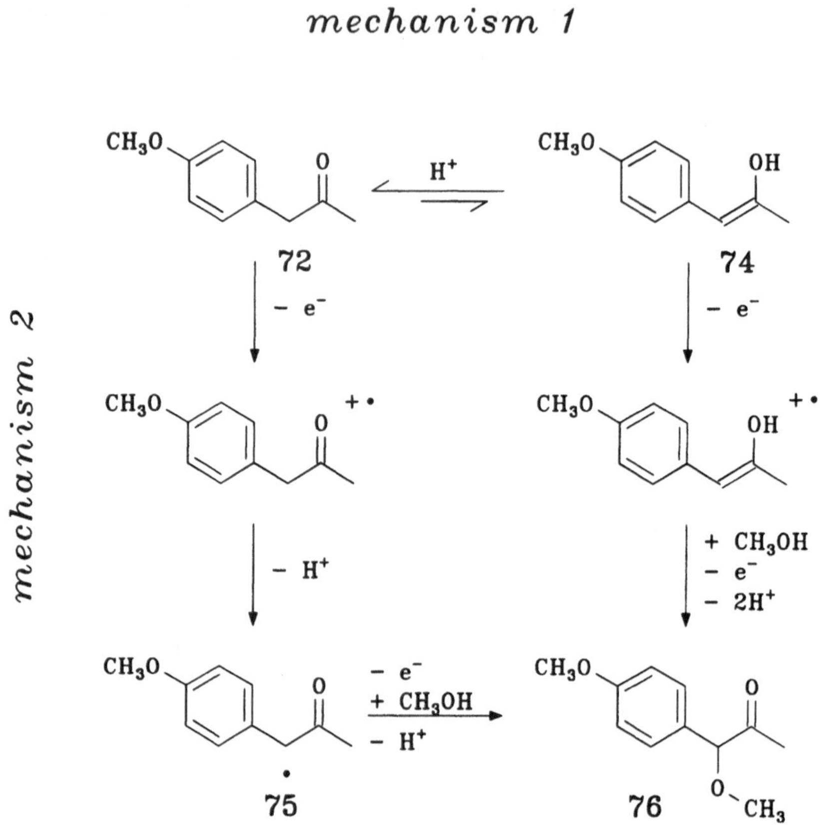

mechanism 1

mechanism 2

Scheme 6.

Table 10. Product distribution in the reaction of **70** with various aminium salts of different oxidation strength in acetonitrile/methanol 9:1 [170]

aminium-salt	$E_{1/2}$/ V	time	71 (%)	77 (%)	78 (%)	70 (%)
TTA$^{+\cdot}$	0.76	33 h	78	< 1	< 1	5
TBPA$^{+\cdot}$	1.06	153 min	30	10	6	29
TDBPA$^{+\cdot}$	1.50	25 s	< 1	7	6	41

For example, addition of 2 equivalents of a nonnucleophilic base, 2,6-di-*tert*-butylpyridine, to the reaction system almost totally suppressed the reaction (Table 9). This proposes that acid catalyzed enolization is important for the α-Umpolung. In support of mechanism 1 the yields of the α-methoxylated products increased with decreasing oxidation strength of the oxidant (Table 10), since endergonic oxidation of the ketone **70** is slowed down (Fig. 2). On the other hand, in the presence of stronger aminium salts the ketone tautomer is oxidized which however does not lead to deprotonation in the benzylic position as outlined in mechanism 2 (Scheme 6) but to carbon–carbon bond cleavage (Scheme 7).

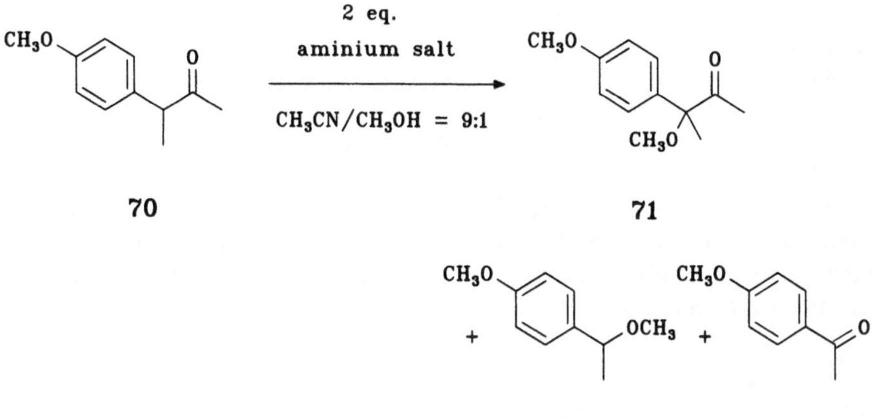

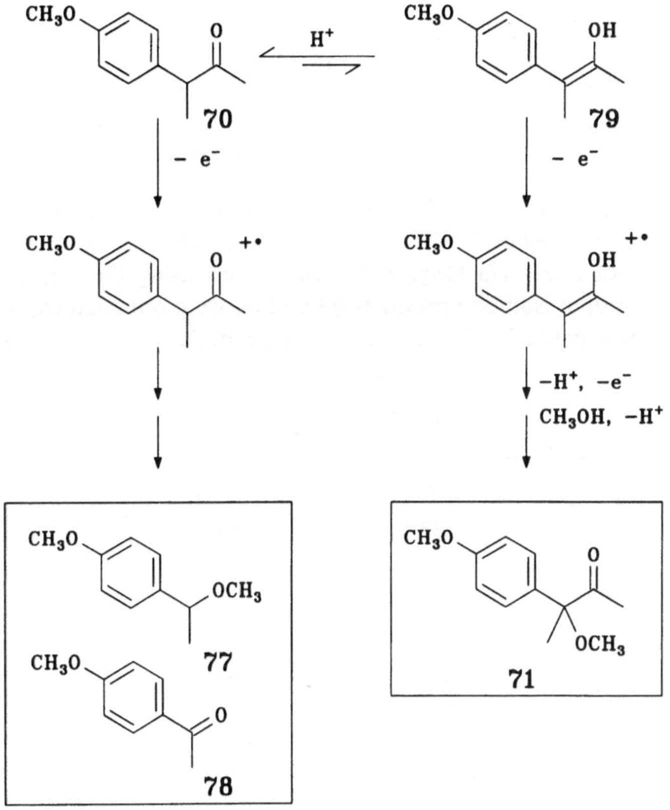

Scheme 7.

Later, it was unambiguously demonstrated in a kinetic study that enoliz-ation is the rate-determining step in the overall reaction of ketone **72** with **FePHEN** [171].

In a paper emphasizing the preparative value, Schulz and Kluge described the α-hydroxylation of ketones in good yields by using 2 equivalents of triarylaminium salts in moist acetonitrile [172]. In contrast to the oxidative functionalization of **68**, **70** and **72** the reaction with cyclohexanone (**64**) and methyl isopropyl ketone (**80**) was run in the presence of a hindered pyridine base. Thus, mechanistically, it cannot rigorously be stated whether enols or the enolates are oxidized (cf Sect. 3.2).

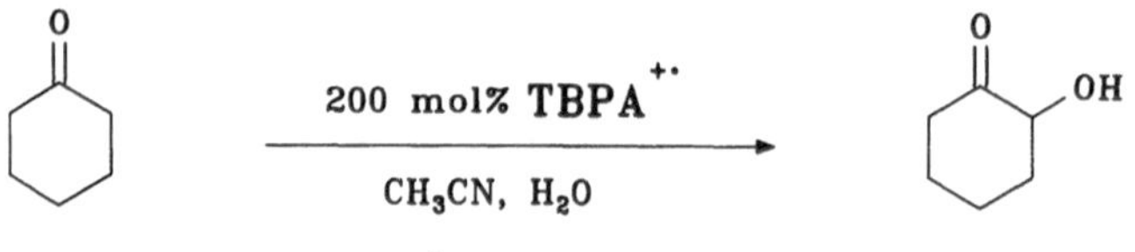

O

64

200 mol% TBPA$^{+\cdot}$

CH$_3$CN, H$_2$O

base

O   OH

67 (71 – 91 %)

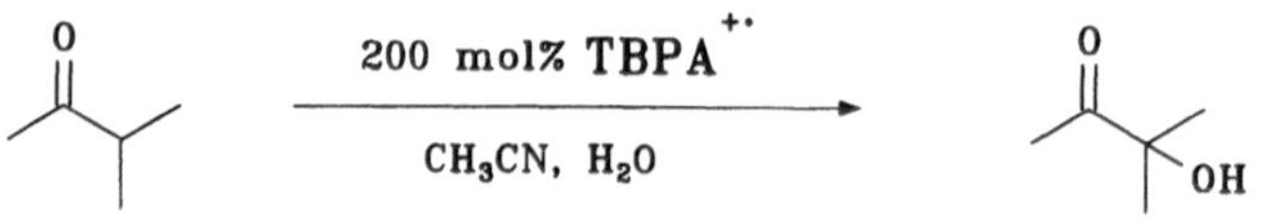

O

80

200 mol% TBPA$^{+\cdot}$

CH$_3$CN, H$_2$O

base

O

OH

81 (51 – 70 %)

In a comparative study the authors demonstrated that aminium salt oxida-tion nicely complements the regioselectivity of other α-hydroxylation methods [173–180], which do not occur via electron transfer. Thus, using the Vedejs reagent [178] two positional isomeric products **83** and **84** were obtained from **82**, whereas aminium salt oxidation leads to only one product.

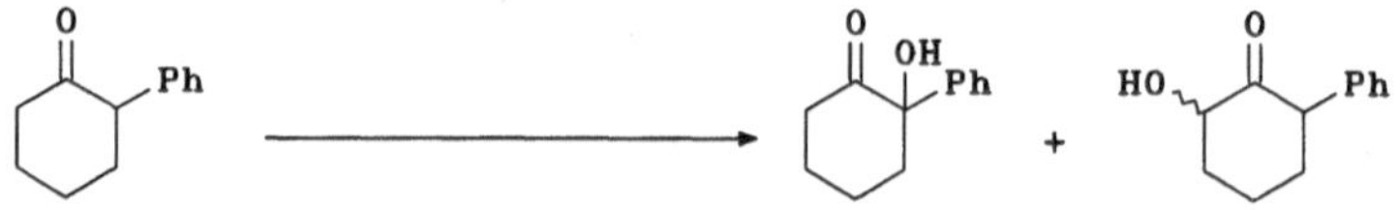

O   Ph

82

O OH Ph

83

+

HO O Ph

84

Schulz 1990

200 mol% TBPA$^{+\cdot}$

Li$_2$CO$_3$, CH$_3$CN, H$_2$O   45%   --

K$_2$CO$_3$, CH$_3$CN, H$_2$O   55%   --

Vedejs 1978

a) MoO$_5$·Py·HMPT, LDA   --   70%

b) MoO$_5$·Py·HMPT, KH   30%   24%

A mechanistic study of the reaction under Schulz conditions later proved that α-bromoketones are intermediates in the reaction which during work-up are transformed to the α-hydroxy ketones [181]. For example, cyclohexanone (64) reacted with 2 equivalents of TBPA$^{+ \cdot}$ to yield 70% of 2-bromocyclohexanone (85). At present it is not clear, whether this reaction is of the electron transfer type, since TBPA$^{+ \cdot}$ is known to dimerize via the ipso position, thus acting as a source of "positive bromine" [182].

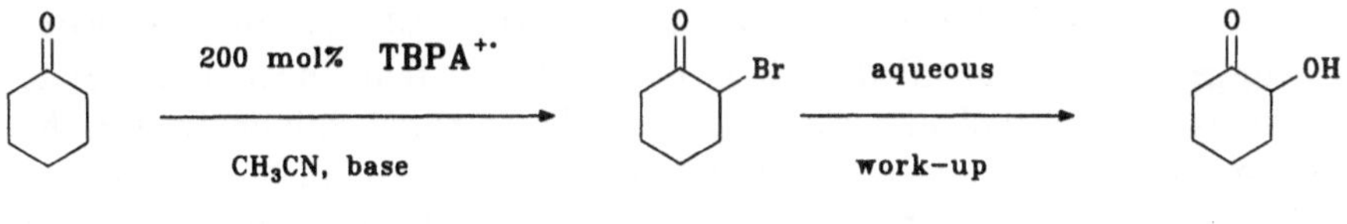

64   **85**   **67**

Consequently, in further oxidation studies of **64** a well defined outer-sphere electron oxidant, i.e. **FePHEN**, was used affording the cyclohexenone as main product indicative of an α-carbonyl cation intermediate [181].

In line with the mechanistic results from above and from the oxidation of stable enols (Sect. 3.1), the following mechanism was postulated (Scheme 8). Considering the multistep nature of the mechanism it is obvious that several, sometimes conflicting requirements have to be accomodated in the reaction system to maximize yields. For example, the redox potential of the oxidant has to be low in order to avoid direct oxidation of the carbonyl compound, but should still allow oxidation of the enol tautomer. In this case the chosen oxidant may turn out to be too weak for fast oxidation of the α-carbonyl radical intermediate, and as a consequence only radical reactions will be triggered.

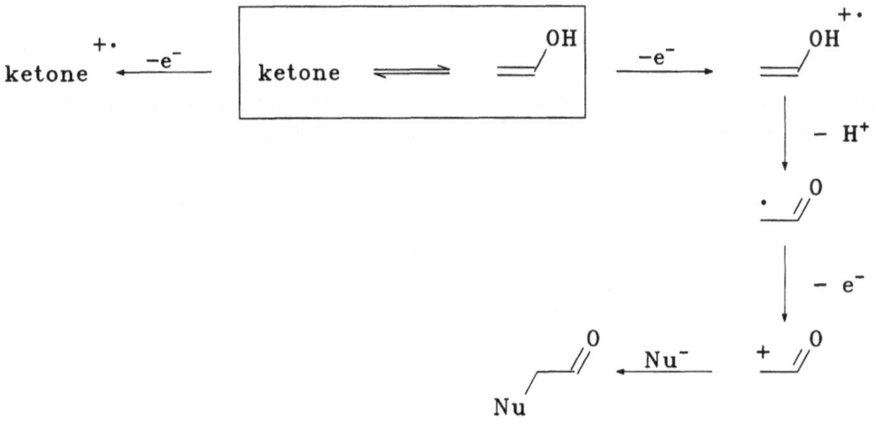

Scheme 8.

Hence, for a rational design of the α-Umpolung reaction some questions have to be answered:

(1) What are the oxidation potentials of the intermediate enols?
(2) If α-carbonyl radicals are intermediates, what are their oxidation potentials?
(3) Under what conditions can one-electron oxidation of radicals compete with other radical reactions?

Quite obviously, direct measurement of the oxidation potentials of the intermediate, highly unstable enols poses a severe experimental problem. From work by Kresge, Capon and others [58–60] it is known that under appropriate conditions even enols of aliphatic ketones can be prepared with lifetimes in the order of minutes. However, no electrochemical data on these systems are available so far. Even for arylacetones, where the enol content is in the order of 10^{-4}%, direct measurement of the oxidation potential is not known to date. Two approaches have been applied to estimate their oxidation potentials [183]. First, using a linear correlation [184] between the experimental oxidation potentials of benzyl radicals [185, 186] and their AM1 calculated ionization potentials (*IPs*) the $E_{1/2}$ of the four enols **86a–86d** were derived from the calculated *IPs*. Secondly, oxidation potentials of silyl enol ethers were used to

Table 11. AM1-calculated ionization potentials and thereof derived estimated oxidation potentials of *para*-substituted enols of arylacetones **86** in comparison with experimental data of silyl enol ethers **87** [171]

86 R	IP_a/eV of **86**	$E_{1/2}/V$ of **86**	E_{pa}/V of **87**
a: *p*-Cl	7.85	1.22	n.k.[a]
b: *p*-H	7.76	1.12	1.02
c: *p*-Me	7.64	1.00	n.k.[a]
d: *p*-MeO	7.43	0.78	0.79

[a] not known.

approximate enol oxidation potentials, since they agreed well for $Mes_2C = C(OSiMe_3)H$ ($E_{pa} = 1.06$ V) and $Mes_2C = C(OH)H$ ($E_{pa} = 1.00$ V). This simple approximation is supported by the similar potentials for **86b** and **87b**, **86d** and **87d**.

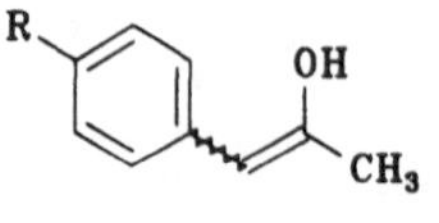

86a–d 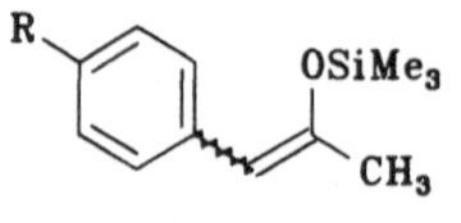   **87a–d**

Since in good hydrogen bond accepting solvents such as DMSO, the equilibrium content of enols of 2-arylpropionaldehydes is relatively high (up to 50%) [187], their oxidation potentials were readily determined by cyclic voltammetry in acetonitrile/DMSO mixtures [171]. A closer analysis, however, indicated that actually enol · DMSO complexes were measured exhibiting some enolate character, since the oxidation potentials found proved to be significantly lower than the ones calculated by the above procedure.

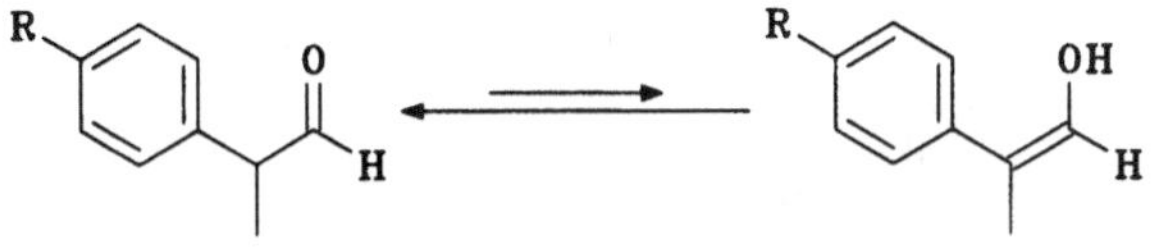

88   89

Although the calculated $E_{1/2}$ (89) should be regarded with appropriate caution, they are still of help for a rational approach to the α-Umpolung reaction since they allow the fine tuning of the potential of the oxidant used. Direct oxidation of the ketone tautomer, therefore, should be largely avoided. After deprotonation of the enol radical cation a highly reactive α-carbonyl radical is formed (Scheme 8). If further oxidation is slow, radical reactions are encountered as demonstrated in several examples by Yoshida [107] and Orliac le Moing [108] in Sect. 3.1. In order to control the chemistry of the intermediate

Table 12. Oxidation potentials, E_{pa}, of *para*-substituted 2-aryl-propionaldehydes **88** and their tautomeric enols **89** in acetonitrile/DMSO [171] compared with calculated $E_{1/2}$ (89). The enol content was taken from Ref. [187]

88/89 R	% enol (89) in DMSO	E_{pa}/V of 88	E_{pa}/V of 89 · DMSO	$E_{1/2}/V^a$ of 89
p-NO$_2$	49.8	2.08	1.00	–
m-NO$_2$	25.3	2.64	0.91	–
p-Br	14.3	2.34	0.74	–
p-Cl	15.6	2.44	0.78	1.38
p-H	8.7	2.42	0.73	1.27
p-MeO	6.3	1.69	0.66	0.93

[a] Calculated from the AM1 adiabatic ionization potentials using the correlation from Wayner [184]

α-carbonyl radicals, their oxidation potentials need to be known. Again, as for the enols, oxidation potentials of α-carbonyl radicals are scarce in the literature. Using the aforementioned correlation of Wayner [184], however, a reasonable estimate for the radicals **90a–d** was possible.

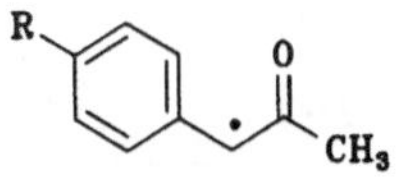

90a–d

211

Although the estimated oxidation potentials of the α-carbonyl radicals **90** are lower than those of the corresponding enols **86**, a strong oxidant is needed to prevent radical type of reactions. According to the rates calculated on the basis of the Marcus theory [188], one-electron oxidation of the radicals can only compete with other radical processes when the electron transfer step is strongly exergonic. Thus, for radicals **90a** and **90b** the one-electron oxidant $TTA^{+\cdot}$ is too weak an oxidant, as reflected by both the yields of the α-Umpolung (Table 14) and the slow electron transfer rates (Table 13).

91a–d   **92a–d**

In summary, these data support the proposed overall mechanistic scheme for the α-Umpolung reaction (Scheme 8), which invokes rapid deprotonation of the enol radical cations $86^{+\cdot}$. Should the other mechanisms (Scheme 3, mech. 4 and 5) be valid for the α-Umpolung reaction of **86** one would expect the yields to go up with less electron rich enols, since both the benzyl radical $ArCH\cdot$-$C(OH)(OMe)CH_3$ and the α-hydroxy radical $ArCH(OMe)$-$C\cdot(OH)CH_3$ should be readily oxidized even for a p-chloroaryl system. In competition experiments

Table 13. Estimated oxidation potentials of α-carbonyl radicals **90a–d** and calculated electron transfer oxidation rates with $TTA^{+\cdot}$ ($E_{1/2} = 0.76$ V), $TBPA^{+\cdot}$ ($E_{1/2} = 1.06$ V) and FePHEN ($E_{1/2} = 1.09$ V) [171]

90 R	$E_{1/2}/V^a$ of 90	$k_{ox}(TTA^{+\cdot})/$ $M^{-1}s^{-1}$	$k_{ox}(TBPA^{+\cdot})/$ $M^{-1}s^{-1}$	$k_{ox}(FePHEN)/$ $M^{-1}s^{-1}$
a: p-Cl	1.11	$2.9\ 10^3$	$3.4\ 10^6$	$1.4\ 10^6$
b: p-H	1.06	$1.1\ 10^4$	$9.2\ 10^6$	$3.9\ 10^6$
c: p-Me	0.84	$1.8\ 10^6$	$4.0\ 10^8$	$2.1\ 10^8$
d: p-MeO	0.58	$2.2\ 10^8$	$1.0\ 10^{10}$	$6.5\ 10^9$

[a] Calculated from the **AM1** adiabatic ionization potentials using the correlation from Wayner [184].

Table 14. Yields of the α-Umpolung reaction of various arylacetones **91a–d** with different oxidants [171]

91 R	$E_{1/2}/V^a$ of 86	yield of 92 with		
		$TTA^{+\cdot}$	$TBPA^{+\cdot}$	FePHEN
a: p-Cl	1.22	<5%	37%	25%
b: p-H	1.12	<3%	15%	43%
c: p-Me	1.00	11%	24%	76%
d: p-MeO	0.78	77%	59%	43%

[a] see table 11.

the different oxidation potentials of the α-carbonyl radicals can be used for increasing the selectivity. For example, the observed selectivity in the reaction of **72** and **91b** with different oxidants originates from selective oxidation of both enols and α-carbonyl radical intermediates [159].

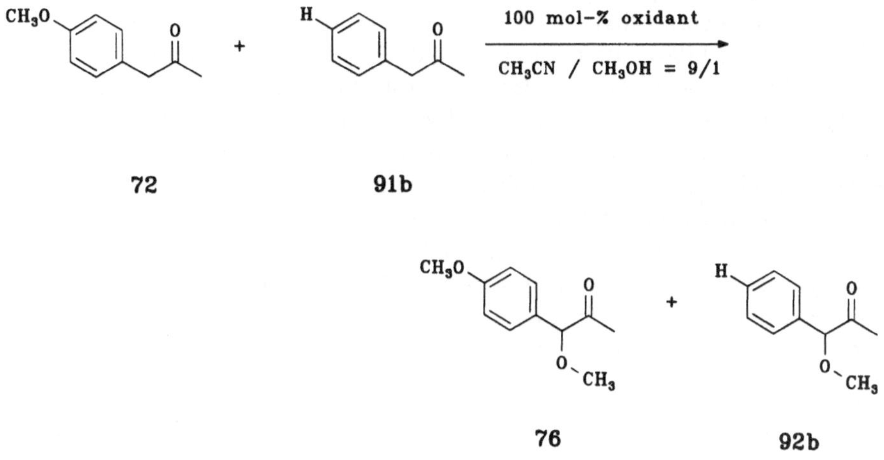

CH₃O

H

100 mol-% oxidant

CH₃CN / CH₃OH = 9/1

72   91b

CH₃O

H

76   92b

Table 15. Results of the competition oxidation of **72** and **91b** as function of the oxidant strength [171]

aminium salt	$E_{1/2}/$ V^a	ratio 76	:	92b
TBPA$^{+\cdot}$	1.06	3		1
TTA$^{+\cdot}$	0.76	20		1
TMPA$^{+\cdot}$	0.52	> 38		1

^{a} from Ref. [189].

4 Reactions of Various Enol Type of Radical Cations

In the following section, the chemistry of other enol type of radical cations will be analyzed with respect to the α-Umpolung reaction. Thus, emphasis was given to reactions of masked enol radical cations that provide α-carbonyl radical and/or α-carbonyl cation intermediates.

4.1 Reactions of Silyl Enol Ether Radical Cations

In terms of one-electron oxidation chemistry, silyl enol ethers are expected to closely resemble enols and enolates. Since they are readily accessible by a number of synthetic procedures, their radical cation chemistry has been studied to more detail. Several reaction modes have been identified so far. O-Si bond cleavage, carbon-carbon bond formation reactions at the stage of either the enol radical cation or the α-carbonyl radical, but little is known about their potential as precursors to α-carbonyl cation chemistry.

Similar to the deprotonation of enol radical cations, silyl enol ether radical cations can undergo loss of "trialkylsilyl cations" (most likely not as ionic silicenium ions [190]). Based on photoinduced electron transfer (PET), Gassman devised a strategy for the selective deprotection of trimethylsilyl enol ethers in the presence of trimethylsilyl ethers [191]. Using 1-cyanonapthalene (1-CN) ($E_{1/2}^{red}* = 1.84$ V) in acetonitrile/methanol or acetonitrile/water trimethylsilyl enol ether 93 ($E_{ox} = 1.29$ V) readily afforded cyclohexanone 64 in 60%. Mechanistically it was proposed that the silyl enol ether radical cation 93$^{+•}$ undergoes O–Si bond cleavage, most likely induced by added methanol [192–194], and that radical 66 abstracts a hydrogen from methanol. Alternatively, back electron transfer from 1-CN$^{-•}$ to 66 would yield the enolate of cyclohexanone which should be readily protonated by the solvent.

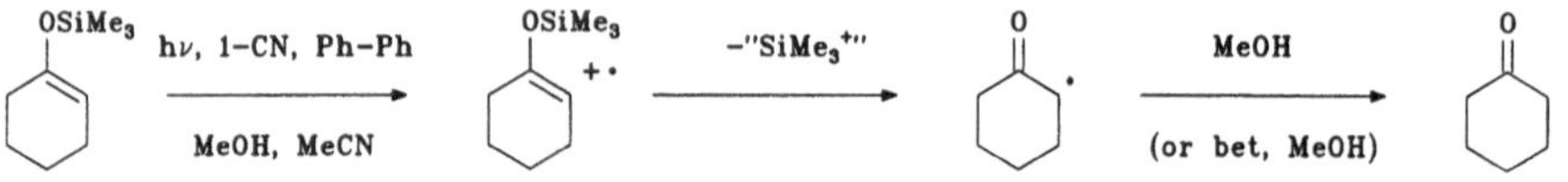

Since simple trimethylsilyl ethers are much more difficult to oxidize ($E_{1/2} > 2.5$ V [191]), the selective removal of the trimethylsilyl group from 94 is possible while leaving intact the silyl group on the unactivated hydroxyl function.

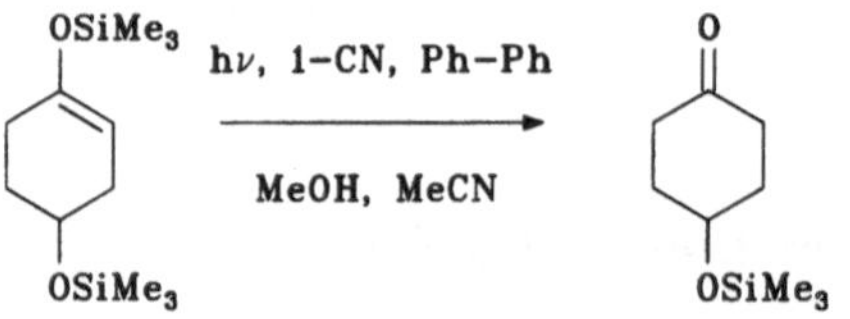

The oxidation of ketones to enones via the reaction of their silyl enol ethers with 2,3-dichloro-5,6-dicyano-1,4-benzoquinone (DDQ) has been suggested originally to proceed via allylic hydride abstraction [195–198]. A recent reinvestigation, however, [199] has established the intermediate formation of a substrate-quinone adduct 96 which was presumably formed from a geminate radical ion pair after electron transfer. Decomposition of the adduct then finally afforded the observed enone product 97. Recently, the critical role of solvent polarity in the formation of 97 from the PET reaction of 93 and chloranil has been identified by time-resolved spectroscopy [200].

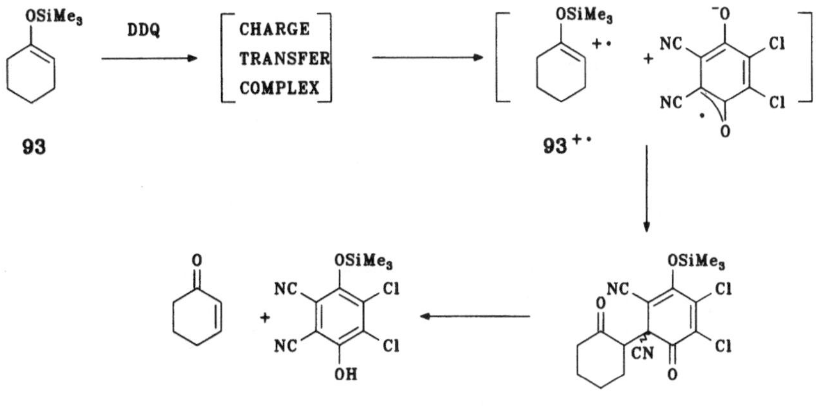

Oxidative coupling of silyl enol ethers as a useful synthetic method for carbon–carbon bond formation has been known for a long time. Several oxidants have been successfully applied to synthesize 1,4-diketones from silyl enol ethers, e.g. Ag_2O [201], $Cu(OTf)_2$ [202], $Pb(OAc)_4$ [203] and iodosobenzene/$BF_3 \cdot Et_2O$ [204]. Although some of these reagents above are known to react as one-electron oxidants, the potential involvement of silyl enol ether radical cations in the above reactions has not been studied. Some recent papers, however, have now established the presence of silyl enol ether radical cations in similar C–C bond formation reactions under well-defined one-electron oxidative conditions. For example, C–C bond formation was reported in the photoinduced electron transfer reaction of 2,3-dichloro-1,4-naphthoquinone (**98**) with various silyl enol ethers **99** [205]. From similar reactions with methoxy alkenes [206, 207] it was assumed that, after photo-excitation, an ion radical pair is formed.

Depending on the steric shielding about the β-carbon in the silyl enol ether the radical cation **99**$^{+\cdot}$ can now attack the radical anion **98**$^{-\cdot}$ at either carbon 2 or at the oxygen providing products **100** or **101**. If R' = R'' = H, then steric effects do not play a role and only C–C bond formation occurs. When either R' or R" are alkyl, steric interactions prohibit formation of the C–C bond and C–O coupling products **101** are obtained exclusively. No coupling products were detected when both R' and R" are alkyl. Although the authors discussed the reaction assuming that C–C or C–O bond formation should occur between radical cation **99**$^{+\cdot}$ and anion radical **98**$^{-\cdot}$, it cannot be excluded that O–Si bond cleavage takes place prior to bond formation.

Electron transfer provides a very efficient methodology to the cross-coupling of 1,2-disubstituted and 1-substituted trimethylsilyl enol ethers as demonstrated by Baciocchi and Ruzziconi [208]. Thus, the synthetically important, un-symmetrical 1,4-diketones **107** were formed in good yields from **102** and **104** using $Ce(NH_4)_2(NO_3)_6$ (**CAN**) as one-electron oxidant.

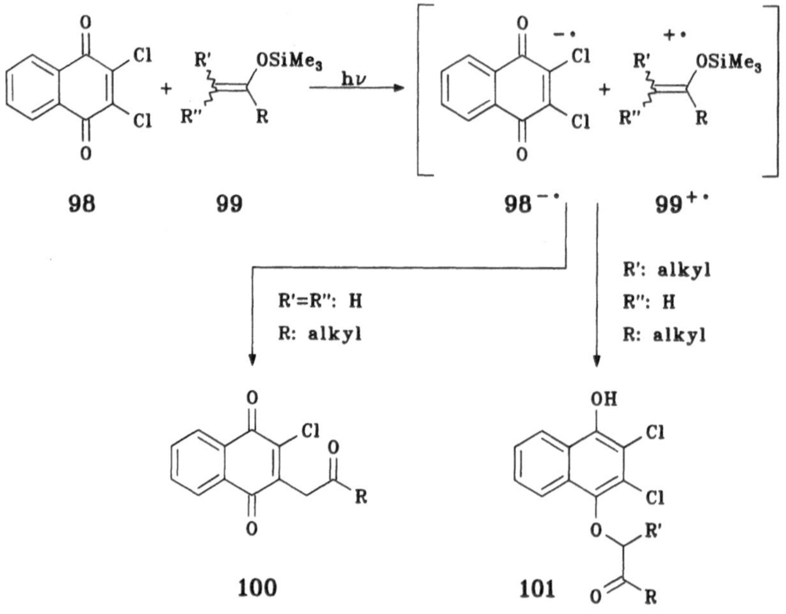

Although the authors do not provide oxidation potentials, it is presumed from oxidation potential considerations that the 1,2-disubstituted silyl enol ethers **102** are oxidized to the radical cation intermediates which after O–Si

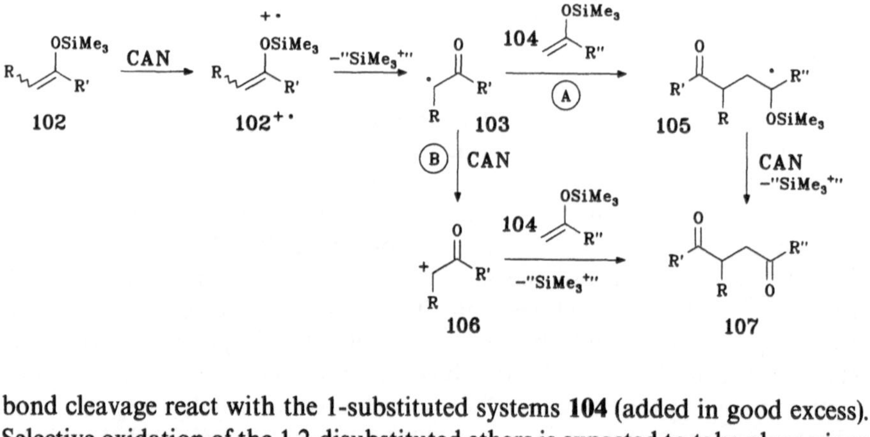

bond cleavage react with the 1-substituted systems **104** (added in good excess). Selective oxidation of the 1,2-disubstituted ethers is expected to take place, since only homodimers thereof were found. It is interesting to note that only radical addition to the silyl enol ether was proposed (path A), but certainly it cannot be excluded that the intermediate α-carbonyl radicals are further oxidized to the α-carbonyl cations prior to C–C bond formation (path B). This may be even more relevant in the recently reported oxidative cross-coupling of trimethyl silyl dienol ethers with silyl enol ethers [209].

A synthetically very interesting oxidative cyclization of δ,ε- and ε,ζ-unsaturated silyl enol ethers using copper(II), **CAN** and **TBPA**$^{+\cdot}$ was recently reported by Snider [210, 211]. Similar products for all three oxidants point to radical cations as common intermediates.

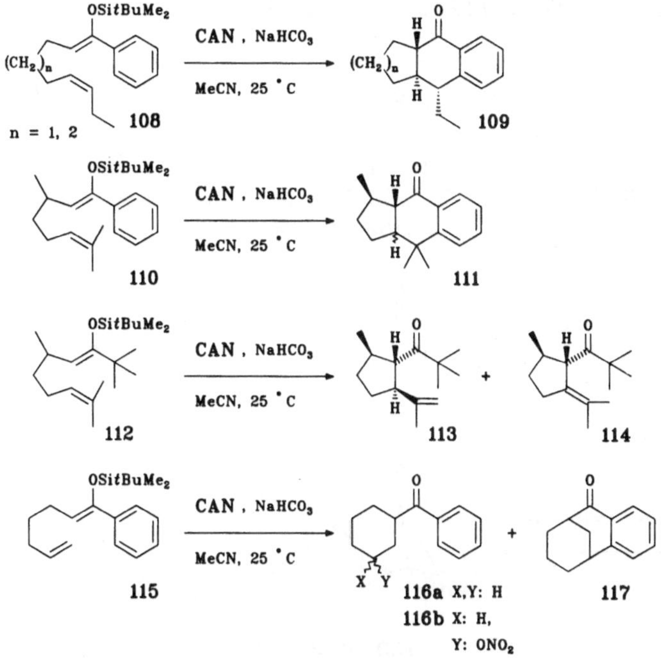

Although several intermediates are possible in the cyclization reaction, strong evidence was presented for the mechanistic proposal, that cyclization takes place at the stage of the silyl enol ether radical cation.

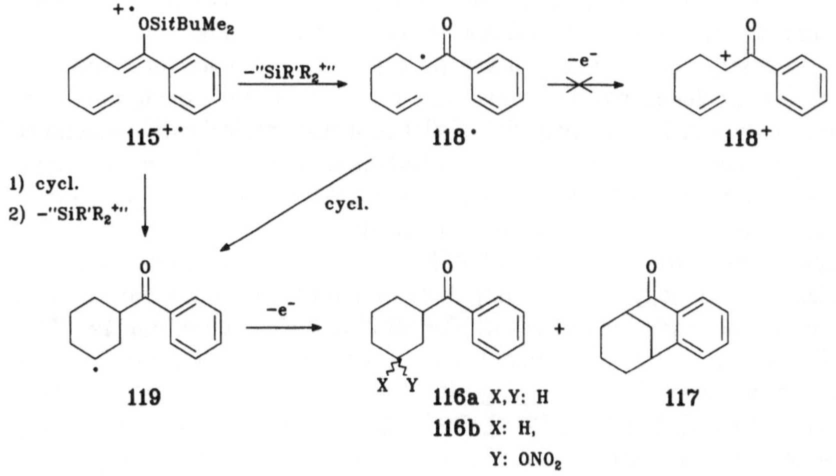

Cyclization of **115**$^{+\cdot}$ via an α-carbonyl radical **118**$^{\cdot}$ would be expected to result in a $3:1$ mixture of 5-*exo* and 6-*endo* products [128], but only 6-*endo* cyclization to **116** and **117** was found. And on the other hand, α-carbonyl cation intermediates, like **118**$^{+}$, are certainly not involved, considering that Mattay could cyclize **115a** to **116a** under PET conditions [212]. Other δ, ε- and ε, ζ-unsaturated silyl enol ethers **120** and **122** could equally be transformed to cyclic

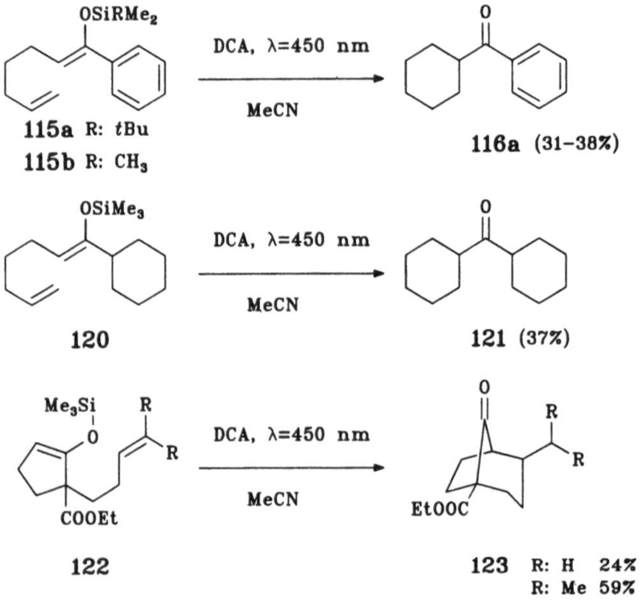

OSiRMe$_2$

DCA, λ=450 nm

MeCN

115a R: *t*Bu
115b R: CH$_3$

116a (31–38%)

OSiMe$_3$

DCA, λ=450 nm

MeCN

120

121 (37%)

Me$_3$Si R
O

DCA, λ=450 nm

MeCN

COOEt

122

EtOOC

123 R: H 24%
         R: Me 59%

ketones (up to 59%) through irradiation in the presence of catalytic amounts of 9,10-dicyanoanthracene as an electron transfer sensitizer. Again, as in Snider's [210, 211] work, it was argued that the cyclization proceeds on the stage of the silyl enol ether radical cation and not on the α-carbonyl radical stage.

In summary, the literature survey provides clear evidence for α-carbonyl radical intermediates but no convincing proof for further oxidation to α-carbonyl cation in the vast majority of silyl enol ether radical cation reactions. This suggests that for most cases, silyl enol ethers are more readily oxidized than the corresponding α-carbonyl radicals. Only in oxidations of β-aryl substituted silyl enol ethers, α-carbonyl cation intermediates have been invoked. For example, one-electron oxidation of **87d** with TTA$^{+\cdot}$ in acetonitrile/MeOH afforded **76** in analogy to the α-Umpolung of ketones via enol radical cations (Scheme 4), and oxidation of **124** with FePHEN provided benzofuran **19** [171].

Other α-Umpolung reactions of ketones via oxidative reactions of silyl enol ethers can be accomplished by using MCPBA as oxidant following the Rubottom method [213], Pb(OAc)$_4$ [214], OsO$_4$ [215], Ni(II)complex/oxygen [216] and iodosobenzene with BF$_3$ [217] or Mn(III) [218] but certainly radical cations are not involved with those. Similarly in the electrochemical generation of α-nitrocycloalkanones (80–95%) [219] from silyl enol ethers the reaction proceeds by anodic formation of NO_2^+ from NO_2 but not via radical cation intermediates.

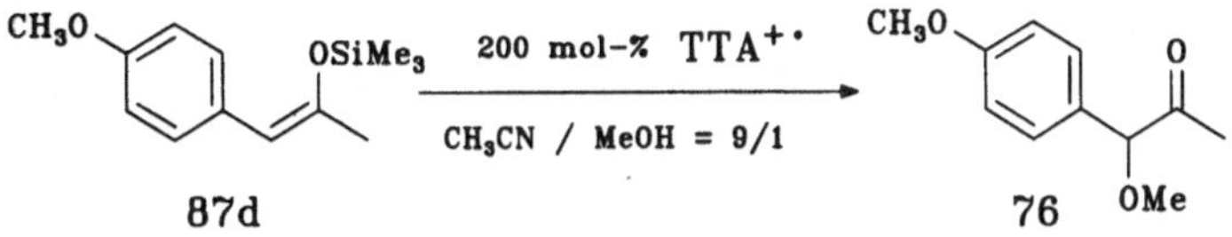

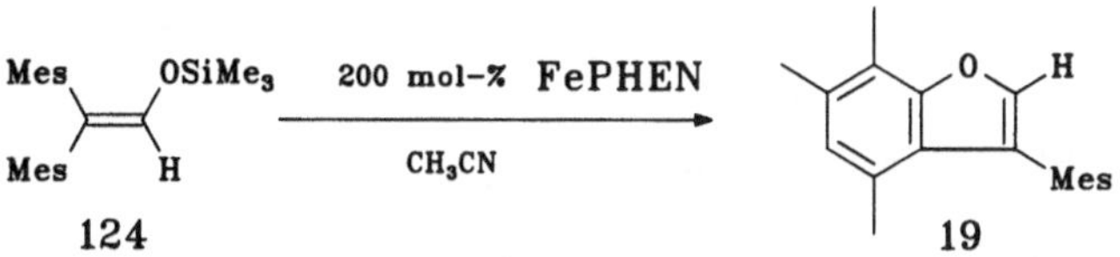

4.2 Reactions of Enol Ester Radical Cations

The reactions of enol ester radical cations formed in anodic oxidations were pioneered by Shono [220–225] almost two decades ago. A reaction mode was identified that formally corresponds to that of enol cation radicals. Depending on the electrolysis conditions enol acetates were either converted to α-acetoxy ketones (high concentration of acetate) or to enone products (absence of acetate). Similarly, α-methoxy ketones were obtained through electrolysis in methanol-Et₄NOTs. Yields for additional reactions not listed here varied between 29% and 90% [222, 223].

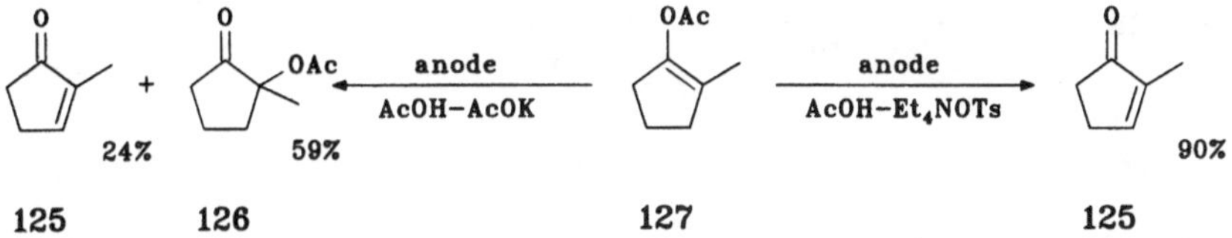

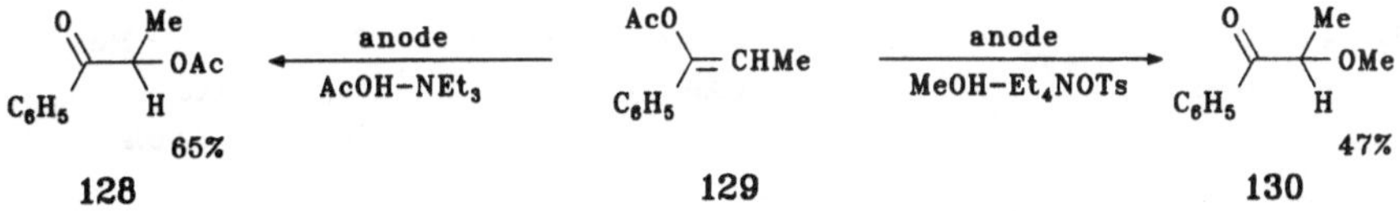

Mechanistically, it was argued that the products are derived from enol acetate radical cations that are either attacked by nucleophiles (A) or by a base (B) (Scheme 9).

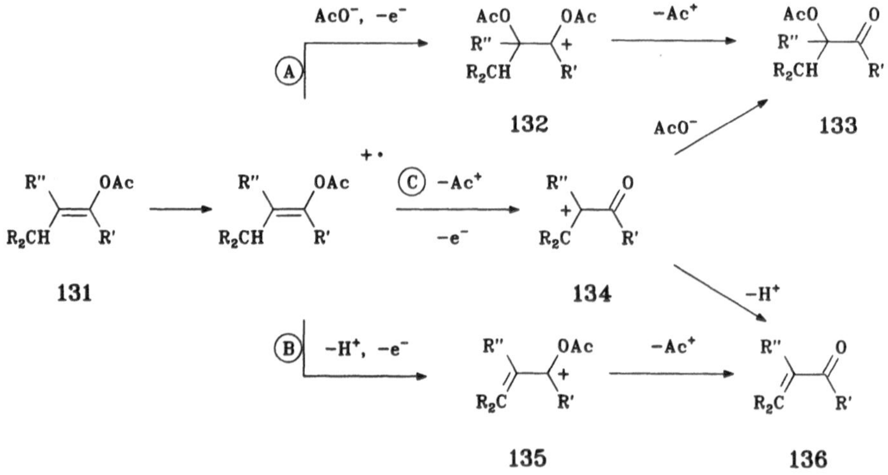

Scheme 9.

Since no evidence was provided in support of the above mechanistic proposals there is a priori no reason to exclude mechanism C as an alternative route to the observed products (see also Scheme 5). Interestingly, route C would also readily explain the observed regioselectivity in the oxidative cyclization reaction of enol acetates [221]. Some years later, however, Laurent and coworkers [226, 227] demonstrated that in the presence of fluoride (CH_3CN/Et_3N, 3HF) enol acetate radical cations partially afforded rearrangement products (e.g. **141**) not compatible with mechanism C. Rather, the products found suggest that fluoride adds directly to enol acetate radical cations providing the most stable radical intermediate (e.g. **140**).

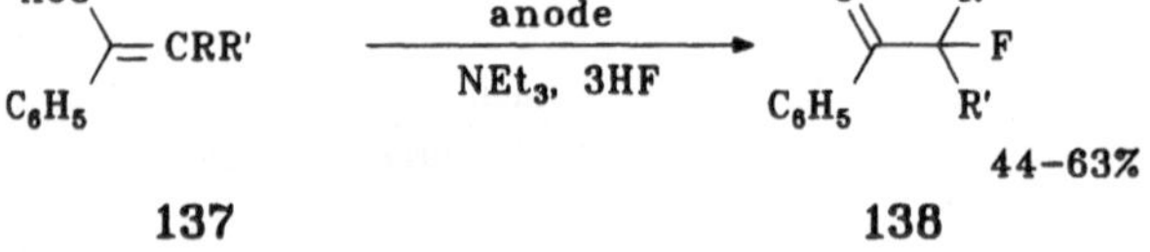

R, R': H,H; H, Me; Me, Me; H, C_6H_5

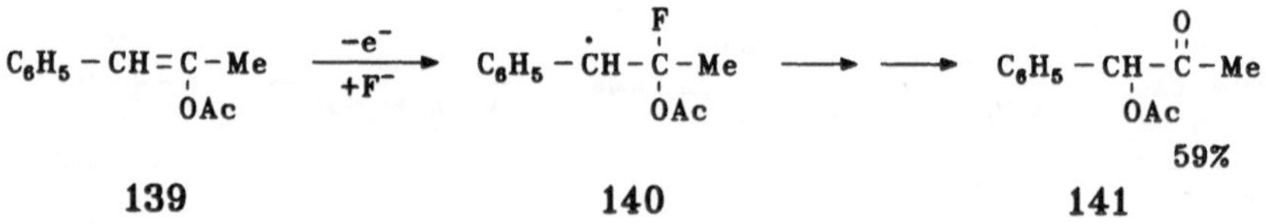

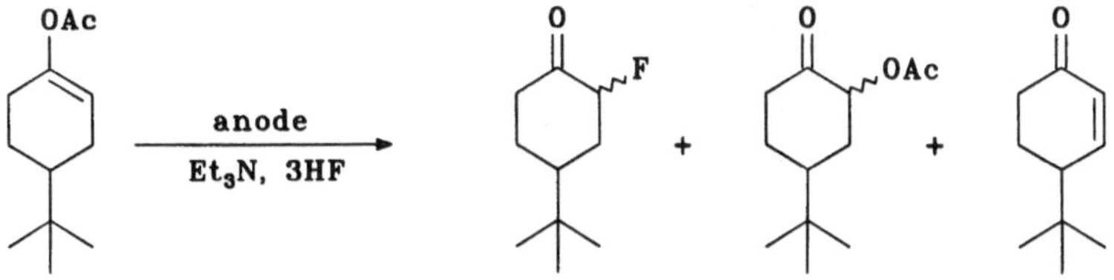

	142		143	144	145
			42%	16%	12%
			(cis/trans =		
			85 : 15)		

These results, however, do not imply that mechanism C is impossible in general [228]. Recently, sterically hindered enol acetates, where nucleophilic attack (mech. A) and deprotonation (mech. B) on the radical cation stage are suppressed, were synthesized and studied by cyclic voltammetry as well as by product analysis [229]. Accordingly, enol acetates **146–149** undergo loss of $CH_3CO^{\cdot}$ upon one-electron oxidation and open up a novel route to α-carbonyl cation chemistry. **150–153** rearrange subsequently to the benzofurans **19–21, 23**. The C–O bond cleavage reaction in **147**$^{+\cdot}$ is rather slow ($k < 10^2$ s^{-1}) as derived from fast-scan cyclic voltammetry studies.

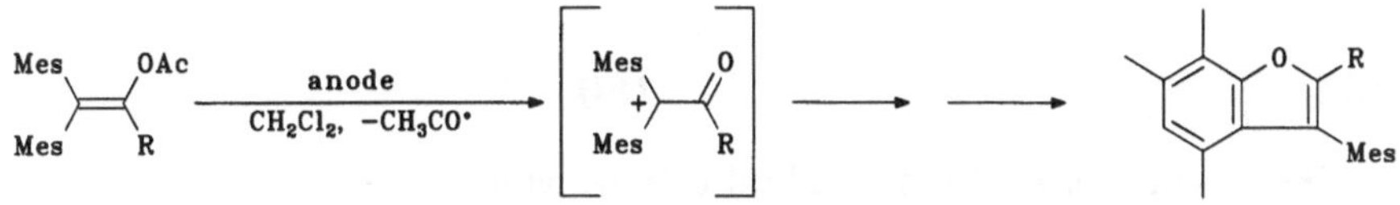

146	R: H	**150**	R: H	**19**	R: H
147	R: tBu	**151**	R: tBu	**20**	R: tBu
148	R: CH$_3$	**152**	R: CH$_3$	**21**	R: CH$_3$
149	R: C$_6$H$_5$	**153**	R: C$_6$H$_5$	**23**	R: C$_6$H$_5$

4.3 Reactions of Enol Ether Radical Cations

Enol ether radical cations undergo a variety of reactions, e.g. carbon–carbon bond formation [206, 230–235], isomerization [236, 237], oxygenation [238–241], cycloaddition [242–245], and they play an important role in the damage to DNA by ionizing radiation [246, 247]. According to ESR studies most of the spin density resides at the β-carbon (89% for EtO$^+$ $= CH$–$CH_2^{\cdot}$ [248]) [234, 249, 250]. The focus of the present brief section, however, will be concerned with their role as potential intermediates in a formal α-Umpolung reaction of ketones.

Under oxidative conditions (anode, aminium salts) in the presence of methanol/base most enol ethers are known to be dimethoxylated, with or without dimerization [235, 251–255], thus opening up a way to formal α-Umpolung products **158**.

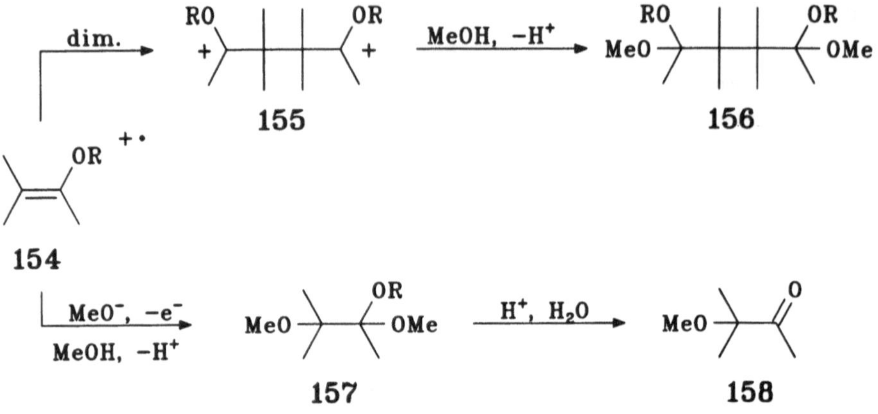

This method could be successfully applied for a straightforward synthetic approach to 2,5,7,10-tetraoxabicyclo[4.4.0]decanes [256]. Some direct [257] and indirect [258] evidence has been found supporting attack of water as a nucleophile at the α-carbon, a reaction mode already identified for enol acetate radical cations [226, 227], although this may not be general.

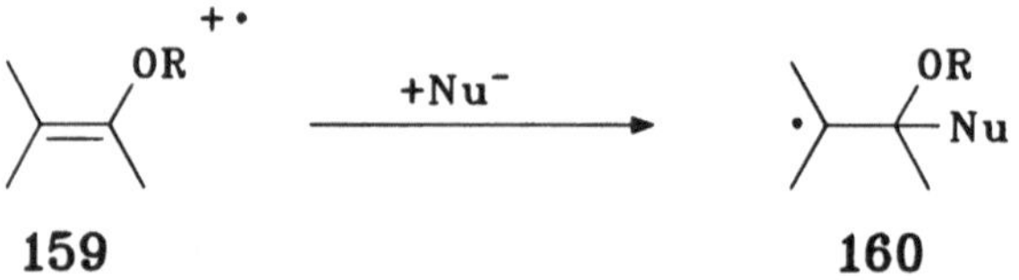

The reaction of enol ether radical cations with hydroxide is six orders of magnitude faster than with water [249], but products have not been identified. In contrast, reaction of $37^{+\cdot}$ and $38^{+\cdot}$ with acetonitrile or methanol is very slow. In a comparative study, it was demonstrated that enol ether reacted much slower than enol ester radical cations with fluoride [227] affording α-fluoroketones only in poor yield. Hitherto, reports in the literature give no evidence for the loss of an alkyl group from enol ether radical cations which would allow to enter α-carbonyl radical and α-carbonyl cation chemistry.

5 Resume

This review has clearly demonstrated that ketones can be functionalized in an α-Umpolung reaction via the corresponding enol radical cation and α-carbonyl cation intermediates. The latter react readily with heteroatom and carbon nucleophiles. Hence, this synthetic approach nicely complements other α-Umpolung strategies for ketones that usually proceed via the reaction of an electron-rich enol derivative with an oxygen donor possessing an electropositive oxygen (dioxirane [259], sulfonyl oxaziridines [173], MoO_5 [178, 179] and others [180, 260]). In general, enol radical cations may be obtained from either direct oxidation of stable enols or by selective oxidation of the enol tautomer of the keto/enol equilibrium. In addition it has been outlined that enol radical cations offer an access to α-carbonyl radical chemistry. Other enol systems like silyl enol ethers, enol esters and enolates similarly may open up after oxidation the chemistry of α-carbonyl radical or α-carbonyl cation intermediates, whereas enol ether oxidative α-functionalization reactions work by another route.

Despite the importance of enol type radical cations in biology [127] relatively little is known to date about oxidation potentials of purely aliphatic enols in solution. To some extent their potentials may be approximated by using those of the corresponding silyl enol ethers [191, 261] (Sect. 3.3). Similarly, almost nothing is known about oxidation potentials of α-carbonyl radicals. As an alternative, ionization potentials of enols and the corresponding α-carbonyl radicals may be used to derive oxidation potentials by known correlations [262, 263]. In Table 16 AM1-calculated IP's of enols and α-carbonyl radicals are listed. Although in some case the calculated values differ significantly from the experimental IP's [264], they are used in order to provide a complete and consistent set of data. Since the solvation terms for enol radical cations and the corresponding α-carbonyl cations should be quite similar the oxidation potential difference is expected to parallel that of the ionization potentials.

From the above data a good part of the observed chemistry can now readily be understood. Only for diarylmethyl and arylmethyl ketones are the ionization potentials of the enols higher than those of the α-carbonyl radicals. Hence, in these cases mostly α-carbonyl cation chemistry is observed. In contrast, the ionization potentials of α-cyanoketones and β-dicarbonyl compounds derived enols are significantly lower than those of the radicals, and as expected radical derived products are now detected [107, 108] (s. Sect. 3.1). Even simple alkyl α-carbonyl radicals are more difficult to oxidize than the corresponding enols, a fact that may play an important role in the reaction of ketones with $TH^{+\cdot}$ [266]. This transformation was originally regarded as an electrophilic substitution, but it could proceed via one-electron oxidation of the enol. Quite obviously, this potential order does not necessarily rule out to trigger α-carbonyl cation chemistry, but then sufficiently strong oxidants are definitely needed. Interestingly, enediols (cf vitamin C, hydroquinone derivatives) show very similar ionization potentials for the enol form and the corresponding α-carbonyl radicals, thus, here again mostly a second electron transfer and rarely radical reactions are observed [267, 268].

$$E_o = 0.16 \ V \ (pH > 6)$$

Table 16. Ionization potentials of enols and α-carbonyl radicals as calculated by AM1 [265]

enol	IP_a [eV]	radical	IP_a [eV]
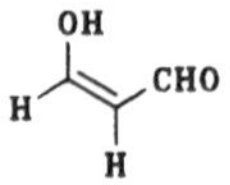	9.47	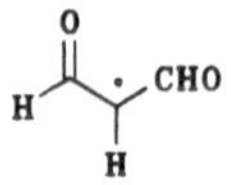	10.31
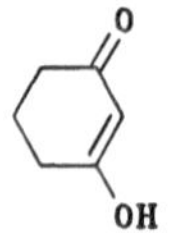	9.13	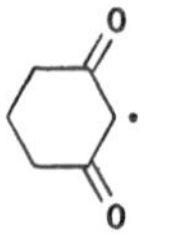	9.92
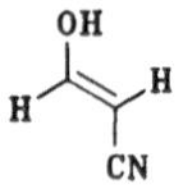	8.82	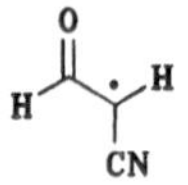	10.04
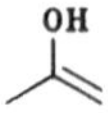	8.73		9.79
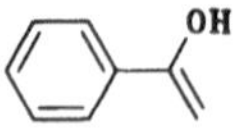	8.48	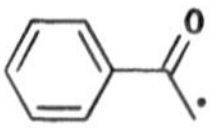	9.47
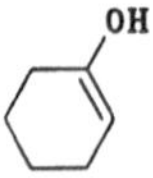	8.18	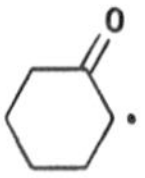	8.79
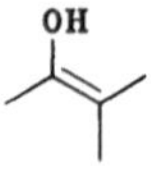	7.94	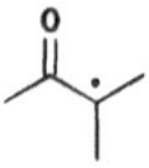	8.43
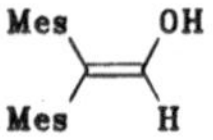	7.84	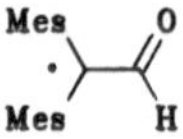	7.24
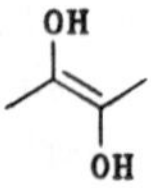	7.96	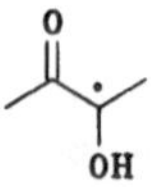	8.04
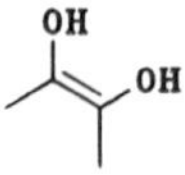	7.81	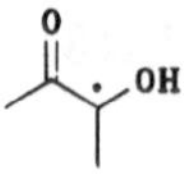	8.05

Selective oxidation of the enol tautomer is not limited to thermal enolization conditions. In a laser-jet study by Wilson [269] enol radical cations are most likely formed after electron transfer trapping of the photoenol of 2-methylbenzophenone with various acceptors as benzoquinones and N-phenyltriazolinedione.

Finally, it is hoped that the present review will stimulate the search for new preparative reactions of enols and related systems under one-electron oxidation conditions, since the richness of their electron transfer chemistry has yet to be uncovered.

Acknowledgements. Our own results in this field have only been made possible by the dedicated efforts of my coworkers, to whom I am very grateful. Their contributions are quoted in the corresponding references. For discussions I am indebted to Prof. M. Schulz and Dr. R. Kluge (Merseburg). Additionally, I want to thank Prof. C. Rüchardt who has been very supportive throughout this work which has been carried out in his laboratories. For generous financial support I am grateful to the Stiftung Volkswagenwerk, Fonds der Chemischen Industrie and the Wissenschaftliche Gesellschaft Freiburg.

6 References

1. Johnson MK, King RB, Kurtz Jr DM, Kutal C, Norton ML, Scott RA (eds) (1990) Electron transfer in biology and the solid state. Inorganic compounds with unusual properties, Advances in chemistry series 226, American Chemical Society, Washington DC
2. Bolton JR, Mataga N, McLendon G (eds) (1991) Electron transfer in inorganic, organic, and biological systems, Advances in chemistry series 228, American Chemical Society, Washington DC
3. Gould ES (1985) Acc Chem Res 18: 22
4. Taube H (1970) Electron transfer reactions of complex ions in solution, Academic, New York
5. Reynolds WL, Lumry RW (1966) Mechanisms of electron transfer, Ronald Press, New York
6. Chanon M, Rajzmann M, Chanon F (1990) Tetrahedron Lett 46: 6193
7, Gilbert A, Baggott J (1991) Essentials of molecular photochemistry, Blackwell, Oxford
8. Eberson L (1987) Electron transfer reactions in organic chemistry, Springer, Berlin
9. Lund H, Baizer MM (eds) (1991) Organic electrochemistry, Marcel Dekker, New York
10. Torii S (1985) Electroorganic syntheses, Kodansha, Tokyo
11. Shono T (1984) Electroorganic chemistry as a tool in organic synthesis, Springer, Berlin Heidelberg New York
12. Yoshida K (1984) Electrooxidation in organic chemistry, Wiley, New York
13. Lund A, Shiotani M (eds) (1991) Radical ionic systems, Kluwer, Dordrecht
14. Neta P, Harriman A (1988) In: Fox MA, Chanon M (eds) Photoinduced electron transfer, Vol. B. Elsevier, Amsterdam, Sect. 2.3
15. Farhataziz, Rodgers MAJ (eds) (1987) Radiation chemistry – principles and applications, VCH, Weinheim
16. Fox MA, Chanon M (eds) (1988) Photoinduced electron transfer, Elsevier, Amsterdam
17. Mattay J (ed) (1990-) Photoinduced electron transfer. A series in topics in current chemistry, Springer, Berlin Heidelberg New York
18. Okamura T, Sancar A, Heelis PF, Begley TP, Hirata Y, Mataga N (1991) J Am Chem Soc 113: 3143
19. Kim JM, Cho IS, Mariano PS (1991) J Org Chem 56: 4943
20. Holland HL (1992) Organic synthesis with oxidative enzymes, VCH, New York
21. Pecoraro VL (ed) (1992) Manganese redox enzymes, VCH, New York
22. Dinnocenzo JP, private communication
23. Bouma WJ, Poppinger D, Radom L (1983) J Mol Struct THEOCHEM 103: 209
24. Haselbach E., private communication
25. Bauld NL (1989) Tetrahedron 45: 5307
26. Dinnocenzo JP, Conlon DA (1988) J Am Chem Soc 110: 2324
27. Roth HD (1992) Top Curr Chem 163: 131
28. Gescheidt G, Lamprecht A, Heinze J, Schuler B, Schmittel M, Kiau S, Rüchardt C (1992) Helv Chim Acta 75: 1607
29. Dinnocenzo JP, Banach TE (1989) J Am Chem Soc 111: 8646

30. Parker VD, Tilset M (1988) J Am Chem Soc 110: 1649
31. Bordwell FG, Bausch MJ (1986) J Am Chem Soc 108: 2473
32. Nicolas AM, de P, Arnold DR (1982) Can J Chem 60: 2165
33. Reynolds DW, Harirchian B, Chiou HS, Marsh BK, Bauld NL (1989) J Phys Org Chem 2: 57
34. Gieseler A, Steckhan E, Wiest O, Knoch F (1991) J Org Chem 56: 1405
35. Schmittel M, von Seggern H (1991) Angew Chem 103: 981, Angew Chem Int Ed Engl 30: 999
36. Bard AJ, Ledwith A, Shine HJ (1976) Adv Phys Org Chem 13: 155
37. Freeman F (1986) In: Mijs WJ, De Jonge CRHI (eds) Organic syntheses by oxidation with metal compounds. Plenum, New York, chapter 5
38. de Klein WJ (1986) In: Mijs WJ, De Jonge CRHI (eds) Organic syntheses by oxidation with metal compounds. Plenum, New York, chapter 4
39. Chanon M, Tobe ML (1982) Angew Chem 94: 27, Angew Chem Int Ed Engl 21: 1
40. Eberson L (1992) New J Chem 16: 151
41. Frenking G, Heinrich N, Schmidt J, Schwarz H (1982) Z Naturforsch 37b: 1597
42. Holmes JL, Terlouw JK, Lossing FP (1976) J Phys Chem 80: 2860
43. Holmes JL, Lossing FP (1980) J Am Chem Soc 102: 1591
44. Holmes JL, Lossing FP (1982) J Am Chem Soc 104: 2648
45. Frenking G, Schwarz H (1982) Z Naturforsch 37b: 1602
46. Frenking G, Schwarz H (1983) Int J Mass Spectr Ion Phys 52: 131
47. Albrecht B, Allan M, Haselbach E, Neuhaus L, Carrupt P-A (1984) Helv Chim Acta 67: 220
48. Frenking G, Schwarz H (1982) Naturwissenschaften 69: 446
49. Frisch MJ, Raghavachari K, Pople JA, Bouma WJ, Radom L (1983) Chem Phys 75: 323
50. Turecek F, Havlas Z (1986) J Org Chem 51: 4066
51. Pedley JB, Naylor RD, Kirby SP (1986) Thermochemical Data of Organic Compounds, Chapman and Hall, London
52. Heinrich N, Schmidt J, Schwarz H, Apeloig Y (1987) J Am Chem Soc 109: 1317
53. Holmes JL, Burgers PC, Terlouw JK (1981) Can J Chem 59: 1805
54. Depke G, Heinrich N, Schwarz H (1984) Int J Mass Spectrom Ion Processes 62: 99
55. Holmes JL, Lossing FP (1980) J Am Chem Soc 102: 3732
56. Heinrich N, Koch W, Frenking G, Schwarz H (1986) J Am Chem Soc 108: 593
57. Rappoport Z (ed) (1990) The chemistry of enols, Wiley, New York
58. Toullec J (1990) In: Rappoport Z (ed) The chemistry of enols. Wiley, New York, p 323
59. Keefe JR, Kresge AJ (1990) In: Rappoport Z (ed) The chemistry of enols. Wiley, New York, p 399
60. Capon B (1990) In: Rappoport Z (ed) The chemistry of enols. Wiley, New York, p 307
61. Mourgues P, Denhez JP, Audier HE, Hammerum S (1993) Org Mass Spect 28: 193
62. Bouchoux G (1988) Mass Spectr Rev 7: 1 and 203
63. Schmittel M, Baumann U (1990) Angew Chem 102: 571, Angew Chem Int Ed Engl 29: 541
64. Schmittel M, Röck M (1992) Chem Ber 125: 1611
65. Splitter JS, Calvin M (1979) J Am Chem Soc 101: 7329
66. Hoppilliard Y, Bouchoux G, Jaudon P (1982) Nouv J Chim 6: 43
67. Bouchoux G, Flament JP, Hoppilliard Y (1984) Int J Mass Spectr Ion Processes 57: 179
68. Apeloig Y, Karni M, Ciommer B, Depke G, Frenking G, Meyn S, Schmidt J, Schwarz H (1984) Int J Mass Spectr Ion Processes 59: 21
69. Heinrich N, Louage F, Lifshitz C, Schwarz H (1988) J Am Chem Soc 110: 8183
70. Gebicki J, Marcinek A, Michalak J, Rogowski J, Bally T, Tang W (1992) J Mol Struct 275: 249
71. Marcinek A, Michalak J, Rogowski J, Tang W, Bally T, Gebicki J (1992) J Chem Soc, Perkin Trans 2 1353
72. Hammerich O, Svensmark B (1991) In: Lund H, Baizer MM (eds) Organic electrochemistry. Marcel Dekker, New York, p 615
73. Rieker A, Beisswenger R, Regier K (1991) Tetrahedron 47: 645
74. Morrow GW, Chen Y, Swenton JS (1991) Tetrahedron 47: 655
75. Angle SR, Arnaiz DO (1992) J Org Chem 57: 5937
76. Swenton JS, Carpenter K, Chen Y, Kerns ML, Morrow GW (1993) J Org Chem 58: 3308
77. Yamamura S, Shizuri Y, Shigemori H, Okuno Y, Ohkubo M (1991) Tetrahedron 47: 635
78. Ayres DC, Loike JD (1990) Lignans: Chemical, biological and clinical properties, Cambridge University Press, Cambridge
79. McDonald PD, Hamilton GA (1973) In: Trahanovsky WS (ed) Oxidation in organic chemistry, Part B. Academic, New York

80. Bordwell FG, Cheng JP (1991) J Am Chem Soc 113: 1736
81. Hapiot P, Pinson J, Yousfi N (1992) New J Chem 16: 877
82. Arnett EM, Amarnath K, Harvey NG, Venimadhavan S (1990) J Am Chem Soc 112: 7346
83. Lind J, Shen X, Eriksen TE, Merényi G (1990) J Am Chem Soc 112: 479
84. Steenken S, Neta P (1982) J Phys Chem 86: 3661
85. Speiser B, Rieker A (1979) J Electroanal Chem 102: 373
86. Bordwell FG, Cheng JP (1989) J Am Chem Soc 111: 1792
87. Dixon WT, Murphy D (1976) J Chem Soc, Faraday Trans 2 1221
88. Evans DH In: Bard AJ, Lund H (eds) Encyclopedia of electrochemistry of the elements. Vol XII, Dekker, New York, p 154, 155
89. Bordwell FG, Harrelson Jr JA (1990) Can J Chem 68: 1714
90. Dorigo AE, McCarrick MA, Loncharich RJ, Houk KN (1990) J Am Chem Soc 112: 7508
91. Green MM, Mayotte GJ, Meites L, Forsyth D (1980) J Am Chem Soc 102: 1464
92. Becker JY, Byrd LR, Miller LL, So YH (1975) J Am Chem Soc 97: 853
93. Danieli B, Palmisano G (1976) Chem Ind 565
94. Becker JY, Miller LL, Siegel TM (1975) J Am Chem Soc 97: 849
95. Soucy P, Ho TL, Deslongchamps P (1972) Can J Chem 50: 2047
96. Boon PJ, Symons MCR, Ushida K, Shida T (1984) J Chem Soc, Perkin Trans II 1213
97. Snow LD, Williams F (1983) Chem Phys Lett 100: 198
98. Snow LD, Williams F (1983) J Chem Soc, Chem Commun 1090
99. Heinrich N, Koch W, Morrow JC, Schwarz H (1988) J Am Chem Soc 110: 6332
100. Bergmark WR, DeWan C, Whitten DG (1992) J Am Chem Soc 114: 8810
101. Akaba R, Niimura Y, Fukushima T, Kawai Y, Tajima T, Kuragami T, Negishi A, Kamata M, Sakuragi H, Tokumaru K (1992) J Am Chem Soc 114: 4460
102. Kitagawa T, Takeuchi K, Murai O, Matsui S, Inoue T, Nishimura M, Okamoto K (1986) J Chem Soc. Perkin Trans II 1987
103. Takeuchi K, Murai O, Matsui S, Inoue T, Kitagawa T, Okamoto K (1983) J Chem Soc, Perkin Trans II 1301
104. Okamoto K, Takeuchi K, Murai O, Matsui S, Inoue T, Kitagawa T (1981) Tetrahedron Lett 22: 2785
105. Okamoto K, Takeuchi K, Murai O, Fujii Y (1979) J Chem Soc, Perkin Trans II 490
106. Röck M, Dissertation Freiburg, in preparation
107. Yoshida J, Sakaguchi K, Isoe S (1988) J Org Chem 53: 2525
108. Orliac-Le Moing MS, Le Guillanton G, Simonet J (1982) Electrochim Acta 27: 1775
109. Ng FTT, Henry PM (1976) J Am Chem Soc 98: 3606
110. Audsley AJ, Quick GR, Littler JS (1980) J Chem Soc, Perkin Trans II 557
111. Cecil R, Littler JS, Easton G (1970) J Chem Soc (B) 626
112. Littler JS, Quick GR, Wozniak D (1980) J Chem Soc, Perkin Trans II 657
113. Yoshida J, Sakaguchi K, Isoe S (1986) Tetrahedron Lett 27: 6075
114. Schäfer H, Al Azrak A (1972) Chem Ber 105: 2398
115. Fuson RC, Rabjohn N, Byers DJ (1944) J Am Chem Soc 66: 1272
116. Fuson RC, Byers DJ, Rabjohn N (1941) J Am Chem Soc 63: 2639
117. Fuson RC, Corse J, McKeever CH (1940) J Am Chem Soc 62: 3250
118. Hart H, Rappoport Z, Biali SE (1990) In: Rappoport Z (ed) The chemistry of enols. Wiley, New York, p 481
119. Röck M, Schmittel M, J Prakt Chem, submitted for publication
120. Takeuchi K, Kitagawa T, Okamoto K (1983) J Chem Soc, Chem Commun 7
121. Bailey PS, Potts III FE, Ward JW (1970) J Am Chem Soc 92: 230
122. Bailey PS, Ward JW, Potts III FE, Chang YG, Hornish RE (1974) J Am Chem Soc 96: 7228
123. Fuson RC, Maynert EW, Tan TL, Trumbull ER, Wassmundt FW (1957) J Am Chem Soc 79: 1938
124. Richards JA, Whitson PE, Evans DH (1975) J Electroanal Chem 63: 311
125. Shin SR, Shine HJ (1992) J Org Chem 57: 2706
126. Gilbert BC, Whitwood AC (1989) J Chem Soc, Perkin Trans II 1921
127. Stubbe JA (1989) Annu Rev Biochem 58: 257
128. Curran DP, Chang CT (1989) J Org Chem 54: 3140
129. Giese B (1986) Radicals in organic synthesis: formation of carbon–carbon bonds, Pergamon, Oxford
130. Hart DJ, Krishnamurthy R (1992) J Org Chem 57: 4457

131. Lai M, Liu L, Liu H (1991) J Am Chem Soc 113: 7388
132. Lenn ND, Shih Y, Stankovich MT, Liu H (1989) J Am Chem Soc 111: 3065
133. Brettle R, Parkin JG (1967) J Chem Soc C 1352
134. VandenBorn HW, Evans DH (1974) J Am Chem Soc 96: 4296
135. Kato S, Dryhurst G (1977) J Electroanal Chem 79: 391
136. Annen K, Hofmeister H, Laurent H, Wiechert R (1983) Liebigs Ann Chem 705
137. Heiba EI, Dessau RM (1974) J Org Chem 39: 3456
138. Heiba EI, Dessau RM (1974) J Org Chem 39: 3457
139. Dombroski MA, Kates SA, Snider BB (1990) J Am Chem Soc 112: 2759
140. Kates SA, Dombroski MA, Snider BB (1990) J Org Chem 55: 2427
141. Santi R, Bergamini F, Citterio A, Sebastiano R, Nicolini (1992) J Org Chem 57: 4250
142. Schäfer HJ (1981) Angew Chem 93: 978, Angew Chem Int Ed Engl 20: 911
143. Schäfer H, Alazrak A (1968) Angew Chem 80: 485, Angew Chem Int Ed Engl 7: 474
144. Brettle R, Parkin JG, Seddon D (1970) J Chem Soc C 1317
145. Torii S, Uneyama K, Onishi T, Fujita Y, Ishiguro M, Nishida T (1980) Chem Lett 1603
146. Pandey G, Krishna A, Bhalerao UT (1989) Tetrahedron Lett 30: 1867
147. Röck M, Schmittel M (1993) J Chem Soc, Chem Commun, 1739
148. Röck M, Gescheidt G, Schmittel M, unpublished results
149. Bordwell FG, Gallagher T, Zhang X (1991) J Am Chem Soc 113: 3495
150. Kern JM, Federlin P (1977) Tetrahedron Lett 837
151. Kern JM, Federlin P (1978) Tetrahedron 34: 661
152. Kern JM, Federlin P (1979) J Electroanal Chem 96: 209
153. Kern JM, Sauer JD, Federlin P (1982) Tetrahedron 38: 3023
154. Ruberu SR, Fox MA (1992) J Am Chem Soc 114: 6310
155. Aurbach I, Ponti PP, Salbeck E, Schmuck H, Daub J (1988) Chem Ber 121: 1101
156. Ashby EC, Argyropoulos JN (1986) J Org Chem 51: 472
157. Kitagawa T, Nishimura M, Takeuchi K, Okamoto K (1991) Tetrahedron Lett 32: 3187
158. Okamoto K, Takeuchi K, Kitagawa T (1982) Bull Soc Chim Belg 91: 410
159. Hudlicky M (1990) Oxidations in organic chemistry, American Chemical Society, Washington DC
160. Larock RC (1989) Comprehensive organic transformations, VCH, New York
161. Banerji KK, Nath P, Bakore GV (1970) Bull Chem Soc Jpn 43: 2027
162. Littler JS (1962) J Chem Soc 832
163. Vinogradov MG, Verenchikov SP, Nikishin GI (1976) J Org Chem (USSR) 12: 2245
164. Heiba EI, Dessau RM (1971) J Am Chem Soc 93: 524
165. der Hertog Jr HJ, Kooyman EC (1966) J Catal 6: 357
166. Midgley G, Thomas CB (1987) J Chem Soc, Perkin Trans II 1103
167. Littler JS (1962) J Chem Soc 827
168. Henbest HB, Jones DN, Slater GP (1961) J Chem Soc 4472
169. Gupta M, Saha SK, Banerjee P (1990) Int J Chem Kin 22: 81
170. Schmittel M, Abufarag A, Luche O, Levis M (1990) Angew Chem 102: 1174, Angew Chem Int Ed Engl 29: 1144
171. Levis M, Schmittel M, Chem Ber, in preparation
172. Schulz M, Kluge R, Sivilai L, Kamm B (1990) Tetrahedron 46: 2371
173. Davis FA, Chen BC (1992) Chem Rev 92: 919
174. Moriarty RM, Epa WR, Penmasta R, Awasthi AK (1989) Tetrahedron Lett 30: 667
175. Moriarty RM, Penmasta R, Awasthi AK, Epa WR, Prakash I (1989) J Org Chem 54: 1101
176. Moriarty RM, Prakash O, Duncan MP, Valid RK, Musallam HA (1987) J Org Chem 52: 150
177. Moriarty RM, Prakash O (1986) Acc Chem Res 19: 244
178. Vedejs E, Engler DA, Telschow JE (1978) J Org Chem 43: 188
179. Vedejs E (1974) J Am Chem Soc 96: 5944
180. Reißig HU (1986) Nachr Chem Tech Lab 34: 328
181. Schulz M, Kluge R (1993) Tetrahedron, submitted for publication
182. Eberson L, Larsson B (1986) Acta Chem Scand B40: 210
183. Schmittel M, Levis M, unpublished results
184. Wayner DDM, Sim BA, Dannenberg JJ (1991) J Org Chem 56: 4853
185. Sim BA, Milne PH, Griller D, Wayner DDM (1990) J Am Chem Soc 112: 6635
186. Wayner DDM, McPhee DJ, Griller D (1988) J Am Chem Soc 110: 132
187. Ahlbrecht H, Funk W, Reiner MT (1976) Tetrahedron 32: 479

188. Marcus RA, Sutin N (1985) Biochem Biophys Acta 811: 265
189. Steckhan E (1987) Top Curr Chem 142: 1
190. Olah GA, Heiliger L, Li X-Y, Prakash GKS (1990) J Am Chem Soc 112: 5991
191. Gassman PG, Bottorff KJ (1988) J Org Chem 53: 1097
192. Dinnocenzo JP, Farid S, Goodman JL, Gould IR, Todd WP, Mattes SL (1989) J Am Chem Soc 111: 8973
193. Baciocchi E, Crescenzi M, Fasella E, Mattioli M (1992) J Org Chem 57: 4684
194. Todd WP, Dinnocenzo JP, Farid S, Goodman JL, Gould IR (1993) Tetrahedron Lett 34: 2863
195. Fevig TL, Elliott RL, Curran DP (1988) J Am Chem Soc 110: 5064
196. Fleming I, Paterson I (1979) Synthesis 736
197. Ryu I, Murai S, Hatayama Y, Sonoda N (1978) Tetrahedron Lett 3455
198. Jung ME, Pan YG, Rathke MW, Sullivan DF, Woodbury RP (1977) J Org Chem 42: 3961
199. Bhattacharya A, DiMichele LM, Dolling UH, Grabowski EJJ, Grenda VJ (1989) J Org Chem 54: 6118
200. Bockman TM, Perrier S, Kochi JK (1993) J Chem Soc, Perkin Trans 2, 595
201. Ito Y, Konoike T, Saegusa T (1975) J Am Chem Soc 97: 649
202. Kobayashi Y, Taguchi T, Morikawa T, Tokuno E, Sekiguchi S (1980) Chem Pharm Bull 28: 262
203. Moriarty RM, Penmasta R, Prakash I (1987) Tetrahedron Lett 28: 873
204. Moriarty RM, Prakash O, Duncan MP (1985) J Chem Soc, Chem Commun 420
205. Maruyama K, Tai S, Imahori H (1986) Bull Chem Soc Jpn 59: 1777
206. Maruyama K, Otsuki T, Tai S (1984) Chem Lett 371
207. Maruyama K, Otsuki T, Tai S (1985) J Org Chem 50: 52
208. Baciocchi E, Casu A, Ruzziconi R (1989) Tetrahedron Lett 30: 3707
209. Paolobelli AB, Latini D, Ruzziconi R (1993) Tetrahedron Lett 34: 721
210. Snider BB, Kwon T (1992) J Org Chem 57: 2399
211. Snider BB, Kwon T (1990) J Org Chem 55: 4786
212. Heidbreder A, Mattay J (1992) Tetrahedron Lett 33: 1973
213. Rubottom GM, Juve Jr HD (1983) J Org Chem 48: 422
214. Rubottom GM, Gruber JM, Marrero R, Juve Jr HD, Kim CW (1983) J Org Chem 48: 4940
215. McCormick JP, Tomasik W, Johnson MW (1981) Tetrahedron Lett 607
216. Takai T, Yamada T, Rhode O, Mukaiyama T (1991) Chem Lett 281
217. Moriarty RM, Prakash O, Duncan MP (1985) Synthesis 943
218. Reddy DR, Thornton ER (1992) J Chem Soc, Chem Commun 172
219. Bloom AJ, Fleischmann M, Mellor JM (1984) Tetrahedron Lett 25: 4971
220. Shono T, Kashimura S (1983) J Org Chem 48: 1939
221. Shono T, Nishiguchi I, Kashimura S, Okawa M (1978) Bull Chem Soc Jpn 51: 2181
222. Shono T, Nishiguchi I, Nitta M (1976) Chem Lett 1319
223. Shono T, Okawa M, Nishiguchi I (1975) J Am Chem Soc 97: 6144
224. Shono T, Nishiguchi I, Yokoyama T, Nitta M (1975) Chem Lett 433
225. Shono T, Matsumura Y, Nakagawa Y (1974) J Am Chem Soc 96: 3532
226. Laurent E, Tardivel R, Thiebault H (1983) Tetrahedron Lett 24: 903
227. Laurent E, Marquet B, Tardivel R, Thiebault H (1986) Bull Soc Chim Fr 955
228. Algarra F, Baldoví MV, García H, Miranda MA, Primo J (1993) Monatsh Chem 124: 209
229. Schmittel M, Trenkle H, unpublished results
230. Moeller KD, Tinao LV (1992) J Am Chem Soc 114: 1033
231. Hudson CM, Marzabadi MR, Moeller KD, New DG (1991) J Am Chem Soc 113: 7372
232. Moeller KD, Marzabadi MR, New DG, Chaing MY, Keith S (1990) J Am Chem Soc 112: 6123
233. Gersdorf J, Mattay J, Görner H (1987) J Am Chem Soc 109: 1203
234. Mattay J. Gersdorf J, Buchkremer K (1987) Chem Ber 120: 307
235. Koch D, Schäfer H, Steckhan E (1974) Chem Ber 107: 3640
236. Klett MW, Johnson RP (1985) J Am Chem Soc 107: 6615
237. Majima T, Pac C, Sakurai H (1979) Chem Lett 1133
238. Lopez L, Troisi L, Rashid SMK, Schaap AP (1989) Tetrahedron Lett 30: 485
239. Curci R, Luigi L, Troisi L, Rashid SMK, Schaap AP (1987) Tetrahedron Lett 28: 5319
240. Ciminale F, Lopez L (1985) Tetrahedron Lett 26: 789
241. Lopez L, Troisi L (1992) Tetrahedron 48: 7321
242. Mattay J, Trampe G, Runsink J (1988) Chem Ber 121: 1991
243. Mlcoch J, Steckhan E (1985) Angew Chem 97: 429; Angew Chem Int Ed Engl 24: 412

244. Pabon RA, Bellville DJ, Bauld NL (1983) J Am Chem Soc 105: 5158
245. Mizuno K, Kagano H, Otsuji Y (1983) Tetrahedron Lett 24: 3849
246. Sonntag C, Hagen U, Schön-Bopp A, Schulte-Frohlinde D (1981) Adv Rad Biol 9: 109
247. Giese B, Burger J, Kang TW, Kesselheim C, Wittmer T (1992) J Am Chem Soc 114: 7322
248. Symons MCR, Wren BW (1984) J Chem Soc, Perkin Trans II 511
249. Behrens G, Bothe E, Koltzenburg G, Schulte-Frohlinde D (1980) J Chem Soc, Perkin Trans II 883
250. Behrens G, Bothe E, Eibenberger J, Koltzenburg G, Schulte-Frohlinde D (1978) Angew Chem 90: 639, Angew Chem Int Ed Engl 17: 604
251. Cariou M, Simonet J, Toupet L (1987) Tetrahedron Lett 28: 1275
252. Le Moing MA, Le Guillanton G, Simonet J (1981) Electrochim Acta 26: 139
253. Shono T, Matsumura Y, Hamaguchi H, Imanishi T, Yoshida K (1978) Bull Chem Soc Jpn 51: 2179
254. Michel M-A, Martigny P, Simonet J (1975) Tetrahedron Lett 3143
255. Belleau B, Au-Young YK (1969) Can J Chem 47: 2117
256. Lopez L, Calò V, Stasi F (1987) Synthesis 947
257. Gilbert BC, Norman ROC, Williams PS (1980) J Chem Soc, Perkin Trans II 647
258. Lopez L, Troisi L (1989) Tetrahedron Lett 30: 489
259. Adam W, Hadjiarapoglou L, Klicic J (1990) Tetrahedron Lett 31: 6517
260. Rozen S, Mishani E, Kol M (1992) J Am Chem Soc 114: 7643
261. Fukuzumi S, Fujita M, Otera J, Fujita Y (1992) J Am Chem Soc 114: 10271
262. Gassman PG, Yamaguchi R (1979) J Am Chem Soc 101: 1308
263. Miller LL, Nordblom DG, Mayeda EA (1972) J Org Chem 37: 916
264. Turecek F (1990) In: Rappoport Z (ed) The chemistry of enols. Wiley, New York, p 95
265. Schmittel M, unpublished results
266. Kim K, Mani SR, Shine HJ (1975) J Org Chem 40: 3857
267. Fuhrhop JH (1982) Bio-organische Chemie, Thieme, Stuttgart
268. Kobayashi H, Akamine H, Okawa Y, Ohno T, Mizusawa S (1991) Electrochim Acta 36: 1649
269. Wilson RM, Hannemann K, Heineman WR, Kirchhoff JR (1987) J Am Chem Soc 109: 4743

Thermal and Light Induced Electron Transfer Reactions of Main Group Metal Hydrides and Organometallics

Wolfgang Kaim

Institut für Anorganische Chemie der Universität, Pfaffenwaldring 55, 70550 Stuttgart, FRG

Table of Contents

In spite of their practical importance for organic synthesis and industrial catalysis, main group metal hydrides and organometallic compounds without active d orbitals at the metal center have received relatively little attention with regard to their obvious potential for electron transfer reactivity. One reason for this neglect lies in frequent complications arising from solvent-dependent self-association

Topics in Current Chemistry, Vol. 169
© Springer-Verlag Berlin Heidelberg 1994

and from carbanion or hydride exchange equilibria. These complications are due to polar metal-to-carbon or metal-to-hydrogen bonds with the simultaneous presence of a coordinatively un-saturated electrophilic metal center and electron-rich carbanionic or hydridic substituents. On the other hand, this special, electronically ambivalent situation favors the coordination-assisted, i.e. inner sphere electron transfer to unsaturated acceptor substrates via potentially photoreactive charge transfer complexes.

1 Scope and Introduction

While there has been an explosive growth recently in the synthesis, physical characterization and application of transition metal organometallic compounds [1–3], the organic and hydride derivatives of the main group metals certainly enjoy a quantitatively wider distribution due to their bulk availability and because of their long recognized usefulness, especially in organic synthesis [1, 4]. Thus, organometallic and hydridic compounds of lithium, magnesium, zinc or aluminium are being used on quite a large scale as catalysts and in the reductive transfer of H or of organic substituents, i.e. in the formation of carbon–hydrogen or carbon–carbon bonds.

$$
\begin{array}{ccc}
\mathrm{R}^{\delta-} & \mathrm{C}^{\delta+} & \mathrm{R}-\overset{|}{\mathrm{C}}- \\
| & + & \| & \longrightarrow & | \\
\mathrm{M}^{\delta+} & | \; \mathrm{E}^{\delta+} & \mathrm{M}-\mathrm{E} \; | \\
\mathrm{L_n} & & \mathrm{L_n} & & \mathrm{R} = \mathrm{H, \; alkyl, \; aryl}
\end{array}
\tag{1}
$$

This reaction, formulated in a concerted way in Eq. (1), describes an organic type of "reduction" which can, at least formally, be separated into elementary steps including single electron transfer (SET) processes [5]. Transition metal chemistry, including the organometallic variety, has no problems with SET mechanistic concepts because odd-electron oxidation states are well established, especially for the $3d$ metals [3, 6, 7]. On the other hand, most main group metals display only one or two stable even-electron oxidation states and one-electron transfer reactions, e.g. of low-valent indium or tin compounds with suitable acceptor substrates, are thus viewed as something of an oddity [8]. While most organic reduction/oxidation reactions are regarded as two-electron processes, Tributsch has pointed out recently the mechanistic dilemma of one-electron vs multi-electron reactivity from the point of electrochemical electron transfer [9]. In fact, there is an increasing number of reactions between organic compounds and organometallic or hydride derivatives of the main group elements (or, more to the point, of metals with non-active d orbitals) which seem to involve single electron transfer steps, i.e. the splitting of electron pairs.

Major reviews of organometallic electron transfer reactivity, including main group metal examples, have appeared in the form of Kochi's book [10] and reviews by Chanon [11] and Kochi [12]. Other review articles focused on the SET reactivity of Grignard reagents [13], on aspects of coordination [14–16] and on the validity of certain mechanistic criteria [17–19]. This article is an attempt to summarize recent developments in organometallic electron transfer reactivity where changes in the occupancy of d orbitals are not involved. The scope of this article thus includes organometallic compounds of metals from group 1 (Li), 2 (Mg), 4 (Ti^{IV}), 12 (Zn, Hg^{II}), 13 (Al) and 14 (Si, Sn^{IV}). Because of their similar electronic structure and uses [20] the main group metal hydrides and their electron transfer behavior will also be treated. The long neglected field of photoinduced – especially visible light-induced – electron transfer reactivity of main group [21] organometallics is introduced here and the largely untapped synthetic potential will be pointed out.

Although the identification of SET process has sometimes been controversial due to the mechanistic alternative between a "normal" (i.e. concerted, see Eq. (1)) but polar $S_N 2$ reaction and an SET process (cf. below [5, 22, 23]) this article will have to refer to all major articles in which SET reactivity from main group organometallics has been claimed.

2 Mechanistic Concepts and Methodical Aspects

The application of organometallic compounds or hydrides of the more electropositive main group metals lithium, magnesium or aluminium in organic synthesis, catalysis and polymerization is due to the polarity of the metal/carbon and metal/hydrogen bonds, i.e. to the considerably unsymmetrical electron distribution $M^{(\delta+)} - C^{(\delta-)}$ or $M^{(\delta+)} - H^{(\delta-)}$, respectively. While there are rarely any "free" species H^- or even R^- (R = alkyl, aryl) in the absence of charge-compensating electrophiles [20, 24], the term "carbanion" is now pervasive in the literature and will be used here with the above caveat in mind. The high density of negative charge at the metal-bound carbon or hydrogen makes these "ligands" susceptible to interactions with electron-deficient species. Typical acceptor substrates should thus be capable of removing one or two electrons from the metal-carbon or metal-hydrogen σ bond, thus contributing to its cleavage.

$$
\begin{array}{c}
R^{\delta-} \\
(\uparrow\downarrow) \\
M^{\delta+} \\
L_n
\end{array}
\xrightarrow{-e^-}
\begin{array}{c}
R\rceil^{+\cdot} \\
(\uparrow) \\
M \\
L_n
\end{array}
\longrightarrow L_n M + R\cdot
\tag{2}
$$

It should be pointed out here that the geometries of molecular radical cations as primary one electron-oxidized forms and of their neutral precursors can be quite different, leading to high reorganization energies in the sense of the Marcus concept [5]. One of the better known examples in the field of main group organometallic compounds is $SnMe_4$ which apparently distorts from a tetrahedral to an EPR-detectable trigonal pyramidal species $SnMe_4^+ \cdot$ on oxidation [12, 25], accompanied by a comparatively large reorganization energy of about 60 kcal/mol [26].

One-electron processes can be conveniently studied in electrochemical experiments where the oxidation of most simple main group organometallics and hydrides occurs at rather low potentials but generally irreversibly (Eq. (2)) [10, 12, 27, 28]. High reactivity of such cleavage products can even affect the electrochemical process through electrode adsorption.

It is important to note that most main group organometallics and hydrides have both donor *and* acceptor properties. While the carbanionic or hydridic ligands R^- or H^- at the negative end of the polar metal-to-carbon or metal-to-hydrogen bond have a tendency to donate electrons (→ "reduction"), many species L_nM-R (R = H, alkyl, aryl) can also act as electron pair-accepting Lewis acids due to the electron deficiency at a coordinatively unsaturated metal center (cf. Eq. (1)). Of the organometallic compounds discussed here, only the tetrasubstituted silanes SiR_4 are coordinatively so inert that they cannot function as Lewis acids. Whereas the larger Sn(IV) and Ti(IV) centers can easily adopt coordination numbers of 5 or 6 with average-sized ligands, the lower-valent metals such as Li(I), Mg(II), Zn(II) or Al(III) tend to achieve the coordination number 4 in most of their compounds.

Optimal substrates for a synergistic [15, 16] inner-sphere [5, 12] interaction with the electronically ambivalent species L_nM-R (R = H, alkyl, aryl) have a similarly ambivalent acceptor/donor make-up: They should have electron pair-donating heteroatoms such as O, N or S in order to form coordinative bonds to the coordinatively unsaturated metal center of the organometallic compound (Eq. (1)), at the same time, a low-lying unoccupied orbital such as the LUMO (π^*) of a π system is required for the acceptor function.

To a first approximation, both the electronegativity of the metal (Li < Mg < Al < Zn ≈ Si ≈ Sn) and the inductive effects of the substituents R determine the ionicity of the bond and thus the electron donating capability of the "carbanion" as measured e.g. by ionization potentials [10, 12]. While not being strictly organometallic compounds, many main group metal hydrides MH_n behave similarly to MR_n in their propensity towards SET reactivity during reduction of substrates. Although hydride ligands are generally considered to be less electron-rich than e.g. alkyl carbanion ligands, the small size of the hydride ligand often lowers the kinetic barrier for inner-sphere SET reactivity as exemplified by reactions with N-heterocycles of the pyridine type such as 2,2'- or 4,4'-bipyridine: AlH_3 [29] and iso-Bu_2AlH [30] give radical products typical for SET processes while AlR_3 forms just Lewis acid/Lewis base complexes with charge transfer character [28, 31].

(3)

Coordinative unsaturation is an essential requirement for the strong promotion and potentially high selectivity of inner-sphere [5, 12] electron transfer through strong anisotropic metal/ligand interactions [15, 16]. On the other hand, many coordinatively very unsaturated organometallics such as RLi, R_2CuLi, RMgHal, R_3Al and corresponding hydrides, amides or alkoxides can thus exist (and react) in solution in a variety of forms, e.g. as solvate complexes or as aggregates with various degrees of oligomerization [1, 24], which complicates the mechanistic situation considerably. Association equilibria have then to be considered for a meaningful thermodynamic or kinetic treatment [32].

Furthermore, one of the most remarkable and synthetically valuable but mechanistically complicating feature of many main group organometallic or hydridic species lies in the ability of carbanionic groups or hydrides to structurally bridge two or more electrophilic centers, thus facilitating R^- or H^- transfer. Even in the absence of a substrate such intermolecular atom or group transfer processes are evident from electrical conduction due to self-dissociation (Eq. (4)), especially in ion-stabilizing solvents [33, 34].

$$2R_nM \rightleftharpoons [R_{n-1}M]^+ + [R_{n+1}M]^- \quad \text{(“at-complex”)} \qquad (4)$$

(all species solvated)   M: e.g. Al (n = 3), Mg (n = 2)

Whereas certain at-complexes such at $[R_2Cu]^-$ [35–37], $[R_3Zn]^-$ or $[RHgI_2]^-$ [35] are capable of single electron transfer because of their still existing coordinative unsaturation and negative charge, the cationic species $[R_{n-1}M]^+$ are very much suited to form complexes, i.e. organometallic radical ion pairs [14, 38], with reduced acceptors after the SET process.

The basic mechanistic features of a single electron transfer reaction between an electron acceptor A and an (organometallic) donor L_nM–R (R = H, alkyl, aryl) have been discussed at various places [5, 10–16, 38, 39] and may be summarized as follows (Eq. (5)):

Acceptor and donor have to approach each other by diffusion in order to allow a first molecular interaction. In the case of outer sphere-processes [5, 40] with coordinatively saturated components this initial interaction can be quite weak, having to rely e.g. on induced dipole forces if the reaction partners are electroneutral. Ionic species (ion pairs), formed via process (Eq. (4)) or otherwise, have the advantage of long-range coulombic interaction [12, 41]. In contrast, coordinatively unsaturated partners can form the precursor complex [5, 38, 40] of an inner sphere-process which allows for some charge shift but no full electron transfer. Such precursor complexes can exhibit conspicuous “charge transfer absorptions”, yielding electronically excited states which are best described as involving a major exchange of charge between donor and acceptor component [42, 43]. Charge transfer excited states (charge transfer as internal redox process) may convert to the ground state via radiative or non-radiative mechanisms, however, they may also eventually lead to chemical reactivity involving initial bond breaking [12, 44].

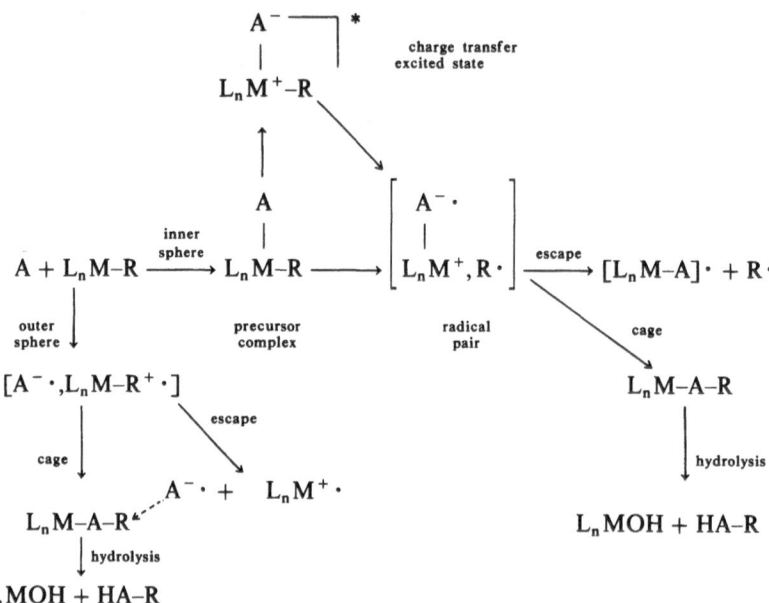

$$(5)$$

The complexes formed after photoinduced or thermal electron exchange in a "solvent cage" can be described within the radical-pair formalism [45, 46]. For instance, neutral even-electron components should yield pairs of radical ions in the case of weak coordinative interactions (outer sphere regime). The alternative is a strong coordination between the reaction partners (inner sphere pathway) with a considerable donor-to-acceptor electron redistribution which, already at this stage, can result in the formal homolysis of a bond within one of the former components (cf. the S_H2-mechanism [11, 54]). Such a bond is likely to be a metal-carbon or metal-hydrogen bond (Eq. (2)), creating an organic or hydrogen radical and a radical complex very close to each other in a solvent cage.

The outer sphere alternative produces oppositely charged ions and is thus energetically more difficult, however, solvation in media of high dielectric constant can facilitate the charge separation. The inner sphere alternative produces neutral species and is favored from the electrostatic perspective.

Although charge separations are energetically always uphill, electron transfer steps can sometimes help to circumvent high activation energy paths as in normally spin-forbidden reactions with triplet dioxygen, $^3O_2$. For instance, the η^3-tris(3-tert-butylpyrazolyl)hydridoborato-organomagnesium compounds L^3MgR, R = Et, iso-Pr, tert-Bu, were found to rapidly insert $^3O_2$ into the Mg–C bond by a mechanism which is likely to involve initial (spin-allowed) SET, Mg–C bond cleavage and peroxymetal radical formation as initiation sequence and subsequent radical activation of $^3O_2$ within a radical chain process (Eq. (6)) [11, 47]. The direct ("concerted") insertion of ground-state $^3O_2$ into M–C bonds is considered to be unlikely since the primary products would be formed in a high-energy triplet state [47].

initiation

$$L^3Mg\text{–}R + {}^3O_2 \longrightarrow [L^3Mg\text{–}R]^+ \cdot + O_2^- \cdot$$

$$[L^3Mg\text{–}R]^+ \cdot \longrightarrow L^3Mg^+ + R\cdot$$

radical chain

$$R\cdot + O_2 \longrightarrow ROO\cdot$$

$$L^3Mg\text{–}R + ROO\cdot \longrightarrow L^3MgOOR + R\cdot$$

termination

$$L^3Mg^+ + O_2^- \cdot \longrightarrow L^3MgOO\cdot$$

$$L^3MgOO\cdot + R\cdot \longrightarrow L^3MgOOR \tag{6}$$

The example in Eq. (6) illustrates nicely that both fundamental physical properties of the electron, charge and spin, play a role in determining the course and energetics of electron transfer reactions.

In either case of alternatives (Eq. (5)), the pairs of radical species can react within the solvent cage to yield a "cage" or "recombination" product which is then typically hydrolized to remove the metal. The alternative to the reaction inside the solvent cage is the "escape" process which can be associated with a spin conversion from the singlet to the triplet state [45, 46]. Main group organometallic radical cations, the most likely primary products of the outer sphere electron transfer, are formed fairly easily as evident from low ionization energies and oxidation potentials [10, 12, 48], however, they are highly reactive in most instances with respect to proton transfer or dissociation (Eq. (2)) [10, 12].

Details of possible electron transfer steps within the cage processes are not easily accessible by physical methods except for fast spectroscopic techniques [12], however, the radical products of the escape pathway may be identified spectroscopically, e.g. by EPR, by kinetic analysis or by a closer look at the product distribution which can reflect typical follow-up reactions of radicals. While organometallic cation radicals tend to cleave the M–C bond because of the σ^1 configuration (Eq. (2)) or lose a proton due to the typically [49] enhanced acidity of cation radicals, the neutral free organic radicals $R\cdot$ usually undergo dimerization, disproportionation or hydrogen abstraction. Of particular value are certain unsaturated free radicals such as 5-hexenyl which are known to cyclize (Eq. (7)) with a defined rate constant (radical clocks) [17–19, 50]. Other indications of intermediate free radical formation in reactions of organometal-lics involve enhanced β-cleavage reactions (see Eq. (10)) [51], loss of stereochemical information [52], typical rearrangements [53], and the initiation of characteristical radical chain processes [11, 35, 54].

$$\tag{7}$$

Reduced acceptor substrates such as π radical anions can form complexes with neutral or cationic metal species which may be quite persistent, especially if the acceptors contain good metal coordination sites such as carbonyl oxygen or imine nitrogen groups (3, 8) [14, 78].

$$\mu\text{-L} + 2\text{MgR}_2 \longrightarrow [\text{R}_2\text{Mg}(\mu\text{-L}^- \cdot)\text{Mg}^+\text{R}] \cdot + \text{R} \cdot \qquad (8)$$

μ-L: bridging acceptor ligand, e.g. 4,4'-bipyridine (3)

Mechanistic evidence for outer sphere SET is not easy to obtain for reactions which lead to conventional products. Instrumental techniques such as EPR or CIDNP often fail to show direct evidence for radical production because of short radical life-times [55]. Spin trapping agents [56] may interfere with the reaction from the start, especially since the widely employed nitroso compounds are good ligands for coordinatively unsaturated organometallics [57]. Even if rather persistent radicals are established via instrumental detection it is not always certain that they arise directly from electron exchange between the precursor molecules or that they are even related to the major reaction [58].

A frequently used indirect method involves cyclizable (cf. (7)) or other "mechanistic probes" which should provide evidence for free radical intermediates and thus for SET [19, 37, 59]. However, Newcomb and Curran have pointed out the pitfalls of such an approach especially if iodide precursors are used [17]. The supposedly radical-indicative reaction may come about albeit slower by a different, nonradical mechanism or the radical formation may occur via a secondary process which is not directly related to the first reaction step. A similar side-route can be made responsible for the appearance of stable radical compounds which may arise via a comproportionation reaction between non-reduced starting material and the doubly reduced species which can be formed from a hydro form (the "normal" product, Eq. (5)) and the usually strongly basic organometallic or hydridic reagents (Eq. (9)) [58]. The ability of strong bases to produce reduced radical species via complicated electron/proton transfer processes has been known for some time in the chemistry of quinones and quaternary salts [60, 61].

$$A \xrightarrow{\underset{\text{process}}{\text{non-SET}}} AH_2$$

$$AH_2 + 2L_nM\text{-R} \longrightarrow (A^{2-})(L_nM^+)_2 + 2H\text{-R} \qquad (9)$$

$$A + (A^{2-})(L_nM^+)_2 \longrightarrow 2(A^- \cdot)(L_nM^+)$$

Much controversy has thus existed around the alternative between a normal but polar S_N2 reaction and a SET process before it was realized that the inner sphere SET variety is in effect very close to the polar S_N2 mechanism [5, 22, 23]. In principle, only the outer sphere SET reaction leading to well separated electron exchange products can be treated simply in terms of individual "free" radical species with their specific reactivity [5, 12]. The "freedom" of radicals in this context refers to their existence within a chemical meaningful period of time and to the full set of motional degrees of freedom.

3 Thermal Processes

3.1 Group 1: Lithium Compounds

Reactions of basic lithium compounds LiR or $LiER_n$ with halides [19, 37, 59], pseudohalides [62] or unsaturated acceptors [63–65] were frequently reported to involve SET steps. The strongly basic character of such lithium compounds may, however, cause the observation of apparent one electron-reduced species such as ketyls or semiquinones via deprotonation of hydrolysis products and ensuing comproportionation reactions (Eq. (9)) [58].

The analysis and interpretation of kinetic isotope effects (KIE) from the carbonyl carbon has been used to postulate SET processes for the first step in the reactions between ketones and MeLi or Me_2CuLi; however, the rate-determining steps within the overall mechanistic scheme in Eq. (5) depend on the steric and electronic properties of the substrate [63].

Organolithium species may form coloured and EPR-active intermediates not only with conventional acceptors such as carbonyl or imine compounds [14–16] but also with unsaturated "inorganic" ions such as ambidentate thiocyanate-SCN; the reaction between n-BuLi and NH_4SCN in the presence of strongly ion-stabilizing hexamethylphosphoric triamide $O=P(NMe_2)_3$ (HMPA) eventually produced $LiNCS \cdot 2$ HMPA via some unusually colored intermediates and thiocyanate-based radical species [62].

Mechanistic probes for radical formation in the form of rapidly cyclizing 5-hexenyl groups (cf. (7)) or the rapidly β-cleaving 2,3,3-trimethyl-2-butoxy (triptoxy) species (Eq. (10)) were used to detect SET processes in the reactions between RLi or RMgHal reagents with acetylenic or vinylic sulfones [64] and peroxybenzoates [51].

$$\cdot OC(Me_2)CMe_3 \longrightarrow OCMe_2 + \cdot CMe_3 \tag{10}$$

Complex lithium-containing reducing reagents such as $LiCuR_2$ [35–37, 63], $LiAlH_4$ [19, 66, 67] or $LiBEt_3H$ [68] were also shown to yield diamagnetic and paramagnetic products which are typical for SET processes. However, the reaction mechanism is further complicated here because of the possible separation between the at-donor anion and the σ-accepting Li^+.

3.2 Group 2: Magnesium Compounds

The majority of main group organometallic compounds for which SET mechanisms were postulated involves Grignard reagents [13, 69–75] in their reactions with carbonyl compounds. Even the formation of these important laboratory reagents RMgHal (which does not fall within the scope of this article) is now believed to involve electron transfer mechanisms and radical formation at the surface of metallic Mg [76, 77]. In contrast to Grignard compounds with their

239

Schlenk equilibrium [13] the simpler diorganomagnesium compounds are often easier to study [14, 78, 79] because of the absence of complications from additional halide ligands. As the organolithium compounds, organomagnesium species often react rapidly at ambient temperatures with substrates so that the charge transfer intermediates are not observed or reported.

The synthetically very useful reactions of Grignard reagents with carbonyl compounds have frequently been studied with diaryl ketones as π acceptors because the ketyl intermediates are sufficiently persistent for EPR detection [69–75]. Relevant studies include kinetic analyses which showed that the rate of disappearance of the ketyl intermediate and the rate of appearance of the alkoxide final products are related [69]. Carbonyl carbon-based kinetic isotope effects confirm the electron transfer character of such reactions even if the rate-determining step may not be the SET process [70]. The structures of long-lived organomagnesium/ketyl complexes from diarylketones and benzil acceptors were studied by Maruyama and coworkers [71–75] via optical absorption and EPR spectroscopy; triplet EPR signals point to an oligomeric structure involving two ketyls bridged by "$(RMgHal)_2^{2+}$". Kinetic analysis from stopped flow experiments points to a complex mechanism in the case of benzil, requiring two molecules of Grignard reagent. Benzophenones and benzil were also used as acceptor substrates in EPR-active SET reactions with aryliminodimagnesium reagents, $ArN(MgBr)_2$ [80, 81].

Based on the kinetic radical pair concept of Fischer [46], Walling has argued that there is no necessity to assume a long-lived cage situation in order to account for the high yield of normal addition (recombination) products in Grignard reactions [82]. In that model, the concentration of the long-lived escaped electron transfer intermediate, the ketyl (complex) will rapidly increase up to a point where the cross reaction with the highly reactive intermediate R · becomes dominant for both species and thus leads to eventual recombination after escape (Eq. (5)). This concept [46, 82] also explains conveniently why ketyl complexes are formed in high concentration as persistent species if sterically crowded and thus slowly recombining acceptors such as dimesitylketone are employed [75].

Mechanistic probes based on typical cleavage, cyclization and chain-propagating reactions of free radicals were used as evidence for SET reactivity from RMgHal to peroxy esters [51] and acetylenic or vinylic sulfones [64, 65].

Several other types of acceptors than carbonyl compounds can undergo SET reactions with organomagnesium reagents. Related to $C=O$ containing compounds are imines which may be part of heterocycles such as 2,2'- or 4,4'-bipyridine [14, 78]; these compounds and nitrosodurene [79] were reacted with dialkylmagnesium species to yield typical EPR-detectable organomagnesium radical complexes (Eq. (8)). The insertion of $^3O_2$ into Mg–C bond of L^3MgR which is likely to have an SET as first step has been described before (Eq. (6)) [47]. 1,2-Diphenyl-substituted cyclopropenylium cations were postulated to react with allylic or benzylic Grignard reagents via SET because the 1,3-diphenyl isomer was observed as major product which implies the intermediate formation of a conjugatively stabilized 3-aryl-substituted cyclopropenylium radical as reduction product [53]. A stable organic cation radical salt, viz., thianthrene perchlorate, was reacted with RMgCl in an SET process to yield thianthrene, follow-up products of R · (RH, RR, alkene, cyclized compounds in the case of 5-hexenyl), and some coupling product (5-alkylthianthrenium cation) [83].

3.3 Group 4: Titanium(IV) Compounds

In the search for more stereoselective organometallic reagents for reductive alkylation the groups of Reetz and Seebach have reported remarkable improvements when organotitanium(IV) compounds were used instead of organolithium or Grignard reagents [84–87]. The high and often particular selectivity of reactions, e.g. of $(RO)_3TiR$, is attributed to the tendency to form strong chelate complexes with suitable substrates in the transition state [84–86]; higher coordination numbers of up to 6 are readily tolerated by the relatively large metal center.

The possibility of electron transfer reactivity is complicated here, however, because Ti(IV) with d^0 configuration can be reduced to the fairly stable d^1 species Ti(III) at not too negative potentials. In fact, an EPR/ENDOR study of the reactions of $(iso\text{-}PrO)_3TiCH_3$ with various acceptor substrates [88] has shown a striking dichotomy of paramagnetic products, based on the ambivalence formulated in Eq. (11).

$$(A^- \cdot)(iso\text{-}PrO)_3Ti^{IV} \longleftrightarrow (A)(iso\text{-}PrO)_3Ti^{III} \qquad (11)$$

With compounds of rather negative reduction potential such as dimesitylketone or 2,2'-bipyridine the reaction produced only Ti(III) species in low amount. On the other hand, good π acceptors such as o-quinones or azo-containing heterocycles yielded large amounts of Ti(IV) complexes of the anion radicals [89] according to Eq. (5) [88]. Similar Ti(IV) semiquinone complexes were reported from the reactions of photogenerated $CpTiCl_2$ with o-quinones [90].

3.4 Group 12: Zinc and Mercury Compounds

In comparison to organmagnesium compounds, the corresponding organozinc species are considerably less reactive. The propensity of zinc for lower coordination numbers is obvious from the existence of R_2Zn as fairly stable molecular species with linear structures [91]. Additional coordination is possible, however, inducing an increased reactivity through bending of the C–Zn–C angle [38, 92]. This enhanced reactivity is mainly caused by the destabilization of the $\sigma(Zn–C)$ orbitals [38] which is also apparent from long-wavelength charge transfer absorptions $\sigma(Zn–C) \rightarrow \pi^*$ in some instances [38, 93–95].

Theoretical, spectroscopic and product distribution studies [38, 95, 96] have been reported for the particularly simple system (Eq. (12)) in which the linear molecules $R'-Zn-R'$ ((Eq. (12 B)) and small chelating 1,4-diaza-1,3-diene acceptors (Eq. (12 A)) react thermally via well-defined, structurally characterized coloured intermediates (Eq. (12 C)), the precursor complexes of Eq. (5), to yield a remarkable spectrum of diamagnetic (Eq. (11 E, G, H)) and paramagnetic products (Eq. (12 F, I)). Reactivity of the precursor and the product distribution were found to be highly dependent on the nature of the carbanionic group R, the more electron rich Zn–C bonds as in $(t\text{-Bu})_2 Zn$ giving rise to higher reactivity, lower charge transfer energies and increased C-alkylation products [38, 95].

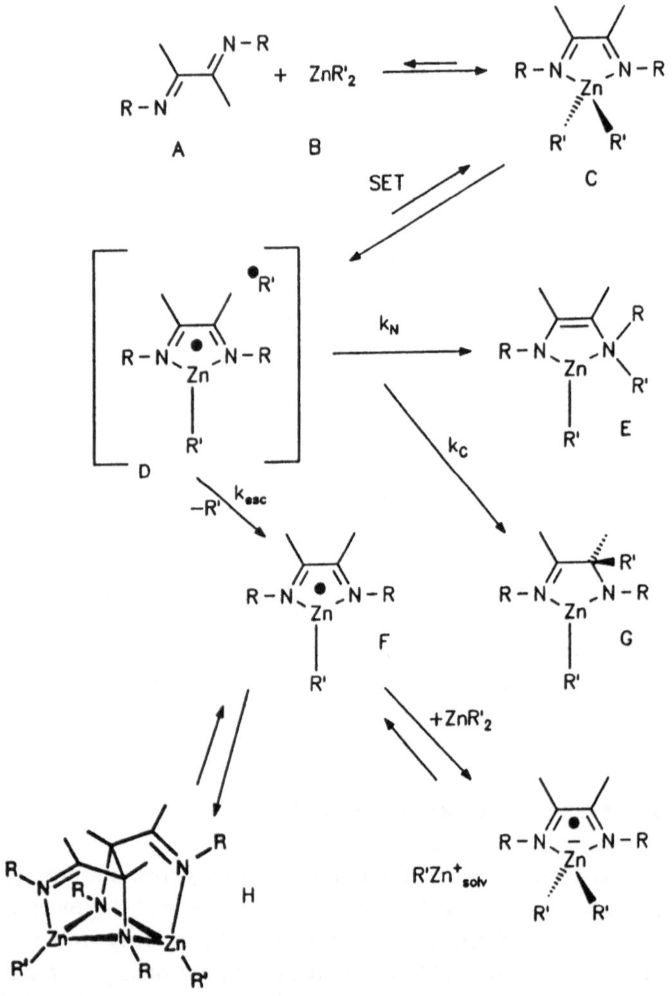

$$(12)$$

The almost optimal orbital correspondence between the antisymmetric HOMO combination of $\sigma(Zn-C)$ and the π^* LUMO of the heterodiene chelate ligand (13) invites the postulation of an intracomplex electron transfer to which one of the labilized ($\sigma^2 \to \sigma^1$) Zn–C bonds reacts with cleavage (Eq. (12 D)) [38]. The recombination of the thus generated alkyl radical with the π stabilized radical complex depends very much on the size and reactivity of R·; the less reactive t-Bu radical attacks further away from the metal site, yielding exclusively (Eq. (12 G)) [95]. The neutral radical complex (Eq. (12 F)) formed as persistent escape product was found to dimerize (C–C coupling, Eq. (12 H)) [96] or formally add a carbanion to yield the anion radical (Eq. (11 I)) of the precursor molecule [38]. The necessary formation of RZn$^+$ species is facilitated by strongly coordinating solvent molecules. The alternative photoinduced reaction of the precursors can yield a different spectrum of paramagnetic and diamagnetic products, indicating a different spin state situation (triplet vs singlet) for the assumed intimate radical pair intermediate (Eq. (11 D)) [38].

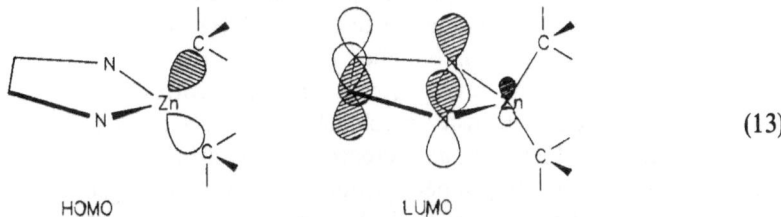

$$(13)$$

HOMO   LUMO

The group transfer regioselectivity exhibited in apparent electron transfer reactions (Eq. (12)) of 1,4-hetero-1,3-dienes with certain organometallics [95, 97] can be exploited to obtain synthetically useful organic compounds [95, 98], including even β-lactams [99].

Although less widely employed than the related dialkylcuprates, zincates such as LiZnR$_3$ can also undergo electron transfer reactions, e.g. with free radicals as chain carriers [35].

According to the "noble" character of Hg the ionization energy of the Hg–C bond in alkylmercury compounds is rather high [10, 12]. On the other hand, the well known tendency of mercury to exist in the non-oxidized elementary form gives rise to reactions which may be interpreted as involving electron transfer processes. For instance, the mercurate [HgI$_2$(t-Bu)]$^-$ was found to react similarly to [Cu(t-Bu)$_2$]$^-$ and [Zn(t-Bu)$_3$]$^-$ with oxidizing radicals [35]. In most instances, thermally [100], chemically [12, 35] or photoinduced [65, 100–103] radical formation from RHgH [100] or RHgHal and the (electron transfer) reactivity of these free radicals determine the chemistry of organomercury compounds. As in the case of RZnR, the linear and thus very open structure of organomercury(II) compounds offers ample opportunities for inner sphere electron transfer with large reorganization energies [12].

3.5 Group 13: Boron and Aluminium Compounds

The less electropositive character of trivalent aluminium and the fairly strong bonds to carbon render the cheap and in-bulk available trialkylaluminium compounds much less suitable for reductive alkylations than e.g. Grignard reagents. Only very strong π acceptors such as α-diimines [96] or o-quinones [104] which are suitable for chelation-assisted electron transfer according to Eqs. (12, 13) [16, 38] seem to undergo reductive SET reactions with trialkylaluminium. However, the substitution of one or more alkyl groups by a hydride ligand enhances the reactivity so that diamagnetic and paramagnetic reduction products can be detected [29, 30]. The latter may also be a product of apparent SET processes from AlR_3 [105] since traces of R_2AlH are notoriously difficult to remove from AlR_3 preparations. Both diisobutylaluminium hydride (DIBAL) [30] and AlH_3 itself [29] were shown to yield well defined radical complexes as products of apparent SET reactions with N-heterocycles such as 2,2'- or 4,4'-bipyridine (3). Studies of kinetic isotopic effects from the carbonyl carbon of benzophenones suggested that DIBAL reacts in hexane via SET in the rate-determining step, in contrast to ether or THF solutions of BH_3, AlH_3, 9-BBN (9-borabicyclo[3.3.1]nonane) or $LiAlH_4$ [106].

Boranes are generally less electron rich than corresponding aluminium analogues and electron transfer mechanisms are usually not considered for reduction reactions. However, an increase of the hydridic character in complex hydrides such as $LiBEt_3H$ ("super hydride") [68] or $LiAlH_4$ [66] allows for the formation of well-characterized radical products in reactions with unsaturated acceptor heterocycles such as (3). Electron transfer mechanisms for the reduction by complex hydrides should be quite intricate because the coordinatively saturated donor moiety (MH_n^-) and the σ acceptor part (e.g. Li^+) can now well separately interact with the coordinating π acceptor substrate.

3.6 Group 14: Silicon and Tin Compounds

The relative strength of the Si–C bond, the less electropositive character of Si and the typical coordinative saturation with a coordination number of 4 are responsible for the fact that organosilicon reagents will generally not engage in reductive alkylation reactions or other potential inner sphere electron transfer processes. However, the rather easy construction of fairly complex organosilicon compounds has allowed to synthesize some special systems which can apparently undergo outer or even inner sphere electron transfer reactivity with extremely strong π acceptors such as tetracyanoethylene (TCNE) [15, 16, 28, 107, 108, 121].

Four types of compounds need to be mentioned in this respect:

i) Compounds with several strategically positioned electron-rich Si–C [107, 109] or Si–O bonds [110, 111] in σ/π-(hyper)conjugative interaction with ethers, amines or carbon π systems [109, 112, 113] such as (14) [107b],
ii) very electron-rich organosilicon-stabilized enamines such as (15) [28, 114],
iii) solvated silyl anions $MSiR_3$, $M = Li, Na, K$, [115], and
iv) disilanes and (cyclo-)polysilanes with destabilized HOMOs [108, 116, 117].

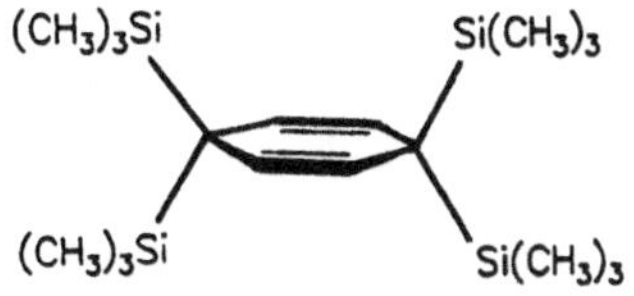

(CH$_3$)$_3$Si   Si(CH$_3$)$_3$

(CH$_3$)$_3$Si   Si(CH$_3$)$_3$

$$(14)$$

$$
\begin{array}{c}
\text{CH}_3 \\
\text{H}_3\text{C}\cdot\text{Si-CH}_3 \\
| \\
\text{N} \\
\\
\text{N} \\
| \\
\text{H}_3\text{C}\cdot\text{Si-CH}_3 \\
\text{CH}_3
\end{array}
$$

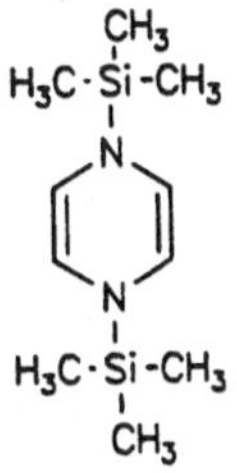

$$(15)$$

The latter compounds have been much studied with respect to (UV-)photo-induced electron transfer [108, 116, 118–121] (see Sect. 4) and under the aspect of information storage (microlithography) [116]. They not only show electon loss to EPR-detectable radical cations [117] but also relatively facile electron uptake to yield radical anions [122] and Si–C cleavage products [123]. Synthetically useful [111, 124] Si–O cleavage is observed after SET oxidation of silyl enol ethers with various Lewis-acidic species (Ce^{IV}, Ti^{IV}, Mn^{III}, Cu^{II}) acting as electrophilic catalysts [124], triorganosiloxy substituents being fairly strong donor substituents [110, 111].

Although more electron rich than organosilane analogs, the tetraorganotin compounds are still relatively poor electron donors in thermal outer sphere electron transfer reactions with acceptors [10, 12]. However, the ability of organotin species to readily increase the coordination number beyond four has made it possible to establish thermal and photoinduced inner sphere SET with sterically undemanding acceptors such as TCNE or I$_2$ via charge transfer precursor compounds [12, 39]. These SET reactions of SnR$_4$ involve tin–carbon bond cleavage with subsequent addition of iodide or insertion of reduced TCNE [39]. The Sn–H bond is so labile [100, 125] that electron transfer is an obvious mechanistic alternative [11]; however, as in the case of mercury hydrides [100], the direct light- and radical-asisted homolysis is the dominating reactivity e.g. of n-Bu$_3$SnH [100, 125a].

Triorganostannates such as LiSnR$_3$ are sufficiently electron rich to be potential candidates for electron transfer to halides [52, 126]. As for the silicon analogues MSiR$_3$ [115], the electron transfer mechanisms for the reduction by such at-complexes may be quite complicated because of the solvent-induced separation between the SiR$_3^-$ donor anion and the σ acceptor cation M$^+$.

Metal–metal bonded species Me$_3$SnMMe$_3$ (M = Sn, Ge, Si) with their de-stabilized HOMO (σ_{SnM}^2) and rather large reorganization energies of about 35 kcal/mol [127a] were found to undergo thermal SET with one-electron accepting acridinium cations via a Sn–M cleavage (Eq. (16)) of the cation radical and a subsequent radical chain process involving further reducing trimethyl-stannyl radical [127b]. In the case of $Me_3MM'Me_3$ (M, M' = Ge, Si) this reaction could be induced by irradiation with visible light [120].

$$[Me_3Sn-MMe_3]^+ \cdot \longrightarrow Me_3Sn \cdot + \, ^+MMe_3 \qquad (16)$$

$$M = Sn, Ge, Si$$

4 Light-Induced Processes

Although there is an extensive literature on the photochemistry and photo-induced electron transfer reactivity of stable and comparatively "free" car-banions [128–130], the photochemistry especially of the more polar main group organometallics and hydrides has received relatively little attention [21, 43, 131, 132]. Among the reasons for this state of affairs is the difficulty of handling such species and their little-defined solvation and aggregation status in solution. Furthermore, the reduction of unsaturated acceptors by organolithium and -magnesium reagents or most hydrides proceeds thermally so rapidly in most instances that additional photochemical activation of possibly very short-lived intermediates is not seriously considered. It may be suspected, however, that the presence of efficient electron transfer chain processes induced by ambient light [133] can go undetected unless control experiments are performed under rigorous exclusion of light. On the other hand, colored charge transfer complexes have long been observed [38, 39, 93–95, 108, 134–136] with the less reactive organo-zinc, -aluminium, -tin or -silicon complexes; in a few instances these were subjected to irradiation with visible light to yield paramagnetic or diamagnetic products which can be attributed to an electron transfer process [38, 39, 108].

The origin of the long-wavelength charge-transfer bands in precursor com-plexes (Eqs. (5, 12)) between main group organometallics and π acceptor sub-strates are due to transitions from the electron rich $\sigma(M–C)$ bond to the π^* orbital of the acceptor [38]. Depending on the definition, such transitions may be referred to as metal-to-ligand or ligand-to-ligand charge transfer (MLCT, LLCT) [21, 38]. Especially the latter view is of interest for organic synthesis because it implies a defined donor/acceptor interaction between two coordinated ligands which can be stimulated to an electron transfer and follow-up reaction by irradiation [12]. The dissociative pathway must dominate the usual non-radiative deactivation of the excited state in order to effect chemical reactivity.

The comparatively poor thermal electron transfer reactivity of organosilanes has triggered fairly extensive and successful efforts to induce such reactions photolytically. Due to the transparence of most organosilanes, even with Si–Si bonds in the visible region, many such reactions were carried out using UV light of fairly high energy [116, 137, 138]. However, the formation of colored charge transfer complexes e.g. of permethylpolysilanes with TCNE [108] or the sensitization with aromatic compounds [121] make it possible to observe light-induced SET reactivity with typical radical production [108] and Si–Si bond cleavage, followed by solvolysis [121].

Although thermal and photoinduced SET reaction products are frequently identical due to converging reaction pathways [12], there are cases where synthetically useful differences can be observed [38] which can be related to different spin states of the intermediate radical pair.

Chelate coordination-assisted SET reactivity as described in Eq. (12) may be invoked in some remarkable insertion reactions of alkylaluminium porphyrins [139] and are not restricted to the main group organometallics. Photoreactions of low-spin d^6 systems with a filled t_{2g} sub-shell such as Co(III) [140] or Pt(IV) [141] with chelating π-acceptor ligands also involve CT absorptions and apparent radical pair formation (spin-trapping, CIDNP [141]) before typical follow-up reactions such as C-alkylation [38, 95, 140] occur.

While the propensity of organometallics to undergo photoinduced electron transfer reactions can lead to (often unwanted) polymerization reactions e.g. of substrates or solvents, there lies an opportunity in the controlled photopolymerization [142] via organometallics.

5 Perspectives

Following a period when single electron transfer was recognized and, perhaps over-enthusiastically, proposed as a potential mechanistic step for almost all reactions which involve main group organometallics and other bases with reducible substrates [19], there is now a more realistic appreciation of the underlying mechanisms [12], the significance of diagnostic criteria [18], and the synthetic opportunities [95].

A major part of the attractivity of SET reactions is related to the possibility of electron transfer catalysis, e.g. within chain processes related to the $S_{RN}2$ mechanism. The wide scope of this kind or reactivity has been pointed out especially by Chanon [11, 143].

In view of the trend to more controlled and stereoselective reactions with readily available, less expensive and environmentally non-problematic reagents, the light-induced inner-sphere electron transfer between M–C bonds of less polar co-ordinating organometallics (Zn, Al) and the organic substrate seems to be a particularly attractive alternative to thermal reactions from organolithium or -magnesium compounds.

Acknowledgements. Support for research on electron transfer phenomena from the Land Baden-Württemberg (Forschungsschwerpunktprogramm), the Volkswagenstiftung (Program: Organometallics in Organic Synthesis), the Deutsche Forschungsgemeinschaft and the Fonds der Chemischen Industrie is gratefully acknowledged. I also wish to thank my coworkers, most notably C. Bessenbacher and T. Stahl for their contributions.

6 References

1. Elschenbroich C, Salzer A (1991) Organometallics. A concise introduction (2nd edn). VCH, Weinheim
2. Wilkinson G, Stone FGA, Abel EW (eds) (1982) Comprehensive organometallic chemistry. Pergamon, Oxford
3. Collman JP, Hegedus LS, Norton JR, Finke RG (1987) Principles and applications of organotransition metal chemistry. University Science Books, Mill Valley (CA)
4. a) Stowell A (1979) Carbanions in organic synthesis. Wiley, New York
   b) Jenkins PR (1992) Organometallic reagents in synthesis. Oxford University Press, New York
5. Eberson L (1987) Electron transfer reactions in organic chemistry. Springer, Berlin Heidelberg New York
6. Trogler WC (ed) (1990) Organometallic radical processes. Elsevier, Amsterdam
7. Müller A, Diemann E, Junge W, Ratajczak H (eds) (1992) Electron and proton transfer in chemistry and biology. Elsevier, Amsterdam
8. Tuck DG (1992) Coord Chem Rev 112: 215
9. Tributsch H (1992) J Electroanal Chem 331: 783
10. Kochi JK (1978) Organometallic Mechanisms and Catalysis. Academic, New York
11. a) Chanon M (1982) Bull Soc Chim France II-197
    b) Chanon M, Rajzmann M, Chanon F (1990) Tetrahedron 46: 6193
12. Kochi JK (1988) Angew Chem 100: 1331; Angew Chem Int Ed Engl 27: 1227
13. Holm T (1983) Acta Chem Scand, Ser B 37: 567
14. Kaim W (1985) Acc Chem Res 18: 160
15. Kaim W, Olbrich-Deussner B (1990) In: Trogler WC (ed) Organometallic radical processes. Elsevier, Amsterdam, p 173
16. Kaim W (1992) In: Müller A, Diemann E., Junge W, Ratajczak H (eds) Electron and proton transfer in chemistry and biology. Elsevier, Amsterdam, p 45
17. Newcomb M, Curran DP (1988) Acc Chem Res 21: 206
18. Newcomb M (1990) Acta Chem Scand 44: 299
19. Ashby EC (1988) Acc Chem Res 21: 414
20. Wiberg E, Amberger E (1971) Hydrides of the elements of main groups I–IV. Elsevier, Amsterdam
21. a) Vogler A, Kunkely H (1990) Top Curr Chem 158: 1
    b) Vogler A, Nikol H (1992) Pure Appl Chem 64: 1311
22. Pross A (1985) Acc Chem Res 18: 212
23. Cho JK, Shaik S (1991) J Am Chem Soc 113: 9890
24. Buncel E, Durst T (eds) (1980, 1984, 1987) Comprehensive carbanion chemistry. Parts A–C. Elsevier, Amsterdam
25. Walther BW, Williams F, Lau W, Kochi JK (1983) Organometallics 2: 688
26. Eberson L, Nilsson M (1990) Acta Chem Scand 44: 1062
27. Bock H, Lechner U (1985) J Organomet Chem 294: 295
28. Baumgarten J, Bessenbacher C, Kaim W, Stahl T (1989) J Am Chem Soc 111: 2126 and 5017
29. Kaim, W (1984) J Am Chem Soc 106: 1712
30. Kaim W (1982) Z Naturforsch 37b: 783
31. a) Thiele KH, Brüser W (1966) Z Anorg Allg Chem 348: 179
    b) Thiele KH, Brüser W (1967) Z Anorg Allg Chem 349: 33
32. Renaud P, Fox MA (1988) J Am Chem Soc 110: 5702
33. a) Strohmeier W (1956) Z Elektrochem Ber Bunsenges 60: 396
    b) Strohmeier W, Hümpfner K (1957) Z Elektrochem Ber Bunsenges 61: 1010

34. Lehmkuhl H, Kobs HD (1965) Tetrahedron Lett. 29: 2505
35. a) Russell GA, Baik W, Ngoviwatchai P, Kim BH (1990) Acta Chem Scand 44: 170
    b) Russell GA, Ngoviwatchai P, Wu YW (1989) J Am Chem Soc 111: 4921
36. Hannah DJ, Smith RAJ, Teoh I, Weavers RT (1981) Aust J Chem 34: 181
37. Ashby EC, Coleman D (1987) J Org Chem 52: 4554
38. Kaupp M, Stoll H, Preuss H, Kaim W, Stahl T, van Koten G, Wissing E, Smeets WJJ, Spek AL (1991) J Am Chem Soc 113: 5606
39. a) Fukuzumi S, Mochida K, Kochi JK (1979) J Am Chem Soc 101: 5981
    b) Fukuzumi S, Kochi JK (1980) J Am Chem Soc 102: 2141
40. Taube H (1984) Angew Chem 19: 315; Angew Chem Int Ed Engl 23: 329
41. Billing R, Rehorek D, Hennig H (1990) Top Curr Chem 158: 151
42. Foster R (1969) Organic charge-transfer complexes. Academic, New York
43. Geoffroy GL, Wrighton MS (1979) Organometallic photochemistry. Academic, New York
44. Saeva FD (1990) Top Curr Chem 156: 59
45. a) Koenig T, Fischer H (1973) In: Kochi J (ed) Free Radicals. Wiley, New York, Vol 1, Chapter 4
    b) Steiner UE, Ulrich T (1989) Chem Rev 89: 51
46. Fischer H (1986) J Am Chem Soc 108: 3925
47. Han R, Parkin G (1990) J Am Chem Soc 112: 3662
48. Barker GK, Lappert M, Redly B, Sharp GJ, Westwood NPC (1975) J Chem Soc, Dalton Trns 1765
49. Schlesener CJ, Amatore C, Kochi JK (1984) J Am Chem Soc 106: 7472
50. Griller D, Ingold KU (1980) Acc Chem Res 13: 317
51. Hendrickson WH, MacDonald WD, Howard ST, Coligado EJ (1985) Tetrahedron Lett 26: 2939
52. Gielen M, Eynde IV (1981) J Organomet Chem 218: 315
53. Padwa A, Goldstein SI, Rosenthal RJ (1987) J Org Chem 52: 3278
54. Russell GA (1989) Acc Chem Res 22: 1
55. Eisch JJ, Kovacs CA, Chobe P (1989) J Org Chem 54: 1275
56. Rehorek D, Janzen EG (1984) J Organomet Chem 268: 135
57. Jaitner O, Huber W, Gieren A, Betz H (1986) Z Anorg Allg Chem 538: 53
58. Newcomb M, Burchill MT (1984) J Am Chem Soc 106: 8276
59. a) Ashby EC, Pham TN (1987) J Org Chem 52: 1291
    b) Ashby EC, Pham TN (1987) Tetrahedron Lett 28: 3197
    c) Ashby EC, Pham TN (1987) Tetrahedron Lett 28: 3183
60. Farrington JA, Ledwith A, Stam MF (1969) J Chem Soc, Chem Commun 259
61. Rieger AL, Edwards JO (1966) J Org Chem 53: 1481
62. Barr D, Doyle MJ, Drake SR, Raithby PR, Snaith R, Wright DS (1988) J Chem Soc, Chem Commun 1415
63. Yamataka H, Fujimura N, Kawafuji Y, Hanafusa T (1987) J Am Chem Soc 109: 4305
64. a) Eisch JJ, Behrooz M, Galle JE (1984) Tetrahedron Lett 25: 4851
    b) Eisch JJ (1984) Pure Appl Chem 56: 35
65. Russell GA, Ngoviwatchai P (1987) Tetrahedron Lett 28: 6113
66. Kaim W (1982) Angew Chem 94: 150; Angew Chem Int Ed Engl 21: 141
67. Lapinte C, Catheline D, Astruc D (1984) Organometallics 3: 817
68. Kaim W, Lubitz W (1983) Angw Chem 95: 915; Angew Chem Int Ed Engl 22: 892; Angew Chem Suppl 1209
69. Zhang Y, Wenderoth B, Su W-Y, Ashby EC (1985) J Organomet Chem 292: 29
70. Yamataka H, Matsuyama T, Hanafusa T (1989) J Am Chem Soc 111: 4912
71. Maruyama K, Katagiri T (1986) J Am Chem Soc 108: 6263
72. Maruyama K, Hayami J, Katagiri T (1986) Chem Lett 601
73. Maruyama K, Katagiri T (1987) Chem Lett 731
74. Maruyama K, Katagiri T (1988) J Phys Org Chem 1: 21
75. Ashby EC, Goel AB (1981) J Am Chem Soc 103: 4983
76. Ashby EC, Oswald J (1988) J Org Chem 53: 6068
77. Walborsky HM (1990) Acc Chem Res 23: 286
78. a) Kaim W (1982) J Am Chem Soc 104: 3833, 7385
    b) Kaim W (1983) J Organomet Chem 241: 157
79. Sobota P, Nowak M, Kramarz W (1984) J Organomet Chem 275: 161

80. Okubo M (1985) Bull Chem Soc Jpn 58: 3108
81. Okubo M, Fukuyama Y, Sato M, Matsuo K, Kitahara T, Nakashima M (1990) J Phys Org Chem 3: 379
82. Walling C (1988) J Am Chem Soc 110: 6846
83. Soroka M, Shine HJ (1986) Tetrahedron 42: 6111
84. Reetz MT (1984) Angew Chem 96: 542; Angew Chem Int Ed Engl 23: 556
85. Reetz MT Organotitanium reagents in organic synthesis, Springer, Berlin, 1986
86. Reetz MT (1991) Angew Chem 103: 1559; Angew Chem Int Ed Engl 30: 1531
87. Weidmann B, Seebach D (1983) Angew Chem 95: 12; Angew Chem Int Ed Engl 22: 31
88. Moscherosch M, Kaim W (1992) J Chem Soc, Perkin Trans 2: 1493
89. Kaim W (1987) Coord Chem Rev 76: 187
90. Vlcek Jr A (1985) J Organomet Chem 297: 43
91. Kaupp M, Stoll H, Preuss H (1990) J Comp Chem 11: 1029
92. Kitamura M, Okada S, Suga S, Noyori R (1989) J Am Chem Soc 111: 4028
93. Coates GE, Gren STE (1962) J Chem Soc 334
94. Noltes JG, van den Hurk JWG (1965) J Organomet Chem 3: 222
95. van Koten G (1988) In: de Meijere A, tom Dieck H (eds) Organometallics in organic synthesis. Springer, Berlin Heidelberg New York, p 277
96. van Koten G, Jastrzebski JTBH, Vrieze K (1983) J Organomet Chem 250: 49
97. a) Klerks JM, Jastrzebski JTBH, van Koten G, Vrieze K (1982) J Organomet Chem 224: 107
    b) van Vliet MRP, van Koten G, Buysingh P, Jastrzebski JTBH, Spek AL (1987) Organometalics 6: 537
98. Stamp L, tom Dieck H (1984) J Organomet Chem 277: 297
99. van Vliet MRP, Jastrzebski JTBH, Klaver WJ, Goubitz K, van Koten G (1987) Rev Trav Chim Pays-Bas 106: 132
100. Giese B (1986) Radicals in organic synthesis: formation of carbon–carbon bonds. Pergamon Press, Oxford
101. Russell GA, Kulkarni SV, Khanna RK (1990) J Org Chem 55: 1080
102. Russell GA, Guo D, Khanna RK (1985) J Org Chem 50: 3425
103. Russell GA, Guo D, Baik W, Herron SJ (1989) Heterocycles 28: 143
104. Razuvaev GA, Abakumov GA, Klimov ES, Gladyshev EN, Bayushkin PY (1977) Izv Akad Nauk SSSR, Ser Khim 1128
105. Ashby EC, Goel AB (1981) J Organomet Chem 221: C15
106. Yamataka H, Hanafusa T (1988) J Org Chem 53: 772
107. a) Hausen HD, Kaim W (1988) Z Naturforsch 43b: 82
     b) Hausen HD, Bessenbacher C, Kaim W (1988) 43b: 1087
108. a) Sakurai H, Kira M, Uchida T (1973) J Am Chem Soc 95: 6825
     b) Sakurai H, Sakamoto K, Kira M (1984) Chem Lett 1213
109. Bock H, Kaim W (1982) Acc Chem Res 15: 9
110. Kaim W (1985) J Organomet Chem 282: 1
111. a) Sato T, Wakahara Y, Otera J, Nozaki H, Fukuzumi S (1991) J Am Chem Soc 113: 4028
     b) Fukuzumi S, Fujita M, Otera J, Fujita Y (1992) J Am Chem Soc 114: 10271
112. a) Yoshida J, Maekawa T, Murata T, Matsunaga S, Isoe S (1990) J Am Chem Soc 112: 1962
     b) Lambert JB, Singer RA (1992) J Am Chem Soc 114: 10246
113. Bock H, Kaim W, Kira M, Osawa H, Sakurai H (1979) J Organomet Chem 164: 295
114. a) Kaim W (1987) Rev Chem Intermed 8: 247
     b) Bessenbacher C, Kaim W (1989) Z Naturforsch 44b: 511
115. a) Sakurai H, Okada A, Kira M, Yonezawa K (1971) Tetrahedron Lett 1511
     b) Sakurai H, Okada A, Umino H, Kira M (1973) J Am Chem Soc 95: 955
     c) Sakurai H, Kira M, Umino H (1977) Chem Lett 1265
116. Miller RD, Michl J (1989) Chem Rev 89: 1359
117. Bock H, Kaim W, Kira M, West R (1979) J Am Chem Soc 102: 1918
118. Nakadaira Y, Komatsu N, Sakurai H (1985) Chem Lett 1781
119. Fukuzumi S, Kitano T, Mochida K (1989) Chem Lett 2177
120. Fukuzumi S, Kitano T, Mochida K (1990) J Chem Soc, Chem Commun 1236
121. Watanabe H, Kato M, Tabei E, Kuwabara H, Hirai N, Sato T, Nagai Y (1986) J Chem Soc, Chem Commun 1986
122. Kirste B, West R, Kurreck H (1985) J Am Chem Soc 107: 3013
123. Allred AL, Smart RT, van Beek DA (1992) Organometallics 11: 4225

124. a) Gassman PG, Bottorff KL (1988) J Org Chem 53: 1097
   b) Baciocchi E, Casu A, Ruzziconi R (1989) Tetrahedron Lett 30: 3707
   c) Bhattacharya A, DiMichele LM, Dolling UH, Grabowski EJJ, Grenda VJ (1989) 54: 6118
   d) Snider RB, Kwon T (1990) J Org Chem 55: 4786
   e) Ali SM, Rousseau G (1990) Tetrahedron 46: 7011
125. a) Giese B (1989) Angew Chem 101: 993; Angew Chem Int Ed Engl 28: 969
   b) Giese B, Kopping B, Chatgilialoglu C (1989) Tetrahedron Lett 30: 681
126. Alnajjar MS, Kuivila HG (1985) J Am Chem Soc 107: 416
127. a) Fukuzumi S, Kitano T, Mochida K (1990) Chem Lett 1741
   b) Fukuzumi S, Kitano T, Mochida K (1990) J Am Chem Soc 112: 3246
128. Krogh E, Wan P (1990) Top Curr Chem 156: 93
129. Tolbert LM (1986) Acc Chem Res 19: 268
130. Fox MA (1979) Chem Rev 79: 253
131. Fox MA, Chanon M (eds) (1988) Photoinduced electron transfer, Vol D. Elsevier, Amsterdam
132. Cox A (1986–1988) In: Photochemistry (specialist periodical report). The royal society of chemistry, London, Chapter 3. Photochemistry.
133. Eberson L, Radner F (1991) Acta Chem Scand 45: 1093
134. Wittig G, Bub O (1950) Liebigs Ann Chem 566: 113
135. Mole T (1963) Aust J Chem 16: 807
136. Ashby EC, Laemmle J, Neumann HM (1968) J Am Chem Soc 90: 5179
137. a) Mizuno K, Ikeda M, Otsuji Y (1985) Tetrahedron Lett 26: 461
   b) Mizuno K, Nakanishi K, Otsuji Y (1988) Chem Lett 1833
   c) Mizuno K, Yasueda M, Otsuji Y (1988) Chem Lett 229
   d) Mizuno K, Kobata T, Maeda R, Otsuji Y (1990) Chem Lett 1821
138. a) Dinnocenzo JP, Farid S, Godman JL, Gould IR, Todd WP, Mattes SL (1989) J Am Chem Soc 111: 8973
   b) Kyushin S, Ehara Y, Nakadaira Y, Ohashi M (1989) J Chem Soc, Chem Commun 279
   c) Kyushin S, Masuda Y, Matsushita K, Nakadaira Y, Ohashi M (1990) Tetrahedron Lett 31: 6395
   d) Sulpizio A, Albini A, d'Alessandro N, Fasani E, Pietra S (1989) J Am Chem Soc 111: 5773
   e) Wakasa M, Sakaguchi Y, Nakamura J, Hayashi H (1992) J Phys Chem 96: 9651
139. a) Kuroki M, Watanabe T, Aida T, Inoue S (1991) J Am Chem Soc 113: 5903
   b) Komatsu M, Aida T, Inoue S (1991) J Am Chem Soc 113: 8492
140. Daikh BE, Finke RG (1992) J Am Chem Soc 114: 2938
141. Hux JE, Puddephatt RJ (1992) J Organomet Chem 437: 251
142. Timpe HJ (1990) Top Curr Chem 156: 167
143. a) Chanon M (1985) Bull Soc Chim France 209
   b) Julliard M, Chanon M (1983) Chem Rev 83: 425

Photoinduced Charge Separation via Twisted Intramolecular Charge Transfer States

Wolfgang Rettig

Institut für Physikalische und Theoretische Chemie, Humboldt-Universität 24 Berlin, Bunsenstr. 1
10117 Berlin

Table of Contents

Topics in Current Chemistry, Vol. 169
© Springer-Verlag Berlin Heidelberg 1994

The principles of Twisted Intramolecular Charge Transfer (TICT) formation are outlined and many examples given. The most recent developments in the understanding of these states both from the theoretical and the phenomenological viewpoint are reviewed. Emphasis is given to the effect of broadened angular distributions and to the possibility of multiple TICT channels. Application possibilities with respect to dye developments and fluorescent probes are outlined as well as various ways of intermolecular reactivity of TICT molecules.

1 Introduction

The discovery of the dual fluorescence of the simple donor–acceptor substituted benzene derivative 4N,N-dimethylaminobenzonitrile (DMABN) by Lippert et al. [1] and the subsequent model compound studies by Grabowski et al. [2–6] including rigidized and pretwisted compounds such as MIN, TMABN and CBQ gave birth to the idea of Twisted Intramolecular Charge Transfer (TICT) states.

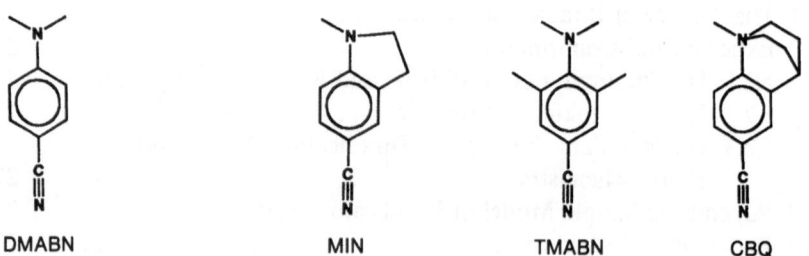

DMABN   MIN   TMABN   CBQ

This was the start of an expanding area of physical, physical-organic and organic chemistry connected with mechanistic and kinetic questions of electron transfer, and with a rationalization of the excited state behavior of many dye systems. The number of applications are growing in various fields such as tailor-made fluorescence dyes [7, 8], sensing of free volume in polymers [9, 10], fluorescent pH or ion indicators [11, 12], fluorescent solar collectors [7], and electron transfer photochemistry [13, 14]. TICT states have been reviewed several times in recent years [2, 8, 15–18] while focusing on various aspects. The present review is intended more as a summary of the important key points than as a comprehensive literature survey which can be found in some of the other reviews. Particular emphasis will be given to the outline of the latest developments with respect to shortcomings and possible extensions of the TICT model (Sects. 5–7).

One of these recent extensions is the finding that excited-state twisting of single and of double bonds can be theoretically described within the same model of biradicaloid states [17, 19, 20] (Sect. 5). Both processes can be viewed as belonging to the area of adiabatic photoreactions [20, 21] defined as occurring on the hypersurface of the lowest (singlet or triplet) excited state and leading to excited products which can be detected e.g. by their luminescence. In other cases, when ground and excited state of the product are energetically very close, excited state twisting can provide a mechanism for efficient and extremely fast nonradiative transitions to the ground state (often called "internal conversion" in a too simplistic language). An outstanding and very relevant example is the primary step of the visual process which involves a concerted combination of several adiabatic photoreaction mechanisms: of single and double bond twist processes, of charge transfer and of proton transfer and which occurs in less than a picosecond [8, 20]. There are also several "in vitro" cases, where TICT reactions are combined, either in series or in parallel, with other adiabatic photoreactions such as excited state proton transfer, or with intra- or inter-molecular excimer or exciplex formation (Sect. 7).

2 The TICT Model

The "Twisted Intramolecular Charge Transfer" (TICT) model was put forward by Grabowski and coworkers [2, 3, 15] to account for the observation that the dual fluorescence of DMABN with its "normal" band (B band) at around 350 nm and its "anomalous" one (A band, around 450 nm in medium polar solvents) depends on the conformational freedom of the dimethylamino (DMA) group: For compounds like MIN, where the DMA group is more or less fixed to a coplanar conformation with the benzonitrile sceleton, and where the lone pair orbital on the amino nitrogen is nearly parallel to the carbon p-orbitals constituting the benzonitrile π-system, only the B band is observed. For the pretwisted compounds TMABN and CBQ, where the nitrogen lone pair is nearly in-plane with the benzonitrile sceleton and perpendicular to the π-orbital system, on the other hand, only the "anomalous" A band is observed. The conclusion regarding DMABN was therefore that a reaction path exists in the excited state leading from the near planar conformation (emitter of the B band) to a photochemical product with an energetic minimum at the perpendicular conformation (emitter of the A band). These two emitters were called the B* state and A* state and were shown to possess a mother–daughter relationship later substantiated by direct kinetic measurements. In many cases, the back-reaction $A^* \to B^*$ also occurs leading to an excited state equilibrium. The ground state of DMABN is known to possess an energy barrier for the perpendicular conformation (the rotational barrier), therefore emission from the perpendicular excited-state minimum occurs to a repulsive potential and is expected to lead to structureless spectra.

The energetic situation, as it can be extracted from the spectra of DMABN and related compounds, is depicted in Fig. 1 which also shows the experimental activation energies for forward and backward reaction. It should, however, be borne in mind that these activation energies involve both the intrinsic barrier and the "dynamic" barrier resulting from the viscous properties of the solvent, and the available kinetic data are more consistent with the assumption that the forward reaction occurs without a significant intrinsic barrier [16, 22]. Another important point to emphasize is that the reaction coordinate is not simply the

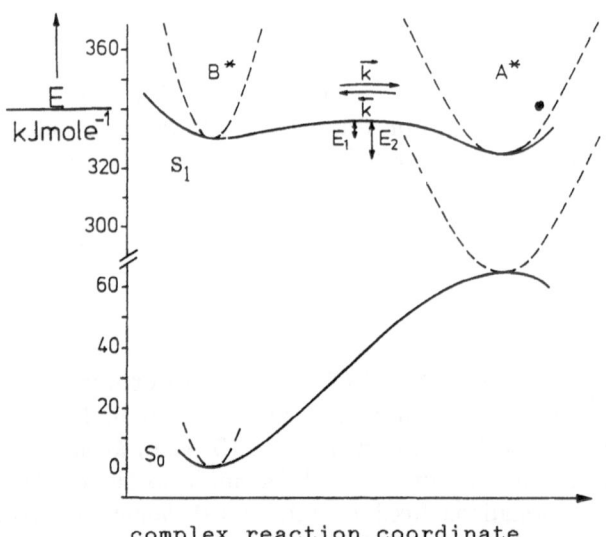

Fig. 1. Potential diagram of the TICT forming reaction for DMABN and derivatives in a medium polar solvent [24]. The reaction coordinate comprises intramolecular twist, relaxation of the surrounding solvent molecules and a further relaxation coordinate, possibly connected with the pyramidalization at the amino group [22] and is therefore multidimensional

intramolecular twisting motion but involves other coordinates, too, such as electron transfer, solvent dipolar relaxation and, most probably, some re-hybridization at the amino nitrogen ("umbrella" motion) [16, 23]. For the perpendicular TICT conformation, donor (dialkylamino group) and acceptor (benzonitrile) π-orbitals are orthogonal (zero overlap) and thus decoupled leading to a maximum for the dipole moment in the excited state (and a minimum in the ground state). This maximum of the dipole moment (near full electron transfer from donor to acceptor) connected with the energetic minimum for the perpendicular conformation are essential ingredients of the so-called "minimum overlap rule" [2]. For the near-planar conformation (B* state), mesomeric interaction between the donor and acceptor π-systems exists and diminishes the dipole moment (in strong donor–acceptor systems the difference between the dipole moments of TICT and B* state may be small, however). The reaction process is summarized in Fig. 2.

Equations 1 and 2 can be used to predict possible new TICT systems [8, 25]. Whether or not the energetic minimum of the A*/TICT state is lower than that of the precursor B* state (inequality Eq. (1) fulfilled) sensitively depends on the electron donor–acceptor properties of the subsystems which can be quantified by ionization (or oxidation) potential IP and electron affinity EA (or reduction potential) of donor D and acceptor A.

$$E_{B*} - E_{TICT} > 0 \tag{1}$$

$$E_{TICT} = IP(D) - EA(A) + C + E_{solv} \tag{2}$$

The B* state responds much less to changes in donor and acceptor proper-ties than the TICT state, and Eq. (1) can often easily be fulfilled by increasing donor and/or acceptor strength. In addition to these two factors which deliver the decisive part of the reaction driving force, polar solvent stabilization E_{solv} and the mutual Coulombic attraction C of the linked donor and acceptor radical anion/cation pair also help to preferentially stabilize the TICT state with respect to the precursor B* state. Figure 3 shows some examples.

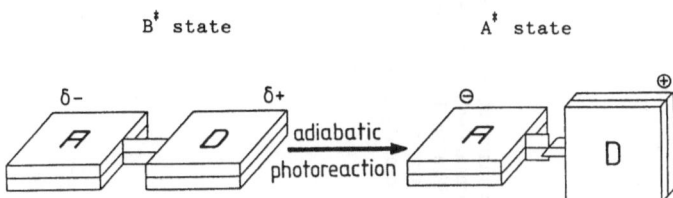

Fig. 2. The TICT model involves a twisted product species with charge transfer or charge shift properties (A* state) formed through an adiabatic photoreaction from the precursor (B*) state which is often (but not always) of near planar conformation

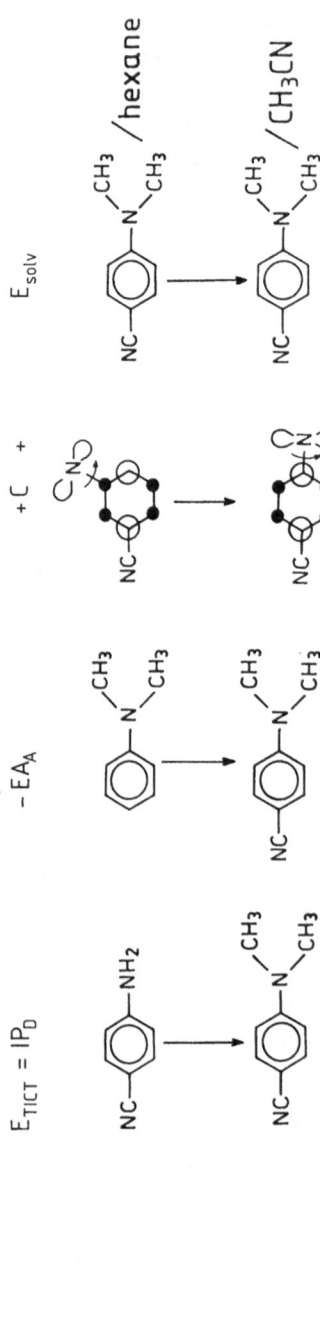

Fig. 3. The main subgroup properties determining the energy of the TICT state: Donor strength (measurable by the ionization potential *IP*) and acceptor strength (measurable by the acceptor electron affinity *EA*). In the charge separated product, coulomb stabilization *C* and polar solvent stabilization E_{solv} further help to stabilize the charge transfer state. Some examples are shown for the effect of changes in the four properties. The upper one is the non-TICT, the lower one the TICT example. In *meta*-dimethylaminobenzonitrile, several factors come together: The coulombic interaction is smaller due to the smaller MO coefficients in meta of the cyano group, but also the energy of the B* state is lowered significantly with respect to DMABN

3 Some Historical Remarks

The history of the TICT model is characterized by the more or less simultaneous proposition of several rivalling mechanisms explaining the dual fluorescence of DMABN. The more important ones are:

1. The simple solvent-induced $^1L_a/^1L_b$ level-inversion originally put forward by Lippert [1] (which does not involve an explicit intramolecular reaction coordinate).
2. Exciplex formation with the solvent [26, 27, 28].
3. Dimer formation in the ground state and/or excimer formation by encounter of excited DMABN with a ground state DMABN molecule [29, 30, 31].
4. Specific solvation with one or several water molecules [32].
5. Rehybridization (pyramidalization or planarization) at the amino nitrogen [33].

A few of these rivalling mechanisms could be rigorously excluded for DMABN under certain conditions but turned out to play an additional minor role in other conditions or for the dual fluorescence of related TICT compounds (see Sect. 4). However, in all cases known so far to the author, the TICT hypothesis (with its extensions, see Sects. 5–7) has proved to be the most general one explaining the experimental findings of many different compounds under a large variety of conditions whereas several of the above mechanisms apply only to specific conditions. All of these mechanisms have their specific limitations. The TICT mechanisms is applicable only to systems with flexible bonds (and its validity is established by comparison with the corresponding bridged system). The following limitations apply to the rivalling mechanisms outlined above: The level inversion mechanism (1) necessitates two energetically close-lying electronic levels or states for the (mostly planar) conformation of the B* state; (2): needs solvents with energetically available lone pair electrons; (3): a concentration dependence is expected; (4): needs the presence of water traces in the solvent and extensive complex formation with water molecules – not evident for biaryls (Sect. 4) and for molecules in the gas phase; (5): applies only to amines.

Unfortunately, many of these mechanisms have been claimed to be the *only* source of the dual fluorescence. From today's viewpoint, it can be said that all these mechanisms are active, for one or the other specific case, and that even more general combinations of TICT with other photoreactions can occur (Sect. 7), but none of these mechanisms *alone* can explain the anomalous fluorescence observed. A consistent picture is obtained, however, when the TICT model is used and the other mechanisms are treated as *additional factors* which are sometimes present, sometimes absent.

Very recently, a further mechanism has been added to the list of additional factors: 6. the encounter of two excited TICT molecules to form a species known as "bicimer". In this case, both a dependence on concentration and on excitation light intensity is expected. This dependence has been found for DCS, an

W. Rettig

elongated stilbene-type derivative of DMABN [34, 35]. Only the model compounds with a flexible dimethylaniline-ethylene bond (like DCS and DCS-B134) show a dual fluorescence with mother–daughter relationship, but not the model compound DCS-B24. Under low-concentration and/or low-excitation-intensity conditions, the dual fluorescence is absent for all compounds (see, however, Sect. 7 for indirect evidence of monomolecular TICT formation in this class of compounds).

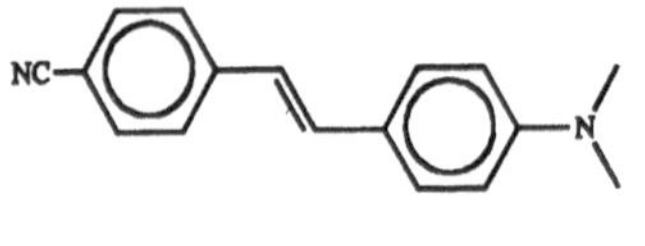

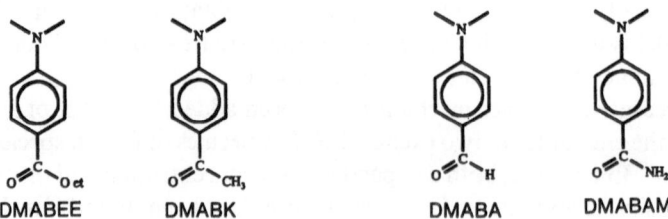

DCS

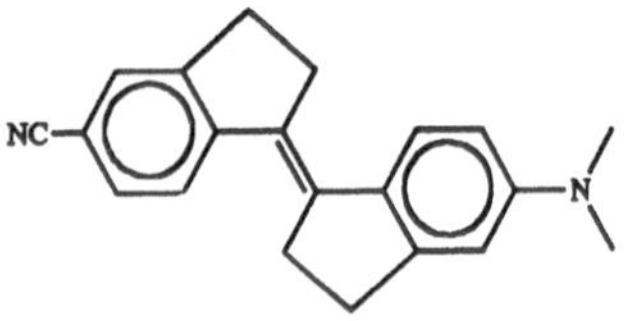

DCS-B24

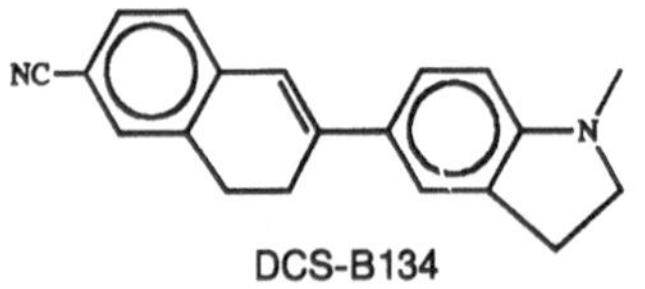

DCS-B134

4 Some Important Classes of TICT Compounds

The important classes of TICT compounds have been reviewed previously [2, 8, 15, 16, 17, 18], and only the most remarkable features and latest developments will be reported here.

4.1 Aromatic Amines

The majority of TICT compounds investigated up to now contain (substituted) amino groups. In some cases, these are directly TICT-active (i.e. the bond linking the amino group twists). In other cases, the amino group merely helps to enhance the donor character of the donor moiety, and twisting occurs around a different (usually C–C) bond. The first set of compounds will be grouped under "aromatic amines", the second one under "aryl-anilines" (Sect. 4.2).

Apart from DMABN, dialkylanilines with other small acceptor substituents exhibit dual fluorescence. Examples are

DMABEE   DMABK   DMABA   DMABAM

260

In many cases, bridged or pretwisted model compounds have established the validity of the TICT model. Instead of acceptor substituents, aza-substitution as in DMAPYR and DMAPYM can also lead to dual fluorescence.

DMAPYR    DMAPYM 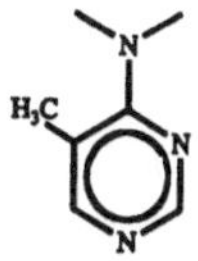   TriMAPYM

Several of these compounds have been studied without surrounding solvent in isolated gas phase conditions (supersonic jet studies) but evidence for TICT emission was found only in a few cases. More successful in terms of dual fluorescence in the jet were studies with respect to microsolvation and self-cluster formation. Both these processes enhance TICT formation. A beautiful example how pretwisting and microsolvation enhance the possibility for TICT formation is shown for a derivative of TriMAPYM in Fig. 4 [36].

Other aromatic amines with TICT-active amino groups include naphthyl-amines like DMANCN, which shows a well-separated dual fluorescence, and DANCA or related dyes which are used as fluorescence probes (sensing micropolarity) in biological investigations. In the latter case, only one (red-shifted) band (or two strongly overlapping bands) can be seen.

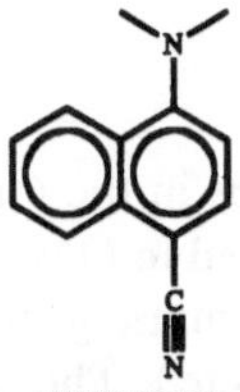

DMANCN 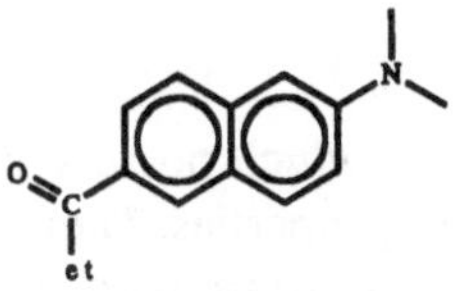   DANCA

If the amino nitrogen bears several aryl groups as in 2,6-TNS, its donor character is increased, and TICT formation is enhanced, Eqs. (1) and (2). This relation is also found in dyes containing amino groups (see Sect. 4.4) and can explain why Fast Acid Violet, a derivative of Rh-FA shows poor fluorescence with respect to its dialkylamino counterpart Rhodamine 3B (Rh-B).

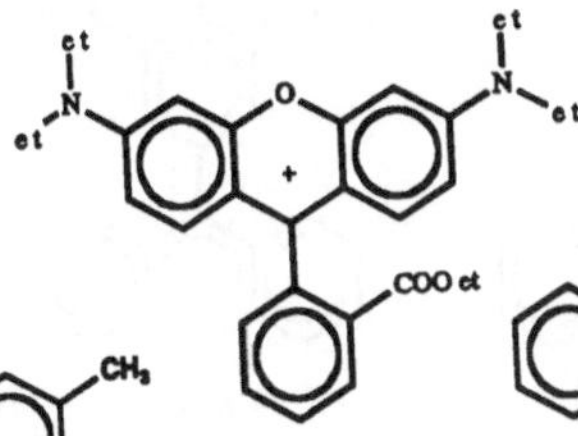

Rh-B

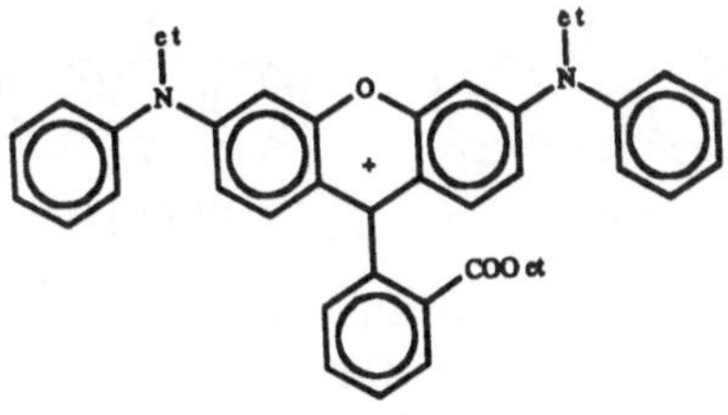

Rh-FA

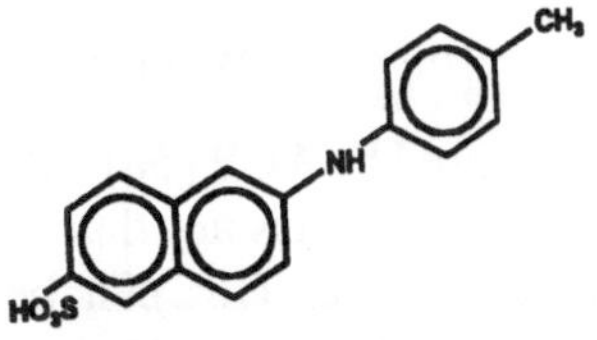

2,6-TNS

261

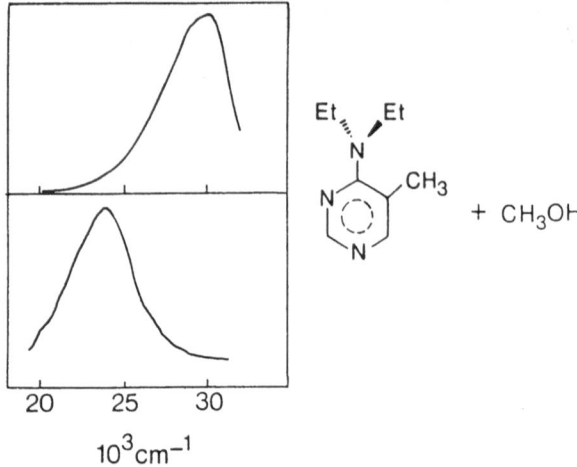

Fig. 4. An *ortho*-methyl group in diethylamino-pyrimidin induces some ground state twist and hence energetically destabilizes the B* state but not yet sufficiently to make the population of the A* state a major process in supersonic jet spectroscopy. Upper panel: dispersed fluorescence spectra of the jet-cooled bare molecule [36]. In clusters with methanol, the TICT state is preferentially lowered, and the majority of the observed red-shifted fluorescence can be assigned to arise from the TICT state (lower panel). This does not occur for the compound without an *ortho*-methyl group.

4.2 Aryl-Anilines

Many dialkylanilines, which possess aromatic groups in the 4-position, exhibit unusual fluorescence properties. They can be compared to DMABN by formally exchanging the cyano group against the larger aromatic group (benzonitrile, naphthalene, anthracene, pyrene, vinylbenzontrile etc.). The best studied example is ADMA, with several model compounds DM-ADMA, MAI, TM-ADMA [18, 37–49]. The behavior of these model compounds leads to the conclusion that the most important TICT channel is that involving the whole dialkylanilino moiety as donor and the anthracene as acceptor unit.

| ADMA | DM-ADMA | MAI | TM-ADMA |

Other compounds recently investigated in this respect are DMACB [50], DS [51], DMAPS [52] and CV [53]. Also here, the TICT-active bond is not that of the amino group but the bond in 4-position of dialkylaniline. This can be concluded by selectively bridging the flexible C–N$_{amino}$ or C–C bond.

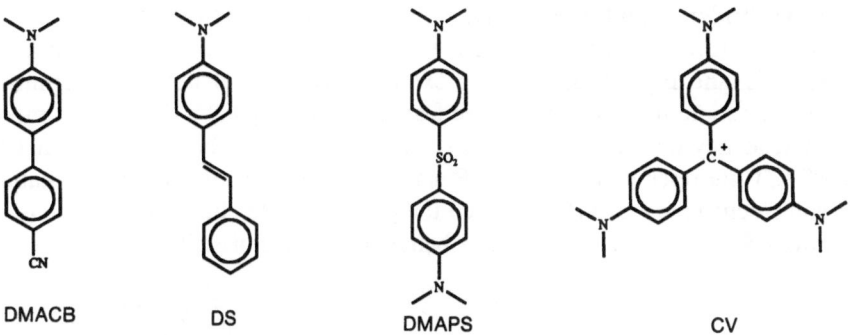

DMACB   DS   DMAPS   CV

4.3 Hetero-Aromatics

If the dimethylamino group in DMABN is exchanged for an *N*-heteroaromatic group like pyrrole, indole, carbazole, compounds such as PBN, CBN etc. result which exhibit strongly redshifted fluorescence. The latter can be rationalized using Eqs. (1) and (2) [17, 54]. Usually, only one band (the "anomalous" one, A*) is observed, with the exception of e.g. *N*-phenyl-pyrrole [56]. This last case shows that even unsubstituted phenyl groups can act as acceptor in TICT molecules in favourable cases (presence of very good donor groups).

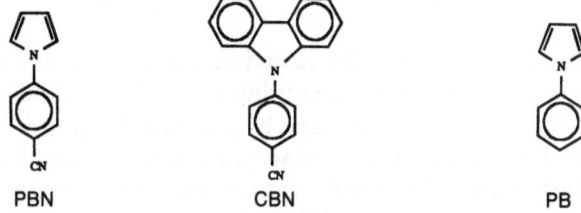

PBN   CBN   PB

By changing the acceptor system, a large variety of TICT compounds is available, only few of which have been studied more closely, e.g. C1N and C9A [54, 55].

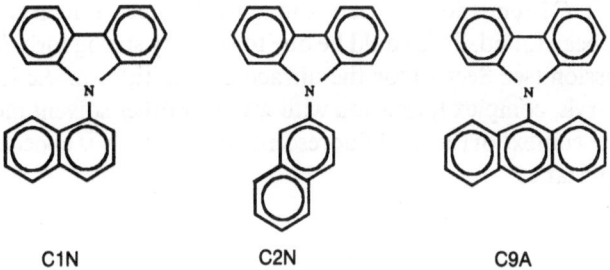

C1N   C2N   C9A

4.4 Biaryls

The above-mentioned C9A leads us directly to the symmetric biaryl compounds like 9,9′-bianthryl BA, the best studied example [8, 17, 25, 57–69]. Although the rigorous bridging and pretwisting studies as in the DMABN or ADMA derivatives have not been conducted yet, the TICT model is a good working hypothesis, and Eqs. (1) and (2) can explain why BA, but not the binaphthyl N1N, shows anomalous fluorescence features, and why C9A shows even more TICT tendency than BA. Equations (1) and (2) have also been used to predict new TICT fluorescing biaryls like BP which has very recently become available. Its solvatochromic redshift confirmed the photoinduced charge separation (see, however, Sect. 5.3 for an anomalous behavior of the emission rate).

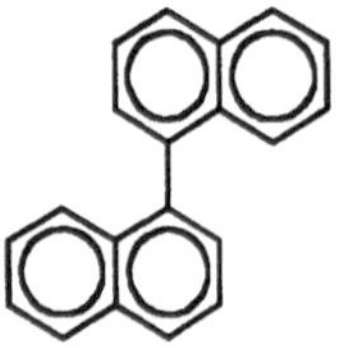

BA has been studied in great detail, and its charge-separated nature in the excited state has been verified by measuring the excited-state dipole moment using different independent methods, and by analysing the transient absorption spectra of B* and A* species. The A* absorption is, as expected from the TICT model, composed of the sum of anthracene radical anion and cation spectra. Detailed supersonic jet studies on BA and derivatives, and solvent clusters thereof, have allowed to quantitatively determine the ground and excited state torsional potentials of the isolated molecules, and to determine the number of polar solvent molecules necessary to induce charge separation. In the isolated molecule, no TICT emission is seen and only indirect evidence for TICT formation has been found. This could be due to the very strong forbiddenness of the TICT emission (see Sect. 6) for the ultracold conditions in the jet.

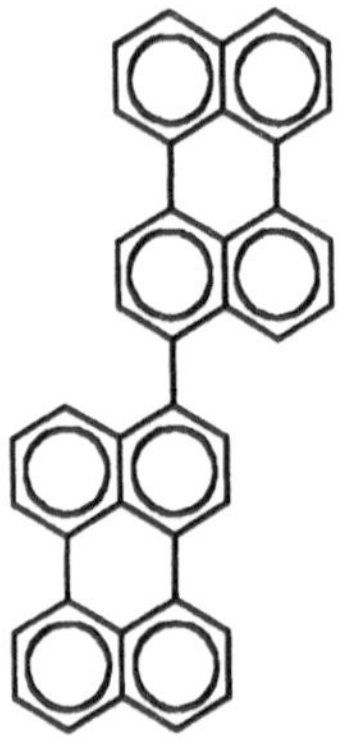

For the biaryls, complex formation with water or other solvent molecules as discussed in the context of the dual fluorescence of DMABN is expected to be of much less importance.

4.5 Dye Systems

As already mentioned in Sect. 4.1, the fluorescence behavior of many dye systems can be rationalized using the TICT model. For example, the fluorescence efficiency of xanthene dyes and derivatives like rhodamine and oxazine dyes correlates with the donor strength of the amino groups as predicted from

Eqs. (1) and (2) and is consistent with the assumption of a nonemissive TICT state: Rh-H shows little nonradiative losses, **Rh-B** medium and Rh-FA – the essential framework of the rhodamine derivative Fast Acid Violet – strongly enhanced ones. Bridging and pretwisting effects point to the involvement of a twisting motion at the dialkylamino bond: for the pretwisted Rh-Pip, enhanced nonradiative losses are observed, which are suppressed for Rh-101 [70, 71, 72]. Of course, other factors like pyramidalization of the nitrogen atom(s) and complexation with the solvent, may additionally play a role [73].

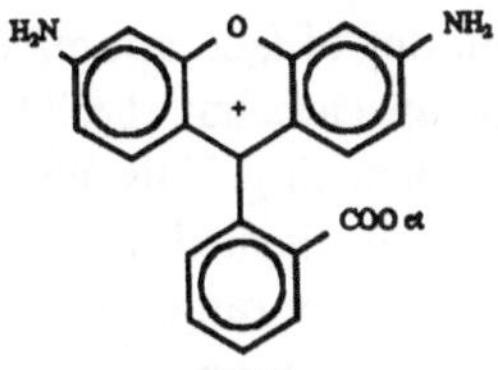

Rh-H

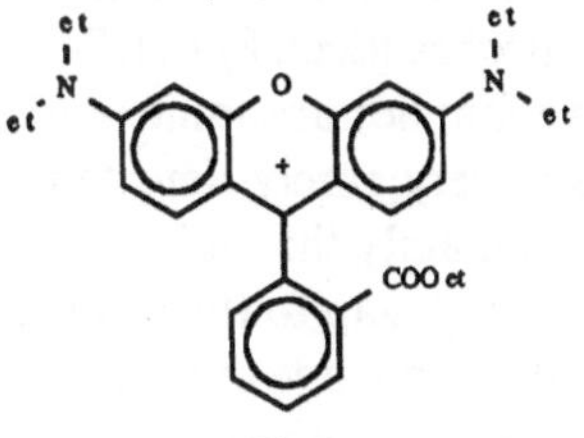

Rh-B

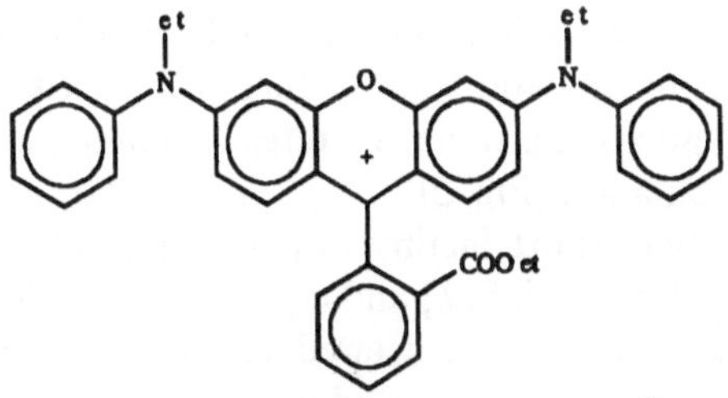

Rh-FA

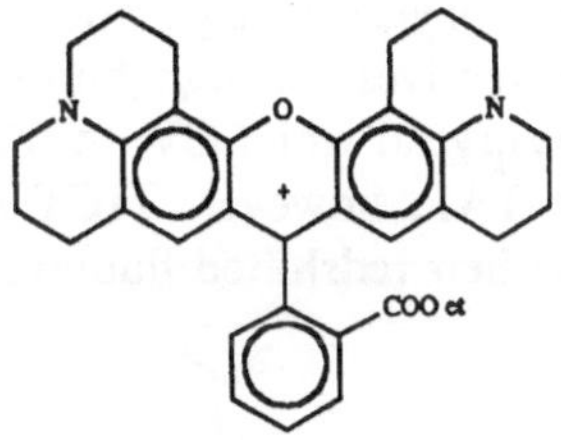

Rh-101

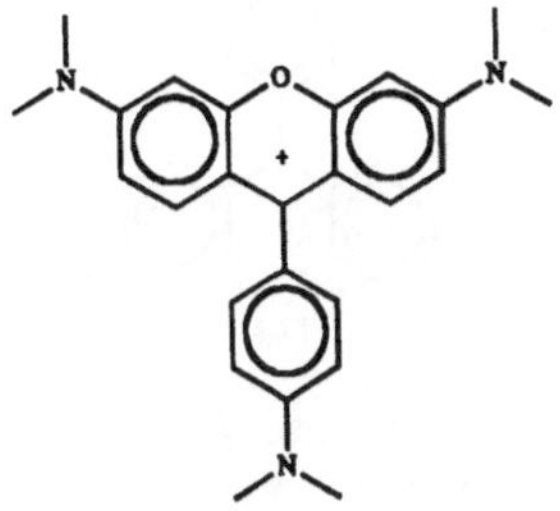

A-Rh

Equations (1) and (2) can be used to construct dyes with new and efficient TICT channels. By substituting the carboxylic acid or ester group in compounds like Rh-H or Rh-B by a dialkylamino group, a very efficient TICT channel is opened in A-Rh, involving the whole dialkylaniline as a donor unit. This

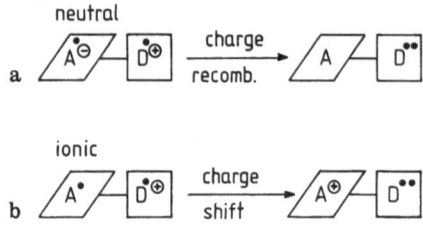

Fig. 5a, b. Decay processes of biradicaloid species: **a** TICT species with charge separation decay nonradiatively (and radiatively) via a charge recombination process; **b** in case of the corresponding ionic species, the twisted conformation leads to charge localization on one of the molecular fragments, and the decay process corresponds to the shift of this charge onto the other fragment

creation of tailor-made nonradiative channels can be used in the construction of fluorescence probes (Sect. 8) [11, 74].

Many dye families are ionic, and as a matter of fact, anomalous redshifted fluorescence is very rarely detected in ionic species, but the TICT formation manifests itself mainly through fluorescence quenching. This may be due 1) to a very narrow energy gap to the ground state for the twisted species such that any emission would be outside the experimental detection range or 2) to fast intrinsic nonradiative losses for the twisted species in spite of a sizable energy gap. The latter can in principle be explained by the charge shift nature of the back-electron transfer (nonradiative transition to the ground state) for ionic systems within recent extensions of the electron transfer theory [75]. Charge shift for ionic systems leads to enhanced nonradiative rates as compared to neutral systems with charge recombination (Fig. 5).

Manifestations of extremely short lifetimes for transient (TICT) states in ionic species have been found in A-Rh [12] and especially in triphenylmethane (TPM) dyes [76–78]. The latter can be viewed as derived from A-Rh by removing the oxygen bridge. This introduces additional TICT channels which manifest themselves through fluorescence quenching. The best known TPM derivative is crystal violet CV. In the neutral but closely related triphenylphosphines like TAP, however, TICT states with long lifetimes can be directly detected by their redshifted fluorescence (Fig. 6) [79].

A similar pair of nonionic/ionic TICT species are ADMA (nanosecond lifetime [43]) and A-Rh (picosecond lifetime [12]).

4.6 Stilbenoid Systems

Very recently, donor–acceptor substituted stilbenes and related dye systems have been studied within the context of possible TICT formation. Examples are

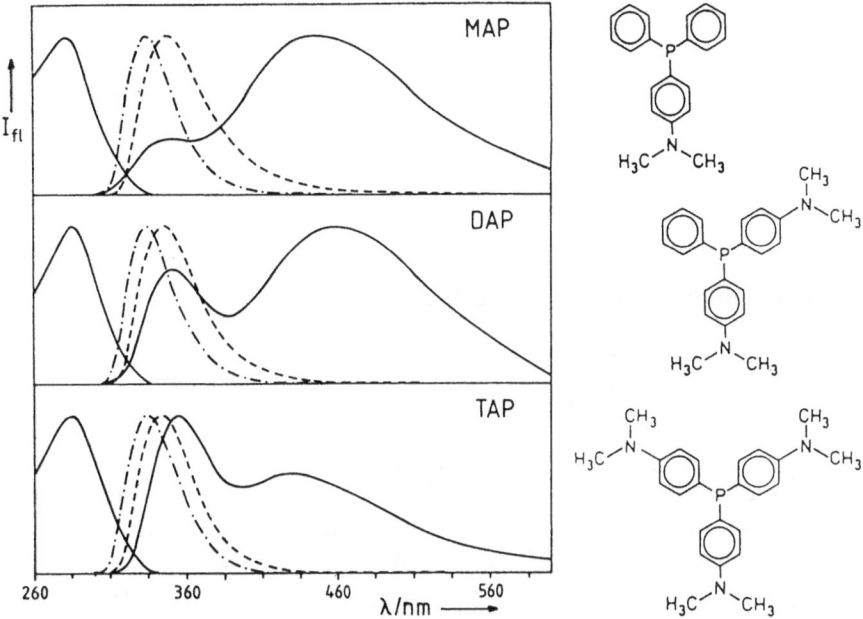

Fig. 6. Fluorescence spectra of triphenylphosphines MAP, DAP, TAP bearing different numbers of dimethylamino donor groups [79]. In the most polar solvent ethanol (————) the redshifted TICT band is clearly developed. The most probable donor moiety is one of the dimethylanilino groups. The rest of the molecule acts as acceptor, which is weakened by the dimethylamino donor groups in the case of the higher substituted compounds DAP and TAP. Therefore, the relative weight of the TICT fluorescence is reduced

DCS, the laser dye DCM, and DPS which can be viewed as an unsymmetrically bridged cyanine dye. In these cases, several bonds can twist, the central double bond being the commonly assumed most active one in parent stilbene. However, selective bridging studies involving compounds like DCS-B24 showed that an intermediate state necessitating single-bond twisting (A* state) is involved in the photophysics of these dyes. The behavior of DCS can be understood on the basis that the A* state is highly luminescent and competes efficiently with the nonradiative deactivation through double-bond twisting (see Sect. 5). In the ionic DPS, the A* state emission is weak because this system seems to possess an intrinsic fast nonradiative channel similar to the other ionic compounds mentioned above.

DCS   DCS-B24

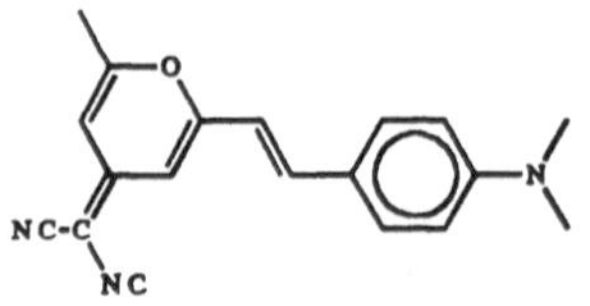

DCM 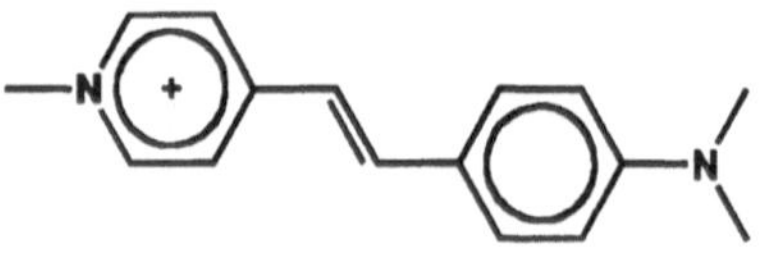   DPS

5 Twisted Single and Double Bonds

5.1 The Theory of Biradicaloid States

In recent years, the theory of TICT states has been merged [16, 19, 20, 80] with the theory of biradicaloid states [81–83] so that twisted single and twisted double bonds can now be understood within one single framework, that of biradicaloid states [8, 16–20, 80]. Typical biradicaloid species possess one or several pairs of nearly degenerate singlet/triplet states, due to a very small exchange integral K for certain orbital combinations. If such a pair happens to constitute the ground state, then one speaks of proper biradicals. Such species are, for example, 1,3-biradicals, but also 90° twisted ethylene. On the other hand, systems with singlet/triplet pairs in the excited state can be called biradicaloid. Every TICT state has its nearly degenerate triplet counterpart. As shown below such biradicaloid systems, if uncharged, involve either charge separation in the ground or in the excited state, and the theoretical model describing them can therefore be termed the "Biradicaloid Charge Transfer" (BCT) model.

A well-studied simple prototype biradicaloid system is twisted ethylene as compared to twisted aminoborane [16, 19, 80]. Figure 7 shows how the localized orbitals for these two species differ: degenerate for ethylene, an energy gap δ for aminoborane. With two π-electrons in the system, three singlet and one triplet state can be constructed [19]. Both electrons on one center ("hole-pair") is a less stable situation for twisted ethylene than one electron on each center ("dot-dot") because electron-electron repulsion is stronger in the former case. This difference is, however, overcompensated for aminoborane with sufficiently large δ (extra stabilization of lowest "hole-pair" state by δ). As seen in Fig. 7, there should be a region with a critical δ, δ_{cr}, where dot-dot and hole-pair states cross. This corresponds to a S_0/S_1 surface touching for 90°, when the twist angle ϕ is allowed to vary, as indicated in Fig. 8.

The surface touchings correspond to photochemical funnels [20] which can lead to ultrafast S_1–S_0 transitions [80] as typically encountered for the energy storage step of retinal pigments in the visual process or in bacterial photosynthesis. The transition along the δ axis in Figs. 7 and 8 can be made smoothly by an external perturbation, e.g. solvent relaxation.

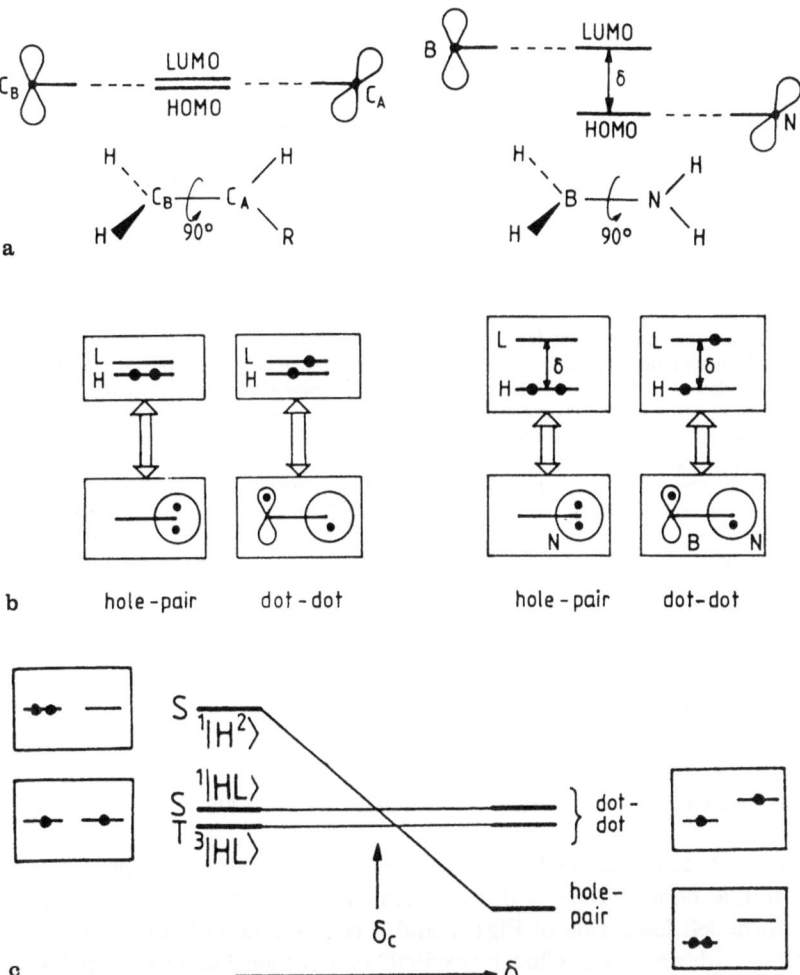

Fig. 7. a Localized frontier orbitals HOMO and LUMO of twisted ethylene slightly symmetry-disturbed by the substituent R and aminoborane. Note the energy gap δ, which is either introduced by donor–acceptor substitution on ethylene or by exchanging the carbon atoms for an atom with stronger (N) or weaker electronegativity (B). **b** Some of the possible electron occupation patterns (configurations) are shown. HOMO and LUMO are abbreviated with H and L. The closed-shell configuration $|H^2\rangle$ with doubly occupied HOMO is called "hole-pair" because an electron pair is localized on one of the two centers. The open-shell configuration $|HL\rangle$ with single occupancy in each orbital and on each center is called "dot–dot". It is always present as a close-lying singlet-triplet pair. **c** The relative energies of these configurations (states) are depicted as a function of δ. For small δ, the configuration $|H^2\rangle$ with the electrons closer to each other possesses higher energy than $|HL\rangle$ (on the left for ethylene). For aminoborane (right hand side) with a large orbital energy difference, δ is sufficiently large to overcompensate the electron repulsion effect, and the lowest state (ground state) possesses a doubly occupied nitrogen orbital ("hole-pair", $|H^2\rangle$), and the excited state is of dot–dot character. Due to the core charges of N and B, $|HL\rangle$ corresponds to charge separation in this case (usual TICT situation). In twisted ethylene with equal core charges of the carbons, $|H^2\rangle$ corresponds to charge separation ("sudden polarization" in the excited state of ethylene). Therefore, for twisted push-pull ethylenes (or donor–acceptor-stilbenes) (right hand side of the figure), the dot–dot excited state although possessing the typical TICT occupation pattern $|HL\rangle$ is of very weakly polar nature

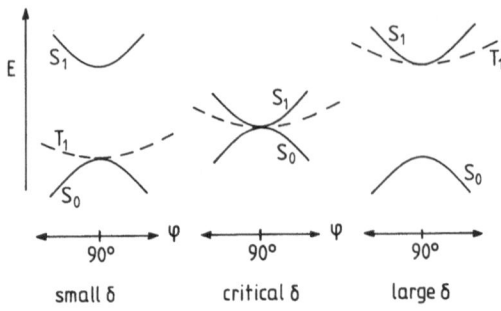

small δ   critical δ   large δ

Fig. 8. Schematic potential surfaces for ground, S_1 (———) and T_1 (– – – –) states in the neighbourhood of the perpendicular geometry. In the simplest case (when perturbation by locally excited states can be neglected), the ground state possesses an energy maximum and the excited states an energy minimum for the perpendicular geometry. This arises because for nonperpendicular conformations, ground and excited states interact and energetically repel each other. For symmetric systems like twisted ethylene (small δ), T_1 is degenerate with S_0 at 90°, and S_1 is of highly polar nature. When δ increases, the energy gap narrows, and a surface touching occurs for the 'critical δ'. For even larger values of δ, S_1 and T_1 are degenerate and of low-polarity nature for systems like double-bond twisted donor–acceptor ethylene or stilbene, but of highly polar nature for the corresponding systems with different core charges like DMABN (cf. the case of aminoborane in Fig. 7)

5.2 Experimental Confirmation

5.2.1 The Photophysics of Stilbene and Related Dye Systems

The BCT model can be applied to larger ethylene derivatives like stilbene. For a twisted double bond species with symmetric subsystems ($\delta = 0$), a situation similar to the left hand side of Figs. 7 and 8 is expected, but complicated by interactions with $\pi\pi^*$-states ("locally excited" or LE states) and by the reduced energy differences between hole-pair and dot–dot states. This arises because of the smaller difference in electron-repulsion integrals for the two electronic structures (the electrons can avoid each other more easily in the larger systems). In fact, CNDO/S calculations predict that the S_0–S_1 energy gap in twisted stilbene is smaller than 0.5 eV [84, 85] whereas for ethylene, this gap amounts to more than 3 eV [16]. Nevertheless, the principle features are expected to be similar: charge separation (high dipole moment) in the S_1 (hole-pair) state and low (or zero) dipole moment in S_0 (dot–dot).

Upon introduction of substituents which produce a sufficient electronic asymmetry δ, the situation of the right hand side of Fig. 8 is expected, with a state reversal, i.e. weakly polar S_1 (dot–dot) and strongly polar S_0 (hole-pair) for the twisted double-bond.

Experimentally, in all cases studied, "intramolecular fluorescence quenching" can be observed due to the adiabatic photoreaction towards the photochemical funnel (twisted double bond, close-lying S_0/S_1, ultrafast nonradiative deactivation). This reaction is very fast in stilbene and 4-cyano-stilbene and quenches nearly all the fluorescence, it is slower in DCS and slowest in DCM. This reduction of nonradiative losses can be understood with the concept of multidimensional photochemistry (simultaneous importance of several BCT/TICT states) as shown in Sect. 7.

5.2.2 The Solvatokinetic Principle

The solvent polarity dependence of rates of chemical reactions, either in the ground or in the excited state, can yield information on the properties of a fleeting species, the so-called transition state. More specifically, it can establish whether a polar surrounding stabilizes to a greater extent the precursor species E or the transition state TS leading to the product P. For uncharged molecules, this solvent stabilization is directly proportional to the square of the solute dipole moment, and therefore, the solvent polarity dependence can tell us on changes of dipole moments and charge distributions. In many cases, the dipolar character of TS is intermediate between that of E and P, such that the comparison of E and TS also allows to draw (qualitative) conclusions regarding the properties of P. This becomes especially important if the reaction is an adiabatic photoreaction with a nonemissive extremely short-lived product P* which may be undetectable. In spite of the undetectability of P*, the solvent polarity dependence of the nonradiative decay rate k_{nr} of the precursor species E* can give information on the polar properties of P*. We will call this behavior solvatokinetic and distinguish between a positive solvatokinetic behavior (the rate constant increases with solvent polarity) and a negative one. Figure 9 exemplifies these two cases. From a positive solvatokinetic behavior, one can directly conclude that the transition state is more polar than the precursor and hence, extrapolating to the product, that $\mu(P^*) > \mu(E^*)$. From a negative solvatokinetic behavior, the opposite can be concluded.

The data in Table 1 refer to the nonradiative decay rate k_{nr} in DCS and DCM and are indicative of the reaction to the photochemical funnel through double-bond twisting. They reveal that k_{nr} is highly polarity dependent, slowest in strongly and fastest in weakly polar solvents (negative solvatokinetic effect). In view of the above, we recognize this as signifying that the funnel state P* is of less polar nature than the precursor state E*.

$$\mu(E^*) << \mu(P^*) \qquad\qquad \mu(E^*) >> \mu(P^*)$$

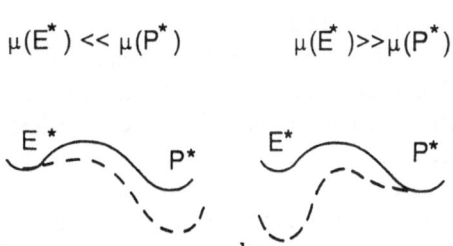

a   b

Fig. 9. Polar solvent influence (----) on the shape of the reactive excited-state surface for **a** product state P* being more polar and **b** less polar than the precursor state E*

Table 1. Intramolecular fluorescence quenching rate constants k_{nr} (10^7 s^{-1}) for DCS and DCM in solvents of different polarity [84]

Solvent	k_{nr} (0 °C) DCS	k_{nr} (20 °C) DCM
dibutyl ether	731	623
diethyl ether	546	283
tetrahydrofuran	240	31
n-butyronitrile	104	12

The behavior of unsubstituted stilbene is the opposite: Comparison of ps fluorescence decay data [86–88] and fluorescence quantum yields [89] as a function of solvent polarity shows that the nonradiative rates increase in parallel with solvent polarity giving evidence that the quenching state P* is more polar than the fluorescing one, E*.

This opposite solvatokinetic behavior related to the opposite dipolar character of P* can be understood by applying the BCT model. As explained for strongly perturbed ethylene derivatives like the donor–acceptor stilbenoid dyes DCS and DCM, the twisted S_1 state (P*, funnel) is expected to possess dot–dot character (right hand side of Figs. 7 and 8) and therefore weakly polar properties in this case, whereas the fluorescing state (E*, near-planar geometry) is highly polar. Increasing the solvent polarity thus lowers the fluorescing state faster than the quenching state (funnel) leading to reduced importance of the quenching channel k_{nr} (negative solvatokinetic behavior). This also allows us to understand why DCM is usable as a laser dye mainly in the most strongly polar solvents like dimethylsulfoxide. Application of the BCT model can also provide a prediction for the solvatokinetic behavior of unsubstituted (or weakly perturbed) stilbene: In this case, the quenching state should be highly polar, the fluorescing state weakly polar. The experiments cited above are in accordance with this expectation and also with the results of corresponding quantum chemical calculations [84, 85].

These effects can be seen even in a more direct way in tetraphenylethylene TPE and its derivatives. As for ethylene and stilbene, a funnel (with an excited state energy minimum) is expected for the twisted double bond geometry, and for unsubstituted TPE with $\delta = 0$, the funnel state P* should be of highly polar (hole-pair) character, similarly as for stilbene. However, in contrast to stilbene, the product P* lives long enough to be detected and characterized by a number of techniques (see for example [90–92]). The longer lifetime is related to a widened S_0/S_1 energy gap at the funnel position. Picosecond calorimetry which can measure the energy content of such transient species directly established that this energy gap decreases with an increase of solvent polarity, in accordance with the highly polar nature of P* and the expected weakly polar nature of the corresponding ground state P. Recent direct dipole moment measurements of P* confirm its polar nature [93].

The relatively large energy gap for TPE, resulting in the long lifetimes (up to nanoseconds) of P* can be contrasted to the picosecond lifetime of P* and the correspondingly smaller gap in stilbene. The strong difference may be related to the fact that TPE resembles ethylene more strongly in its electronic structure: The mutual steric hindrance of the phenyl substituents leads to their twisting partially out of conjugation with the central carbon π-orbitals. Thus, in the twisted biradicaloid structure of TPE, the active electrons are more confined to the central C-atoms resulting in a larger coulombic repulsion and hence hole-pair/dot–dot energy gap (see also the explanation to Fig. 7c). For stilbene, where steric hindrance is less, the orbitals on the phenyl groups overlap much better with those on the adjacent C-atoms, and the biradicaloid state species are better described by two close-lying benzyl radicals with reduced electron density at the central C-atoms and hence reduced coulomb repulsion and hp/dd energy gap. This view is in accordance with the very short lifetimes (low fluorescence quantum yields) observed for planarized model compounds of TPE like TPE-B.

TPE   TPE-B   TPE-DA

Upon introduction of donors and acceptors in an unsymmetrical way (TPE-DA), the energy gap should at first narrow, then widen again for stronger donor/acceptor substituents, according to the BCT model. For weak substituents, the narrowing of the gap has already been shown through the shortening of the lifetimes [92, 93].

5.2.3 The Yin-Yang Principle in Biradicaloid Photo- and Thermochemistry

Up to now, we have mainly discussed the properties of the excited state at the biradicaloid geometry. However, from Fig. 7, it is clear that the corresponding ground state possesses the opposite electronic structure. Thus, if the excited state is of hole-pair character, then the ground state is of dot–dot character and vice-versa (Fig. 10). For the discussion of double-bond twisting, this signifies that the electronic nature of the excited-state funnel is the complement to that of the top of the ground state barrier. The latter is the transition state for thermal *cis-trans* isomerization, and its properties will determine the solvatokinetic behavior of this ground state reaction. We can therefore expect opposite solvatokinetic behavior for ground and excited-state *cis-trans* isomerization reactions.

Unfortunately, due to very high ground state barriers, comparative solvatokinetic measurements in the stilbene series are difficult. It may therefore suffice here to cite two examples of related compounds which differ in the solvatokinetic behavior of the ground state reaction (Fig. 11).

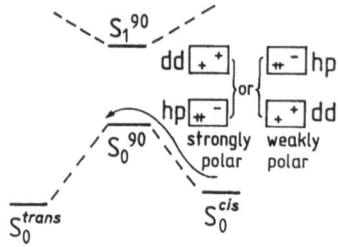

Fig. 10. The electronic characters of the excited state funnel (S_1^{90}) and of the ground state barrier to *cis-trans* isomerization (S_0^{90}) of stilbenoid or similar compounds are opposite and complementary

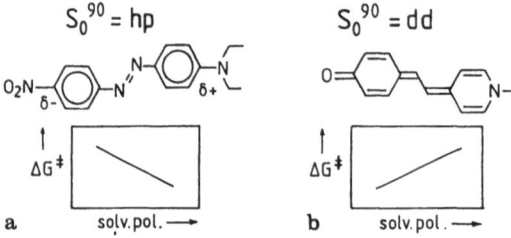

Fig. 11. a Example of a positively solvatokinetic ground state isomerization reaction (the activation barrier decreases with increasing solvent polarity) for an aza-dervative of a donor–acceptor stilbene [94]. As outlined above, related donor–acceptor–stilbenes exhibit negative solvatokinetic behavior for the excited-state reaction. **b** For the merocyanine dye shown, the ground state *trans-cis*-isomerization across the central bond shows negative solvatokinetic behavior [95], the expectation for the excited state reaction is positive solvatokinetic behavior

5.3 Beyond the Simple Model of Biradicaloid States

The model of biradicaloid states as outlined and applied above is strictly valid only for cases where no other low-lying states are important. In this respect, ethylene, aminoborane (and probably also DMABN) are good examples. Then, the model clearly predicts a potential minimum in the excited state and a maximum in the ground state for a twist angle of 90°, and strongly forbidden emission from this perpendicular conformation. Upon introducing larger aromatic systems, however, low-lying π, π^*-states become more and more important and interact with the biradicaloid states, for geometries deviating from 90°. As the angular dependence of the π, π^*-states is opposite to that of the biradicaloid states (i.e. they gain stability upon twisting away from 90°), the interaction of both types of states can lead to complicated excited-state potentials with a very flat angular dependence and, in principle, minima for geometries other than 90° but nevertheless possessing a large degree of the features of classical TICT states with a 90° minimum, especially sizeable charge separation. Due to the strong coupling with π, π^*-states, these strongly interacting TICT states offer the possibility for allowed TICT emission and thus also for large fluorescence quantum yields, a feature which is highly desirable in the application of these dyes e.g. as fluorescence probes, tracers and labels.

A general and simple theory describing the interaction of these two types of states is not yet available. In principle, quantum chemical calculations with configuration interaction for many different twist angles can yield a quantitative answer, but in order to understand and interprete the results, detailed configuration analysis is necessary.

Three examples may serve to illustrate these points. Both BA and BP show dual fluorescence which can be explained with the TICT model. Both compounds possess symmetry-equivalent aromatic moieties in the ground state (zero dipole moment), and the observed redshift of the fluorescence spectra with increasing solvent polarity implies large excited-state dipole moments with a reduction of the symmetry (nonequivalence of radical cation/anion molecular moieties). Yet BA and BP differ strongly in the allowedness of the TICT transition, that of BA being forbidden (effective radiative lifetime in acetonitrile $\tau_r = \tau_f/\phi_f \approx 150$ ns) but that of BP ($\tau_r \approx 7$ ns) being even more allowed than for perylene [96]. The allowed TICT-nature of BP goes along with a reduction of the observed charge separation which is also interpretable within the above state coupling model. If charge transfer were quantitative, the elongated nature of BP should lead to a very large TICT dipole moment (42 Debye), twice the size as for BA (21 D). Experimentally, ca. 18 D are found for both compounds.

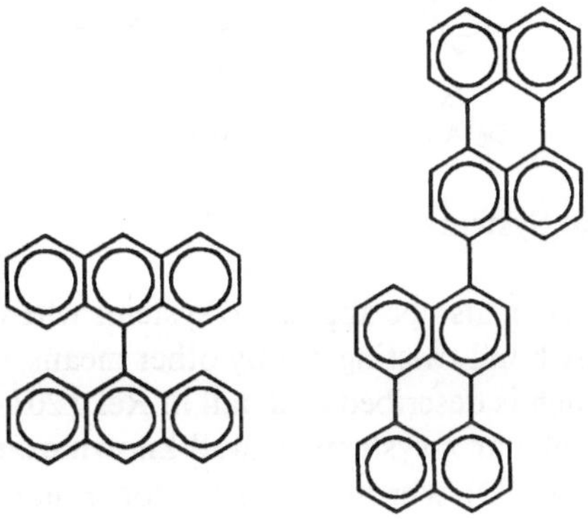

BA   BP

A further example is DCS-B34 which, according to the behavior of the other derivatives of DCS (see Sect. 7), populates to a large extent a TICT state available through twisting the single bond connecting the dialkylanilino group. Nevertheless, the fluorescence quantum yields are very high (80%) and the short lifetimes are indicative of an allowed emissive nature. The same applies to DCS, if at low temperature the fluorescence loss process (twisting of the double bond) is made unimportant [35].

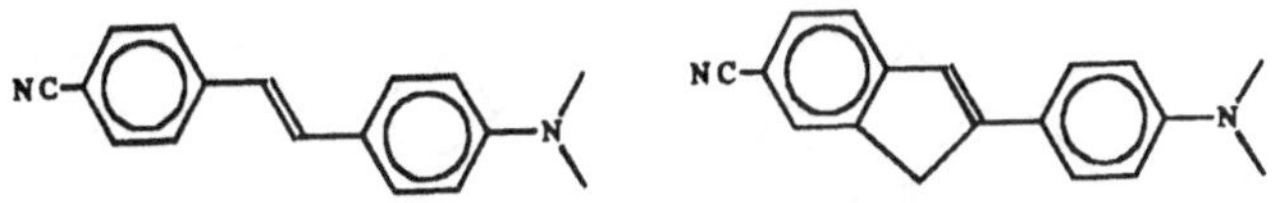

DCS   DCS-B34

Recently, detailed quantum chemical calculations comparing DMA-CB and CN-ADMA have been made [50]. The results can be summarized (Fig. 12) by stating that 1) the angular dependence of the CT state can be very flat and does not necessarily show an energy minimum for 90° twist; 2) solvation leads to stronger energy changes and may be a more important relaxation coordinate than twisting; 3) nevertheless, the perpendicular conformation is always preferentially lowered by solvation, because it is connected with a maximum of the dipole moment; 4) the allowedness of the TICT emission strongly depends on the symmetry of the nearby π, π^*-states: In CN-ADMA, only a low-lying π, π^*-state energetically far apart can couple through the twisting vibration away from 90°, and little transition moment is introduced (forbidden TICT emission) in contrast to DMA-CB (strong coupling with energetically close-lying π, π^*-states). The latter is predicted to possess a highly emissive TICT state even for twist angles quite close to 90°.

DMA-CB 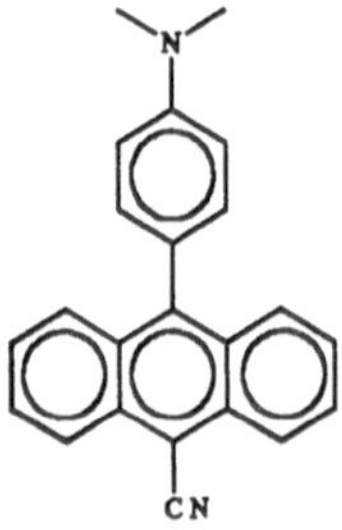   CN-ADMA

5.4 Planar Biradicaloid Systems

The biradicaloid model can also be applied to systems where orbital decoupling is brought about not by bond twisting but by other means. One such possibility is σ-bond breaking which is described in detail in Ref. [20]. Orbital decoupling can also be present in planar π-systems. A well-known example is the antiaromatic cyclobutadiene (CBD) which possesses, for equal bond lengths, two degenerate noninteracting frontier orbitals. In push-pull cyclobutadienes, one of the orbitals is pulled down by the acceptor, the other one pushed upwards by the donor substituents (Fig. 13a), and the orbital energy gap δ depends on the substituent strength. Thus, according to the biradicaloid model, the donor–acceptor-character of the substituents can modulate the S_1–S_0 energy gap, and the nature of the excited state is predicted to switch from hole-pair to dot–dot upon increasing δ (Fig. 13b). Relaxation pathways may involve in-plane distortions (Fig. 13c) [20, 97].

CBD 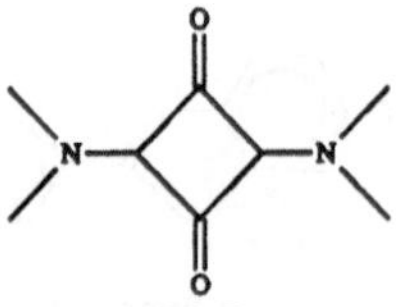   SQ-D 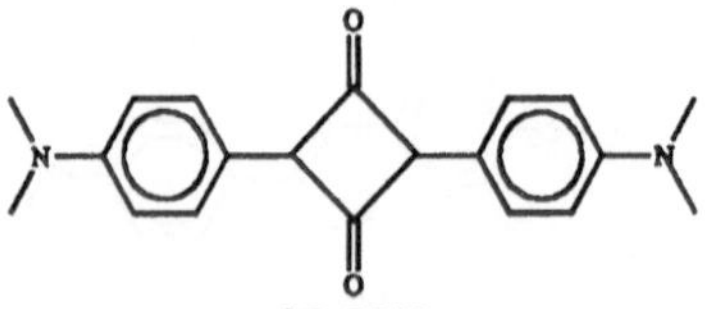   SQ-DMA

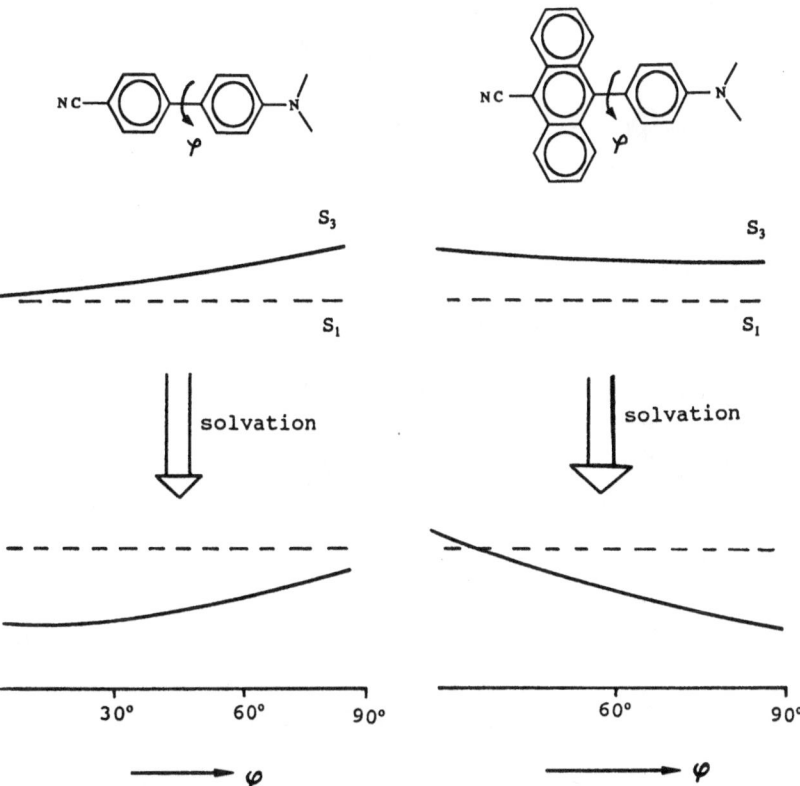

Fig. 12. Upper part: Schematic drawing according to calculated gas phase potential energy curves for the twist coordinate in dimethylaminocyano-biphenyl (*left*) and dimethylaminocyano-phenylanthracene (*right*) [50]. The lowest singlet state S_1 is of locally excited nature (small dipole moment), S_3 is of charge transfer nature with pure TICT character at 90° but without the usual energy minimum for the biphenyl derivative. Solvation (*lower part*) preferentially lowers the CT state. It becomes S_1 in sufficiently polar solvents, and the energetic lowering is biased towards 90° because the dipole moment possesses a maximum for the perpendicular conformation

Due to its antiaromaticity, cyclobutadiene CBD is not stable under ordinary conditions, but derivatives thereof are, for example from the squaric acid family. Compounds SQ-D and SQ-DMA even possess the possibility for a planar biradicaloid state (within the CBD unit) and a perpendicular one (through twisting of the substituents). The "tuning" of δ has also been proposed as a means of generating new long-wavelength absorbing dye systems [98].

6 The Role of Pretwisting and Rotational Distribution Functions

If the major relaxation coordinate in a TICT system is the torsional or twisting motion, and if the TICT equilibrium geometry is close to 90°, then the geometry

W. Rettig

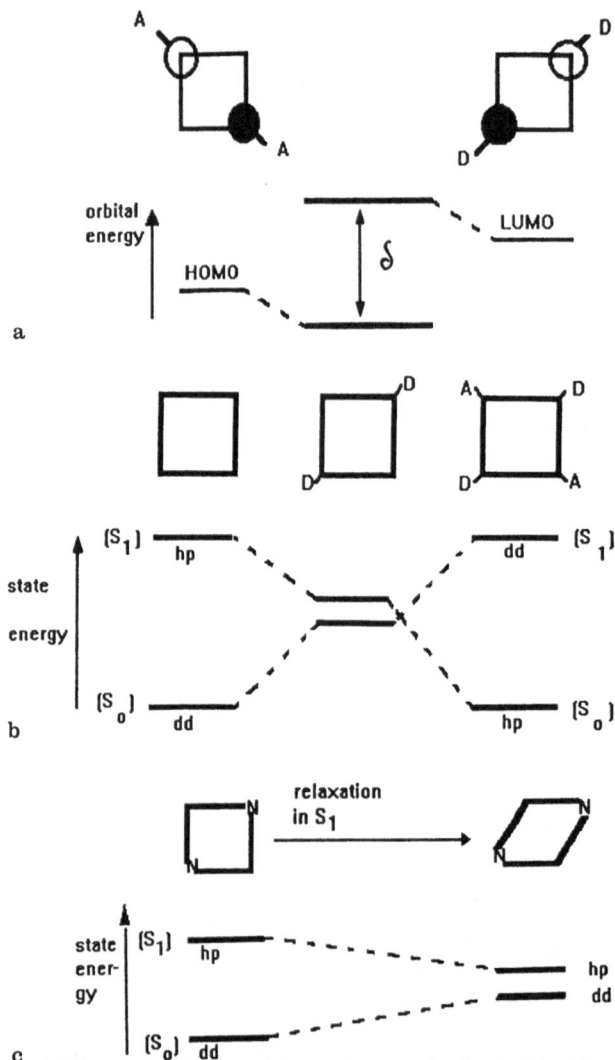

Fig. 13a–c. Push-pull cyclobutadienes as biradicaloid species [20]. **a** The two degenerate frontier orbitals localize upon introduction of donor and acceptor substituents as shown and energetically split by the energy gap δ. **b** Ground and S_1 states can therefore be assigned hole-pair (hp – one doubly occupied, one unoccupied frontier orbital) and dot–dot character (dd – two singly occupied frontier orbitals) similarly as in the case of twisted ethylene (Fig. 7). The energetic order is determined by the interplay of electron repulsion and the orbital energy gap δ which depends on the substituents. **c** In-plane relaxational deformations in S_1 can lead to an approach of S_0 and S_1 and thus to fluorescence red shifts or even to photochemical funnels

of the B* state (or more precisely of the Franck–Condon state, the "starting point" for the relaxation to both B* and A* states) is an important factor determining the kinetics of TICT formation. For pretwisted compounds, energy barriers separating B* and TICT state are expected to be reduced [36, 99], but even for the barrierless case, pretwisting should lead to accelerated kinetics due to the reduced angular gap to be crossed [100]. This has been verified in a number of cases. For amino group twisting, pyrrolidino and piperidino substituents provide a good pair for comparison. Due to increased steric repulsion, the piperidino substituent is somewhat twisted away from planarity in the ground (and B*) state. Correspondingly, TICT formation kinetics is found to be enhanced by a factor of about 10 in medium polar solvents [99, 100]. This is found for quite different systems such as PIPBN [100], Rh-PIP [70], Oxaz-PIP [72]. Pretwisting effects are also found in the pair PYRBN/DMPYRBN [10].

PIPBN 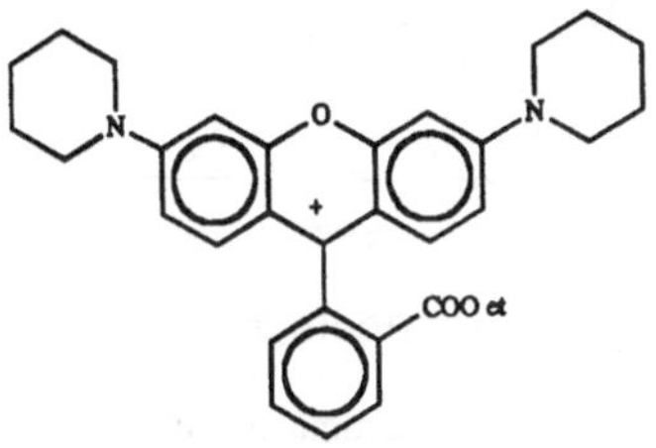   Rh-Pip 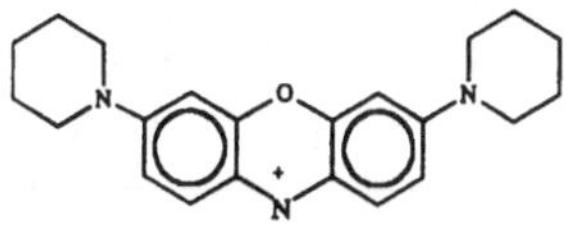   Oxaz-Pip

PYRBN    DMPYRBN

In TICT systems which are in a conformation close to perpendicularity already in the ground state, solvation effects are generally more important than twisting effects and determine the kinetics. Similar to pretwisting, presolvation can be present, i.e. the presolvated system in the Franck–Condon state is advanced with respect to the solvation coordinate leading to the TICT state. Such a situation can arise for example for perpendicularly twisted compounds with zero and nonzero ground state dipole moment. For zero dipole moment (e.g. symmetric compounds like BA) the surrounding solvent molecules are randomly oriented whereas a nonvanishing ground state dipole moment prepolarizes the solvent distribution. Examples are BA, BACl and CBA which possess a perpendicular ground state minimum [61], and for which the substituted compounds with a ground state dipole moment show faster TICT formation than the parent molecule BA [58]. In C9A, which is nonperpendicular in the ground state and hence less "pretwisted" than BA [55] but possesses a nonvanishing dipole moment, hence is presolvated in contrast to BA, presolvation

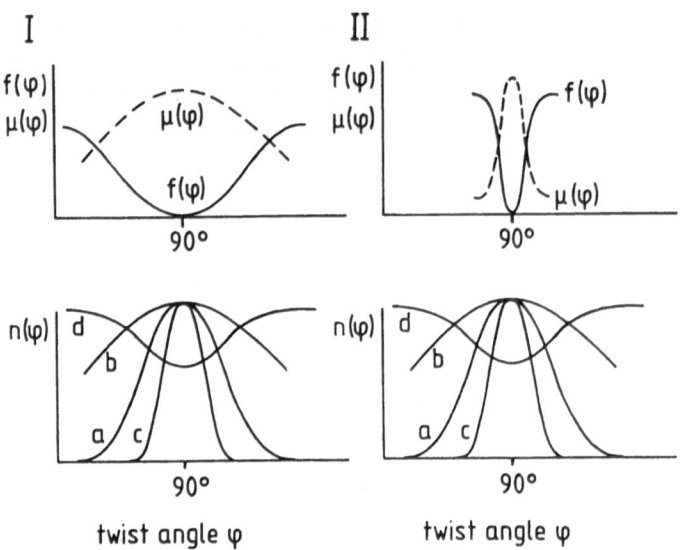

Fig. 14a–d. Broad (**I**) and narrow (**II**) maxima and minima of dipole moment μ and oscillator strength f as a function of twist angle ϕ for a TICT state, together with different forms of rotational distribution functions n(ϕ) (normalized to the maximum – instead of equal surface area – for better representation here): An assumed TICT distribution (*a*) is expected to be broadened by decreasing solvent polarity or increasing temperature (*b*) and expected to be reduced in width by bulky substituents (sterical confinement effect regarding the range of near-perpendicular twist angles), (*c*). (*d*) covers the case of strong TICT/LE interactions (Sect. 5.3) which can lead to TICT state potentials with minima different from 90° twist. For narrow peaks of μ and f (case **II**), broadening of the distribution function (*b* and *d*) is expected to have little influence on the integral, Eq. 3, whereas sterical confinement effects (*c*) are expected to lead to stronger changes for case **II** than for case **I**

and pretwisting effects can compete, and it depends on the nature of the solvent (interplay of solvent viscosity and dielectric relaxation time) which of them is more important [96].

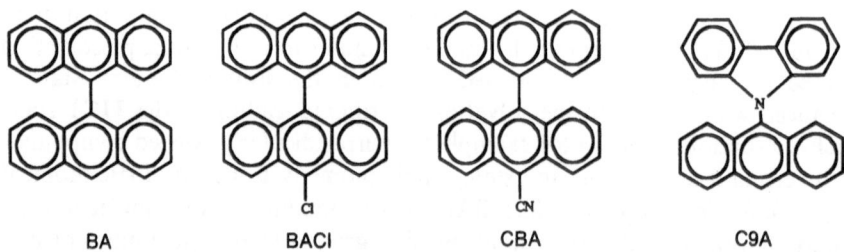

Conformational effects not only determine TICT formation kinetics but can also play a role in determining the TICT state properties. It has to be borne in mind that properties such as the radiative rate constant k_f (or oscillator strength f) and the dipole moment μ can strongly depend on the twist angle especially in the neighborhood of 90°, because k_f minimizes and μ maximizes for the perpendicular conformation [2, 50, 101]. But the observed TICT properties are given by the contribution of all twist angles populated around 90°, and hence the observable TICT emission can be regarded as deriving mainly from twist angles different from 90° or, in other words, corresponds to a "hot" fluorescence from vibrationally excited TICT levels [2, 8, 23]. The TICT properties can furthermore be expected to depend on the width of the rotational distribution around 90° or, more generally, on the shape of the rotational distribution function, if the maximum occurs for values different from 90°. Distribution function effects have also been discussed in the context of the normal (B) emission of DMABN to explain the anomalously low f values of planarized DMABN model compounds [99, 102]. Any observed property p (dipole moment, oscillator strength, radiative or nonradiative rate, or other properties depending on the twist angle ϕ) has to be regarded as mean value $\langle p \rangle$ resulting from the folding of $p(\phi)$ with the angular distribution function $n(\phi)$ according to Eq. 3. Some cases are exemplified in Fig. 14.

$$\langle p \rangle = \int p(\phi) \cdot n(\phi) \cdot d\phi \qquad (3)$$

The width of the TICT state rotational distribution function can depend on solvent polarity. In fact, the solvent-polarity induced increase of the TICT dipole moment [18, 103–105] can be understood by such a "nuclear polarizability" effect [105] brought about by the energetic deepening of the TICT potential and the corresponding angular narrowing of the TICT rotational distribution function in more polar solvents. Bulky ortho-substituents can also influence the shape of the TICT potential and generally lead to a "confinement effect" i.e. to a narrowing of the rotational distribution function around 90°. This

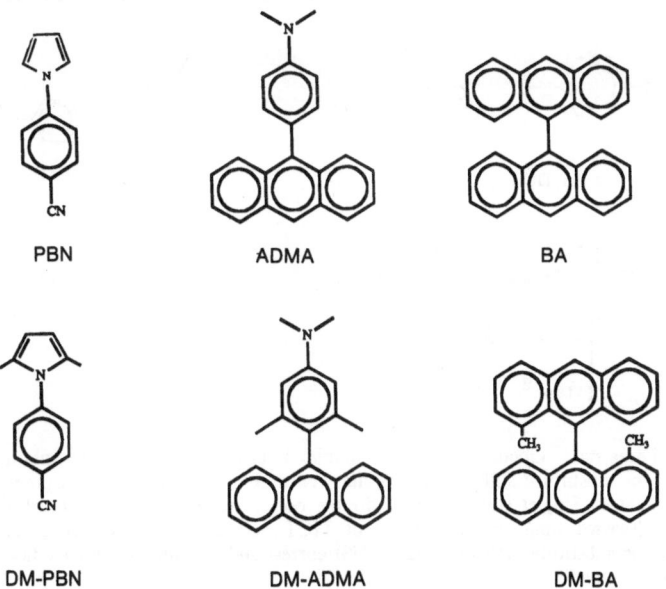

PBN   ADMA   BA

DM-PBN   DM-ADMA   DM-BA

is expected to lead to increased effective values for the dipole moment and to reduced values for the radiative rate constant of the TICT state. Three examples for such sterically confined pairs are given below. The first two examples have been experimentally studied [8, 39, 41, 43, 49, 56], the third is based on prediction.

7 Multiple Adiabatic Photochemistry

7.1 Multidimensional Photochemistry in Stilbenoid Dye Systems

As evident from the molecular structures (dimethylanilino group with very good donor properties), both DCS and DCM are likely to possess comparatively low-lying TICT states reachable by twisting one of the single bonds adjacent to the central double bond. These TICT-channels can compete with the double-bond-twist channel leading to the biradicaloid funnel state P* with fast nonradiative deactivation (Sect. 5.2). For a closer discussion of these multidimensional effects, let us assume that the only relevant TICT state is that reached by twisting the anilino group (called A* state) (quantum chemical calculations reveal, however, that twisting motions around the other single bond adjacent to the central double bond can also be an important TICT channel [106]). Then, the situation can be represented by a three-state kinetic scheme (Scheme 1) [35, 84, 106], with the primarily reached essentially planar Franck–Condon state E*, and involving the competing adiabatic photoreactions k_{EP} and k_{EA} along different coordinates (double vs single bond twist) equivalent to different dimensions.

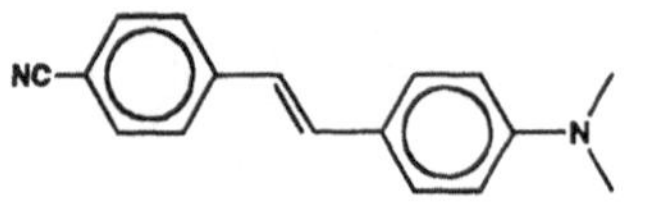

DCS 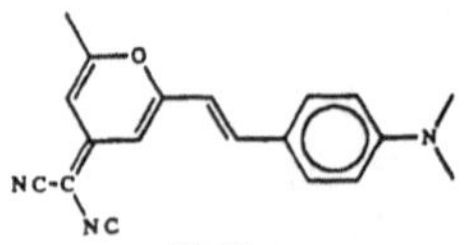   DCM

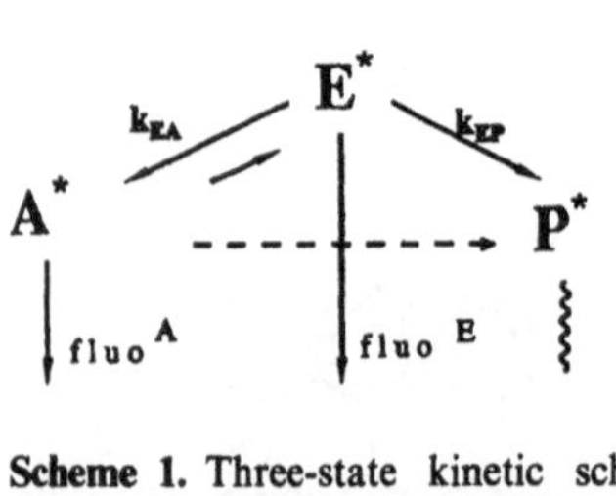

Scheme 1. Three-state kinetic scheme as minimal basis to explain the photophysics of donor–acceptor substituted stilbenes. E* (primary excited π, π* state, essentially planar geometry) and P* ("Phantom Singlet state", twisted double bond) are the "classical" states discussed for stilbene, A* (twisted single bonds, state(s) of TICT nature, up to 4 different possibilities in donor–acceptor substituted stilbenes like DNS) correspond to motions along different reaction coordinates than for P*

For DCS, k_{EA} is much larger than k_{EP} [35] and therefore competes efficiently with the direct nonradiative decay process towards P*. If the TICT state A* is highly fluorescent, this competition will correspond to an effective reduction of the fluorescence quenching via the funnel P*. Thus, most of the nonradiative deactivation observed for DCS is thought to occur, after initial nearly quantitative population of A*, by the indirect route A* → E* → P*, (or by a direct multidimensional pathway involving the simultaneous twisting of single and double bonds) [106]. A further consequence of multidimensionality are complicated lifetime behaviors. DCS possesses a lifetime maximum at low temperature (2.3 ns at – 90 °C in ethanol) corresponding mainly to A* fluorescence and to a freezing-out of the two-step quenching reaction A* → E* → P* which shortens the lifetimes at higher temperature [35, 96]. At still lower temperatures and strongly increased viscosities, the initial TICT population reaction E* → A* is frozen out, too, and the lifetimes decrease again reaching a value of 1.7 ns at 77 K (E* fluorescence only). The availability of the A* state for the flexible compounds (DCS) acts like a stabilizing factor and leads to increased fluorescence lifetimes and quantum yields with respect to bridged model compounds like DCS-B24 which are unable to populate the A* state [35]. Selective bridging has also been used to show that DS populates a TICT state through dimethylanilino twisting (Fig. 15) [51] and that nitro group twisting is also a possible TICT channel in DNS [107].

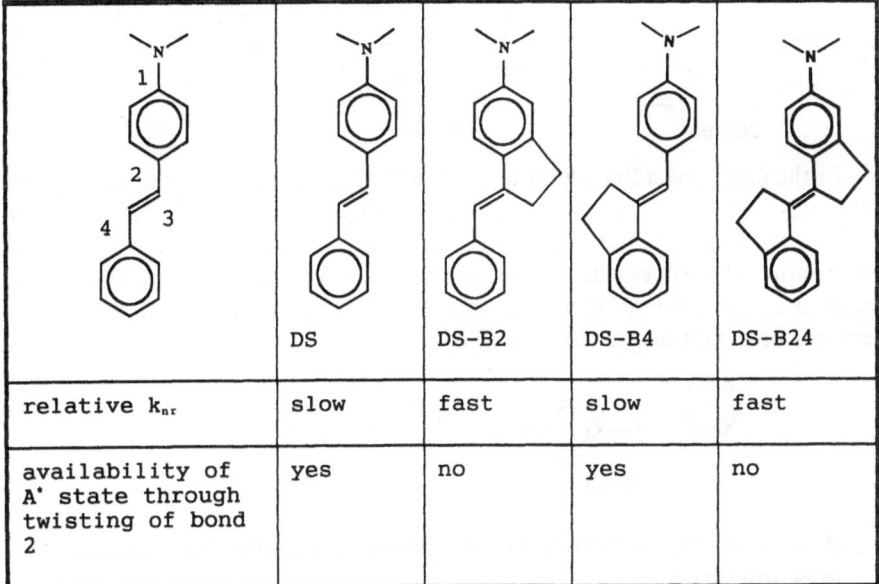

	DS	DS–B2	DS–B4	DS–B24
relative k_{nr}	slow	fast	slow	fast
availability of A* state through twisting of bond 2	yes	no	yes	no

Fig. 15. Nonradiative deactivation pattern of different selectively bridged derivatives of DS. When bond 2 is free to rotate, nonradiative decay is slowed down due to the formation of a TICT state hindering the access to the funnel state. Bond 4 is not TICT-active [51]

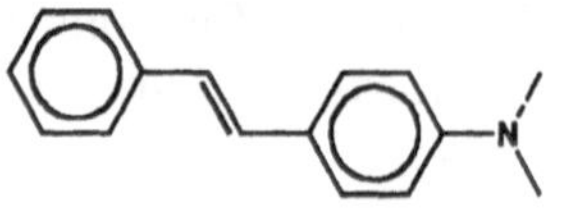

DS

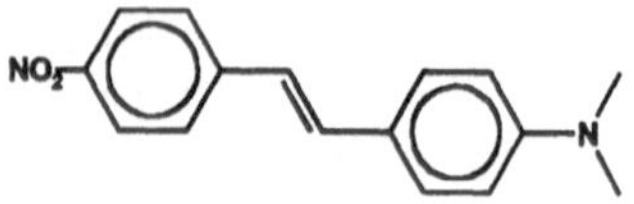

DNS

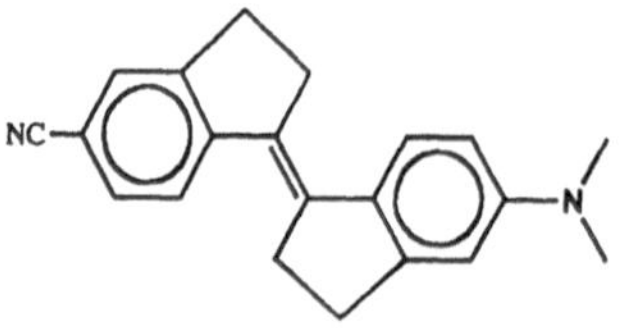

DCS-B24

Intramolecular bridging can also be used to suppress the P* state selectively. This has dramatic photophysical consequences: DCS-B34 shows fluorescence quantum yields of about 80% and no sign of intramolecular fluorescence quenching [35] but some indication of a temperature-dependent $A^* \rightleftarrows E^*$ equilibrium. The high quantum yield of this compound shows that the TICT emission is allowed in this case. The same can be concluded for DCS from the sizeable quantum yields at the lifetime maximum near $-90\,°C$ in ethanol [35]. A possible explanation are narrow minima in k_f or f (case II of Fig. 14). First results for compounds like DM-DS-B34 sterically confined to the neighbourhood of a perpendicular conformation show a large reduction in fluorescence quantum yield with respect to the unconfined case (DS-B34).

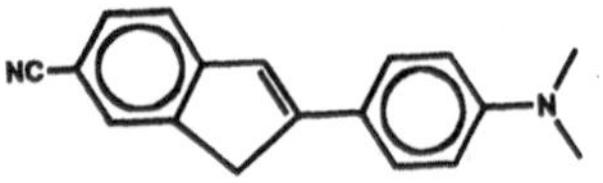

DCS-B34

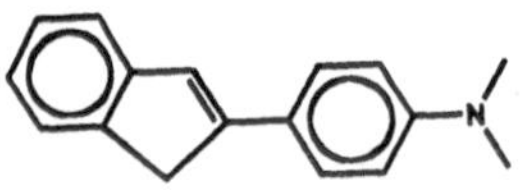

DS-B34

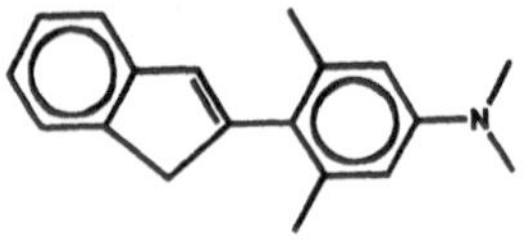

DM-DS-B34

In the family of stilbazolium dyes DASPMI, DPS etc., a similar three-state mechanism is active leading in this case to single-bond twisted TICT species with a high intrinsic nonradiative rate [108, 109] probably connected with the ionic nature of the dyes (charge shift, see Sect. 4.5). This results in similar lifetime maxima as a function of temperature as for DCS but sizeably shifted to lower temperatures with respect to DCS [96].

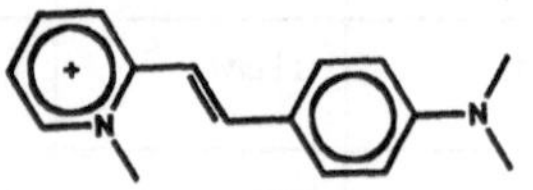

DASPMI

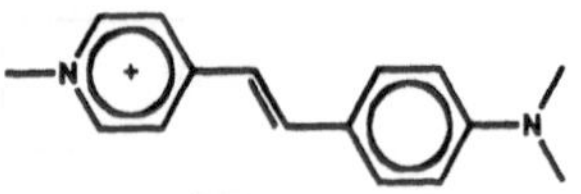

DPS

7.2 Reversal of Charge Shift Directions through Competing TICT Channels in Benzopyrylium Dyes

Benzopyrylium dyes are also positively charged molecules, and consequently, TICT formation corresponds to shift and localization of that charge [53].

Benzopyrylium dyes with two flexible and TICT-active donor substituents at the two ends of the molecule therefore show opposite charge shift directions as indicated in Fig. 16.

These two channels compete kinetically and lead to nonradiative TICT species. In bridged compounds, where TICT formation is not possible, the quenching is absent. The competition in the flexible compounds can be governed by tuning the electron accepting properties of the benzopyrylium system with the aid of the other (non-twisted) substituent [106, 110]. Some examples are shown in Fig. 17. Remarkable is the "anti-loose-bolt" behavior, i.e. the compounds OO, NO and NN with two flexible groups show reduced nonradiative

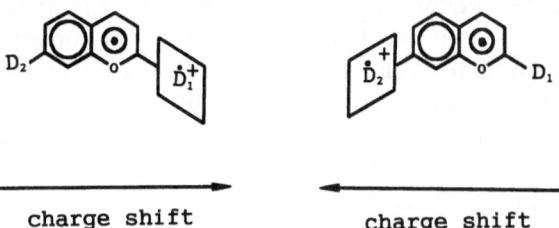

charge shift   charge shift

Fig. 16. Benzopyrylium dyes with multiple TICT channels. Depending on which of the two donor substituents twists, charge shift occurs at different ends of the molecule

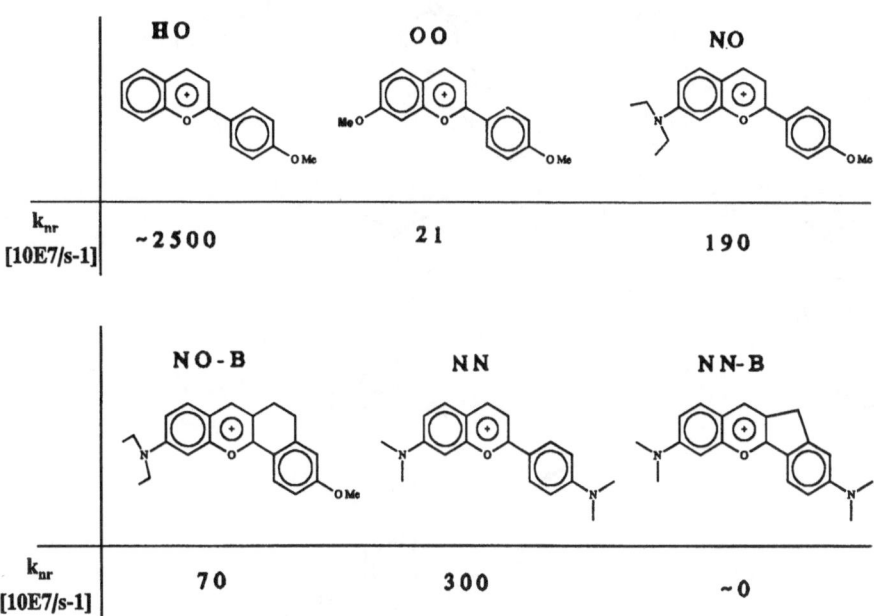

Fig. 17. Relationship between molecular structure and nonradiative decay rate constants for benzopyrylium dyes at room temperature. The observed dependencies are predicted by the TICT model [110]

losses with respect to compound HO with only one flexible group. In NO, both TICT channels are active and are of similar importance. In NO-B, only the dialkylamino group twisting remains active. Its activity is strongly reduced in NN-B because the benzopyrylium acceptor strength is weakened by the stronger and rigidized donor D_1. It can therefore be concluded, that the main TICT activity in NN occurs through dimethylanilino (D_1) twisting.

7.3 Combination of TICT with other Types of Adiabatic Photochemical Reactions

Different photochemical channels leading either to TICT or to other products can be combined in one and the same molecule. An example involving competition between "Excited State Intramolecular Proton Transfer" (ESIPT) and TICT formation (Scheme 2) is the molecule Ka1 [111]. In this case, three fluorescence bands can be expected in principle (the precursor state E* and the two product species ESIPT and TICT). The product channels can be selectively blocked in the model compounds Ka2 and Ka3.

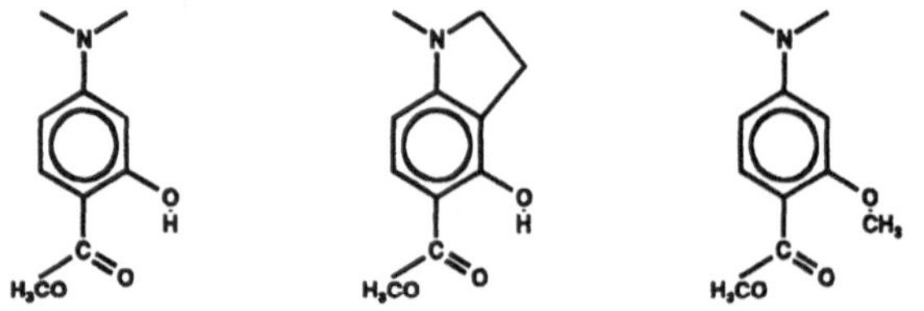

Ka1   Ka2   Ka3

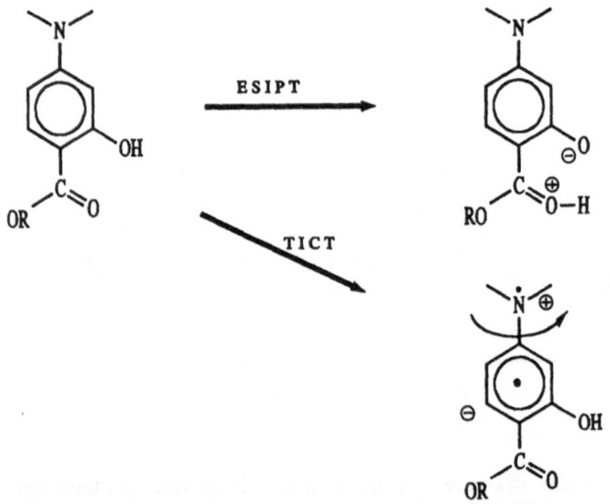

Scheme 2. Example for molecules possessing competing TICT and ESIPT photoreaction processes

In the molecules BO1 and BO2, ESIPT and TICT are consecutively coupled (Scheme 3). Both compounds show exclusively the strongly redshifted ESIPT fluorescence but the fluorescence quantum yield of BO1 is by orders of magnitude lower than that of BO2 [112, 113]. This can be rationalized by the fluorescence quenching action of the TICT state available through twisting the central bond in the ESIPT tautomer. In the case of BO1, this TICT state is calculated to be of much lower energy than for BO2 explaining the viscosity-dependent nonradiative decay and the large difference between both compounds [113].

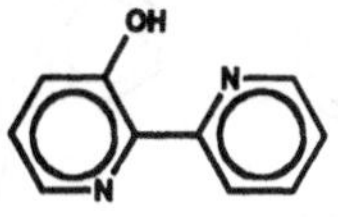

BO1 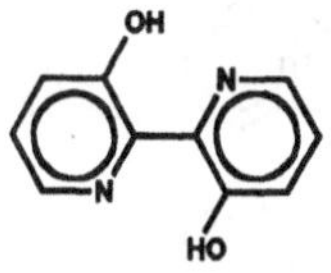   BO2

TICT formation can also be combined with excimer or exciplex formation. An example is compound Bi-DMABK, which, in addition to the TICT state, can also form head-to-head intramolecular excimers even at very low concentration and therefore exhibits triple fluorescence [114]. At sufficiently high concentrations, DMABN [115, 116] and its derivatives like DMABME [117–121, 126] are also expected to form, in addition to TICT, (intermolecular) excimers, ground state dimers or larger aggregates.

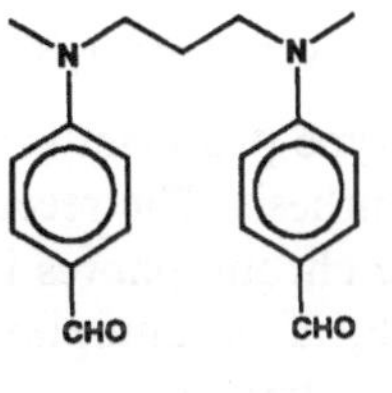

BI-DMABK    DMABN    DMABME

Also in methylene-linked biaryl systems like A1A, two conformers (conformation A and B) preferentially prone for excimer (A) and TICT formation (B) can be discussed. TICT formation in B should be related to that in bianthryl BA and lead to a large excited state dipole moment. Hence, a solvent-polarity induced redshift of the TICT fluorescence component is expected. The excimer

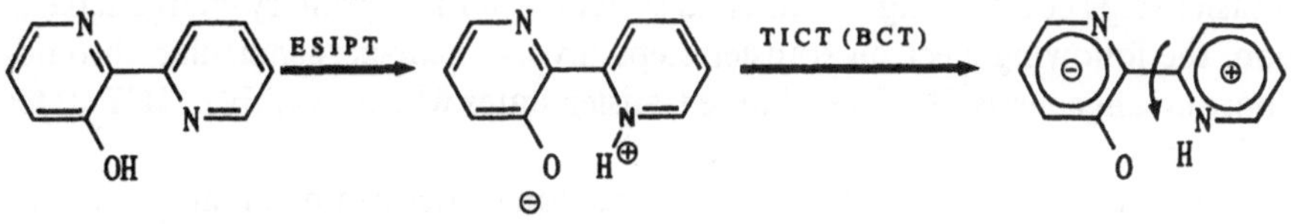

ESIPT   TICT (BCT)

Scheme 3. Example for molecules possessing consecutive TICT and ESIPT photoreaction processes

as well as the anthracene-type (excitonic) state, on the other hand, correspond to states with very low dipole moment and should not show a strong redshift. Experimentally, the fluorescence of A1A can be decomposed into an excitonic and an excimer component similar to the case of the related dianthrylethane system [122], but low-temperature measurements also yield evidence for the contribution of a TICT component [123, 124]. The latter is strongly enhanced and responsible for the major part of the fluorescence at low temperature in the compound A(OH)A [124].

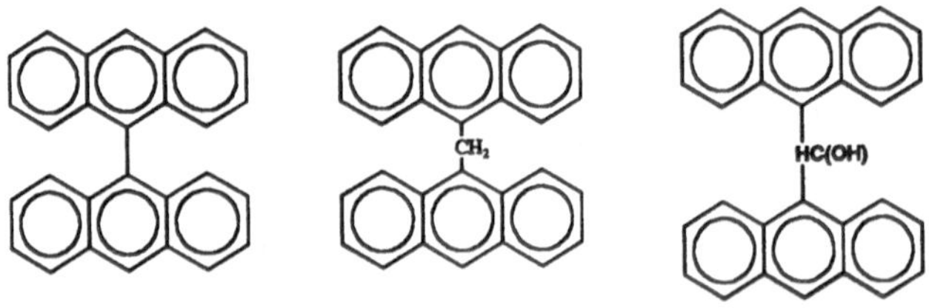

BA   A1A   A(OH)A

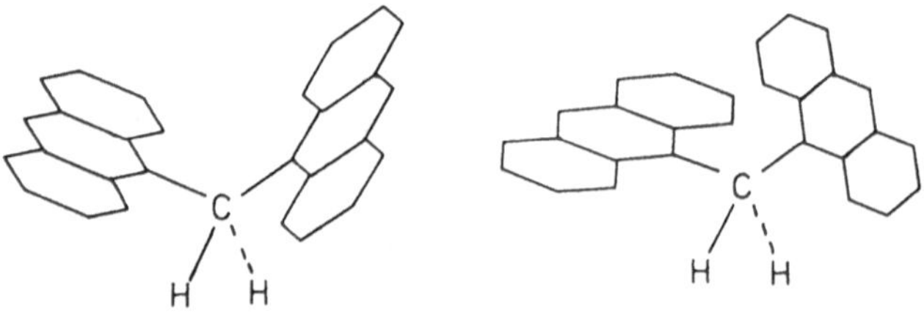

conformation A   conformation B

This aspect of coupling excimer and TICT-type processes is also important for understanding the primary step of photosynthesis. The recent X-ray determination of the relative mutual position of the chromophores in the bacterial reaction center [125] has shown that the central chromophore dimer ("the special pair") where the initial charge separation takes place on a subpicosecond time scale possesses two nearly coplanar and only partially overlapping chromophores (Fig. 18). This overlap (and the corresponding interaction) is small enough to allow for charge separation (and symmetry breaking); in symmetric systems with large overlap and interaction, symmetry breaking and charge separation are rendered much more difficult [8, 17]. On the other hand, the interaction is strong enough to lead to redshifted absorption spectra of the special pair which assure that it is the lowest trap in the sequence of energy transfer steps constituting the antenna system. After this primary charge separation, the following electron transfer steps involve near-perpendicular chromophore arrangements [133] i.e. charge transfer states which resemble TICT states and possess much smaller interaction.

The situation in the photosynthetic reaction center can be mimicked by the dianthryl model compounds. In BA with perpendicular ground state conformation the interaction is small enough for charge separation/symmetry breaking to

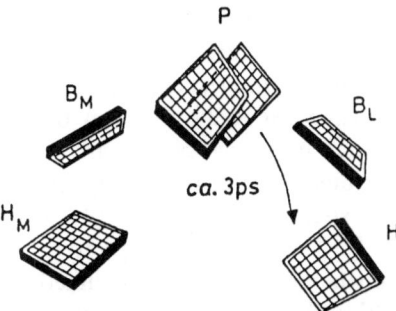

P

B_M

ca. 3ps

B_L

H_M

H_L

Fig. 18. Spatial arrangement of the chromophores in the photo-synthetic reaction system (adapted from [127]). The *top* shows the partially overlapping special pair *P*; *left* and *right* – electron transfer branches incorporate chromophores in a nearly perpendicular arrangement

occur; in A2A, where a sandwich-type conformation is possible, the orbital overlap is larger leading to a stronger interaction, and no symmetry breaking is observed [8, 128]. Compounds with intermediate coupling like A1A can possibly bridge the gap between the two cases.

CH 2
CH 2

NC-

NC-

-N

-N

A2A

DCB

DCH

Unless the two chromophores are chemically linked as in Bi-DMABK, A1A and A2A, the excimer formation occurs through a bimolecular encounter and is therefore concentration dependent. Recently, such behavior was also discovered for DCS and its derivatives [34, 35], and even more strongly for DCB and DCH [129, 130]. However, in these cases, the observed effects were also dependent on the square of the laser intensity indicating a biphotonic mechanism [34, 35, 129, 130]. Thus, the complexes formed involve the encounter of two excited molecules and are termed "bicimers" [131]. TICT relaxation channels can be strongly slowed down in these species to become conveniently observable by picosecond spectroscopy [132].

8 The Chemical Reactivity of TICT States

8.1 TICT States as Contact Radical Ion Pairs

Because of the orthogonality of donor and acceptor π-orbitals in the ideal TICT conformation, the electron transfer from donor to acceptor has to be virtually complete [2], possibly somewhat reduced by σ-back donation effects [134].

Therefore, a molecule in the TICT state can be regarded as a rigidly linked radical anion-radical cation pair. Experimental proof for this expectation can be gained from transient absorption spectroscopy. In the simplest case, the absorption spectrum of the TICT excited state is expected to be the sum of the individual ion spectra. This was indeed found in a few cases, with some perturbations which can be explained by the interaction of the closely-spaced radical ions. Thus, the transient absorption spectra of DMABN [135], of DMABK (or DMABA) [137] and of BA [58] resemble the spectrum of benzonitrile and acetophenone radical anion (the absorption of the dimethyl-amino radical cation is expected to be situated in the UV region and could not be observed) and to the sum of anthracene anion and cation absorption spectra, respectively.

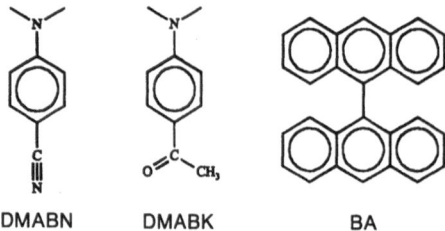

DMABN	DMABK	BA

On the other hand, the perpendicularity of the TICT state is only a model, and in reality, thermal motion will lead to a more or less broad distribution around the 90° twist angle, or the TICT potential may even possess a more complicated shape with minima differing from 90° [17, 50, 96] (leading to twist angle distributions of type d in Fig. 14, see above). Then, strong mixing of the CT with the LE wavefunction will result, and the corresponding absorption spectra may deviate considerably from the sum of radical anion/cation spectra and exhibit a pronounced dependence on solvent polarity. This was found for the biaryl C9A [136] which is electronically closely related to BA, but possesses less sterical strain and thus more conformational freedom away from perpendicularity. Whereas BA has a perpendicular ground state potential minimum [61], this is loosened to a shallow double minimum potential for C9A [55]. Similar behavior could be expected for C1N, a TICT compound which is closely related to C9A. The fluorescence quenching properties for C1N and C9A as discussed below show that although there may be nonideal TICT behavior, the chemical reactivity is nevertheless that of a radical anion/cation pair.

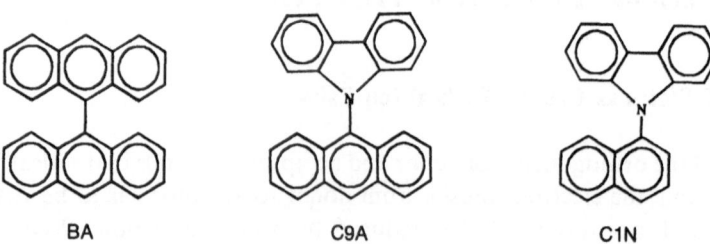

BA	C9A	C1N

8.2 Reactivity and Interaction with Inorganic Ions

Relatively soon after the announcement of the TICT model, the latter was tested by observing the fluorescence quenching behavior with inorganic ions. In particular, the halide anions fluoride, chloride, bromide and iodide were tested with respect to the two TICT compounds DMABN and PBN [139]. While PBN responded in the expected manner, with the strongest quenching observed for the iodide and no quenching for the fluoride ion, the order for DMABN was different, with fluoride being a stronger quencher than chloride, and with no quenching observed for the bromide ion. This was traced back to the different distribution of the positive charge for the two molecules in the TICT state: For DMABN, it is concentrated on the lone pair of the amino nitrogen, for PBN, the positive charge is distributed over the entire pyrrole heterocycle. The latter thus behaves as a soft Lewis acid with respect to the anion (viewed as base), whereas

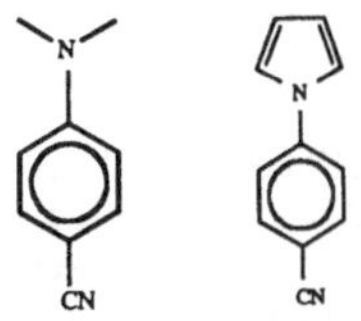

DMABN   PBN

DMABN corresponds to a hard acid favouring the interaction with a hard base, the fluoride anion.

8.3 Reactivity with Organic Molecules

Recently, TICT compounds were used to initiate electron-transfer-induced organic reactions [13]. If an organic electron donor transfers its electron to the cationic end of the excited TICT molecule (to the donor moiety of the TICT molecule in the ground state) the positive charge is transferred and radical cation sensitized reactions can be started in such a way. Likewise, electron acceptors can react with the "acceptor" end of the excited TICT molecule, and radical anionic sensitized reactions such as the dechlorination of pentachloro-benzene can be performed (Fig. 19). Although the reactivity of the latter reaction is not outstanding and although it was used only as a mechanistic test, this procedure shows a possible way to the photodestruction of ecologically harmful materials, in many cases chlorinated organic compounds like chlorobenzenes or chloro-s-triazines, by using TICT molecules as catalysts.

The original study was done using C1N as TICT catalyst [13, 14]. A later study [138] compared the reactivity of C1N with the related TICT compounds C9A and CBN which possess a differently sized acceptor group. Donors were used as quenchers, and the good correlation of the quenching rate constants k_q with the donor oxydation potential establish the expected electron transfer

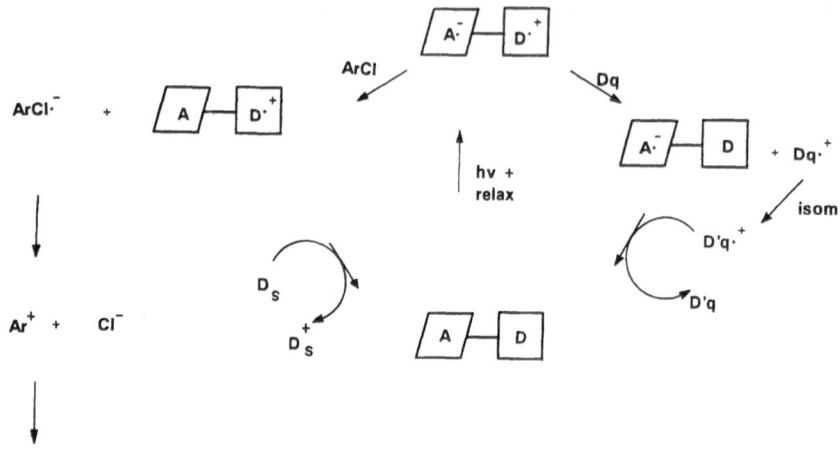

Fig. 19. Possible scheme for use of TICT compounds as catalysts in photoreduction (e.g. photo-dechlorination of organic compounds ArCl). D_s is a sacrificial donor to reduce the TICT cation radical back to the starting material in the ground state. TICT compounds can also be used as photooxydation catalysts as shown for the example of norbornadiene(Dq)-isomerization [13]. In this case, direct back electron transfer in the ground state takes place as shown in the figure

Table 2. Bimolecular fluorescence quenching rate constants k_q (10^9 l mol^{-1} s^{-1}) for different TICT molecules with various electron donors [138]. The oxidation potential of the donor (in Volts) is also given

Quencher	E_{ox}	C1N	C9A	CBN
N,N-dimethylaniline	0.78	11.5	13.4	13.1
		12.4[a]		
1,2,4,5-tetramethoxybenzene	0.81	6.0	8.3	7.4
1,2,4-trimethoxybenzene	1.12	1.9	4.3	5.1
1,4-dimethoxybenzene	1.34	< 0.1	0.1	0 1
1,2,3-trimethoxybenzene	1.42	< 0.1	< 0.1	< 0.1
hexamethylbenzene	1.46	< 0.1	< 0.1	< 0.1

[a] [13]

mechanism (Table 2). The reactive end of the TICT molecule is the carbazole (radical cation) unit in all three cases, and the fact that the correlation holds for both anilines and methoxybenzenes as electron donating quenchers and further-more that isomeric trimethoxybenzenes show strongly different quenching rates indicates that specific short range interactions (like 2 center-3-electron bonds, or specific solute-solvent exciplexes, see below) are not operative here. This is probably related to the rather expansively distributed positive charge on the carbazole moiety in the TICT state (compare the case of PBN above). Although the acceptor end of the TICT molecule is not directly involved in the intermolecular electron transfer, it nevertheless influences the quenching rate constant k_q (Table 2). Thus C9A and CBN are found somewhat more efficient than C1N. This may be related to differently-sized contributions of the (nonpolar) LE-wavefunction to the TICT state (see the discussion of conformational effects above).

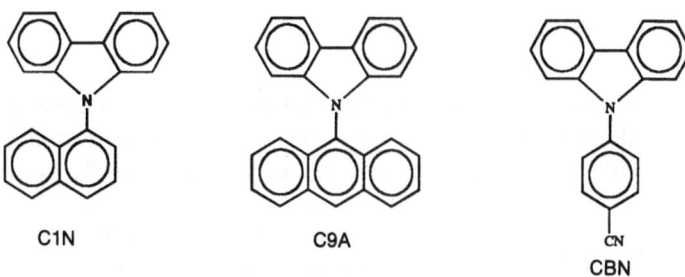

C1N   C9A

CN
CBN

8.4 Specific Interactions with TICT Molecules

There are a number of studies which indicate that certain TICT molecules undergo specific interactions with solvent or quencher molecules. One of the early proposals to explain the dual fluorescence of DMABN was in fact that of a solute-solvent exciplex (excited complex) [26–28]. The first specific interaction documented as such, however, was that of DMABN (and model compounds) with saturated amines, where a mechanism necessitating a close approach was postulated. This was concluded from the lack of correlation of the quenching reactivity with the ionization potential of the amine quencher. However, a correlation was found between the reactivity pattern and the sterical demands of the saturated amine [140, 141]. The complexes formed were non-emissive and were proposed to involve a two-center-three-electron bond between the aromatic and the saturated amino group.

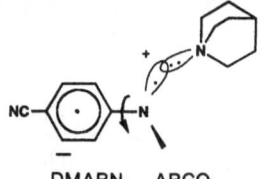

DMABN. . .ABCO

The first emissive exciplex-type species were reported for DMABEE derivatives in chlorinated alkane solvents [142]. In this case, a significant blueshift (40 nm) of the TICT band with respect to the expected position (concluded by comparison with other nonchlorinated solvents of the same and of different polarity) was observed. A further case of an unexpected sizeable solvent-specific shift of the TICT band has recently been observed for amino-substituted triphenylphosphines like MAP in *n*-butyl chloride as solvent [79]. In this case, the specific shift occurs to the red.

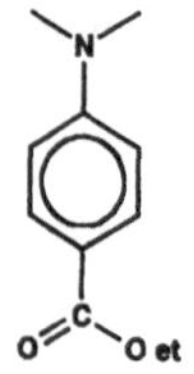

DMABEE 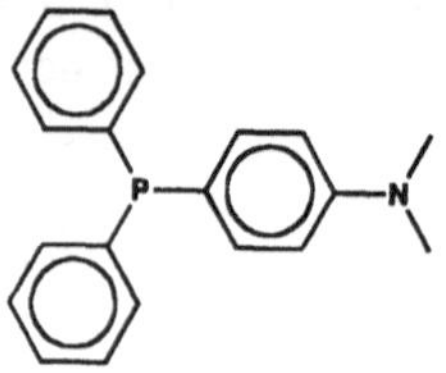   MAP

Taken together, these results indicate that in some cases (preferably in TICT molecules with amino group lone pairs, the "hard Lewis acid case" [139], see Sect. 8.2) specific complexes with solvents bearing energetically available lone pair electrons may be formed. But except for solvents bearing amino groups (where the N ∴ N two-center-three-electron bond can form) the energetical changes upon exciplex formation seem to be small, and exciplex formation alone as claimed in [26, 143, 144] cannot be responsible for the observed large redshifts of the TICT band, but it seems to be an additional small factor influencing the energetic position of the TICT band in some cases and explaining the observed scatter in the solvatochromic plots. The same applies to the proposition [32, 148] that TICT fluorescence is uniquely the result of ground state aggregates with water molecules (as solvent impurity) believed to induce pretwisted conformations at the dimethylamino group. The complication by aggregates with water molecules cannot be excluded in a multitude of cases, but the different experimental results (e.g. the parent-daughter kinetic relationship) viewed as a whole are not explainable by this model alone.

An interesting new approach of using TICT compounds in intermolecular interactions are molecules with a crown ether function. These molecules can form complexes with cations, e.g. Ca^{2+}, and the complex can have different emission properties in the complexed and the uncomplexed state. In this way, specific ion probes usable in medicine, biology and analytical chemistry can be developed. The mechanism of cation sensing and the TICT-crown molecules studied to date [145–147] are collected in Fig. 20.

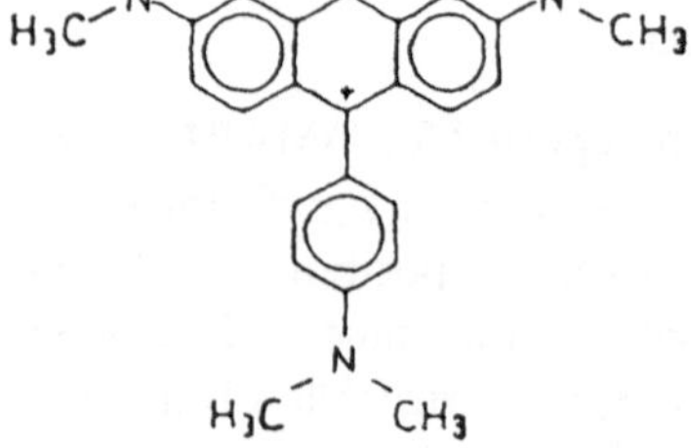

ARh 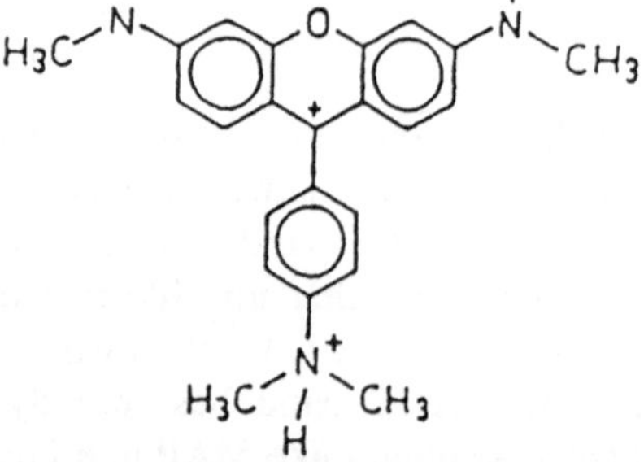   Arh/H⁺

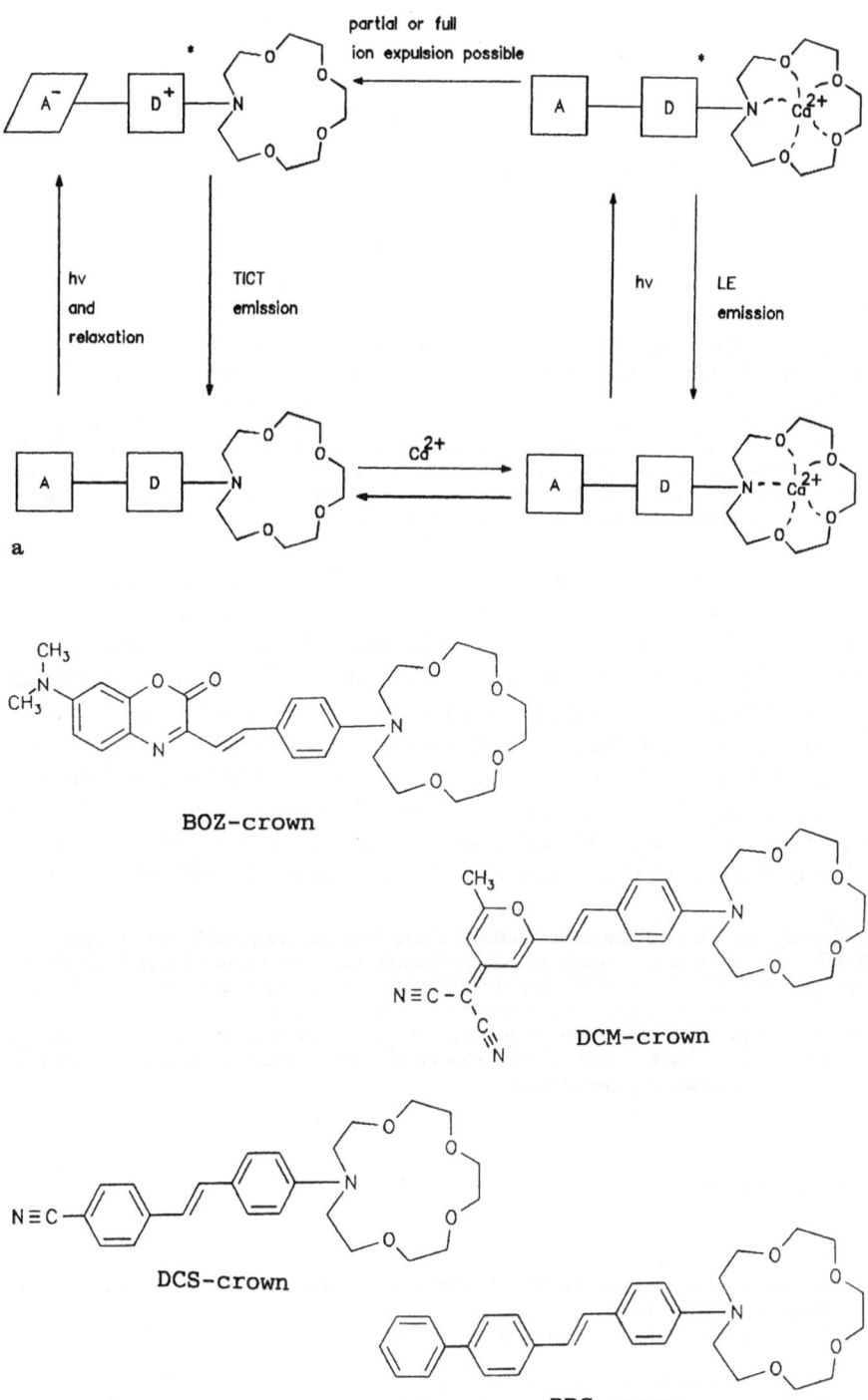

Fig. 20. (continued)

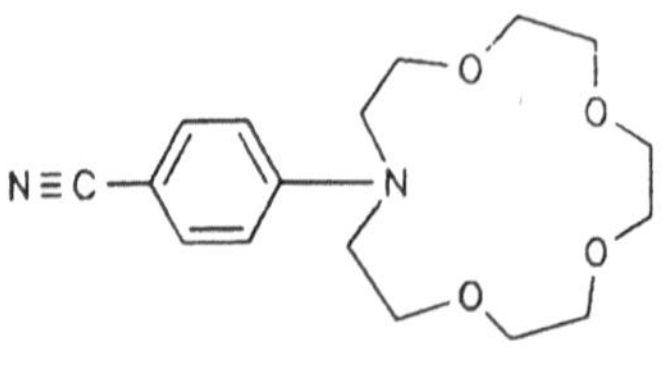

DMABN-crown

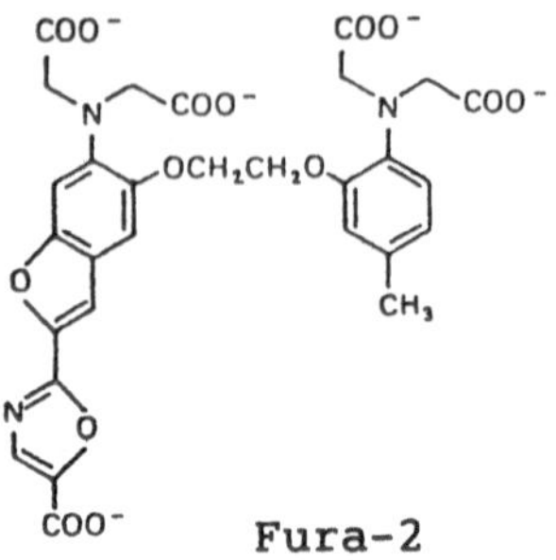

b

Fura-2

Fig. 20. a In a favourable TICT-crown ether compound, the uncomplexed species exhibits TICT emission in polar solvents, whereas for the complexed species, the ion (here Ca^{2+}) diminishes the donor strength of the amino group and raises the TICT energy far enough such that only blue-shifted LE emission can occur. In some cases, expulsion of Ca^{2+} in the excited state has been found. **b** A selection of TICT-crown ether molecules studied to date. The last molecule (Fura-2) has different complexing groups especially suited for Calcium. Its chromophore system, however, is related to the other compounds and can be viewed as a hetero-derivative of a bridged donor–acceptor–stilbene. It is commercially available and used as a biological fluorescent probe [147]

A similar approach can be used to generate pH-sensing fluorescent dyes by introducing an amino group into highly fluorescent compounds leading to an efficient TICT channel. An example is ARh, where the unprotonated dimethyl-anilino group is an efficient TICT donor and leads to a fast nonradiative channel quenching the fluorescence nearly completely [11, 12]. When the amino group is protonated in acidic media (Arh/H^{+}), however, this channel is closed because an ammonium group acts as an acceptor, and highly brilliant fluorescence, comparable to that of other rhodamine dyes, results. These fluorescent indicators are much more sensitive than ordinary indicator dyes and can be used to measure pH values of micro-domains (like cells in biological material, microdroplets in fog).

Acknowledgment. Many of the results outlined above have become possible due to support by BMFT (fluorescence kinetics using BESSY synchrotron radiation, projects 05 314 FAI5, 05 414 FAB1 and 05 5KT FAB) and DFG (Heisenberg fellowship to the author), and due to the skill of my coworkers. I also want to thank my collaborators from other laboratories abroad and in Germany, who provided the stimulus for many discussions and helped to develop the ideas presented here. I am especially grateful to Professor Z.R. Grabowski for his open-minded collaboration with the new aspects of TICT formation outlined here.

9 References

1. Lippert E, Lüder W, Boos H (1962) In: Mangini A (ed) Advances in molecular spectroscopy. Pergamon, Oxford, p 443
2. Grabowski ZR, Rotkiewicz K, Siemiarczuk A, Cowley DJ, Baumann W (1979) Nouv J Chim 3: 443
3. Rotkiewicz K, Grellmann KH, Grabowski ZR (1973) Chem Phys Lett 19: 315
4. Rotkiewicz K, Grabowski ZR, Krowczynski A, Kühnle W (1976) J Lumin 12: 877
5. Grabowski ZR, Rotkiewicz K, Rubaszewska W, Kirkor-Kaminska E (1978) Acta Phys Pol A 54: 767

6. Grabowski ZR, Rotkiewicz K, Siemiarczuk A (1979) J Lumin 18: 420
7. Rettig W (1991) Nachr Chem Tech Lab 39: 298
8. Rettig W (1986) Angew Chem Int Ed Engl 25: 971
9. Al-Hassan KA, Rettig W (1986) Chem Phys Lett 126: 273
10. Rettig W, Fritz R, Springer J (1991) In: Honda K (ed) Photochemical processes in organized molecular systems. Elsevier, Amsterdam, p 61
11. Rettig W (1993) In: Wolfbeis, OS (ed) Fluorescence spectroscopy. New methods and applications, Springer, Berlin Heidelberg New York, p 31
12. Plaza P, Jung ND, Martin MM, Meyer YH, Vogel M, Rettig W (1992) Chem Phys 168: 365
13. Habib Jiwan JL, Soumillion JP (1992) J Photochem Photobiol A: Chem 64: 145
14. Soumillion JP (1993) In: Topics in Current Chemistry, Springer, Berlin Heidelberg New York (in press)
15. Grabowski ZR, Dobkowski J (1983) Pure Appl Chem 55: 245
16. Lippert E, Rettig W, Bonačić-Koutecký V, Heisel F, Miehé JA (1987) Adv Chem ʾhys 68: 1
17. Rettig W (1988) In: Liebman JF, Greenberg A (eds) Modern models of bonding and delocalization; Series: Molecular structure and energetics, vol 6, chapt 5, VCH Publishers, New York, p 229
18. Rettig W, Baumann W (1992) In: Rabek, JF (ed) Photochemistry and Photophysics, vol 6, CRC Press, Boca Raton, p 79
19. Bonačić-Koutecký V, Koutecký J, Michl J (1987) Angew Chem 99: 216; (1987) Angew Chem Int Ed Engl 27: 170
20. Michl J, Bonačić-Koutecký V (1990) Electronic aspects of organic photochemistry. J Wiley, New York
21. Turro NJ, McVey J, Ramamurthy V, Lechtken P (1979) Angew Chem 91: 597
22. Rettig W (1992) In: Mataga N, Okada T, Masuhara H (eds) Dynamics and mechanisms of photoinduced electron transfer and related phenomena. Elsevier, Amsterdam, p 57
23. van der Auweraer M, Grabowski ZR, Rettig W (1991) J Phys Chem 95: 2083
24. Rettig W, Lippert E (1980) J Mol Struct 61: 17
25. Zander M, Rettig W (1984) Chem Phys Lett 110: 602
26. Visser RJ, Varma CAGO (1980) J Chem Soc Faraday Trans II 76: 453
27. Visser RJ, Weisenborn PCM, Varma CAGO (1984) Chem Phys Lett 104: 38
28. Chandross EA (1975) In: Gordon M, Ware WR (eds) The exciplex, Academic Press, New York p 187
29. Khalil OS, Hofeldt RH, McGlynn SP (1973) Spectrosc Lett 6(3): 147
30. Khalil OS (1975) Chem Phys Lett 35: 172
31. Khalil OS, Hofeldt RH, McGlynn SP (1973) J Lumin 6: 229
32. Cazeau-Dubroca C, Ait Lyazidi S, Nouchi G, Peirigua A, Cazeau P (1986) Nouv J Chim 10: 337 New J Chem
33. Schuddeboom W, Jonker SA, Warman JM, Leinhos U, Kühnle W, Zachariasse KA (1992) J Phys Chem, 96: 10809
34. Gilabert E, Lapouyade R, Rullière C (1991) Chem Phys Lett 185: 82
35. Lapouyade R, Czeschka K, Majenz W, Rettig W, Gilabert E, Rullière C (1992) J Phys Chem 96: 9643
36. Herbich J, Perez Salgado F, Rettschnick RPH, Grabowski ZR, Wojtowicz H (1990) J Phys Chem 2: 1
37. Okada T, Fujita T, Kubota M, Masaki S, Mataga N, Ide R, Sakata Y, Misumi S (1972) Chem Phys Lett 14: 563
38. Okada T, Kawai M, Ikemachi T, Mataga N, Sakata Y, Misumi S, Shionoya S (1984) J Phys Chem 88: 1976
39. Okada T, Mataga N, Baumann W, Siemiarczuk A (1987) J Phys Chem 91: 4490
40. Baumann W, Petzke F, Loosen K-D (1979) Z Naturforsch A 34: 1070
41. Detzer N, Baumann W, Schwager B, Fröhling J-C, Brittinger C (1987) Z Naturforsch A 42: 395
42. Baumann W, Schwager B, Detzer N, Okada T, Mataga N (1988) J Phys Chem 92: 3742
43. Siemiarczuk A, Grabowski ZR, Krowczynski A, Asher M, Ottolenghi M (1977) Chem Phys Lett 51: 315
44. Tominaga K, Walker GC, Kang TJ, Barbara PF, Fonseca T (1991) J Phys Chem 95: 10485
45. Barbara PF, Tominaga K, Walker GC (1992) In: Mataga N, Okada T, Masuhara H (eds) Dynamics and mechanisms of photoinduced electron transfer and related phenomena. Elsevier Amsterdam, p 21
46. Siemiarczuk A, Koput J, Pohorille A (1982) Z Naturforsch A 37: 598

47. Siemiarczuk A (1984) Chem Phys Lett 110: 437
48. Kajimoto O, Hayami S, Shizuka H (1991) Chem Phys Lett 177: 219
49. Herbich J, Kapturkiewicz A (1991) Chem Phys 158: 143; (1993) Chem Phys 170: 221
50. Klock AM, Rettig W (1993) Polish J Chem 67: 1375
51. Létard JF, Lapouyade R, Rettig W (1993) J Am Chem Soc 115: 2441
52. Rettig W, Chandross EA (1985) J Am Chem Soc 107: 5617
53. Vogel M, Rettig W (1985) Ber Bunsenges Phys Chem 89: 962
54. Rettig W, Zander M (1982) Chem Phys Lett 87: 229
55. Monte C, Roggan A, Subaric-Leitis A, Rettig W, Zimmermann P (1993) J Chem Phys 98: 2580
56. Rettig W, Marschner F (1983) Nouv J Chim 7: 425; (1990) New J Chem 14: 819
57. Nakashima N, Murakawa M, Mataga N (1976) Bull Chem Soc Jpn 49: 854
58. Mataga N, Yao H, Okada T, Rettig W (1989) J Phys Chem 93: 3383
59. Migira M, Okada T, Mataga N, Sakata Y, Misumi S, Nakashima N, Yoshihara K (1981) Bull Chem Soc Jpn 54: 3301
60. Rettig W, Zander M (1983) Ber Bunsenges Phys Chem 87: 1143
61. Subaric-Leitis A, Monte C, Roggan A, Rettig W, Zimmermann P, Heinze J (1990) J Chem Phys 93: 4543
62. Lueck H, Windsor MW, Rettig W (1990) J Phys Chem 94: 4550
63. Kang TJ, Jarzeba W, Barbara PF, Fonseca T (1990) Chem Phys 149: 81
64. Baumann W, Spohr E, Bischof H, Liptay W (1987) J Lumin 37: 227
65. Visser R-J, Weisenborn PCM, van Kan PJM, Huizer BH, Varma CAGO (1985) J Chem Soc Faraday Trans II 81: 689
66. Kajimoto O, Yamasaki K, Arita K (1986) Chem Phys Lett 125: 184
67. Hara K, Arase T (1984) Chem Phys Lett 197: 178
68. Toublanc TB, Fessenden RW, Hitachi A (1989) J Phys Chem 93: 2893
69. Honma K, Arita K, Yamasaki K, Kajimoto O (1991) J Chem Phys 94: 3496
70. Vogel M, Rettig W, Sens R, Drexhage KH (1988) Chem Phys Lett 147: 452
71. Vogel M, Rettig W, Sens R, Drexhage KH (1988) Chem Phys Lett 147: 461
72. Vogel M, Rettig W, Fiedeldei U, Baumgärtel H (1988) Chem Phys Lett 148: 347
73. López Arbeloa I, Rohatgi-Mukherjee KK (1986) Chem Phys Lett 129: 607; López Arbeloa F, López Arbeloa T, Gil Lage E, López Arbeloa I, De Schryver FC (1991) J Photochem Photobiol A: Chem 56: 313
74. Letard JF, Lapouyade R, Rettig W (1993) Mol Cryst Liq Cryst 236: 41
75. Kakitani T, Mataga N (1985) J Phys Chem 89: 8; (1986) J Phys Chem 90: 993
76. Martin MM, Breheret E, Nesa F, Meyer YH (1989) Chem Phys 130: 279
77. Martin MM, Plaza P, Meyer YH (1991) J Phys Chem 95: 9310
78. Ben-Amotz D, Harris CB (1985) Chem Phys Lett 119: 305; (1987) J Chem Phys 86: 4856, 5433
79. Vogel M, Rettig W, Heimbach P (1991) J Photochem Photobiol A: Chem 61: 65
80. Bonačić-Koutecký V, Michl J (1985) J Am Chem Soc 107: 1765
81. Borden WT (ed) (1982) Biradicals, Wiley, New York
82. Salem L (1982) Electrons in chemical reactions, Wiley, New York
83. Turro NJ (1986) Angew Chem 98: 872
84. Rettig W, Majenz W (1989) Chem Phys Lett 154: 335
85. Strehmel B, Rettig W, Majenz W (1993) Chem Phys 173: 525
86. Sundström V, Gillbro T (1985) Ber Bunsenges Phys Chem 89: 222
87. Hicks JM, Vandersall MT, Sitzmann EV, Eisenthal KB (1987) Chem Phys Lett 135: 413
88. Hicks J, Vandersall M, Babarogic Z, Eisenthal KB (1985) Chem Phys Lett 116: 18
89. Rettig W, Majenz W, Herter R, Létard JF, Lapouyade R (1993) Pure and applied chem (in press)
90. Schilling CL, Hilinski EF (1988) J Am Chem Soc 110: 2296
91. Morais J, Ma J, Zimmt MB (1991) J Phys Chem 95: 3885
92. Ma J, Zimmt MB (1992) J Am Chem Soc 114: 9723
93. Warman JM (1993) Pure Appl Chem (in press)
94. Shin D-M, Whitten DG (1988) J Am Chem Soc 110: 5206
95. Abdel-Halim ST, Abdel-Kader MH, Steiner UE (1988) J Phys Chem 92: 4324
96. Rettig W, Majenz W, Lapouyade R, Vogel M (1992) J Photochem Photobiol A: Chem 65: 95
97. Bonačić-Koutecký V, Schöffel K, Michl J (1989) J Am Chem Soc 111: 6140
98. Fabian J, Zahradnik R (1989) Ang Chem 101: 693
99. Rettig W, Gleiter R (1985) J Phys Chem 89: 4676
100. Rettig W (1991) Ber Bunsenges Phys Chem 95: 259

101. Lippert E, Ayuk AA, Rettig W, Wermuth G (1981) J Photochem 17: 237
102. Rettig W, Rotkiewicz K, Rubaszewska W (1984) Spectrochim Acta A 40: 241
103. Baumann W (1981) Z Naturforsch A 36: 868
104. Rettig W, Braun D, Suppan P, Vauthey E, Rotkiewicz K, Luboradzki R, Suwinska K (1993) (to be published)
105. Baumann W, Bischof H, Fröhling J-C, Brittinger C, Rettig W, Rotkiewicz K (1992) J Photochem 64: 49
106. Rettig W, Majenz W, Lapouyade R, Haucke G (1992) J Photochem Photobiol A: Chem 62: 415
107. Lapouyade R, Kuhn A, Letard JF, Rettig W (1993) Chem Phys Lett (in press)
108. Ephardt H, Fromherz P (1991) J Phys Chem 95: 6792
109. Fromherz P, Dambacher KH, Ephardt H, Lambacher A, Müller CO, Neigl R, Schaden H, Schenk O, Vetter T (1991) Ber Bunsenges 95: 1333
110. Haucke G, Czerney P, Steen D, Rettig W, Hartmann H (1993) Ber Bunsenges Phys Chem (in press)
111. Gormin D, Kasha M (1988) Chem Phys Lett 153: 574
112. Bulska H, Grabowska A, Grabowski ZR (1986) J Lumin 35: 189
113. Vollmer F, Rettig W (1990) Contribution to the 8th Int Conf Photochem conversion and storage of solar energy, Palermo, 15–20 July 1990, abstracts p 267
114. Dähne S, Freyer W, Teuchner K, Dobkowski J, Grabowski ZR (1980) J Lumin 22: 37
115. Nakashima N, Mataga N (1973) Bull Chem Soc Jpn 46: 3016
116. Rotkiewicz K, Leismann H, Rettig W (1989) J Photohem Photobiol A: Chem 49: 247
117. Howell R, Jones AC, Taylor AG, Phillips D (1989) Chem Phys Lett 163: 282
118. Phillips D, Howell R, Taylor AG (1992) Proc Indian Acad Sci (Chem Sci) 104: 153
119. In supersonic jet experiments with mass selection, ground state dimers, trimers and higher aggregates of DMABME could clearly be identified (Ref. 120)
120. Dedonder-Lardeux C, Jouvet C, Martrenchard S, Solgadi D (1992) (private communication)
121. Even in these aggregates, a TICT relaxation might be possible and lead to the observed redshifted emission spectra (Ref. 116, 118, 120)
122. Scholes GD, Ghiggino KP, Wilson GJ (1991) Chem Phys 155: 127
123. Paeplow B, Rettig W (1992) (unpublished results)
124. Cornelißen C, Rettig W (1992) (unpublished results)
125. Deisenhofer J, Epp O, Miki K, Huber R, Michel H (1984) J Mol Biol 180: 385; Deisenhofer J, Michel H (1989) Science 245: 1463
126. Revill JAT, Brown RG (1992) Chem Phys Lett 188: 433
127. Windsor MW (1986) J Chem Soc, Faraday Trans II, 82: 2237
128. Hayashi T, Suzuki T, Mataga N, Sakata Y, Misumi S, (1977) J phys Chem 81: 420
129. Viallet JM (1993) PhD thesis, Bordeaux
130. Seifert H, Rettig W (1993) (to be published)
131. Locke RJ, Lim EC (1987) Chem Phys Lett 134: 107; Locke RJ, Modiano SH, Lim EC (1988) J Phys Chem 92: 1703
132. Gilabert E, Lapouyade R, Rullière C (1991) Chem Phys Lett 185: 82
133. Rettig W (1992) Proc Indian Acad Sci (Chem Sci) 104: 89
134. Rettig W, Bonačić-Koutecký V (1979) Chem Phys Lett 62: 115
135. Okada T, Mataga N, Baumann W (1987) J Phys Chem 91: 760
136. Mataga N, Yao H, Okada T, Rettig W (to be published)
137. Rullière C, Grabowski ZR, Dobkowski J (1987) Chem Phys Lett 137: 408
138. Schopf G, Bendig J, Rettig W (to be published)
139. Kolos R, Grabowski ZR (1982) J Mol Struct 84: 251
140. Wang Y (1988) J Chem Soc Faraday Trans II 84: 1809
141. Wang Y (1985) Chem Phys Lett 116: 286
142. Guo RK, Kitamura N, Tazuke S (1990) J Phys Chem 94: 1404
143. Visser RJ, Varma CAGO, Konijnenberg J, Bergwerf P (1983) J Chem Soc 79: 347
144. Weisenborn PCM, Huizer AH, Varma CAGO (1988) Chem Phys 126: 425
145. Bourson J, Valeur B (1989) J Phys Chem 93: 3871
146. Létard JF, Lapouyade R, Rettig W (1993) Pure and applied chem (in press)
147. Cobbold PH, Rink TJ (1987) Biochem J 248: 313
148. Cazeau-Dubroca C, Ait Lyazidi S, Cambou P, Peirigua A, Cazeau P, Pesquer M (1989) J Phys Chem 93: 2347

Addition and Cycloaddition Reactions via Photoinduced Electron Transfer

Kazuhiko Mizuno and Yoshio Otsuji

Department of Applied Chemistry, College of Engineering, University of Osaka Prefecture, Sakai, Osaka 593, Japan

Table of Contents

Topics in Current Chemistry, Vol. 169
© Springer-Verlag Berlin Heidelberg 1994

This review article deals with addition and cycloaddition reactions of organic compounds via photoinduced electron transfer. Various reactive species such as exciplex, triplex, radical ion pair and free radical ions are generated via photoinduced electron transfer reactions. These reactive species have their characteristic reactivities and discrimination among these species provides selective photoreactions. The solvent and salt effects and also the effects of electron transfer sensitizers on photoinduced electron transfer reactions can be applied to the selective generation of the reactive species. Examples and mechanistic features of photoaddition and photocycloaddition reactions that proceed via the following steps are given: reactions of radical cations with nucleophiles; reactions of radical anions with electrophiles; reactions of radical cations and radical anions with neutral radicals; radical-radical coupling reactions; addition and cycloaddition reactions via triplexes; three-component addition reactions.

1 Introduction

Addition and cycloaddition reactions are one of the most fundamental reactions for carbon–carbon, carbon–heteroatom, and heteroatom–heteroatom bond formations in organic chemistry. They are put into three categories depending on the mechanistic modes of reactions; that is, ionic, free radical and concerted electrocyclic reactions. The other classification is based on the reactive species involved. The reactive species for addition and cycloaddition reactions are cations, anions, free radicals, radical cations and radical anions. All of these reactive species can be generated by using photochemical methods. Ionic addition reactions are usually accomplished under acidic or basic conditions in the dark. Although some reactions can be carried out under milder conditions by applying advanced synthetic methodologies, acidic or basic conditions are usually necessary for generation of ionic species in the ground state chemistry.

Photoinduced electron transfer reactions are a useful method, particularly for generation of radical ion species [1–10]. Radical ions can also be generated by electrochemical method or by redox reactions using metal compounds with low or high redox potentials. One-electron oxidizing agents such as triphenylmethyl (trityl) cation and tris (p-bromophenyl)aminium cation are also useful for generation of organic radical cations. If charge and spin densities are localized on different atoms in radical ions, these species would have both ionic and radical characters. These radical ions become a good candidate not only for ionic reactive species that react with nucleophiles or electrophiles, but also for radical species that react with neutral free radicals.

Photoinduced electron transfer reactions that occur between neutral electron donor molecules and neutral electron acceptor molecules have several characteristic features: (1) a radical cation and a radical anion are produced as a pair, (2) radical ion species are produced under neutral and mild conditions, and (3) the polarity inversion (umpolung) of original electron donor and electron acceptor molecules arises through their conversion into radical ion species. As a result, the radical cation $D^{+\bullet}$ can interact with another D or a different electron donor molecule D' to yield a dimer radical cation $D_2^{+\bullet}$ or heterodimer radical cation DD′⁺•. These reactive species become useful intermediates for selective photoaddition reactions.

This review article deals primarily with addition reactions of nucleophiles, electrophiles, and neutral radicals to photochemically generated radical ions of organic compounds and some organometallic compounds. Photocyclodimerizations of electron-rich alkenes, photo-Diels-Alder reactions between alkenes and alkadienes via dimer or heterodimer radical cations, and photocycloadditions via triplexes are also included.

2 Generation of Reactive Species via Photoinduced Electron Transfer

Photoinduced electron transfer reaction between an electron donor molecule (D) and an electron acceptor molecule (A) can be initiated by the photoexcitation of either a D or A molecule. Various reactive species are generated in the course of this process. Important reactive species involved in photoinduced electron transfer reactions are shown in Scheme 1.

An exciplex $[D^{\delta+} \cdots A^{\delta-}]^*$ is formed by interaction of an electronically excited molecule A* (or D*) with a ground state molecule D (or A). Exciplex formation is favored in nonpolar solvents. In general, an exciplex has a relatively rigid and oriented structure. Highly regio- and stereoselective photocycloaddition reactions between A and D can often be achieved via this intermediate.

A triplex $[D \cdots D \cdots A]^*$ or $[D \cdots A \cdots A]^*$ is formed by interaction of an exciplex with another D or A within its lifetime. If an interacting molecule is different from the original D or A, a triplex of the type, $[D' \cdots D \cdots A]^*$ or $[D \cdots A \cdots A']^*$ will be produced. In some photoreactions, triplexes become important reactive intermediates for photocycloadditions.

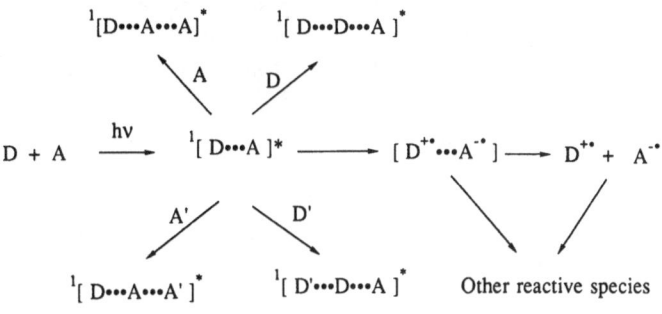

A,A' : Electron acceptor   D,D' : Electron donor

Scheme 1.

In some cases, formation of exciplexes and triplexes can be evidenced by fluorescence quenching experiments. Exciplex emission is usually observed in less polar solvents. However, in many cases the light-emissive exciplexes are photochemically unreactive. Triplex emission is rarely observed [11–14]. Caldwell demonstrated a termolecular interaction in excited molecules by means of exciplex quenching [15–17]. The photoreaction of 9-cyanophenan-threne with trans-1-methoxy-4-(1-propenyl)benzene in benzene gives a (2 + 2)cycloadduct in a regio- and stereospecific manner (Scheme 2). This photo-reaction is accompanied by exciplex emission in a longer wavelength region than fluorescence of 9-cyanophenanthrene. Both the exciplex emission and the formation of the photocycloadduct are quenched by dimethyl acetylenedi-carboxylate at nearly identical rates.

The character of an exciplex depends upon the nature of solvents and the redox properties of the components. Radical ion species are generally produced in polar solvents. In polar solvents, an exciplex often dissociates to free radical ions via a radical ion pair. In some cases, a radical ion pair is directly produced by photoinduced electron transfer from D to A* or from D* to A without intervention of an exciplex.

An electronically excited molecule has a strong electron-donating or elec-tron-accepting ability (Fig. 1). A single-electron transfer can occur from D to A to generate radical ions. The free energy change associated with formation of a radical ion pair via photoinduced electron transfer can be calculated by the Rehm-Weller equation [18]:

$$\Delta G = E^{ox}(D) - E^{red}(A) - E_{0-0} - e^2/\varepsilon r \qquad (1)$$

where $E^{ox}(D)$, $E^{red}(A)$ and E_{0-0} are the oxidation potential of D, the reduction potential of A and the 0–0 excitation energy of the molecule to be excited, respectively. The term $e^2/\varepsilon r$ expresses the coulombic interaction energy between $D^{+\bullet}$ and $A^{-\bullet}$ in a solvent of dielectric constant ε of distance r apart. When ΔG is negative, the formation of radical ion pair $[D^{+\bullet} \cdots A^{-\bullet}]$ takes

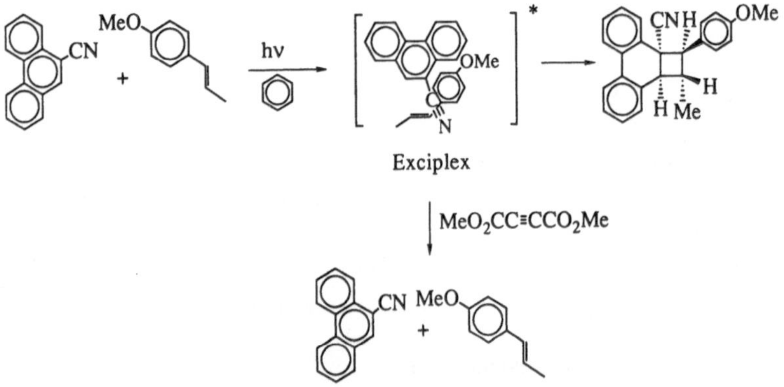

Exciplex

Scheme 2.

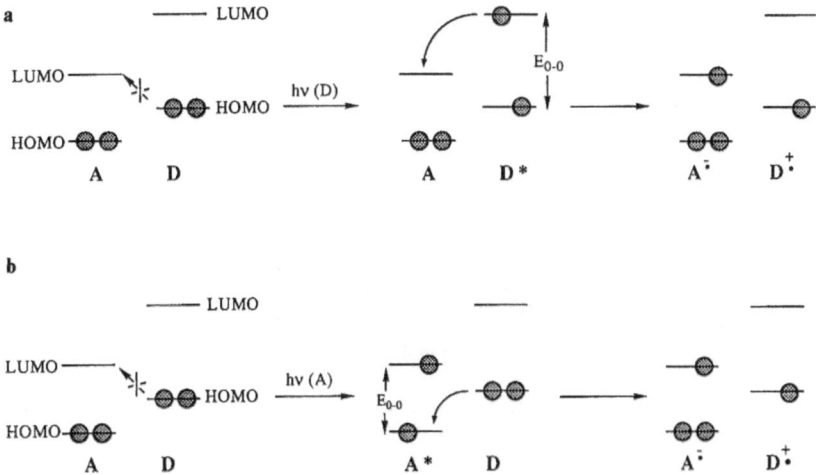

Fig. 1. Schematic orbital representation of photoinduced electron transfer processes: (**a**) Excitation of electron donor (D), (**b**) Excitation of electron acceptor (A)

place as an exothermic process. If the excited molecule emits fluorescence, the rate of forward electron transfer can be estimated from the fluorescence quenching rate.

Various theories have been proposed for prediction of rate constants for electron transfer quenching k_q. Most of the theories are based on the relationship between the free energies of activation for photoinduced electron transfer and the free energy changes, ΔG, associated with radical ion pair formation through photoinduced electron transfer.

The Marcus theory predicts that the k_q values should increase up to a maximum and then decrease with increasing the oxidation potentials of D, or with decreasing the reduction potentials of A [19]. The region where k_q decreases with increasing the ΔG values is called the "Marcus inverted region". The bell-shaped curve "a" in Fig. 2 illustrates this theory.

However, in the usual fluorescence quenching, the Rehm-Weller treatment (the curve "b") well accounts for the forward electron transfer rate constants from D (or D*) to A* (or A), and the inverted region is not usually observed in the rate constants for the emission quenching. More sophisticated theories than the Marcus theory have been advanced in recent years [20–21]. However, we will not discuss this subject further.

Recent studies of photoinduced electron transfer reactions have demonstrated that the reactivity of a radical ion pair is different from that of free radical ions. Farid showed that in the photoreaction of 1,1-diphenylethene the reactivity of the radical ion pair can be distinguished from that of the free radical cation by kinetic measurements [22]. Mizuno and Otsuji showed that the reactivity feature of a radical ion pair in solvent cage and that of solvent separated radical

K. Mizuno and Y. Otsuji

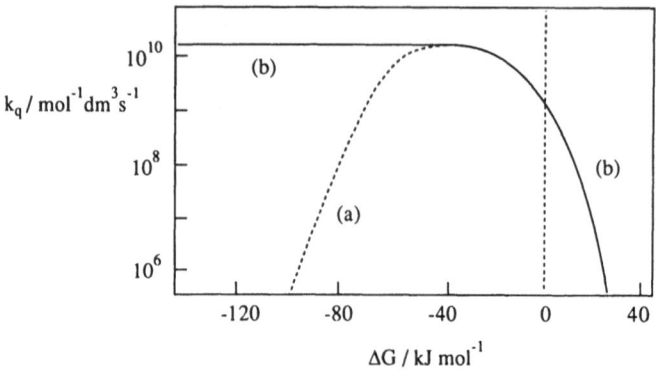

Fig. 2. Plot of the fluorescence quenching rate constants k_q vs the free energy changes ΔG for the electron transfer process: (a) Marcus relationship, (b) Rehm-Weller relationship

ions (free radical ions) can be differentiated by utilizing the salt effect on the photoreaction [23]. The mechanistic scheme for the salt effect and the other additive effect on the photoreaction is illustrated in Scheme 3.

Photoinduced electron transfer reactions involve several unavoidable problems in their nature: (1) many reactive species are generated simultaneously in the reaction system and (2) back-electron transfer from $A^{-\bullet}$ to $D^{+\bullet}$ usually occurs very rapidly and ionic species decay to their ground state molecules A and D. For the development of efficient and selective organic photoreactions it would therefore be important to find a method for controlling the nature of

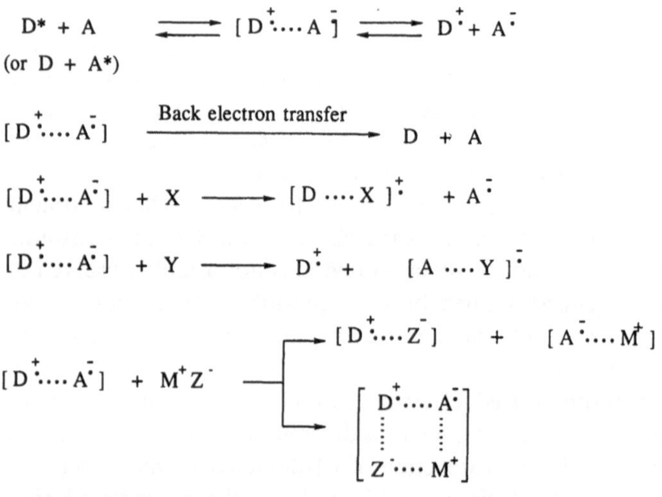

$$D^* + A \rightleftarrows [\overset{+}{D}\cdots \overset{-}{A}] \rightleftarrows \overset{+}{D} + A^{-}$$
(or $D + A^*$)

$$[\overset{+}{D}\cdots \overset{-}{A}] \xrightarrow{\text{Back electron transfer}} D + A$$

$$[\overset{+}{D}\cdots \overset{-}{A}] + X \longrightarrow [D\cdots X]^{+} + A^{-}$$

$$[\overset{+}{D}\cdots \overset{-}{A}] + Y \longrightarrow \overset{+}{D} + [A\cdots Y]^{-}$$

$$[\overset{+}{D}\cdots \overset{-}{A}] + M^{+}Z^{-} \longrightarrow
\begin{cases}
[\overset{+}{D}\cdots \overset{-}{Z}] + [A\cdots M^{+}] \\[6pt]
\begin{bmatrix} \overset{+}{D}\cdots \overset{-}{A} \\ \vdots \quad \vdots \\ Z\cdots M^{+} \end{bmatrix}
\end{cases}$$

D ; Electron donor   A ; Electron acceptor   X, Y ; Additive $M^{+}Z^{-}$; Salt

Scheme 3.

306

reactive species to be generated and also for suppressing the back-electron transfer.

These problems can be solved in part by utilizing medium effects such as solvent and additive effects [24–27]. For example, photoreactions involving exciplexes and radical ion pairs as reactive species are suppressed and those involving free radical ions are accelerated by adding some salts to the reaction system [23].

Aromatic hydrocarbons also affect the efficiency of photoinduced electron transfer reactions. In such cases, aromatic hydrocarbons often act as redox photosensitizers or co-sensitizers. The role of these sensitizers is shown in Scheme 4. The primary photoinduced electron transfer occurs from ArH* to A (or ArH to A*) to give $ArH^{+\bullet}$ and $A^{-\bullet}$. The succeeding secondary electron transfer from D to $ArH^{+\bullet}$ produces $D^{+\bullet}$ and ArH in which $D^{+\bullet}$ is a real reactive

$$\mathrm{ArH} \xrightarrow{h\nu} \mathrm{ArH}^*$$

$$\left(\text{or } \mathrm{A} \xrightarrow{h\nu} \mathrm{A}^*\right)$$

$$\mathrm{ArH}^* + \mathrm{A} \longrightarrow \mathrm{ArH}^{+\bullet} + \mathrm{A}^{-\bullet}$$

$$\left(\mathrm{ArH} + \mathrm{A}^*\right)$$

$$\mathrm{ArH}^{+\bullet} + \mathrm{D} \longrightarrow [\,\mathrm{ArH} \cdots\cdots \mathrm{D}\,]^{+\bullet}$$

$$[\,\mathrm{ArH} \cdots\cdots \mathrm{D}\,]^{+\bullet} \longrightarrow \mathrm{ArH} + \mathrm{D}^{+\bullet}$$

A; Electron acceptor D; Electron donor ArH; Aromatic hydrocarbon

Scheme 4.

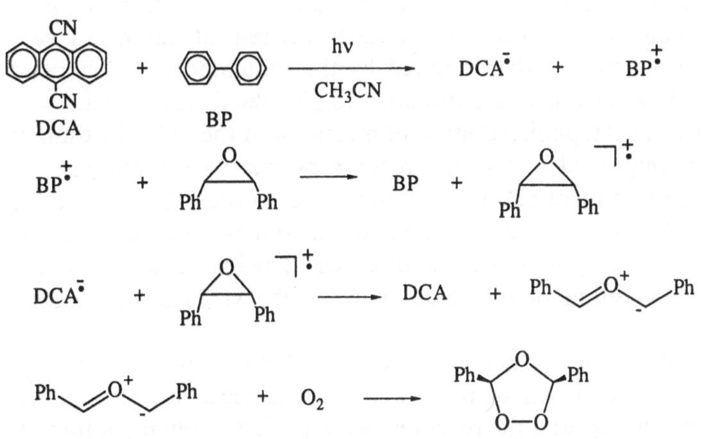

Scheme 5.

intermediate. The secondary electron transfer process from D to ArH$^{+\bullet}$ often takes place even in an unfavorable endothermic process. Pac proposed that such a process may proceed via π-complex [ArH $\bullet\bullet\bullet$ D]$^{+\bullet}$ formation [28–30].

Redox photosensitization or co-sensitization by aromatic hydrocarbons has been utilized for enhancement of the efficiency of photoinduced electron transfer reactions. For example, the efficiency of the 9,10-dicyanoanthracene-sensitized photooxygenation of 1,2-diphenyloxirane in acetonitrile is enhanced appreciably by adding biphenyl as a co-sensitizer, giving 3,5-diphenyl-1,2,4-trioxolane in good yield [31–32]. This photoreaction does not take place in the absence of biphenyl. Schaap proposed that in this photoreaction the primary electron transfer reaction occurs from biphenyl (BP) to ^{1}DCA* to produce biphenyl radical cation BP$^{+\bullet}$ and DCA$^{-\bullet}$. The secondary electron transfer from the oxirane to BP$^{+\bullet}$ produces BP and the radical cation of the oxirane which is converted into the trioxolane (Scheme 5).

3 Addition of Nucleophiles to Radical Cations

3.1 Addition to Alkenes

A variety of photoaddition reactions of nucleophiles to electron-rich substrates in the presence of electron acceptors were developed in the early 1970s to 1980s. Arnold and his coworkers reported that the photoaddition of methanol to 1,1-diphenylethene occurs in the presence of methyl p-cyano-benzoate in an anti-Markownikoff manner [33]. Photoaddition of nucleophiles such as water, alcohols, acetic acid, and cyanide anion to electron-rich arylalkenes were also reported [34–37]. The photoaddition reaction proceeds via the mechanism as shown in Scheme 6.

The possibility of generation of a radical cation can be predicted by use of the Rehm-Weller equation. A nucleophile adds to the radical cation to give a neutral radical. The radical is then reduced by the simultaneously generated radical anion to an anion which is protonated to give the final product.

The efficiency of nucleophilic addition of methanol to the radical cation of arylalkenes can be improved by use of a redox sensitizer system, and also by utilizing solvent and additive effects. Pac and his coworkers found that the aromatic hydrocarbon-electron acceptor sensitizer system, such as phenanthrene-p-dicyanobenzene system, acts as an excellent redox sensitizer for the anti-Markownikoff addition of methanol to 1,1-diphenylethene and indene (Scheme 7) [29].

Mizuno and Otsuji found that the efficiency of 9,10-dicyanoanthracene (DCA)-sensitized photoaddition of methanol to arylalkenes varies with the structure of the arylalkenes and the reaction media [37]. The photoaddition of methanol to 1,1-diphenylpropene occurs more efficiently in benzene than in

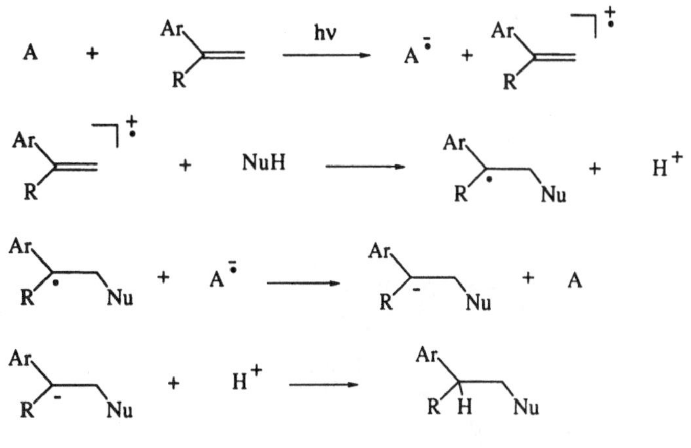

A : Electron acceptor   Nu : Nucleophile

Scheme 6.

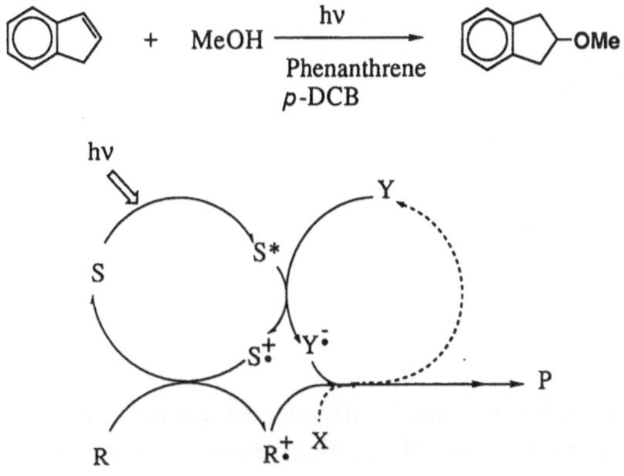

S; Phenanthrene   R; Arylalkene   X; MeOH
Y; p-Dicyanobenzene (p-DCB)   P; Product

Scheme 7.

acetonitrile. However, the photoaddition of methanol to 9-ethylidenefluorene occurs less efficiently in benzene than in acetonitrile. The former photoreaction is suppressed by adding salts such as Bu_4NClO_4 and $Mg(ClO_4)_2$, but the latter reaction is facilitated by adding these salts (Scheme 8). These results can be explained by assuming that in the former reaction, methanol adds to the exciplex which is formed between DCA and 1,1-diphenylpropene and that in the latter reaction methanol adds to the free radical cation of the alkene. These results suggest that the nature of reactive species involved in photoinduced electron transfer mediated reactions depends also on their stability as well as on

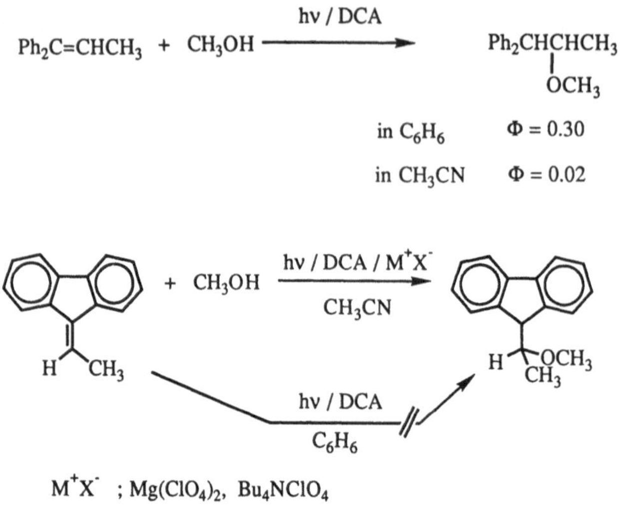

$$\text{Ph}_2\text{C}=\text{CHCH}_3 \ + \ \text{CH}_3\text{OH} \xrightarrow{\ \ hv\,/\,DCA\ \ } \text{Ph}_2\text{CHCHCH}_3$$

$$\underset{\ \ \ \ \ \ \ \ \ \ \ \ }{\quad}\overset{|}{\text{OCH}_3}$$

in C_6H_6 $\Phi = 0.30$

in CH_3CN $\Phi = 0.02$

M⁺X⁻ ; Mg(ClO₄)₂, Bu₄NClO₄

Scheme 8.

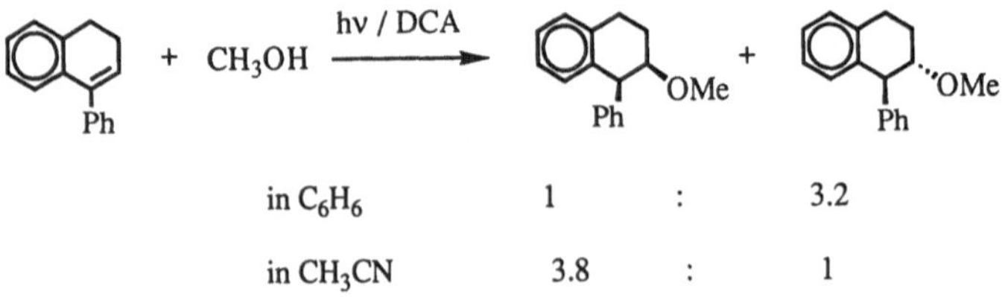

in C_6H_6 1   :   3.2

in CH_3CN 3.8   :   1

Scheme 9.

the reaction media. A one-electron removal from 9-ethylidenefluorene affords a stable, delocalized radical cation. In such a case, the radical cation appears to be forced to become the reactive intermediate.

The stereoselectivity in the photoaddition of methanol to 1,2-dihydro-4-phenylnaphthalene in the presence of DCA depends upon solvents. In benzene, the formation of the *trans*-adduct predominates and in this case the reaction occurs via a *cis*-addition of methanol to the alkene-DCA exciplex. In acetonitrile the formation of the *cis*-adduct predominates and in this case the reaction occurs via a *trans*-addition of methanol to the free radical cation of the alkene (Scheme 9).

The intramolecular version of photoaddition of alcohols to alkenes was also reported. Mizuno and Otsuji reported the DCA-sensitized photocyclization of ω-hydroxy-1,1-diphenylalkenes (n = 3–5) via exciplex in benzene [38]. The long chain ω-hydroxyalkenes (n = 6, 10) do not give the corresponding cyclized ethers. Gassman reported the photocyclization of unsaturated carboxylic acids by the use of a sensitizer system consisting of sterically hindered electron acceptors and biphenyl to give γ-lactones (Scheme 10) [39].

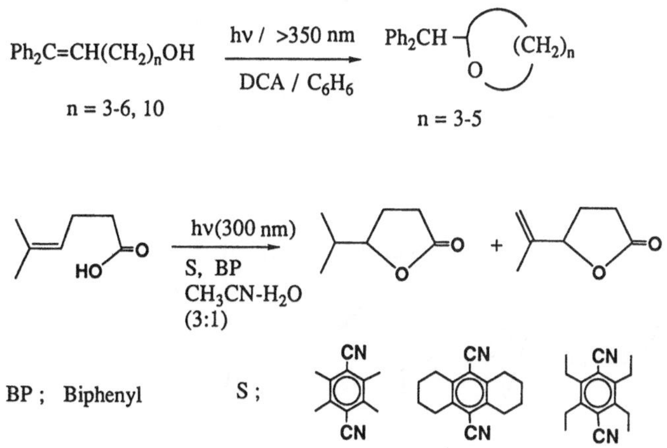

$$Ph_2C=CH(CH_2)_nOH \xrightarrow[DCA\ /\ C_6H_6]{hv\ /\ >350\ nm} Ph_2CH-\underset{O}{\underset{\big|}{\diagup}}(CH_2)_n$$

n = 3-6, 10   n = 3-5

BP ; Biphenyl   S ;

Scheme 10.

3.2 Addition to Small Ring and Strained Molecules

Nucleophiles also add to radical cations of arylcyclopropanes to give anti-Markownikoff-type adducts [40–43]. There are two possible structures for the radical cations of arylcyclopropanes; i.e., ring opened 1,3-radical cation and ring closed cyclopropane radical cation. Dinnocenzo reported that the photoinduced electron transfer reaction of chiral *trans*-1-methyl-2-phenylcyclopropane in methanol gives chiral 3-methoxy-1-phenylbutane via the radical cation [44]. This result indicates that the radical cation in the ring closed form is attacked by methanol. The fact that this cyclopropane does not isomerize under the photochemical reaction conditions supports this conclusion (Scheme 11).

Gassman reported the photoaddition of nucleophiles to radical cations of highly strained aliphatic polycyclic molecules such as tricyclo[2.2.1.0^{5,6}]hexane and related compounds. The radical cation of bicyclo[1.1.0]butane is postulated as a key intermediate (Scheme 12) [45–46].

Two electron photooxidation of arylalkenes and arylcyclopropanes can be achieved by carrying out the photoreaction in the presence of DCA and Cu(II) ions [47]. In this photoreaction, a primary photoinduced one-electron oxidation of alkenes by $^1DCA^*$ is followed by one-electron oxidation of the resulting

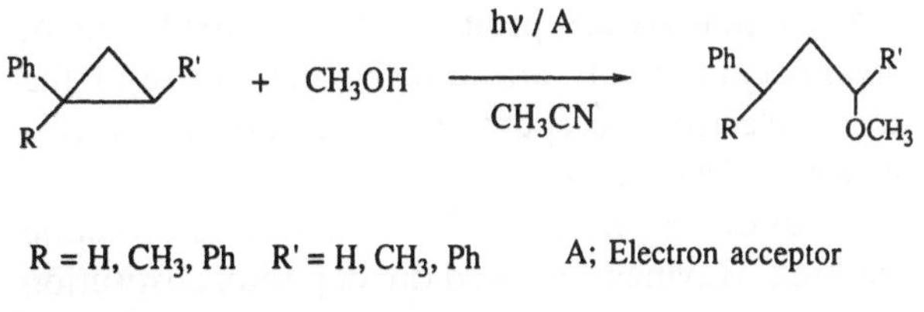

R = H, CH$_3$, Ph   R' = H, CH$_3$, Ph   A; Electron acceptor

Scheme 11.

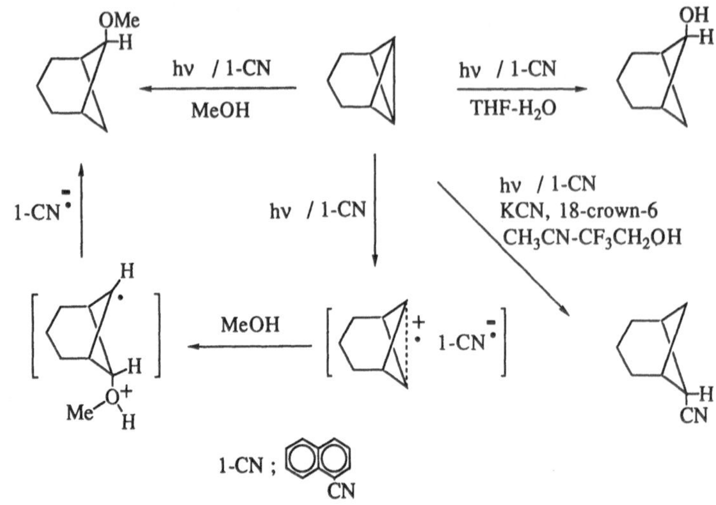

Scheme 12.

neutral radicals by Cu(II) ion. A direct two-electron photooxidation of arylalk-enes and arylcyclopropanes upon irradiation in the presence of Cu(II) or Fe(III) ions was also reported [48–51]. The photooxidation occurs via the mechanism similar to that of electrochemical anodic oxidation [52].

3.3 Addition to Aromatic Rings

Nucleophiles add also to radical cations of aromatic compounds that are generated by photoinduced electron transfer. The photo-Birch reduction of naphthalene, phenanthrene, anthracene and their derivatives occurs when the aromatic compounds are irradiated in the presence of NaBH$_4$ and electron accepting aromatic cyano compounds [53]. This photoreaction gives the corresponding dihydro compounds in high yields (Scheme 13). The photocyanation of aromatic compounds with NaCN occurs under similar reaction conditions to give aromatic cyano and dihydroaromatic cyano compounds (Scheme 14) [54]. In these photoreactions, aqueous acetonitrile can be used as solvent. The use of the aqueous solvent allows to dissolve nucleophilic reagents. Since the radical cations of aromatic compounds do not react with water, this is of practical advantage for carrying out nucleophilic photoaddition reactions. Under aerobic conditions, aromatized cyano compounds are produced by the oxidation of radical intermediates. The cyano compounds produced in the course of the photoreaction serve also as an electron acceptor. The mechanism for this photoaddition reaction is shown in Scheme 14.

The efficiency and selectivity of direct photoaddition reactions to aromatic compounds are usually not so high. Havinga reviewed direct photosubstitution reactions on aromatic compounds [55].

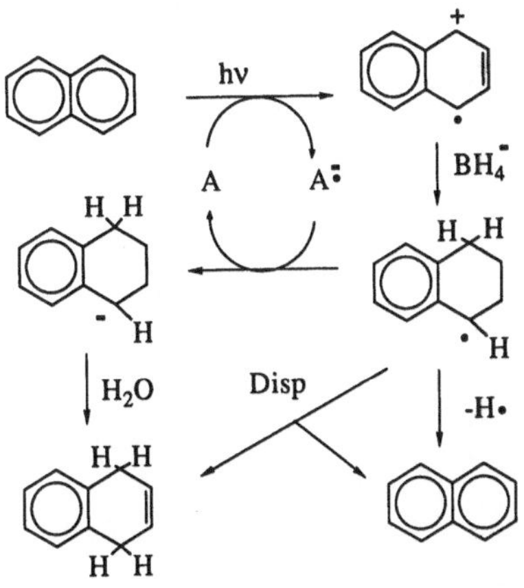

A : Electron acceptor   Disp : Disproportionation

Scheme 13.

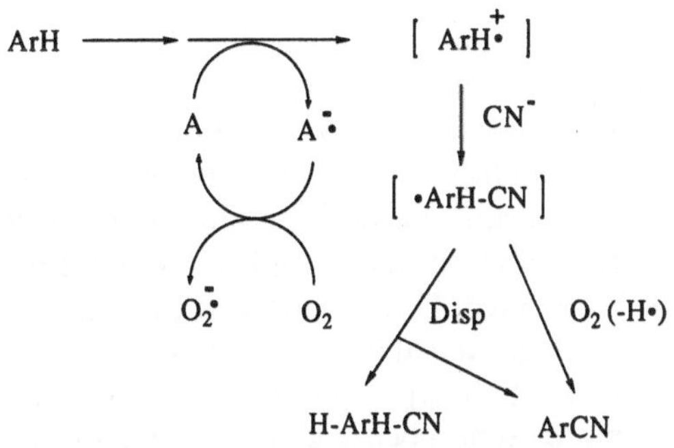

A ; Electron acceptor   Disp ; Dispropotionation
ArH ; Aromatic hydrocarbon

Scheme 14.

The photoamination of aromatic hydrocarbons and arylalkenes by ammonia and primary aliphatic amines occurs via photoinduced electron transfer in the presence of m-dicyanobenzene (Scheme 15) [56–58]. In this photoreaction, secondary amines are less reactive than ammonia and primary amines, and the high concentration of the aminated products retards the photoreaction. This is due to the fact that the oxidation potentials of the aminated products produced by the photoreaction are usually much lower than those of the starting ammonia and primary amines. The secondary amines quench the reactive species much faster than the primary amines. Yasuda extended this photoreaction to intramolecular cyclization reactions [59].

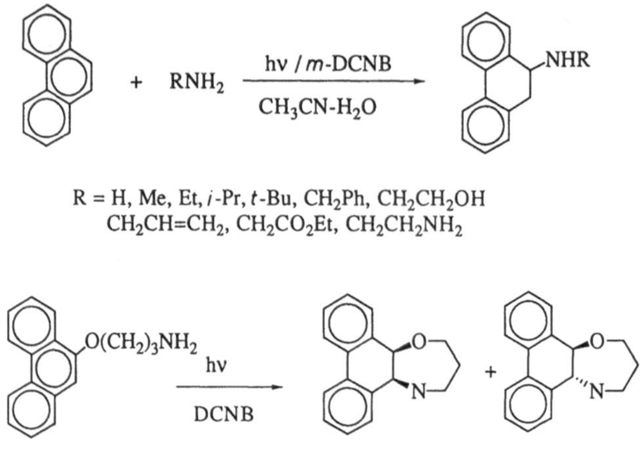
+ RNH2
hν / m-DCNB
CH3CN-H2O
NHR
R = H, Me, Et, i-Pr, t-Bu, CH2Ph, CH2CH2OH
CH2CH=CH2, CH2CO2Et, CH2CH2NH2
O(CH2)3NH2
hν
DCNB
O
N
+
O
N

Scheme 15.

Photoinduced electron transfer between amines and aromatic hydrocarbons occurs to generate radical cations of amines and radical anions of aromatic hydrocarbons. Pac and Sakurai reported the photoaddition of *N,N*-dimethylaniline to anthracene via photoinduced electron transfer [60]. In benzene, the $(4\pi + 4\pi)$ photocyclodimer of anthracene is produced as a sole isolable product, although an emission due to the exciplex formed from anthracene and *N,N*-dimethylaniline is observed. In acetonitrile, the addition of dimethylaniline to anthracene occurs via their radical ions to give 9,10- dihydro-9-(4′-dimethylaminophenyl)anthracene as the major product. However, the photoamination on anthracene takes place even in benzene when *N*-methylaniline is used as an electron donor. Sugimoto and his coworkers reported the intramolecular photoaddition of anilines to aromatic hydrocarbons to give cyclic amino compounds (Scheme 16) [61–63].

The photoreactions of aliphatic amines with aromatic hydrocarbons have also been reported by several groups. With tertiary amines, deprotonation occurs from the radical cations of amines at the α-carbon to generate carbon radicals which react with the radical anion of aromatic hydrocarbons. With secondary amines, deprotonation from the radical cations of amines occurs both at the α-carbon and at the nitrogen atom, so that the reaction becomes complicated [64–65].

Lewis and his coworkers reported the intramolecular photoaddition of aliphatic secondary amines to arylalkenes to give 5- and 6- membered cyclic amines (Scheme 17) [66–67]. Ohashi reported that the other type of photoaddition reaction takes place when 1,4-dicyanobenzene and triethylamine are used as substrates. This photoreaction can be explained in terms of the deamination of the initially produced diethylaminoethylated compound (Scheme 18) [68].

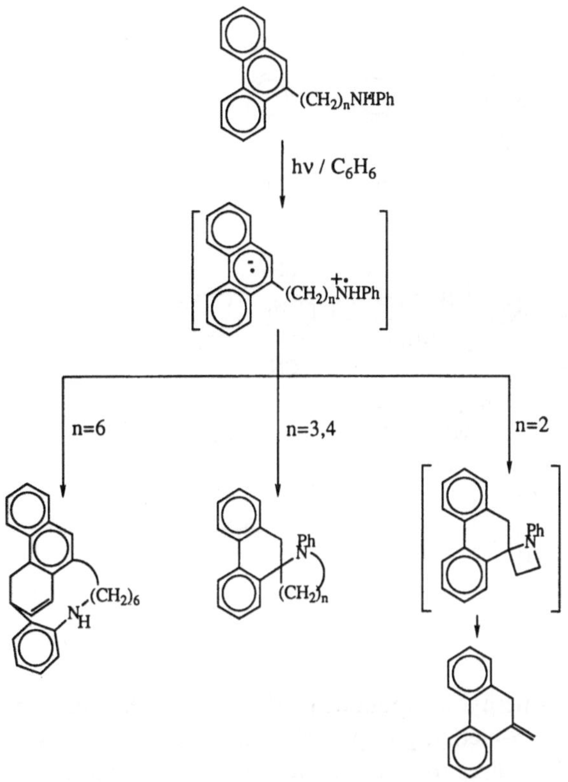

Scheme 16.

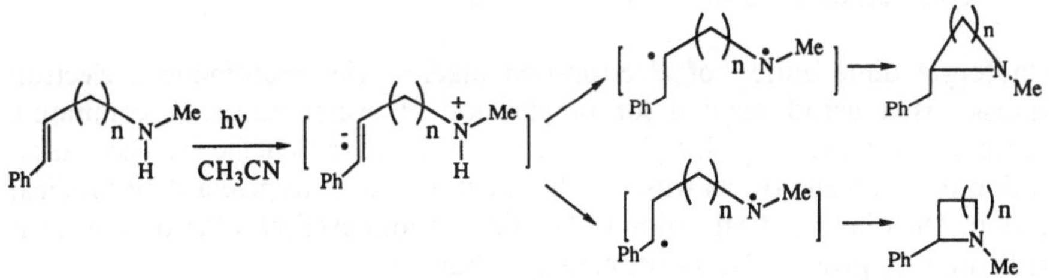

Scheme 17.

Yonemitsu and his coworkers found that the intramolecular photoaddition of chloroacetoamide derivatives with an electron donating aromatic ring in molecules gives the cyclic amide compounds in water [69]. However, the photoreaction in organic solvents proceeds via the exciplex to give the different cyclized products.

Kanaoka and his coworkers developed a method for preparation of sulfur-containing cyclic compounds by utilizing the intramolecular photoinduced

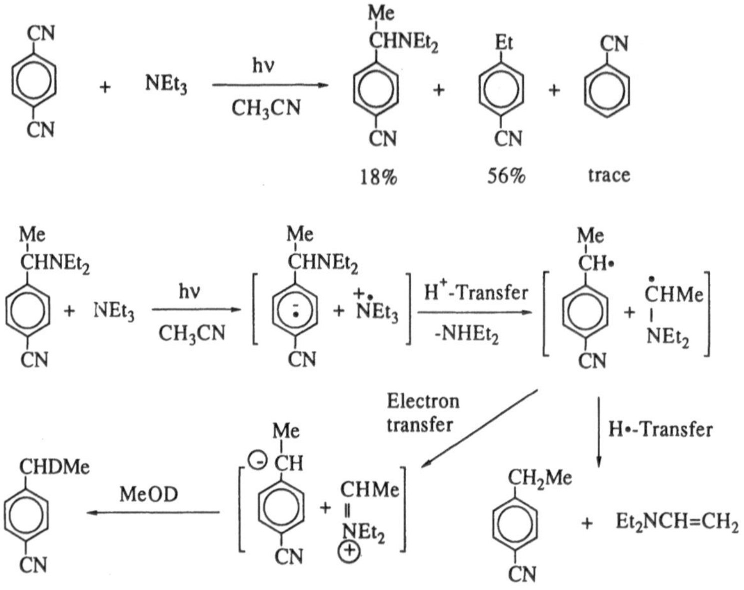

Scheme 18.

electron transfer reaction of *S*-methyl compounds [70]. In this case, deprotonation from the α-carbon of the *S*-methyl group in the radical cation occurs to give the $S\text{-CH}_2^{\bullet}$ radical which cyclizes to give the cyclic thio compounds.

3.4 Cycloaddition via Dimer Radical Cations

Photocyclodimerization of electron-rich alkenes via photoinduced electron transfer is a useful method for preparation of cyclobutanes. A commonly accepted mechanism for this photoreaction is shown in Scheme 19. The radical cation of an alkene reacts with another neutral alkene to give a dimer radical cation. The dimer radical cation (1,4-radical cation) cyclizes, after one-electron reduction, to give a 1,2-disubstituted cyclobutane.

The photocyclodimerization of *N*-vinylcarbazole, which was reported by Ledwith and Shirota, can be accounted for by this mechanism [71–73]. A chain process is involved in this photoreaction, and the quantum yield exceeds unity (the maximum quantum yield is 66). The hole transfer from the cyclobutane radical cation to a neutral *N*-vinylcarbazole is a key reaction for the chain process. Similar photodimerizations of electron-rich alkenes such as aryl vinyl ethers [74–76], indenes [29, 77, 78], styrenes [79–80] and enamines [71] have been reported by several groups. The DCA-sensitized photodimerization of phenyl vinyl ether gives *cis*- and *trans*-1,2-diphenoxycyclobutanes. This photoreaction also involves a chain process although the chain length is short [75].

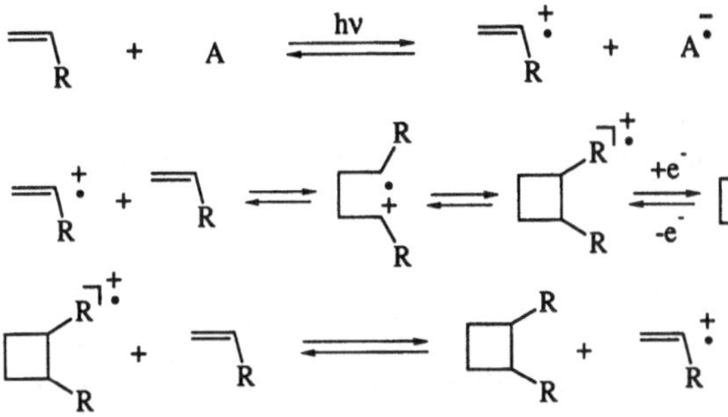
hv
+e⁻
-e⁻

A; Electron acceptor R; Aryl, Aryloxy

Scheme 19.

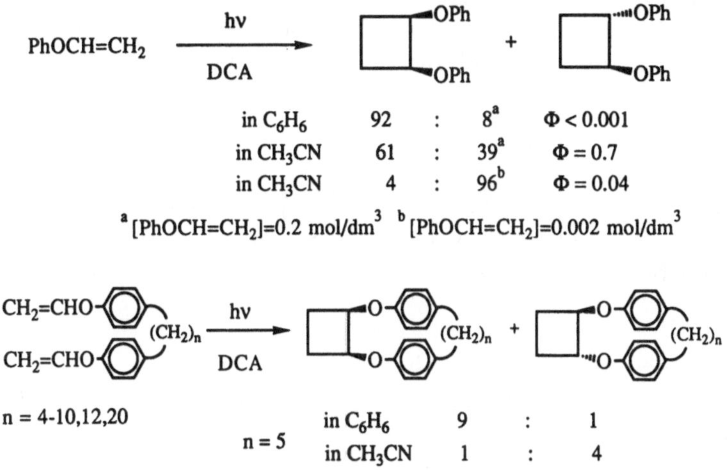
PhOCH=CH2
hv
DCA
OPh
OPh
OPh
OPh
in C6H6 92 : 8a Φ < 0.001
in CH3CN 61 : 39a Φ = 0.7
in CH3CN 4 : 96b Φ = 0.04
a [PhOCH=CH2]=0.2 mol/dm3 b [PhOCH=CH2]=0.002 mol/dm3
CH2=CHO
CH2=CHO
(CH2)n
hv
DCA
(CH2)n
(CH2)n
n = 4-10,12,20
n = 5
in C6H6 9 : 1
in CH3CN 1 : 4

Scheme 20.

The stereoselectivity in the DCA-sensitized inter- and intra- molecular ($2\pi + 2\pi$) photocycloaddition depends on the nature of solvents, the reaction temperature, and the concentration of the vinyl ether [76]. In benzene, formation of the *cis* isomer predominates and the reaction proceeds via a triplex. Involvement of triplexes in photoreactions will be discussed again in Sect. 3.6. In acetonitrile, the product ratio depends on the concentration of the vinyl ether and the reaction temperature. Formation of the *trans* isomer predominates at low concentration of the vinyl ether and at high reaction temperature ($> 100°C$). At the high concentration of the vinyl ether, the *cis* and *trans* isomers are obtained in a 3:2 ratio (Scheme 20).

Transition metal complexes and semi-conductors can also be used as electron accepting photosensitizers for the photocyclodimerization of phenyl vinyl ether [81–82].

1,4-Dimer radical cation intermediates can sometimes be trapped by the reaction with cyano compounds and molecular dioxygen, giving pyridine derivatives and 1,2-dioxane derivatives, respectively (Scheme 21) [75, 83]. In some cases, $(4 + 2)$ cycloadducts are obtained via the same intermediates [22,23].

Another mechanism for the $(2\pi + 2\pi)$ photocycloaddition of alkenes via electron transfer is the reaction that proceeds via a triplet state which is produced by a back-electron transfer from a radical anion of the electron acceptor to a radical cation of the electron donor. The triplet state alkenes generated by this way can undergo the cyclodimerization (Scheme 22). Farid showed that the DCA-sensitized $(2\pi + 2\pi)$ photocyclodimerization of 1,2-diphenylcyclo-propene-3-carboxylate occurs via the triplet state of the cyclopropene in acetonitrile [84]. In this photoreaction, two types of the $(4\pi + 2\pi)$ photocycloaddition reactions take place between DCA and the cyclopropene depending upon solvents. One type of the cycloadduct is produced in benzene via exciplex and the other type of the photocycloadduct is produced in

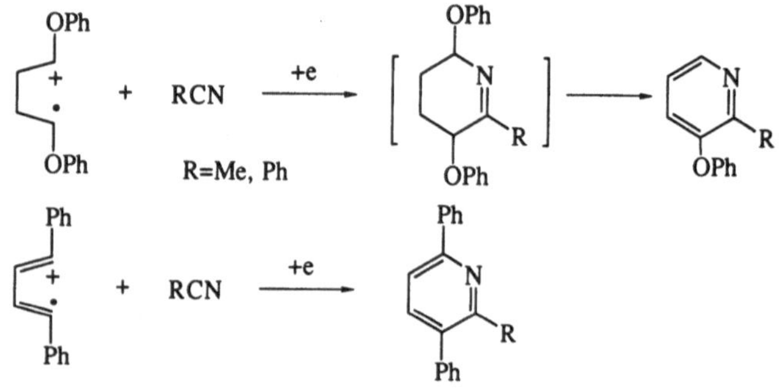

Scheme 21.

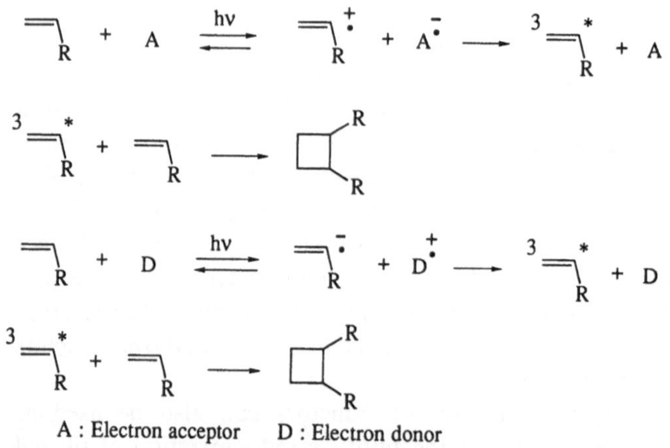

A : Electron acceptor   D : Electron donor

Scheme 22.

acetonitrile via a radical ion pair (Scheme 23). Shirota reported the $(2\pi + 2\pi)$ photocycloaddition of 2-vinylnaphthalene in the presence of tertiary amines via the triplet alkene which is generated by a back electron transfer (Scheme 24) [85, 86].

The DCN-sensitized photocyclodimerization of 1,3-cyclohexadiene involves two different reactive species. The $(2\pi + 2\pi)$ photocycloadducts are formed from the excited triplet state of 1,3-cyclohexadiene that is generated by a back

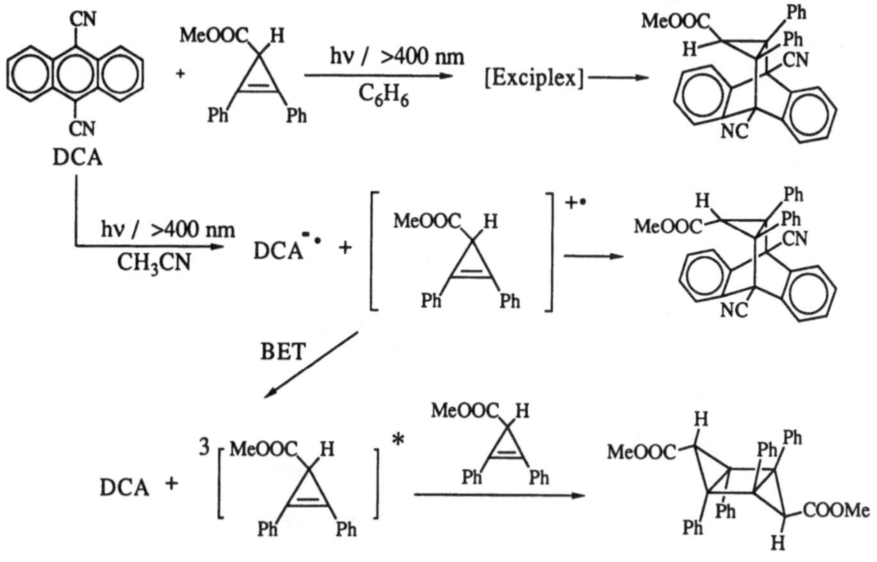

Scheme 23.

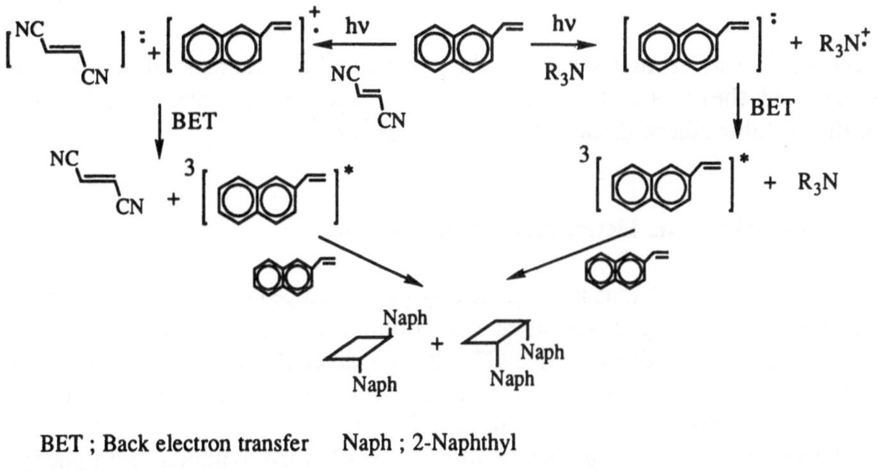

BET ; Back electron transfer   Naph ; 2-Naphthyl

Scheme 24.

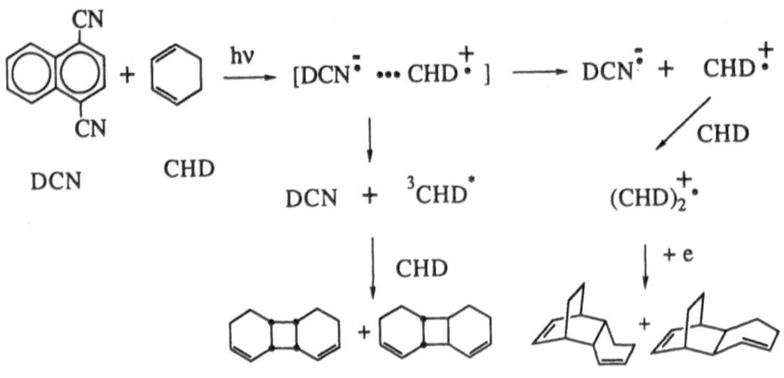

$$\text{DCN} + \text{CHD} \xrightarrow{h\nu} [\text{DCN}^{\overline{\cdot}} \cdots \text{CHD}^{+\cdot}] \longrightarrow \text{DCN}^{\overline{\cdot}} + \text{CHD}^{+\cdot}$$

Scheme 25.

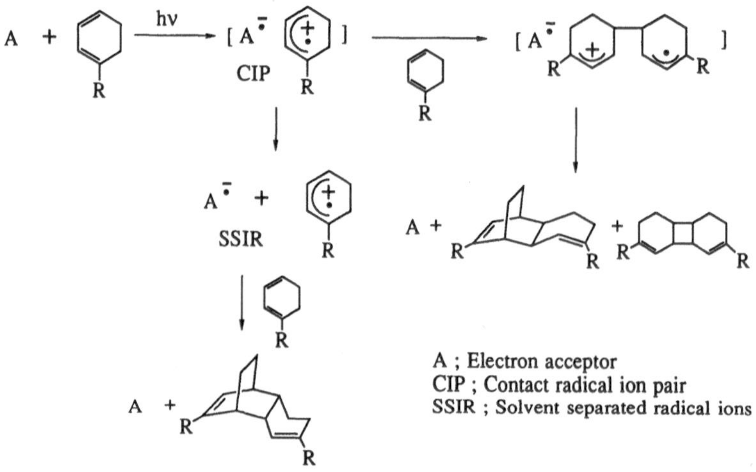

A ; Electron acceptor
CIP ; Contact radical ion pair
SSIR ; Solvent separated radical ions

Scheme 26.

electron transfer from $\text{DCN}^{-\cdot}$ to the cyclohexadiene radical cation. On the other hand, the $(4\pi + 2\pi)$ cycloadducts are produced via a dimer radical cation of the cyclohexadiene (Schemes 25 and 26) [87, 88].

3.5 Cycloaddition via Heterodimer Radical Cations

A radical cation $D^{+\cdot}$ that is generated by photoinduced electron transfer from an electron donating unsaturated compound can react with another D to produce a dimer radical cation $D_2^{+\cdot}$. If two different electron-donating unsaturated compounds are present in the reaction system, a heterodimer radical cation $DD'^{+\cdot}$ may be produced (Scheme 27) [89].

Farid reported that the crossed addition of 1,1-dimethylindene with phenyl vinyl ether occurs upon irradiation of their acetonitrile solution in the presence

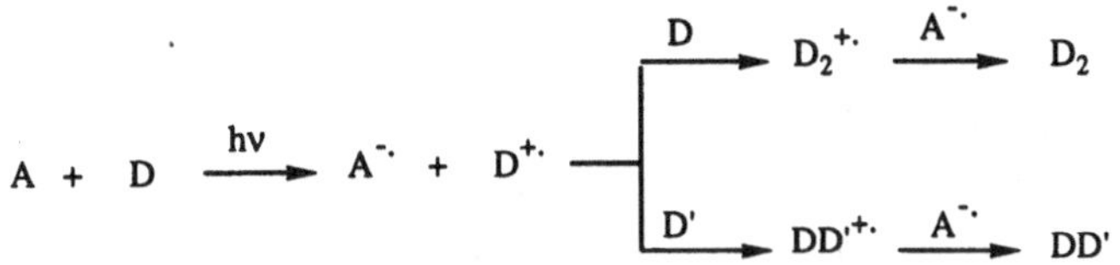

A + D $\xrightarrow{\text{hv}}$ A$^{-\cdot}$ + D$^{+\cdot}$

D $\longrightarrow$ D$_2{}^{+\cdot}$ $\xrightarrow{\text{A}^{-\cdot}}$ D$_2$

D' $\longrightarrow$ DD'$^{+\cdot}$ $\xrightarrow{\text{A}^{-\cdot}}$ DD'

A ; Electron acceptor   D; Electron donor

Scheme 27.

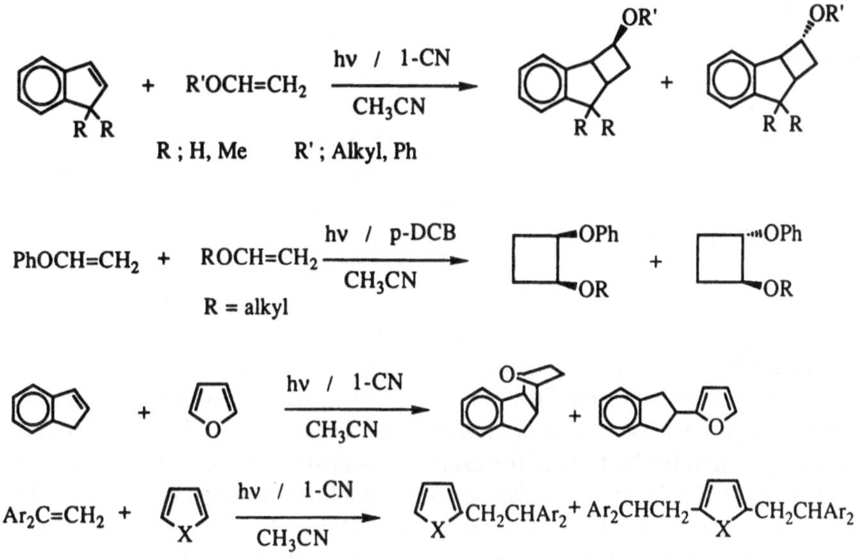

$\qquad$ + R'OCH=CH$_2$ $\xrightarrow[\text{CH}_3\text{CN}]{\text{hv / 1-CN}}$ $\qquad$ +

R ; H, Me   R' ; Alkyl, Ph

PhOCH=CH$_2$ + ROCH=CH$_2$ $\xrightarrow[\text{CH}_3\text{CN}]{\text{hv / p-DCB}}$ $\qquad$ +

R = alkyl

$\qquad$ + $\qquad$ $\xrightarrow[\text{CH}_3\text{CN}]{\text{hv / 1-CN}}$ $\qquad$ +

Ar$_2$C=CH$_2$ + $\qquad$ $\xrightarrow[\text{CH}_3\text{CN}]{\text{hv / 1-CN}}$ CH$_2$CHAr$_2$ + Ar$_2$CHCH$_2$ $\qquad$ CH$_2$CHAr$_2$

Ar ; Aryl   X ; O, NMe   1-CN ; 1-Cyanonaphthalene   p-DCB ; 1,4-Dicyanobenzene

Scheme 28.

of an electron acceptor (Scheme 28) [90]. Mizuno and Otsuji showed many examples of crossed photocycloaddition and photoaddition via heterodimer radical cations [89, 91–93]. An important feature of these photoreactions is that the photoreactions take place efficiently only when the oxidation potentials of two substrates are close to each other (within 0.3–0.4 V). Photo-Diels-Alder reaction via DD'$^{+\cdot}$ was reported by various groups (Scheme 29) [92, 94, 95]. Bauld reported the Diels-Alder reaction via a radical cation by use of the tris (p-bromophenyl)aminium salt [96–98].

3.6 Cycloaddition via Triplexes

Photocycloaddition via triplexes is a new area in organic photochemistry. As mentioned in Sect. 3.4, phenyl vinyl ether undergoes photocyclodimerization

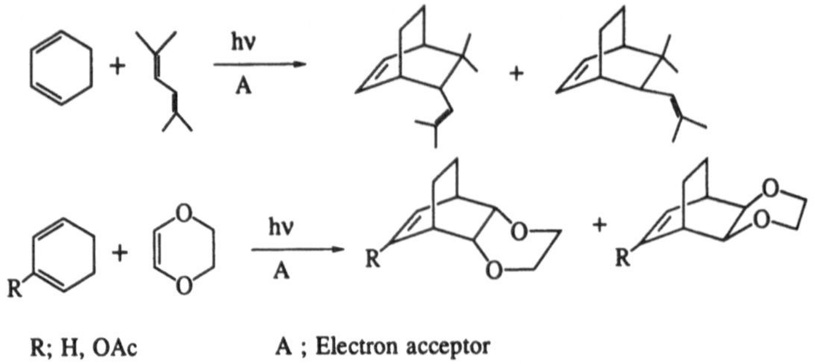

R; H, OAc   A ; Electron acceptor

Scheme 29.

to give two stereoisomeric head-to-head cyclobutanes. This photoreaction proceeds via the dimer radical cation in acetonitrile and via triplex in benzene. However, alkyl vinyl ethers hardly undergo photocyclodimerization by any of direct, electron transfer and triplet sensitized photoreactions. Mizuno and Otsuji found that 1,4-dicyanonaphthalene (DCN) can be used as a sensitizer for this photocyclodimerization (Scheme 30) [99]. This reaction takes place efficiently in aromatic solvents such as benzene and toluene, but does not occur in acetonitrile. Similar photocycloaddition of allyltrimethylsilane in the presence of DCN was also reported [100]. Pac and Das found that DCN forms an emissive exciplex with benzene and toluene and the lifetimes of the exciplexes are much longer than the lifetime of the excited singlet state of DCN in cyclohexane [14, 101, 102]. The exciplex emission from DCN and toluene is quenched by electron-rich alkenes such as ethyl vinyl ether, although triplex emission is not observed. Mizuno and Otsuji proposed that triplex, which is formed from DCN, toluene and alkyl vinyl ether, is a key intermediate for the photocyclodimerization of alkyl vinyl ethers.

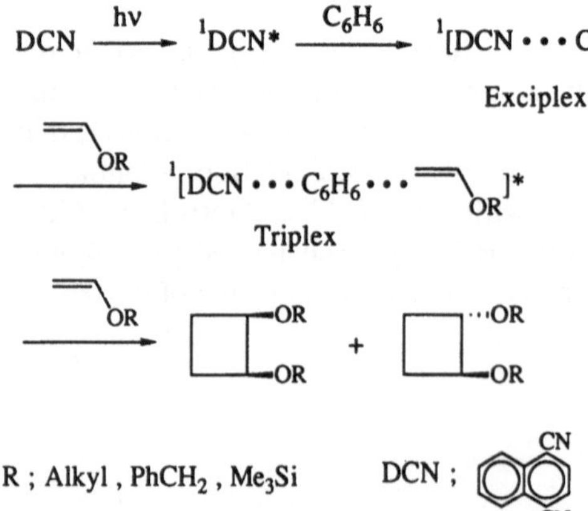

R ; Alkyl , PhCH$_2$, Me$_3$Si   DCN ;

Scheme 30.

Cyclobutano crown ethers and 3-silabicyclo [3.2.0] compounds can be synthesized by the intramolecular version of this photocyclodimerization (Scheme 31) [103, 104].

Schuster proposed a triplex as a key intermediate for the inter- and intramolecular photo-Diels Alder reaction in nonpolar solvents in the presence of an electron accepting sensitizer [105–107]. This photoreaction can be applied to an enantioselective photo-Diels Alder reaction between 1,3-cyclohexadiene and propenylbenzene by the use of a chiral sensitizer (Scheme 32) [108].

3.7 Oxygenation by $O_2^{-\bullet}$

Photooxygenation initiated by photoinduced electron transfer is of considerable interest [109, 110]. Foote and his coworkers demonstrated that the DCA-sensitized photooxygenation of arylalkenes is initiated by photoinduced electron transfer and proceeds via the mechanism shown in Scheme 33 [111, 112].

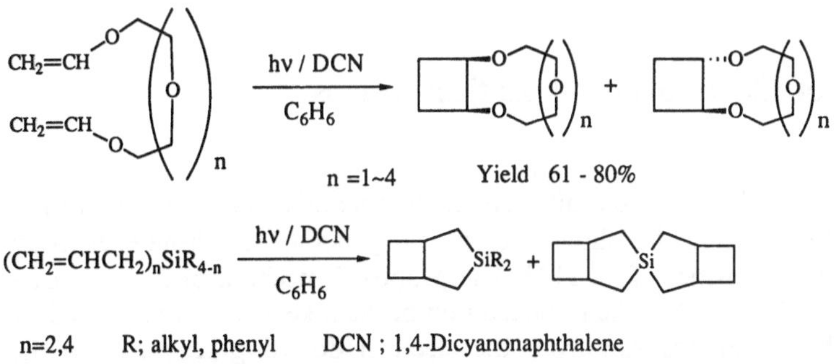

n = 1~4   Yield 61 - 80%

n=2,4   R; alkyl, phenyl   DCN ; 1,4-Dicyanonaphthalene

Scheme 31.

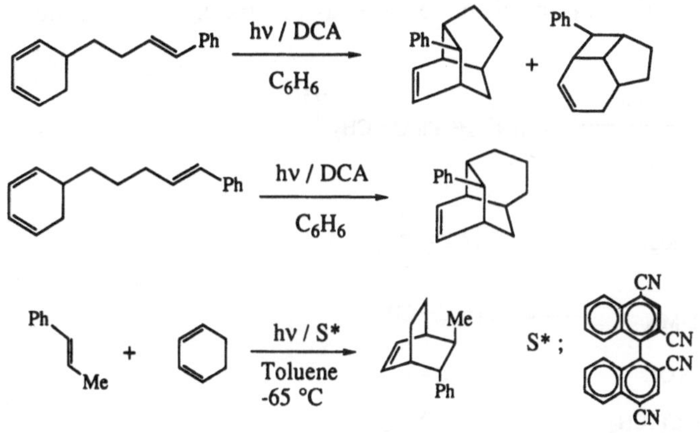

Scheme 32.

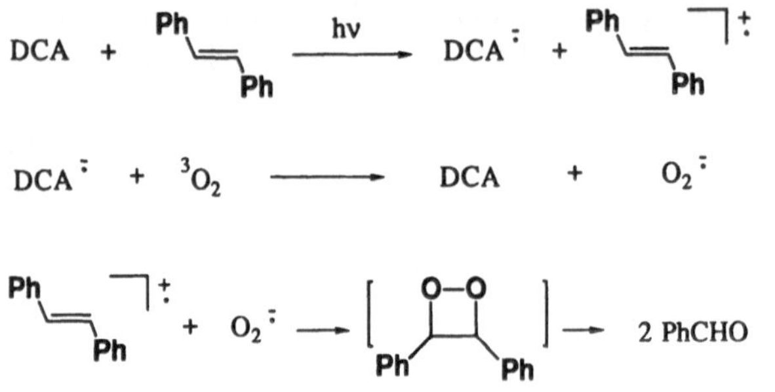

Scheme 33.

The radical cations of arylalkenes are oxidized through a nucleophilic attack of $O_2^{-\bullet}$ which is generated by a secondary electron transfer from $DCA^{-\bullet}$ to $^3O_2$. The other mode of the photooxygenation is the attack of $^3O_2$ toward the radical cations of arylalkenes. Details will be discussed in the latter section.

4 Addition of Electrophiles to Radical Anions

Arnold reported that the photoaddition of methanol and trifluoroethanol to 1,1-diphenylethene occurs in the photoreaction using methoxynaphthalenes as an electron donating sensitizer [35, 113]. The proposed mechanism for this reaction is shown in Scheme 34. The radical anion of the alkene is first produced by photoinduced electron transfer from the electron donating sensitizer to the alkene and it is protonated in a Markownikoff fashion to form the 1,1-diphenylethyl radical. The resulting radical is then oxidized by the radical cation of the electron donating sensitizer to generate the cation of the alkene. Finally, a nucleophilic attack of alcohol on this cation affords the alkoxylated product.

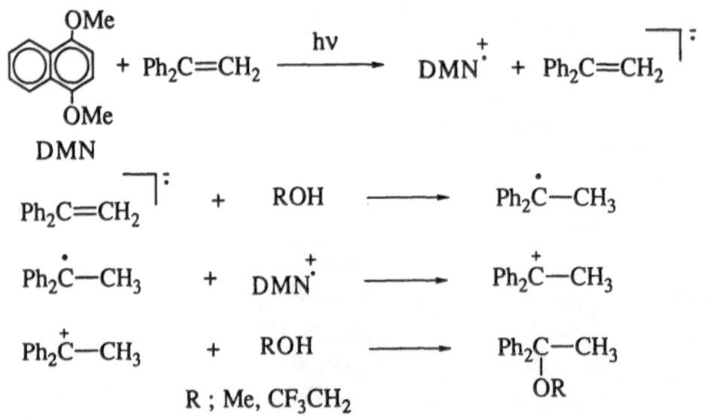

Scheme 34.

The radical anions of cyanoaromatic compounds, cyanoalkenes and aromatic hydrocarbons, which are generated by photoinduced electron transfer, can be protonated by protic solvents or by radical cations of amines to form their neutral radicals. Disproportionation of the radicals yields the reduction products [114].

Pac and his coworkers reported the $Ru(bpy)_3^{2+}$-sensitized photoreduction of electron deficient alkenes by 1-benzyl-1,4-dihydronicotinamide (BNAH) in methanol [116–118]. In this photoreaction, BNAH is first oxidized by $*Ru(bpy)_3^{2+}$ to give $BNAH^{+\bullet}$ and $Ru(bpy)_3^+$. The radical anions of the electron-deficient alkenes, which are generated by photoinduced electron transfer from $Ru(bpy)_3^+$ to the alkene, are then protonated to give the reduction products (Scheme 35).

Tazuke reported the carboxylation of the radical anions of aromatic hydrocarbons that are generated by photoinduced electron transfer from the tertiary amines to the excited singlet aromatic hydrocarbons (Scheme 36) [119]. Toki and his coworkers reported the photofixation of CO_2 with styrene using tertiary amines as electron donors [120]. Tomioka reported the photoaddition of tertiary amines to electrophilic cyclopropanes [115].

The DCA-sensitized photooxygenation of biphenyl is reported by Mizuno and Otsuji [121–123]. In this photooxygenation, both DCA and biphenyl are oxidized by $^3O_2$ to give anthraquinone and benzoic acid (Scheme 37). Although the detailed mechanism of this reaction is still obscure, it is likely that the radical anion of DCA is oxidized with $^3O_2$ after converted to either the anion or the radical. The radical cation of biphenyl is oxidized with $^3O_2$ to give benzoic acid.

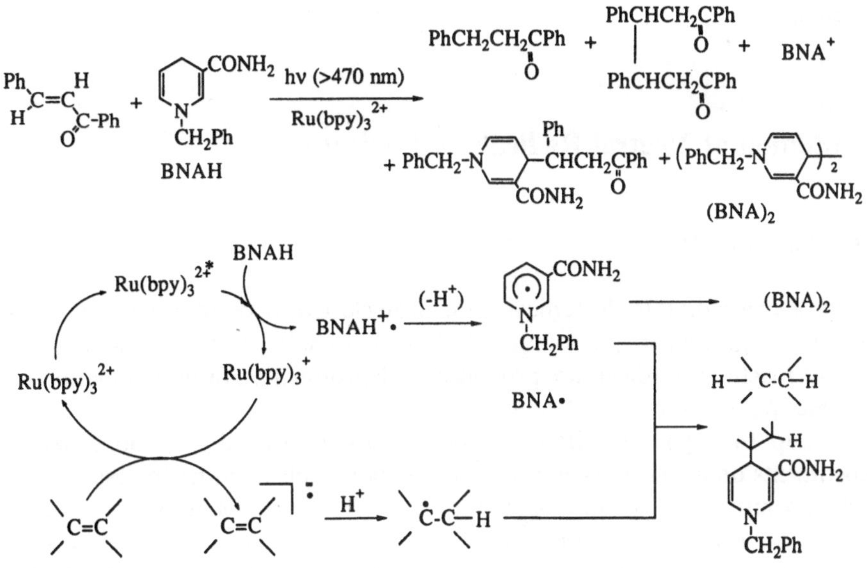

Scheme 35.

325

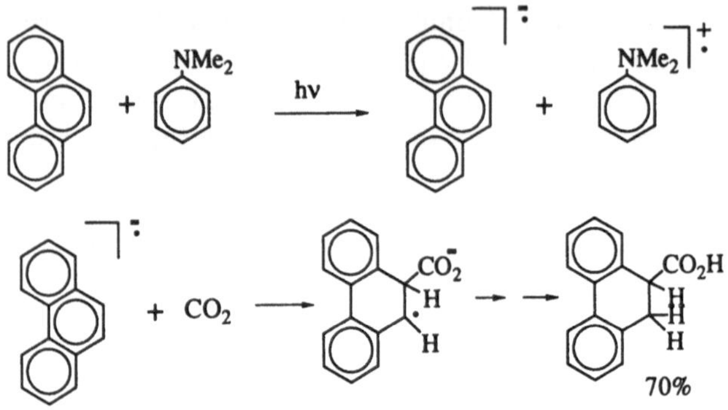

Scheme 36.

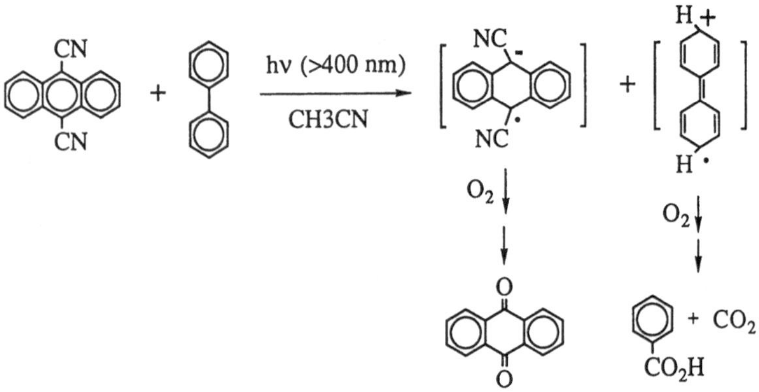

Scheme 37.

5 Addition of Neutral Radicals to Radical Ions

5.1 Addition to Alkenes

Addition of a radical to the radical anion of an electron acceptor is an important elementary process in photoinduced electron transfer reactions and also in organic reactions. Radicals are produced by deprotonation or by elimination of cationic species from radical cations.

Group 14 organometallic compounds bearing alkyl, allylic, and benzylic substituents on metals have lower oxidation potentials than the corresponding carbon compounds. The photoreaction of these compounds in the presence of an electron acceptor generates the radical cations of group 14 organometallic compounds and the radical anion of the electron acceptor. The photochemical processes involving these reactive species are shown in Scheme 38. The radical

R-MR′$_3$ + A $\xrightarrow{\text{hv}}$ [R-MR′$_3$$^{+\bullet}$••• A$^{\bullet-}$]

[R-MR′$_3$$^{+\bullet}$••• A$^{\bullet-}$] $\longrightarrow$ R-MR′$_3$$^{+\bullet}$ + A$^{\bullet-}$

R-MR′$_3$$^{+\bullet}$ + Nu$^-$ $\longrightarrow$ R• + NuMR′$_3$

R• + A$^{\bullet-}$ $\longrightarrow$ $\longrightarrow$ Product

R : alkyl, allyl, arylmethyl   R′ : alkyl
A : Electron acceptor   M : Si, Ge, Sn

Scheme 38.

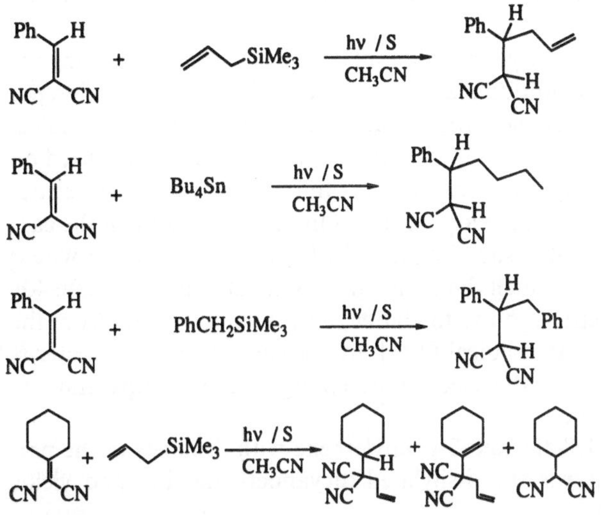

S; Phenanthrene

Scheme 39.

cations of group 14 organometallic compounds are cleaved to give the carbon radicals and the group 14 organometallic cations such as R_3Si^+, R_3Ge^+, and R_3Sn^+. This process is often assisted by attack of nucleophiles such as alcohols and nitriles toward group 14 metals. The radicals thus generated react with the radical anion of the electron acceptor to produce anions which undergo further chemical reactions. The typical examples of photoreactions of group 14 organometallic compounds are shown in Scheme 39.

The photoreaction of 1,1-dicyano-2-phenylethene with allyltrimethylsilane in the presence of phenanthrene in acetonitrile affords 2,2-dicyano-3-phenyl-1-pentene in good yield [124]. Phenanthrene acts as a redox sensitizer and a key process in this photoreaction is a one-electron transfer from the excited singlet phenanthrene (Phen) to 1,1-dicyano-2-phenylethene to give the radical cation of

Phen and the radical anion of the alkene. Secondary electron transfer from allylsilane to Phen$^{+\bullet}$ produces the radical cation of allylsilane and neutral Phen. The radical cation of allylsilane is cleaved by assistance of acetonitrile to generate an allyl radical. The allyl radical adds to the radical anion of the alkene to give the allylated anion which is converted into the product upon protonation. Alkyl and arylmethyl radicals can be generated in a similar manner from tetraalkyl tin compounds and arylmethylsilanes, respectively [124]. These radicals add regioselectively to the β-position to the cyano groups in the radical anions of alkenes.

However, in the case of the photoreaction of alkylidenepropanedinitriles, a different regioselectivity is observed. The photoreaction of cyclohexylidenepropanedinitrile with allyltrimethylsilane in the presence of Phen in acetonitrile gives the α-allylated compounds as major products along with the reduction products (Scheme 39).

The difference between the aryl and alkyl substituted dinitriles regarding the regioselectivity in the photoaddition reactions can be attributed to the difference in the electronic structures of their radical anions. In the case of the radical anions of the aryl substituted dinitriles, the radical center would be localized on the carbon to which the aryl groups are attached. In this case, the radical center is stabilized by conjugation with the aryl groups. On the other hand, in the case of the radical anions of the alkyl substituted dinitriles, the radical center would be localized on the carbon to which the cyano groups are attached (Scheme 40). The unusual electronic distribution in the latter radical anions comes from the capto-dative stabilization of the radical [126]. In this case, the radical center is stabilized by interaction with the electron-accepting cyano groups and the electron-donating carbanion.

The photoreaction of 1,1-dicyano-2-phenylethene with disilanes in the presence of phenanthrene affords the β-silylated dicyanoethanes in good yields

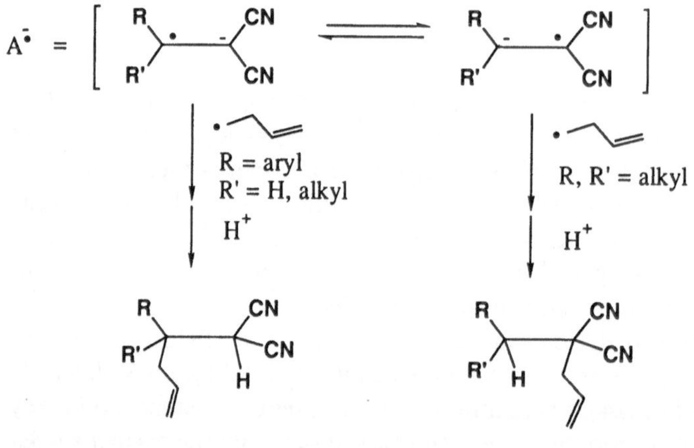

Scheme 40.

[127]. When unsymmetrical disilanes are used as substrates, more bulky silyl groups are predominantly added to the dicyanoethenes. A key step of this photoreaction is also the reaction of the radical anion of the dicyanoethene with silyl radicals which are generated by solvent-assisted cleavage of the radical cations of disilanes (Scheme 41) [125].

Metal cation-catalyzed photoadditions of radical species to radical cations of electron rich alkenes have been reported. Lewis found that the radical cation of norbornene generated by photoinduced electron transfer from this alkene to Ag(I) reacts with acetonitrile to produce 2-cyanomethylnorbornane (Scheme 42) [128].

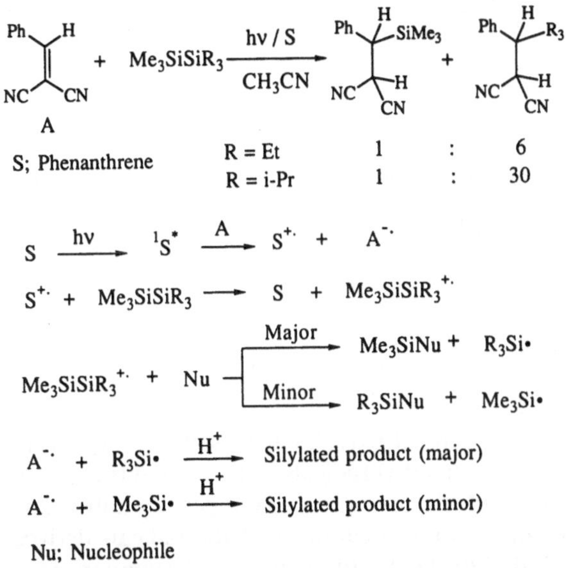

Scheme 41.

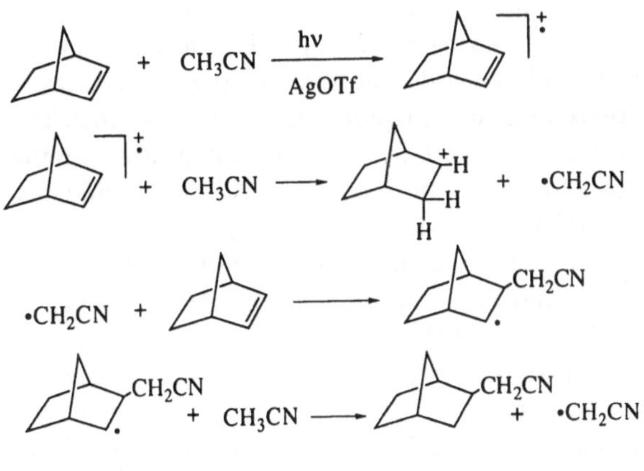

Scheme 42.

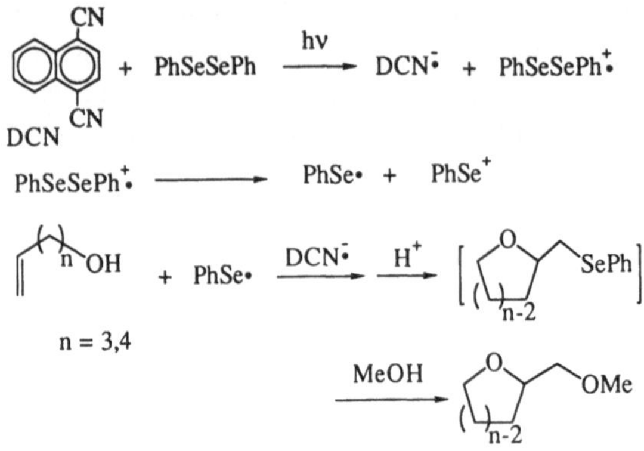

Scheme 43.

Pandey reported the 1,4-dicyanonaphthalene-sensitized cyclization of $CH_2 = CH(CH_2)_nOH$ (n = 3, 4) in the presence diphenyldiselenide [129]. This photoreaction is initiated by the addition of the PhSe• radical to the alkenes as shown in Scheme 43.

5.2 Addition to C=X Bonds

Photoinduced electron transfer reactions of pyrrolidinium salts with electron-rich alkenes generate the neutral iminomethyl radicals =N–C• and the radical cations of the alkenes. Mariano and his coworkers developed the intra- and intermolecular photoadditions of the radical cations and the radicals derived from them to the =N–C• radical [130–132]. Although they are summarized in his review article [3, 133–135], some examples are shown in Scheme 44. The photoaddition of allylic or benzylic radicals to the =N–C• radical can be achieved by the photoreaction of iminium compounds with group 14 organometallic compounds bearing allylic and benzylic groups. The mechanism for generation of the radicals has already been discussed in the preceding section.

α-Silyl amine adds to cyclohexenone upon irradiation in polar solvents. The radical cation of α-silyl amine cleaves to two types of radicals, depending on the nature of solvents. In acetonitrile formation of $Et_2N(SiMe_3)CH•$ predominates, but in methanol formation of $Et_2NCH_2^+$ predominates [136].

Kubo, Maruyama and Takuwa reported the photoaddition of allylic radicals generated from allylic silane and tin compounds to the radical anions of carbonyl compounds (Scheme 45) [137–139].

Photoinduced electron transfer from amines to the excited ketones generates the radical cations of amines and the radical anions of ketones [140]. The radical anions of ketones are usually reduced to the corresponding alcohols. However, in some cases the coupling reaction between the radical anions and

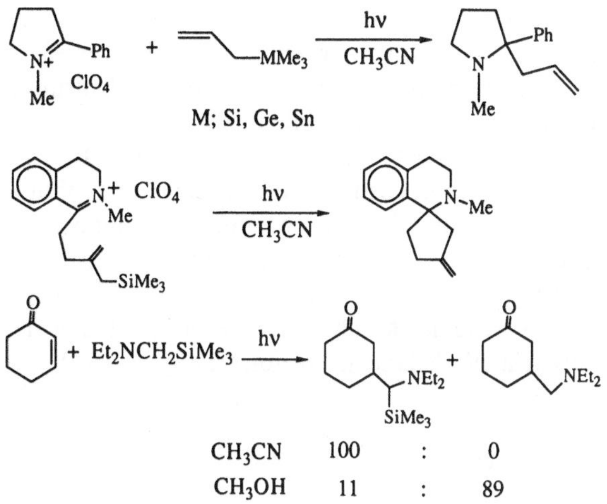

Scheme 44.

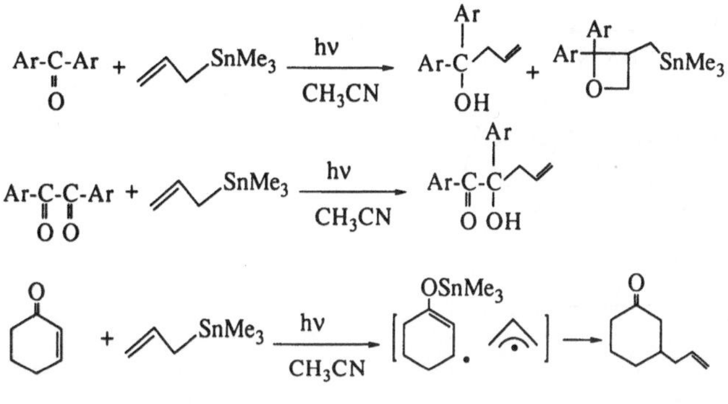

Scheme 45.

the radical cations takes place, giving addition products of aryl group to the carbonyl group [141, 142].

5.3 Addition to Aromatic Rings

Photoaddition of neutral radicals R$^\bullet$ to aromatic rings occurs efficiently, especially for aromatic compounds having electron-withdrawing substituents. The addition of R$^\bullet$ to the radical anions of aromatic compounds affords the anion intermediates. The anion intermediates are converted to reductive addition products by protonation and to substitution products by elimination of an anion moiety. The addition of R• to the radical cations of aromatic compounds affords the cation intermediates. The cation intermediates are converted to reductive addition products by addition of a nucleophile and to substitution products by deprotonation or elimination of a cationic moiety.

The photoaddition of radicals to benzene rings usually gives substitution products as major products. The photosubstitution and reductive photo-addition occur competitively in the case of polycyclic aromatic compounds. The reductive addition becomes predominant over the substitution with increasing the number of the component rings.

Some examples are shown in Scheme 46. The photoreaction of dicyano aromatic compounds with allyltrimethylsilane in acetonitrile gives both the

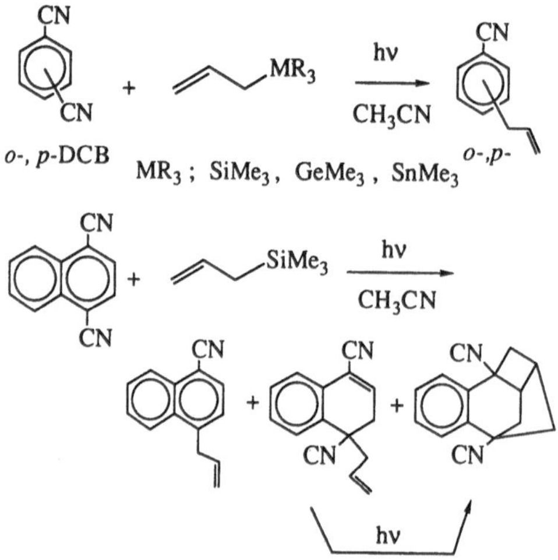

Scheme 46.

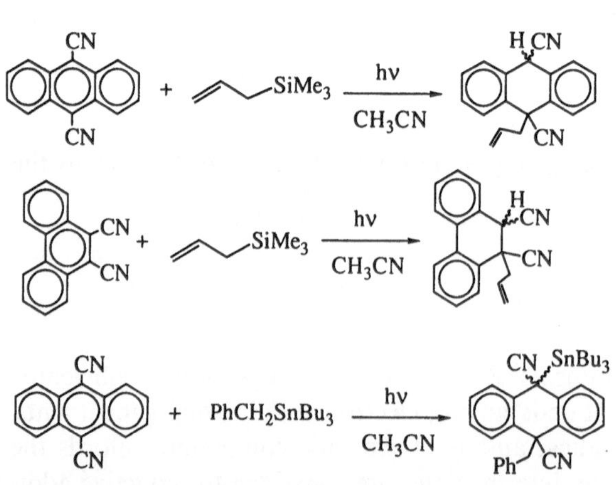

Scheme 47.

photosubstitution products and the reductive photoaddition products, the product ratios depending on the number of benzene rings. In the case of dicyanobenzenes, only the substitution products are produced [143, 144]. In the case of 1,4-dicyanonaphthalene, both 1-allyl-4-cyanonaphthalene and 1-allyl-1,4-dicyano-1,2-dihydronaphthalene are produced [145]. In the case of tricyclic ring compounds such as 9, 10-dicynoanthracene and 9, 10-dicyanophenanthrene, only the reductive photoallylation products are produced (Scheme 47) [145–146]. The photosubstitution and photoaddition reactions via the addition of alkyl and silyl radicals to the radical anions of polycyanoaromatic compounds and the neutral radicals of cationic heteroaromatic compounds also occur when alkyl borates, silanes, stannanes and disilanes are used as substrates (Schemes 48 and 49) [147–150]. The mechanism for these photoreactions is shown in Scheme 50.

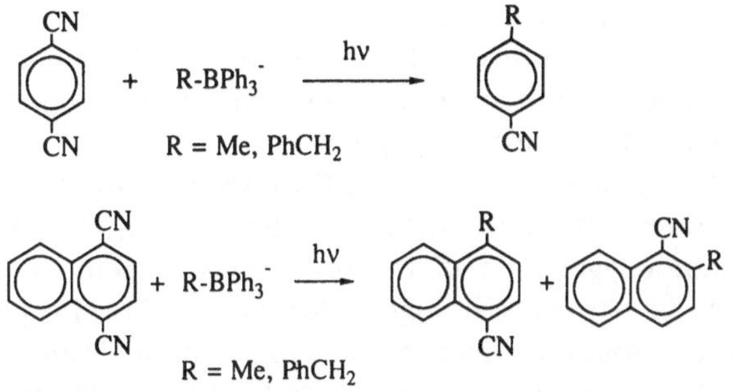

Scheme 48.

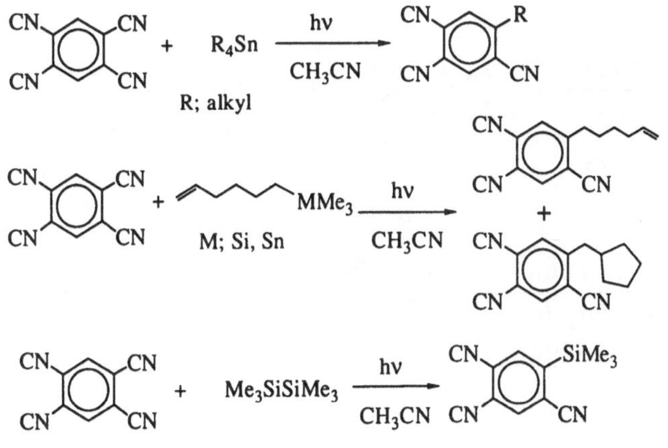

Scheme 49.

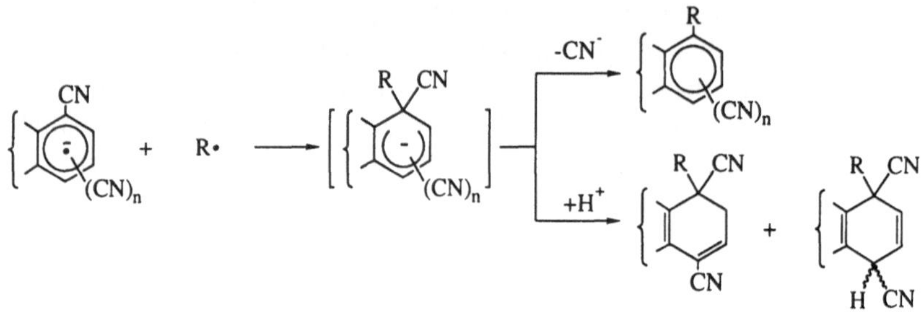

Scheme 50.

5.4 Oxygenation by $^3O_2$ and Insertion of NO

In some cases, photooxygenation of organic molecules by molecular dioxygen occurs via photoinduced electron transfer and proceeds through a radical addition of dioxygen to the radical cations and radical anions of organic compounds. The photooxygenation of this type has been reviewed by Lopez [109]. In this section, some selected examples are shown.

The DCA-sensitized photooxygenation of 1,1-diarylethylenes in acetonitrile gives 3,3,6,6-tetraaryl-1,2-dioxanes in good yields (Schemes 51 and 52) [151–153]. Intramolecular photooxygenation of 1,1,ω,ω-tetrakis (4-methoxyphenyl)-alkadienes affords bicyclic peroxides in high yields (Scheme 51) [154]. The 1,4-radical cations generated from the dimer radical cations of the alkenes are trapped by $^3O_2$, giving 1,2-dioxanes. The DCA-sensitized photooxygenation of N-vinylcarbazole in acetonitrile gives 3,6-bis (9-carbazolyl)-1,2-dioxane via the 1,4-radical cation (Scheme 53) [155].

The DCA-sensitized photooxygenation of electron-rich 1,2-diarylcyclopropanes affords 3,5-diaryl-1,2-dioxolanes in high yields (Scheme 54) [156]. In

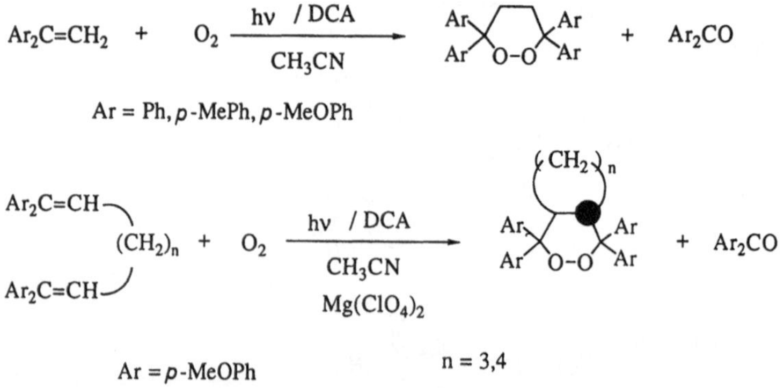

Ar = Ph, p-MePh, p-MeOPh

Ar = p-MeOPh   n = 3,4

Scheme 51.

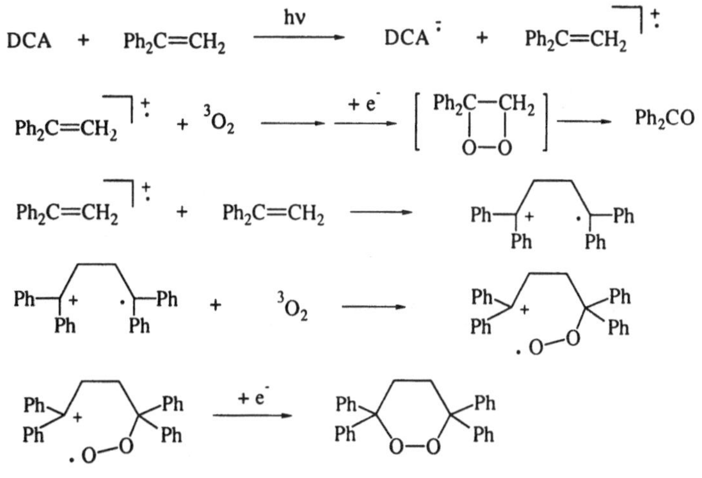

$$DCA + Ph_2C=CH_2 \xrightarrow{h\nu} DCA^{-\cdot} + [Ph_2C=CH_2]^{+\cdot}$$

$$[Ph_2C=CH_2]^{+\cdot} + {}^3O_2 \xrightarrow{+e^-} \left[\underset{O-O}{Ph_2C-CH_2} \right] \longrightarrow Ph_2CO$$

$$[Ph_2C=CH_2]^{+\cdot} + Ph_2C=CH_2 \longrightarrow Ph\underset{Ph}{-}\!\!\diagdown\!\!+ \cdot\underset{Ph}{-}Ph$$

$$Ph\underset{Ph}{-}\!\!\diagdown\!\!+ \cdot\underset{Ph}{-}Ph + {}^3O_2 \longrightarrow \underset{Ph}{\overset{Ph}{}}\!\!\diagdown\!\!+\!\!\underset{\cdot O-O}{\overset{Ph}{}}Ph$$

$$\underset{Ph}{\overset{Ph}{}}\!\!\diagdown\!\!+\!\!\underset{\cdot O-O}{\overset{Ph}{}}Ph \xrightarrow{+e^-} \underset{Ph}{\overset{Ph}{}}\diagdown\!\!\underset{O-O}{}\!\!\diagup\underset{Ph}{\overset{Ph}{}}$$

Scheme 52.

Scheme 53 (diagram):

$$\text{carbazole vinyl} \xrightarrow[\text{CH}_3\text{CN}]{h\nu \ / \ DCA} \text{cyclobutane}(Ar, Ar) \xrightarrow[\text{CH}_3\text{CN}]{h\nu \ / \ DCA \ / \ O_2} \text{dioxane}(Ar, Ar)$$

$$Ar = \text{carbazolyl}$$

$$\left[\text{cyclobutane cation radical (Ar, Ar)} \right] \xrightarrow{O_2} \ +e^-$$

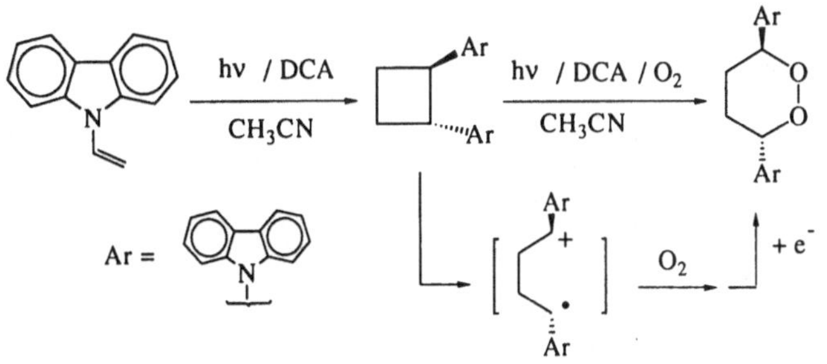

Scheme 53.

the case of 1,2-bis (4-methoxyphenyl) cyclopropanes, the quantum yield for formation of 1,2-dioxolanes exceeds unity. The mechanism for this photo-oxygenation is shown in Scheme 55. The key step is the attack of $^3O_2$ on the radical site of the 1,3-radical cation produced by a ring opening of the cyclopropane ring. The peroxy radical cations are converted into 1,2-dioxolanes by one-electron transfer from neutral 1,2-diarylcyclopropanes or from $DCA^{-\cdot}$. When the electron transfer occurs from 1,2-diarylcyclopropanes, the photo-reaction becomes a chain reaction.

The DCA-sensitized photooxygenation of less electron-rich cyclopropanes in the absence of $Mg(ClO_4)_2$ does not afford 1,2-dioxolanes. However, the photoreaction in the presence of $Mg(ClO_4)_2$ gives 1,2-dioxolanes in moderate yields [157]. In this photoreaction, $Mg(ClO_4)_2$ suppresses not only a back electron transfer from $DCA^{-\cdot}$ to the radical cations of cyclopropanes, but also the decomposition of the 1,2-dioxolanes produced.

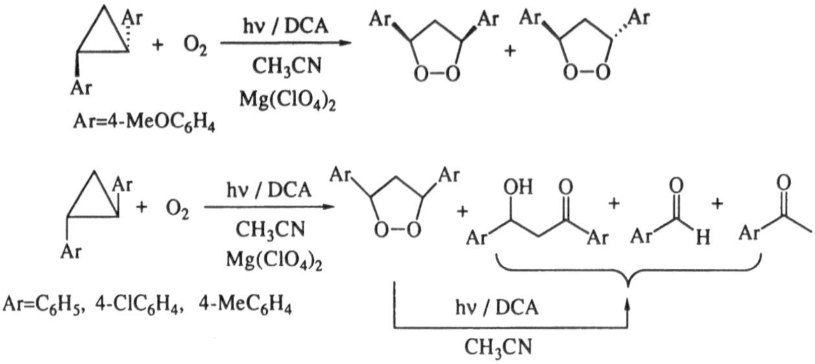

K. Mizuno and Y. Otsuji

Ar=4-MeOC$_6$H$_4$

Ar=C$_6$H$_5$, 4-ClC$_6$H$_4$, 4-MeC$_6$H$_4$

hv / DCA

CH$_3$CN

Scheme 54.

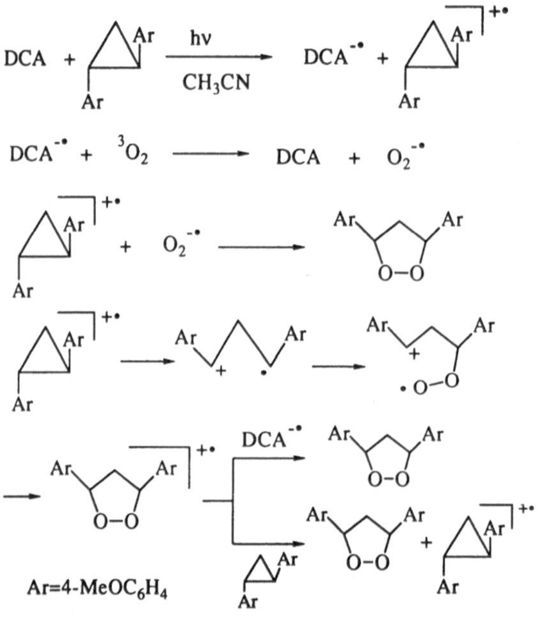

Ar=4-MeOC$_6$H$_4$

Scheme 55.

The photooxygenation of 3-alkyl-1,2-diarylcyclopropanes occurs with a high stereoselectivity in the presence of Mg(ClO$_4$)$_2$ (Scheme 56) [158]. Shim reported the DCA-sensitized photooxygenation of vinyl cyclopropanes in the presence of Mg(ClO$_4$)$_2$ and biphenyl [159].

The photooxygenation of 2,2-diarylmethylenecyclopropanes proceeds via trimethylenemethane radical cation intermediates, giving 1,2-dioxolanes [160, 161]. When 1,1,2,2-tetracyanoethene (TCNE) is used as an electron acceptor, the (3 + 2) cycloaddition between methylenecyclopropanes and

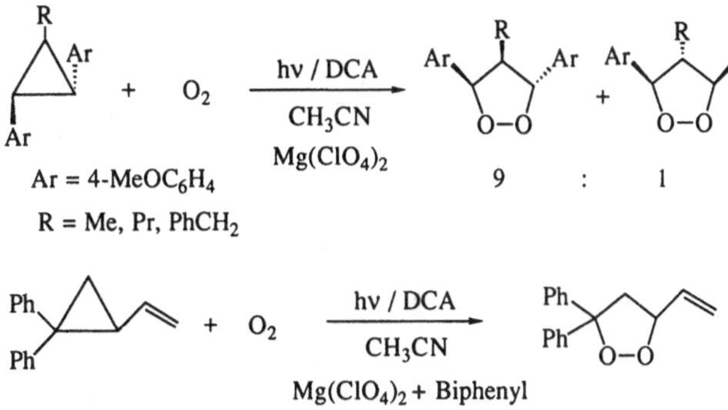

Scheme 56.

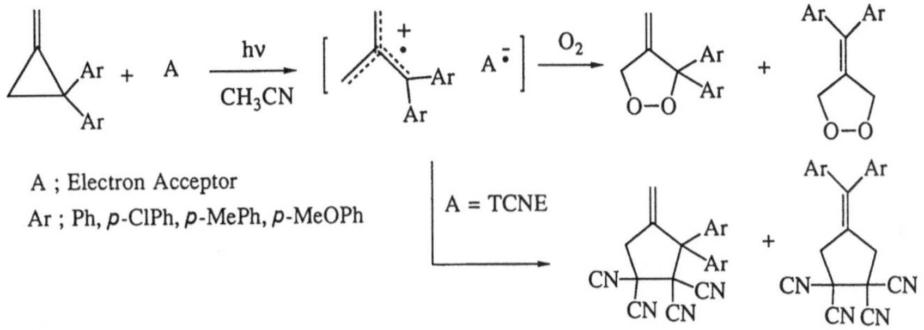

Scheme 57.

TCNE takes place (Scheme 57). Tomioka reported the (3 + 2) photocycloaddition between 1,1,2-triarylcyclopropanes and vinyl ethers in the presence of p-DCB [162]. Mizuno and Otsuji reported the (4 + 2) photocycloaddition between 1,2-diarylcyclopropanes and DCA [23]. The 1,4-radical cation produced as an intermediate of the Cope rearrangement of 1,5-dienes via photoinduced electron transfer can be trapped by molecular dioxygen, giving bicyclic dioxanes (Scheme 58) [163]. This photooxygenation takes place in a stereospecific manner.

The photooxygenation of three- and four-membered heterocyclic compounds affords five- and six-membered heterocyclic peroxides (Scheme 59) [164, 165]. In the case of the photooxygenation of aziridines, the *cis* and *trans* isomer ratio of the five-membered peroxides depends on the bulkiness of substituents on the nitrogen atom.

Nitrogen oxide NO also inserts into the cyclopropane ring of 1,2-diarylcyclopropanes under photoinduced electron transfer reaction conditions, giving 3,5-diarylisoxazolines (Scheme 60).

The (3 + 2) photocycloaddition occurs between three-membered heterocyclic compounds and electron deficient alkenes in the presence of 1,4-dicyanonaphthalene to afford five-membered heterocyclic compounds

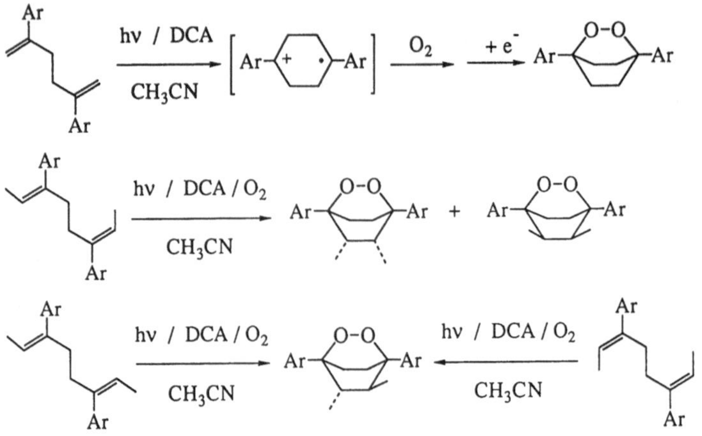

Ar; Ph, *p*-MeOC$_6$H$_4$

Scheme 58.

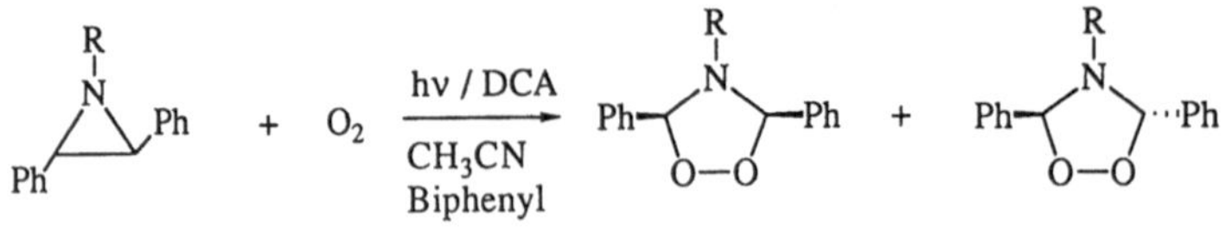

R = H	> 99	:	< 1
R = Me	85	:	15
R = PhCH$_2$	60	:	40
R = *t*-Bu	< 1	:	> 99

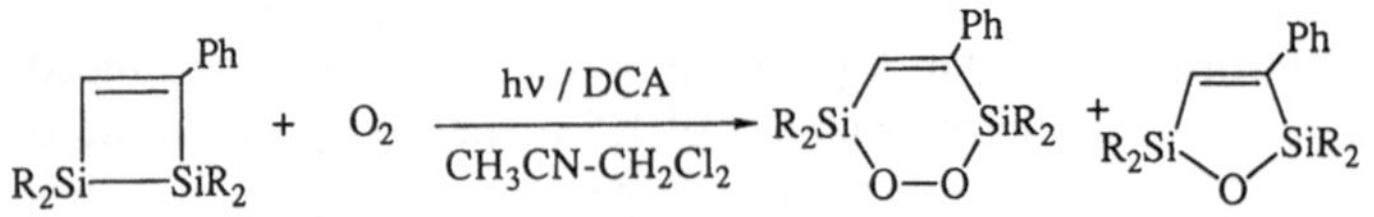

R = mesityl

Scheme 59.

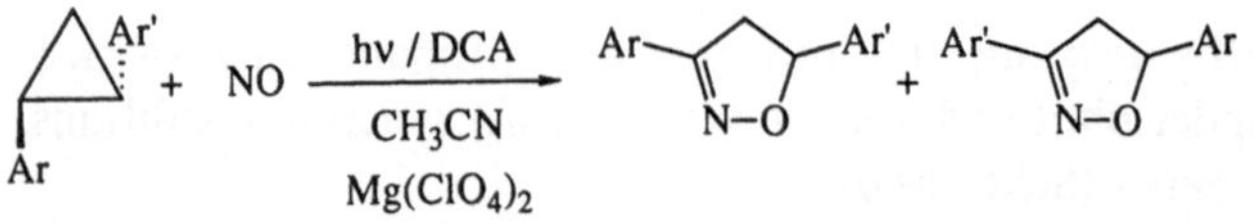

Ar, Ar' = Aryl

Scheme 60.

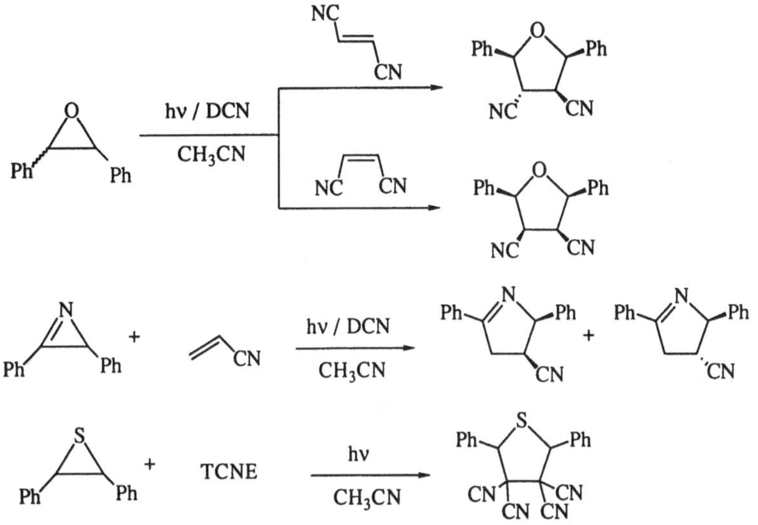

Scheme 61.

(Scheme 61) [167–169]. In these photoreactions, dicyanonaphthalene acts as an effective sensitizer for generation of the radical cations of the heterocyclic compounds.

6 Three-Component Addition

The photoreaction of electron-rich alkenes and cyclopropanes with cyanoaromatic compounds and cyanoarylalkenes in the presence of nucleophilic solvents affords the solvent-incorporated adducts (Scheme 62). Three mechanisms have been proposed for these three-component addition reactions.

In the photoreaction of 2-cyanonaphthalene with 2,3-dimethyl-2-butene in methanol, McCullough proposed that a key step of this photoreaction is a radical coupling between the radical anion of 2-cyanonaphthalene and the radical cation of 2,3-dimethyl-2-butene [170]. The zwitter ionic intermediate thus produced reacts then with methanol (Scheme 63). In the photoreaction of 9-cyanophenanthrene with 2,3-dimethyl-2-butene in methanol, Pac proposed that a nucleophile adds first to the radical cation of the alkene to produce the radical species which then reacts with the radical anion of 9-cyanophenanthrene [171]. Protonation of the anion gives the solvent-incorporated adduct (Scheme 64). In the photoreaction of 1-cyano-2,3-diphenylethene with 2,5-dimethyl-2,4-hexadiene in methanol, Lewis proposed that a key step of this photoreaction is a

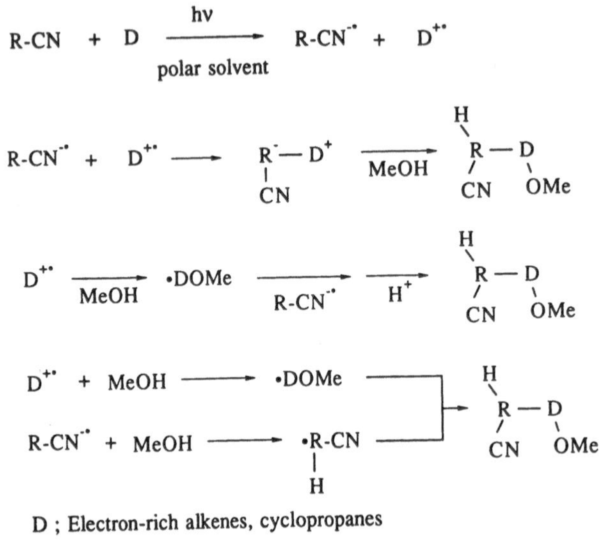

$$\text{R-CN} + \text{D} \xrightarrow[\text{polar solvent}]{h\nu} \text{R-CN}^{-\bullet} + \text{D}^{+\bullet}$$

$$\text{R-CN}^{-\bullet} + \text{D}^{+\bullet} \longrightarrow \underset{\underset{\text{CN}}{|}}{\text{R}^{-}} \text{—D}^{+} \xrightarrow{\text{MeOH}} \underset{\underset{\text{CN}}{/}}{\overset{\overset{\text{H}}{\diagdown}}{\text{R}}} \text{—} \underset{\text{OMe}}{\overset{\diagdown}{\text{D}}}$$

$$\text{D}^{+\bullet} \xrightarrow{\text{MeOH}} \bullet\text{DOMe} \xrightarrow[\text{R-CN}^{-\bullet}]{} \xrightarrow{\text{H}^{+}} \underset{\underset{\text{CN}}{/}}{\overset{\overset{\text{H}}{\diagdown}}{\text{R}}} \text{—} \underset{\text{OMe}}{\overset{\diagdown}{\text{D}}}$$

$$\text{D}^{+\bullet} + \text{MeOH} \longrightarrow \bullet\text{DOMe}$$

$$\text{R-CN}^{-\bullet} + \text{MeOH} \longrightarrow \underset{\underset{\text{H}}{|}}{\bullet\text{R-CN}} \longrightarrow \underset{\underset{\text{CN}}{/}}{\overset{\overset{\text{H}}{\diagdown}}{\text{R}}} \text{—} \underset{\text{OMe}}{\overset{\diagdown}{\text{D}}}$$

D ; Electron-rich alkenes, cyclopropanes

Scheme 62.

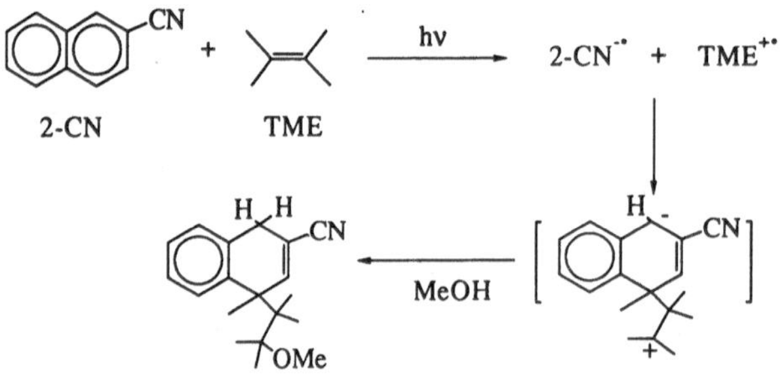

Scheme 63.

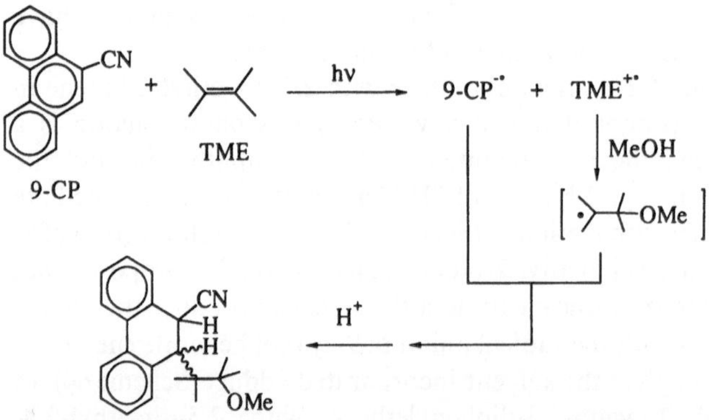

Scheme 64.

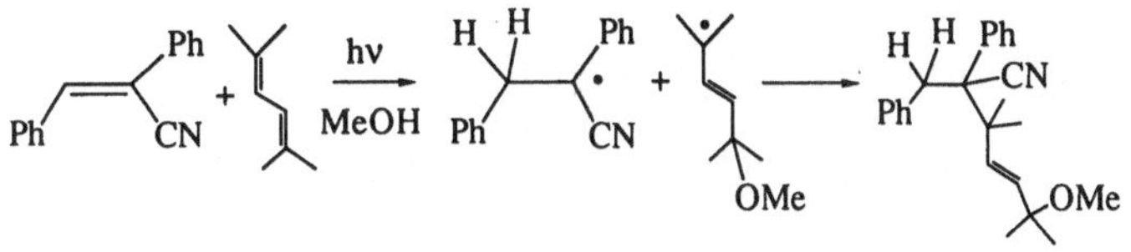

Scheme 65.

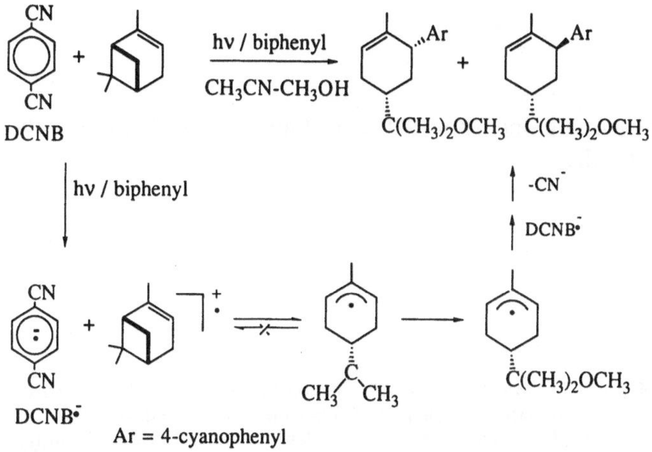

Ar = 4-cyanophenyl

Scheme 66.

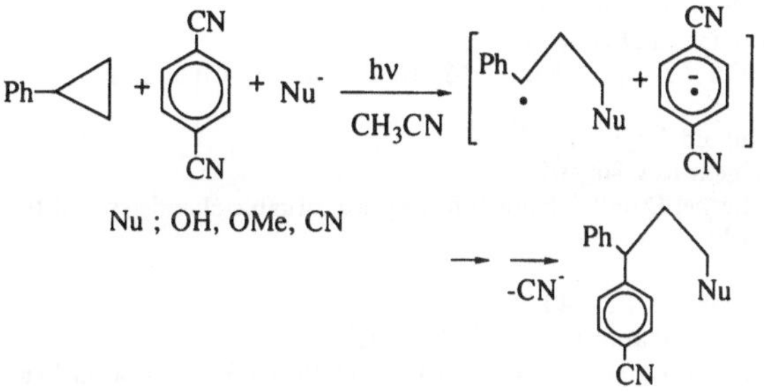

Nu ; OH, OMe, CN

Scheme 67.

radical coupling between the protonated radical anion of 1-cyano-2,3-diphenylethene and the methanol-incorporated radical of 2,5-dimethyl-2,4-hexadiene which is produced from the reaction of the radical anion of the diene with methanol (Scheme 65) [172].

The three-component photoaddition reactions of this type have also been reported by several groups [29,40–42,173–178]. These examples are shown in Schemes 66,67 and 68. These photoreactions involve, in most cases, a coupling

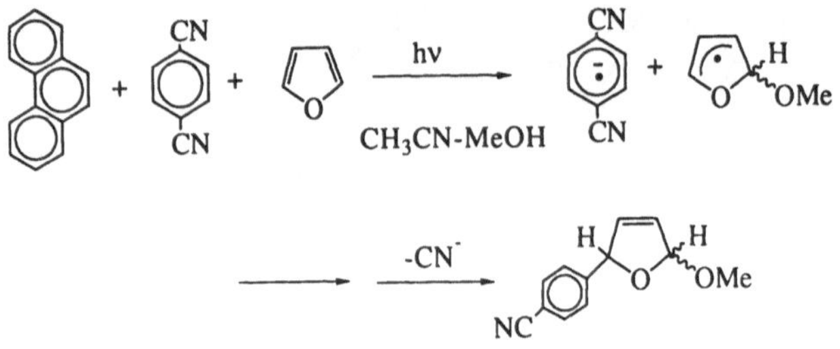

K. Mizuno and Y. Otsuji

Scheme 68.

reaction between a nucleophile-incorporated radical and a radical anion of the electron accepting molecule.

7 References

1. Chanon M, Hawley MD, Fox MA (1988) In: Fox MA, Chanon M (eds) Photoinduced electron transfer, Part A, Elsevier, Amsterdam, p 1; Chanon M, Eberson L (1988) Ibid, p 409
2. Roth H (1990) In: Mattay J (ed) Photoinduced electron transfer I, Topics in Current Chemistry, vol 156. Springer, Berlin Heidelberg New York, p 1; Mattay J (1991) In: Mattay J (ed) Photoinduced electron transfer III, Top Curr Chem, vol 159, p 219
3. Mariano PS, Stavinoha JL (1984) In: Horspool WM (ed) Synthetic organic photochemistry. New York, p 145
4. Mattes SL, Farid S (1983) Org Photochem 6: 233
5. Mattay J (1987) Angew Chem Int Ed Engl 26: 825
6. Chanon M (1985) Bull Soc Chim Fr 209; Chanon M, Rajzmann M, Chanon F (1990) Tetrahedron 46: 6193
7. Kochi JK (1988) Angew Chem Int Ed Engl 27: 1227
8. Kavarnos GJ, Turro NJ (!986) Chem Rev 86: 401
9. Davidson RS (1983) In: Gold V, Bethel D (ed) Advances in Physical organic chemistry, vol 19, Academic, London New York, p 1
10. Mattay J (1989) Synthesis 233
11. Beens H, Weller A (1968) Chem Phys Lett 2: 140
12. Saltiel J, Townsend DE, Watson BD, Shannon P (1975) J Am Chem Soc 97: 5688
13. Yang NC, Minsek DW, Johnson DG, Larson JR, Petrich JW, Gerald III R, Wasieleski MR (1989) Tetrahedron 45: 4669
14. Masaki Y, Yanagida S, Pac C (1988) Chem Lett 1305; Masaki Y, Uehara Y, Yanagida S, Pac C (1990) Ibid. 1339; (1992) Ibid. 315
15. Caldwell RA, Creed D (1978) J Am Chem Soc 100: 2905
16. Caldwell RA, Creed D (1980) Acc Chem Res 13: 45
17. Caldwell RA, Creed D, DeMarco DC, Melton LA, Ohta H, Wine PH (1980) J Am Chem Soc 102: 2369
18. Rehm D, Weller A (1970) Isr J Chem 8: 259
19. Marcus RA (1956) J Chem Phys 24: 966; Marcus RA (1959) Can J Chem 37: 155; Marcus RA (1965) J Chem Phys 43: 58
20. Gould IR, Young RH, Farid S (1991) In: Honda K (ed) Photochemical processes in organized molecular systems. Elsevier, Amsterdam, p 19; Miller JR (1991) Ibid, p 41
21. Mataga N (1992) In: Mataga N, Okada T, Masuhara H (eds) Dynamics and mechanisms of photoinduced electron transfer and related phenomena. Elsevier, Amsterdam, p 1; Kakitani T (1992) Ibid, p 71

22. Mattes SL, Farid S (1986) J Am Chem Soc 108: 7356
23. Mizuno K, Ichinose N, Otsuji Y (1992) J Org Chem 57: 1855; Mizuno K, Hiromoto Z, Ohnishi K, Otsuji Y (1983) Chem Lett 1059; Ichinose N, Mizuno K, Hiromoto Z, Otsuji Y (1986) Tetrahedron Lett 27: 5619
24. Santamaria J (1988) In: Fox MA, Chanon M (eds) Photoinduced electron transfer, Part B. Elsevier, Amsterdam, p 483
25. Pac C, Ishitani O (1988) Photochem Photobio 1988: 47
26. Loupy A, Tchoubar B, Astruc D (1992) Chem Rev 92: 1141
27. Mizuno K, Otsuji Y (1988) Yuki Gousei Kagaku Kyoukaishi (J Synth Org Chem Japan) 47: 916
28. Majima T, Pac C, Sakurai H (1980) J Am Chem Soc 102: 5265
29. Majima T, Pac C, Sakurai H (1981) J Am Chem Soc 103: 4499; Pac C, Nakasone A, Sakurai H (1977) Ibid. 99: 5806
30. Pac C (1986) Pure & Appl Chem 58: 1249
31. Mattes SL, Farid S (1982) Acc Chem Res 15: 80
32. Schaap AP, Lopez L, Gagnon SD (1983) J Am Chem Soc 105: 663
33. Neunteufel RA, Arnold DR (1973) J Am Chem Soc 95: 4080
34. Shigemitsu Y, Arnold DR (1975) J Chem Soc, Chem Commun 407; Maroulis AJ, Shigemitsu Y, Arnold DR (1978) J Am Chem Soc 100: 535
35. Maroulis AJ, Arnold DR (1979) Synthesis 819
36. Maroulis AJ, Arnold DR (1979) J Chem Soc, Chem Commun 351
37. Mizuno K, Nakanishi I, Ichinose N, Otsuji Y (1989) Chem Lett 1095
38. Mizuno K, Tamai T, Tani K, Otsuji Y (to be published)
39. Gassman PG, De Silva SA (1991) J Am Chem Soc 113: 9870
40. Rao VR, Hixson SS (1979) J Am Chem Soc 101: 6458
41. Mizuno K, Ogawa J, Otsuji Y (1981) Chem Lett 741
42. Mazzocchi PH, Somich C, Edwards M, Morgan T, Ammon HL (1986) J Am Chem Soc 108: 682
43. Ichinose N, Kitamura N, Masuhara H (1991) J Chem Soc, Chem Commun 985; Tomioka H, Inoue O (1988) Bull Chem Soc Jpn 61: 1404
44. Dinnocenzo JP, Todd WP, Simpson TR, Gould IR (1990) J Am Chem Soc 112: 2462
45. Gassman PG, Olson KD, Walter L, Yamaguchi R (1981) J Am Chem Soc 103: 4977; Gassman PG, Smith JL (1983) J Org Chem 48: 4438
46. Gassman PG (1988) In: Fox MA, Chanon M (eds) Photoinduced electron transfer, Part C. Elsevier, Amsterdam, p 70
47. Mizuno K, Yoshioka K, Otsuji Y (1983) Chem Lett 941
48. Murai S, Tsutsumi S (1966) Bull Chem Soc Jpn 34: 198
49. Mizuno K, Ogawa, Kagano, Otsuji Y (1981) Chem Lett 437
50. Kojima M, Sakuragi H, Tokumaru K (1985) Bull Chem Soc Jpn 58: 521
51. Asai M, Matsui H, Tazuke S (1974) Bull Chem Soc Jpn 47: 864
52. Shono T, Matsumura Y (1970) J Org Chem 35: 4157
53. Mizuno K, Pac C, Sakurai H (1975) J Chem Soc, Chem Commun 553; Yasuda M, Pac C, Sakurai H (1981) J Chem Soc Perkin Trans I 746
54. Mizuno K, Okamoto H, Pac C, Sakurai H (1975) J Chem Soc, Chem Commun 839; Yasuda M, Pac C, Sakurai H (1981) J Org Chem 46: 788
55. Cornelisse J, Havinga E (1975) Chem Rev 75: 353
56. Yasuda M, Yamashita T, Matsumoto T, Shima K, Pac C (1985) J Org Chem 50: 3667
57. Yasuda M, Yamashita T, Shima K, Pac C (1987) J Org Chem 52: 753
58. Yasuda M, Shima K (1991) Reviews on Heteroatom Chem 4: 27
59. Yasuda M, Kubo J, Shima K (1990) Heterocycles 31: 1007; Yasuda M, Hamasuma, Yamano K, Kubo J, Shima K (1992) Heterocycles 34: 965
60. Pac C, Sakurai H (1969) Tetrahedron Lett 3829; Yasuda M, Pac C, Sakurai H (1981) Bull Chem Soc Jpn 54: 2352
61. Sugimoto A, Sumida R, Tamai N, Inoue H, Otsuji Y (1981) Bull Chem Soc Jpn 54: 3500; Sugimoto A, Sumi K, Urakawa K, Ikemura M, Sakamoto S, Yoneda S, Otsuji Y (1983) Bull Chem Soc Jpn 56: 3118
62. Sugimoto A, Yamano J, Suyama K, Yoneda S (1989) J Chem Soc Perkin Trans 1 483; Sugimoto A, Yoneda S (1982) J Chem Soc Chem Commun 376
63. Sugimoto A, Hiraoka R, Inoue H, Adachi T (1992) J Chem Soc, Perkin Trans, 1, 1559; Sugimoto A, Hiraoka R, Fukada N, Kosaka H, Inoue H (1992) Ibid 1, 2871

64. Gilbert A (1984) In: Horspool WM (ed) Synthetic organic photochemistry. New York, p 1
65. Albini A, Sulpizio A (1988) In: Fox MA, Chanon M (eds) Photoinduced electron transfer, Part C. Elsevier, Amsterdam, p 88
66. Lewis FD, Reddy GD, Schneider S, Gahr M (1989) J Am Chem Soc 111: 6465
67. Lewis FD, Reddy GD, Schneider S, Gahr M (1991) J Am Chem Soc 113: 3498
68. Ohashi M, Miyake K, Tsujimoto K (1980) Bull Chem Soc Jpn 53: 1683
69. Naruto S, Yonemitsu O, Kanamaru N, Kimura K (1971) J Am Chem Soc 93: 4053
70. Kanaoka Y (1978) Acc Chem Res 11: 407; Sato Y, Nakai H, Mizoguchi T, Hatanaka Y, Kanaoka Y (1976) J Am Chem Soc 98: 1976
71. Ledwith A (1972) Acc Chem Res 5: 133
72. Shirota Y, Mikawa H (1978) J Macromol Sci Rev Macromol Chem C16: 129
73. Shirota Y (1988) In: Fox MA, Chanon M (eds) Photoinduced electron transfer, Part D, Elsevier, Amsterdam, p 441
74. Kuwata S, Shigemitsu Y, Odaira Y (1972) Chem Commun 2; Kuwata S, Shigemitsu Y, Odaira Y (1973) J Org Chem 38: 3803
75. Evans TR, Wake RW, Jaenicke O (1975) In: Gordon M, Ware WR (eds) The exciplex. Academic, New York, p 345
76. Mizuno K, Kagano H, Kasuga T, Otsuji Y (1983) Chem Lett 133; Mizuno K, Otsuji Y (1986) Chem Lett 683
77. Farid S, Shealer SE (1973) J Chem Soc Chem Commun 677
78. Yasuda M, Pac C, Sakurai (1980) Bull Chem Soc Jpn 53: 502; Cedheim L, Eberson L (1976) Acta Chem. Scand. B30: 527
79. Lewis FD, Kojima M (1988) J Am Chem Soc 110: 8664; Kojima M, Sakuragi H, Tokumaru K (1991) Tetrahedron Lett 22: 2889
80. Asanuma T, Gotoh T, Tsuchida A, Yamamoto M, Nishijima Y (1977) J Chem Soc Chem Commun 485; Yamamoto M, Asanuma T, Nishijima Y (1975) J Chem Soc Chem Commun 53; Asanuma T, Yamamoto M, Nishijima Y (1975) J Chem Soc Chem Commun 56
81. Mizuno K, Ogawa J, Kamura M, Otsuji Y (1979) Chem Lett 731
82. Ilyas M, de Mayo P (1985) J Am Chem Soc 107: 5093
83. Mattes SL, Farid S (1980) J Chem Soc Chem Commun 126
84. Brown-Wensley KA, Mattes SL, Farid S (1978) J Am Chem Soc 100: 4162
85. Shirota Y, Nishikata A, Aoyama T, Saimatsu J, Oh S-C, Mikawa H (1984) J Chem Soc Chem Commun 64
86. Yamaguchi K, Oh S-C, Shirota Y (1986) Chem Lett 1445
87. Mattay J, Trampe G, Runsink J (1988) Chem Ber 121: 1991
88. Mella M, Fasani E, Albini A (1991) Tetrahedron 47: 3137; Jones CR, Allman BJ, Mooring A, Spahic B (1983) J Am Chem Soc 105: 652
89. Mizuno K, Kaji R, Otsuji Y (1977) Chem Lett 1027
90. Farid S, Hartman SE, Evans TR (1975) In: Gordon M, Ware WR (eds) The exciplex. Academic, New York, p 327
91. Mizuno K, Kaji R, Okada H, Otsuji Y (1978) J Chem Soc, Chem Commun 594
92. Mizuno K, Ishii M, Otsuji Y (1981) J Am Chem Soc 103: 5570
93. Mizuno K, Ueda H, Otsuji Y (1981) Chem Lett 1237
94. Maroulis AJ, Arnold DR (1979) J Chem Soc Chem Commun 351
95. Arnold DR, Borg RM, Albini A (1981) J Chem Soc Chem Commun 138
96. Bellville DJ, Wirth DD, Bauld NL (1981) J Am Chem Soc 103: 718
97. Bauld NL, Bellville DJ, Pabon PA, Chelsky R, Green G (1983) J Am Chem Soc 105: 2378
98. Bauld NL, Bellville DJ, Harirchian B, Lorenz KT, Pabon PA, Reynolds DW, Wirth DD, Chion HS, Marsh BK (1987) Acc Chem Res 20: 371
99. Mizuno K, Hashizume T, Otsuji Y (1983) J Chem Soc, Chem Commun 772
100. Mella M, Fasani E, Albini A (1992) J Org Chem 57: 6210
101. Davis HF, Chattopadhyay SK, Das PK (1984) J Phys Chem 88: 2798
102. Pac C, Ohtsuki S, Yanagida S, Sakurai H (1986) Bull Chem Soc Jpn 59: 1133
103. Mizuno K, Hashizume T, Otsuji Y (1983) J Chem Soc, Chem Commun 977
104. Nakanishi K, Mizuno K, Otsuji Y (1991) J Chem Soc, Chem Commun 90
105. Calhoun GC, Schuster GB (1986) J Am Chem Soc 108: 8021
106. Wolfle I, Chan S, Schuster GB (1991) J Org Chem 56: 7313
107. Hartsough D, Schuster GB (1989) J Org Chem 54: 3
108. Kim J, Schuster GB (1990) J Am Chem Soc 112: 9635

109. Lopez L (1990) In: Mattay J (ed) Photoinduced electron transfer I, Topics in current chemistry, vol 156. Springer, Berlin Heidelberg, New York
110. Fox MA (1988) In: Fox MA, Chanon M (eds) Photoinduced electron transfer, Part D. Elsevier, Amsterdam, p 1
111. Eriksen J, Foote CS, Parker TL (1977) J Am Chem Soc 99: 6455
112. Eriksen J, Foote CS (1980) J Am Chem Soc 102: 6083
113. Arnold DR, Maroulis AJ (1977) J Am Chem Soc 99: 7355
114. Lewis FD (1986) Acc Chem Res 19: 401; Pienta NJ (1988) In: Fox MA, Chanon M (eds) Photoinduced electron transfer, Part C, Elsevier, Amsterdam, p 421
115. Tomioka H, Miyagawa H (1988) J Chem Soc Chem Commun 1183
116. Pac C, Ihama M, Yasuda M, Miyauchi Y, Sakurai H (1981) J Am Chem Soc 103: 6495
117. Ishitani O, Ihama M, Miyauchi Y, Pac C (1985) J Chem Soc, Perkin Trans 1 1527
118. Pac C, Miyauchi Y, Ishitani O, Ihama M, Yasuda M, Sakurai H (1984) J Org Chem 49: 26
119. Tazuke S, Ozawa H (1975) J Chem Soc, Chem Commun 273
120. Toki S, Hida S, Takamuku S, Sakurai H (1984) Nippon Kagaku Kaishi 152
121. Mizuno K, Ichinose N, Tamai T, Otsuji Y (!985) Tetrahedron Lett 26: 5823
122. Mizuno K, Tamai T, Nakanishi I, Ichinose N, Otsuji Y (1988) Chem Lett 2065
123. Tamai T, Mizuno K, Hashida I, Otsuji Y (1991) Photochem Photobiol 54: 23
124. Mizuno K, Ikeda M, Otsuji Y (1988) Chem Lett 1507; Mizuno K, Nakanishi K, Tachibana A, Otsuji Y (1991) J Chem Soc, Chem Commun 344
125. Dinnocenzo JP, Farid S, Goodman JL, Gould IR, Todd WP, Mattes SL (1989) J Am Chem Soc 111: 8973; Nakadaira Y, Otani S, Kyushin S, Ohashi M, Sakurai H, Funada Y, Sakamoto K, Sekiguchi A (1991) Chem Lett 601
126. Viehe HG (1979) Angew Chem Int Ed Engl 18: 917; (1985) Acc Chem Res 18: 148
127. Mizuno K, Nakanishi K, Chosa J, Nguyen T, Otsuji Y (1989) Tetrahedron Lett 28: 3689; Mizuno K, Nakanishi K, Chosa J, Otsuji Y, to be published
128. Bruno JW, Marks TJ, Lewis FD (1982) J Am Chem Soc 104: 5579
129. Pandey G, Soma Sekhar BBV, Bhalerio UT (1990) J Am Chem Soc 112: 5650
130. Ohga K, Mariano PS (1982) J Am Chem Soc 104: 617
131. Ohga K, Yoon UC, Mariano PS (1984) J Org Chem 49: 213
132. Ahmed-Schofield R, Mariano PS (1985) J Org Chem 50: 5667; (1987) Ibid 52: 1478
133. Mariano PS (1983) Tetrahedron 39: 3485
134. Mariano PS (1983) Acc Chem Res 16: 130
135. Yoon UC, Mariano PS (1992) Acc Chem Res 25: 233
136. Hasegawa E, Xu W, Mariano PS, Yoon UC, Kim JU (1988) J Am Chem Soc 110: 8099; Xu W, Jeon YT, Hasegawa E, Yoon UC, Mariano PS (1989) J Am Chem Soc 111: 413; Xu W, Zhang XM, Mariano PS (1991) J Am Chem Soc 113: 8863
137. Kubo Y, Imaoka T, Shiragami T, Araki T (1986) Chem Lett 1749
138. Maruyama K, Imahori H, Osuka A, Takuwa A, Tagawa H (1986) Chem Lett 1719
139. Takuwa A, Nishigaichi Y, Yamashita K, Iwamoto H (1990) Chem Lett 639; Takuwa A, Nishigaichi Y, Iwamoto H (1991) Ibid 1013
140. Cohen SG, Parola A, Parsons Jr GH (1973) Chem Rev 73: 141
141. Pac C, Sakurai H, Tosa T (1970) Chem Commun 1311
142. Pac C, Mizuno K, Tosa T, Sakurai H (1974) J Chem Soc Perkin I 561
143. Mizuno K, Ikeda M, Otsuji Y (1985) Tetrahedron Lett 26: 461; Mizuno K, Nakanishi K, Otsuji Y (1988) Chem Lett 1833; Mizuno K, Kobata T, Maeda R, Otsuji Y (1990) Ibid 1821
144. Mizuno K, Toda S, Otsuji Y (1987) Chem Lett 203; (1987) Nippon Kagaku Kaishi 1183
145. Mizuno K, Terasaka K, Ikeda M, Otsuji Y (1985) Tetrahedron Lett 26: 5819; Mizuno K, Terasaka K, Yasueda M, Otusji Y (1988) Chem Lett 145; Mizuno K, Nishiyama T, Terasaka K, Yasuda M, Shima K, Otsuji Y (1992) Tetrahedron (in press)
146. Eaton DF (1980) J Am Chem Soc 102: 328; (1981) J Am Chem Soc 103: 7235
147. Mella M, Fasani E, Albini A (1992) J Photochem Photobiol A: Chem 65: 383; (1992) J Org Chem 57: 6210
148. Lan JY, Schuster GB (1985) J Am Chem Soc 107: 6710; (1986) Tetrahedron Lett 4261
149. Kyushin S, Ehara Y, Nakadaira Y, Ohashi M (1989) J Chem Soc Chem Commun 279; Kyushin S, Nakadaira Y, Ohashi M (1990) Chem Lett 2191; Kyushin S, Masuda Y, Matsushita K, Nakadaira Y, Ohashi M (1990) Tetrahedron Lett 31: 6395
150. Fukuzumi S, Kuroda S, Tanaka T (1986) J Chem Soc Chem Commun 1553; Fukuzumi S, Kitano T, Mochida K (1989) Chem Lett 2177

151. Mattes SL, Farid S (1986) J Am Chem Soc 108: 7356
152. Gollnick K, Schanatterer A (1984) Tetrahedron Lett 25: 185 and 2735
153. Haynes RK, Probert MKS, Wilmot ID (1978) Aust. J Chem 31: 1737
154. Tamai T, Mizuno K, Hashida I, Otsuji Y (1993) Tetrahedron Lett 34: 2641
155. Mizuno K, Murakami K, Kamiyama N, Otsuji Y (1983) J Chem Soc Chem Commun 462
156. Mizuno K, Kamiyama N, Ichinose N, Otsuji Y (1985) Tetrahedron 41: 2207; Mizuno K, Kamiyama N, Otsuji Y (1983) Chem Lett 477
157. Tamai T, Mizuno K, Hashida I, Otsuji Y (1992) J Org Chem 57: 5338
158. Ichinose N, Mizuno K, Tamai T, Otsuji Y (1990) J Org Chem 55: 4079
159. Shim SC, Song JS (1986) J Org Chem 51: 2817
160. Takahashi Y, Miyashi T, Mukai T (1983) J Am Chem Soc 105: 6511; Miyashi T, Takahashi Y, Mukai T, Roth HD, Schilling MLM (1985) Ibid 107: 1079
161. Miyashi T, Kamata M, Mukai T (1986) J Chem Soc Chem Commun 1577; Miyashi T, Kamata M, Mukai T (1987) J Am Chem Soc 109: 2780
162. Tomioka H, Kobayashi D, Hashimoto A, Murata S (1989) Tetrahedron Lett 30: 4685
163. Miyashi T, Konno A, Takahashi Y (1988) J Am Chem Soc 110: 3676; Miyashi T, Ikeda H, Konno A, Okitsu O, Takahashi Y (1990) Pure & Appl Chem 62: 1531
164. Schaap AP, Siddiqui S, Gagnon SD, Prasad G, Palmono E, Lopez L (1984) J Photochem 25: 167; Schaap AP, Prasad G, Siddiqui S (1984) Tetrahedron Lett 25: 3035
165. Akasaka T, Sato K, Kako M, Ando W (1991) Tetrahedron Lett 32: 6605
166. Mizuno K, Ichinose N, Tamai T, Otsuji Y (1992) J Org Chem 57: 4669; Ichinose N, Mizuno K, Yoshida K, Otsuji Y (1988) Chem Lett 723
167. Albini A, Arnold DR (1978) Can J Chem 56: 2985
168. Muller F, Mattey J (1991) Angew Chem Int Ed Engl 30: 1336
169. Kamata M, Miyashi T (1989) J Chem Soc Chem Commun 557
170. McCullough JJ, Wu WS (1972) J Chem Soc Chem Commun 1136; McCullough JJ, Miller RC, Fung D, Wu WS (1975) J Am Chem Soc 97: 5942; McCullough JJ, Miller RC, Wu W-S (1977) Can J Chem 55: 2909
171. Mizuno K, Pac C, Sakurai H (1974) J Am Chem Soc 96: 2993
172. Lewis FD, DeVoe RJ, MacBlane DB (1982) J Org Chem 47: 1392
173. Lewis FD, DeVoe RJ (1982) Tetrahedron 38: 1069
174. Borg RM, Arnold DR, Cameron TS (1984) Can J Chem 62: 1785; Arnold DR, Snow MS (1988) Can J Chem 66: 3012; Arnold DR, Du X (1989) J Am Chem Soc 111: 7666
175. Kubo Y, Suto M, Araki T, Mazzocchi PH, Klingler L, Shook D, Somich C (1986) J Org Chem 51: 4404; Maruyama K, Kubo Y (1985) Ibid 50: 1426; Maruyama K, Kubo Y, Machida M, Oda K, Kanaoka Y, Fukuyama K (1978) Ibid 43: 2303
176. Mariano PS, Stavinoha JL, Pepe G, Meyer EF (1978) J Am Chem Soc 100: 7114
177. Mazzocchi PH, Minamikawa S, Wilson P (1985) J Org Chem 50: 2681
178. Yamada S, Kimura Y, Ohashi M (1977) J Chem Soc Chem Commun 667

Photophysical and Photochemical Properties of Fullerenes*

Christopher S. Foote

Department of Chemistry and Biochemistry, University of California, Los Angeles, California 90024-1569, USA

Table of Contents

---

* Portions of this work were presented at the NATO workshop on Fullerenes, Agia Pelagia, Crete, June 1993 and will appear in the report of this workshop [1].

Topics in Current Chemistry, Vol. 169
© Springer-Verlag Berlin Heidelberg 1994

Christopher S. Foote

On irradiation with visible and UV light, fullerenes C_{60} and C_{70} give high yields of their triplet states. Fluorescence is very weak from both, and phosphorescence has been observed only from C_{70}. The triplet states are also formed efficiently by energy transfer from triplet sensitizers with energies above 35 kcal/mol. Triplet fullerenes transfer energy efficiently to oxygen, giving singlet molecular oxygen, but they react and quench singlet oxygen very inefficiently, making them excellent photosensitizers. The fullerenes C_{60} and C_{70} are easily reduced but very difficult to oxidize. Electron transfer to triplet C_{60} and C_{70} occurs readily from electron donors to produce the radical anions. C_{60} can also be oxidized by excited high-potential photosensitizers to the radical cation. Some dihydrofullerenes also give high yields of the triplet state and singlet oxygen on irradiation. Photoreaction of C_{60} with electron-rich compounds appears to be an efficient route to difunctionalized adducts.

1 Introduction

The recent production in gram quantities of the intriguing soccerball-shaped molecule, "buckminsterfullerene" (C_{60}) and in lesser quantities, the rugby-ball-shaped C_{70}, stimulated intense worldwide activity [2–6]. The physical and photophysical properties of C_{60} and C_{70} have been well characterized during the last two years. The chemistry of the fullerenes has exploded, particularly during the last year, and a large number of functionalized dihydrofullerenes and some polyadducts have been characterized. (The term "dihydrofullerene" refers to any fullerene in which both carbons of one double bond have been converted to sp^3 hybridization by attachment of two substituents, not necessarily hydrogen; thus all monoadducts of C_{60} are dihydrofullerenes). This field has already outstripped an excellent recent review [7]. This chapter summarizes the photochemical and photophysical properties of C_{60} and C_{70} and the small amount that is known of these properties of dihydrofullerenes. The coverage is limited to work in liquid solution (with a few useful exceptions), and while extensive, is not exhaustive. A review of the extensive work in solid and gas phases, and in particular, the very interesting superconducting properties of alkali metal compounds, has recently appeared [8]. Where numerical data are given, I have attempted to select those that I believe are most reliable, but original references should be consulted for details.

2 Photophysical Properties

2.1 Introduction

The basic photophysical properties of C_{60} and C_{70} were first described by Arbogast et al. [9, 10]. These studies have been greatly expanded and extended [11–30]. Scheme 1 summarizes the photophysical processes reported.

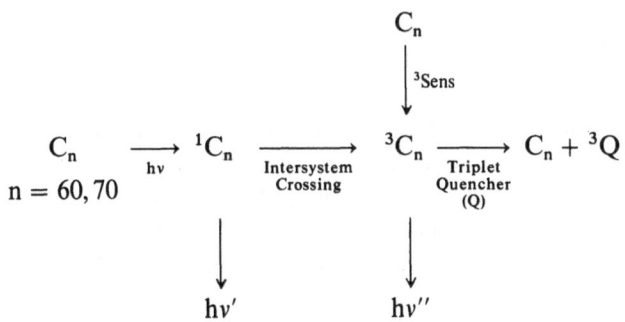

Scheme 1. Photophysical Processes in C_{60} and C_{70}

2.2 Electronic Absorption and Fluorescence

Both C_{60} and C_{70} absorb strongly in the UV ($\varepsilon_{max} \sim 10^5$ M^{-1} cm^{-1}) and much more weakly in the visible region (λ_{max} 540 nm, $\varepsilon_{max} \sim 710$ M^{-1} cm^{-1} for C_{60} and λ_{max} 468 nm, $\varepsilon_{max} \sim 15\,000$ M^{-1} cm^{-1} for C_{70}) because of symmetry prohibitions [5, 18]. The visible absorption of C_{60} is allowed only because of vibronic interactions and is responsible for the beautiful magenta color of these solutions. Because of its lower symmetry, the visible absorption of C_{70} is much stronger and solutions have a reddish-orange color.

No fluorescence emission was originally detected from C_{60} [9]. Very weak fluorescence (λ_{max} 720 nm, quantum yield (Φ_F) $\sim 10^{-5} - 10^{-4}$) has since been reported by several groups; there is disagreement about the quantum yield [29, 31–34]. Weak, structured fluorescence from C_{70} ($\Phi_F \sim 10^{-4}$) occurs at 77 K in a glass with λ_{max} 682 nm and is much broader at room temperature [10, 28, 35, 36]. The very low fluorescence yields of C_{60} and C_{70} probably result both from the very short lifetime of the singlet state and from the symmetry-forbidden nature of the lowest-energy transition of C_{60}.

Christopher S. Foote

2.3 Excited States

2.3.1 The S_1 State

Transient absorption studies have characterized the singlet excited states (S_1) of C_{60} (λ_{max} 513, 759, 885 nm, lifetime ~ 1.3 ns) and of C_{70} (λ_{max} 675, 840 nm, lifetime ~ 700 ps) [11, 12, 30, 37, 38]. The very short singlet lifetimes of both compounds are caused mainly by the high rate of intersystem crossing to the triplet.

2.3.2 The Triplet State

Triplet quantum yields (Φ_T) are near 100% for both C_{60} and C_{70} [12–17]. The high yield is a consequence of the very high rate of intersystem crossing (much faster than for typical aromatic hydrocarbons) and the inefficient fluorescence. These values were determined by photothermal and spectroscopic methods and agree with lower limits determined from singlet oxygen quantum yields [9, 10]. (see below). A cyclodextrin complex also produces the triplet state efficiently in water [39].

The triplet-triplet absorption spectra and extinction coefficients of C_{60} (λ_{max} 330, 750 nm, ε 40 000, 20 000 $M^{-1} cm^{-1}$) and C_{70} (λ_{max} 335, 480, 690, 970 nm, ε 35 000, 12 000, 2250, 3800 $M^{-1} cm^{-1}$) have been measured by several groups [9–13, 15, 17, 19, 22, 26, 27, 30, 38]. The details of the spectra in the UV (and for C_{70} at 480 nm, where there is intense ground state absorption), vary considerably because of the difficulty in correcting the spectra accurately for ground-state depletion. The triplet lifetimes in fluid solution, limited by triplet-triplet annihilation and ground-state quenching in solution at room temperature, are quite short (50–100 µs); in glasses, they are fairly long, but much lower than those of planar aromatic hydrocarbons.

The triplet energy levels (E_T) of both C_{60} and C_{70} were estimated to be close to 35 kcal/mol by triplet-triplet energy transfer [9, 10]. Both compounds quench the triplet states of sensitizers with $E_T \geq 42$ kcal/mol with essentially diffusion-controlled rate constants, but quenching of the tetraphenylporphine (TPP) triplet ($E_T = 33$ kcal/mol) [40] is significantly slower, (~ 2–$4 \times 10^7 M^{-1} s^{-1}$), probably because energy transfer with this sensitizer is slightly endothermic. Energy transfer from C_{60} and C_{70} triplets to acceptors with $E_T \leq 28$ kcal/mol is diffusion-controlled. Wasielewski et al. measured C_{70} phosphorescence at 77 K, which also confirmed the triplet energy of C_{70} [37]. Phosphorescence of C_{60} has not been observed despite the high triplet yield. However, heavy-atom-induced phosphorescence can be observed, and gives a triplet energy of 36.3 kcal/mol [23]. These values are in good agreement with those determined by photothermal [14–16, 21] and spectroscopic [18, 30] methods.

ESR Studies of triplet C_{60} have been carried out and suggest the electrons are on opposite sides of the molecule on the average [37]. EPR studies give much insight into rotational dynamics of the triplet [25, 41–43]. EPR studies of triplet C_{70} have also been reported [25, 37]. This technique has been used to measure triplet-triplet annihilation and quenching by nitroxide radicals [44]. (Such quenching has also been studied by flash photolysis and is very fast) [45].

2.4 Summary

The short lifetimes of both the singlet and triplet states, high triplet yields, low fluorescence and phosphorescence yields and high intersystem crossing rates of the two fullerenes are probably a result of the unique geometry of these compounds. The nearly spherical shape causes a relatively large spin-orbit coupling [46] which results in efficient intersystem crossing, compared, for example, to planar aromatic hydrocarbons. The high symmetry, which makes the lowest energy transition forbidden and is responsible for the weak absorption of the visible bands, is also partly responsible for the weak fluorescence and phosphorescence.

The small S-T splitting in C_{60} and C_{70} ($\sim 10\,\text{kcal/mol}$ in C_{60}, [18, 23] $\sim 7.0\,\text{kcal}$ in C_{70} [33]) is probably a result of the large diameter of the molecules and the resulting small electron-electron repulsion energy. Overall, the behavior of C_{70} is similar to that of C_{60}, but the lower symmetry relaxes the forbiddenness of some absorption and emission processes.

3 Singlet Oxygen

The triplet states of C_{60} and C_{70} transfer energy efficiently to $^3O_2$ to give singlet molecular oxygen ($^1O_2$) [9, 10]. The yield of $^1O_2$ is conveniently measured by the intensity of its luminescence at 1270 nm in comparison to that of a standard [47]. The quantum yield ($\Phi_{^1O_2}$) is nearly quantitative for C_{60} and slightly lower for C_{70} [9, 10]. The C_{70} value would imply that the energy transfer from C_{70} to oxygen goes with slightly less than unit efficiency, since the triplet yield is near unity (see above). Subsequent work has confirmed the C_{70} value [27]. Singlet oxygen yields have also been measured by photothermal measurements; [14, 16, 21] there is disagreement about the wavelength-dependence of the yield from C_{60}, but because of the relatively large experimental errors in measuring quantum yields, all values are nearly within the experimental error of 1.0. These values also represent lower limits for the quantum yield of triplet production (Φ_T). The values of Φ_T estimated in this way agree well with estimates obtained by photothermal and spectroscopic methods (see above).

$$^3C_n + {}^3O_2 \longrightarrow C_n + {}^1O_2$$

C_{60} quenches singlet oxygen with an approximate rate constant $k_q(^1O_2) = (5 \pm 2) \times 10^5\ M^{-1}s^{-1}$, as measured by the small shortening of the lifetime of the singlet oxygen luminescence with a nearly-saturated solution of C_{60} in C_6D_6 [9]. A somewhat lower value has since been reported [48]. Quenching by C_{70} was difficult to measure because of the low solubility; it has since been measured in C_6F_6, in which singlet oxygen has a very long lifetime, and is $(2.8 \pm 0.7) \times 10^6\ M^{-1}s^{-1}$, considerably larger than for C_{60} [49]. The mechanism of quenching by C_{60} and C_{70} is unknown. Chemical reaction is not an important cause because negligible loss of starting material or formation of new products (by HPLC or UV-visible absorption spectra) occurs following hundreds of laser pulses under O_2 at either excitation wavelength. However, longer irradiation under oxygen leads to production of $C_{60}O$, the epoxide, and other materials by an unknown mechanism [50]. Most of the quenching of singlet oxygen appears to be thermal, probably also involving the spin-orbit coupling of the fullerenes in a collision complex, causing deactivation of singlet oxygen [9].

$$C_n + {}^1O_2 \longrightarrow C_n + {}^3O_2$$

In any case, the relative photochemical stability of both C_{60} and C_{70} and their low fluorescence yields make these compounds particularly useful as photosensitizers. For example, they have been used to prepare some novel oxygen-metal adducts by reaction of singlet oxygen with certain organometallic complexes [51].

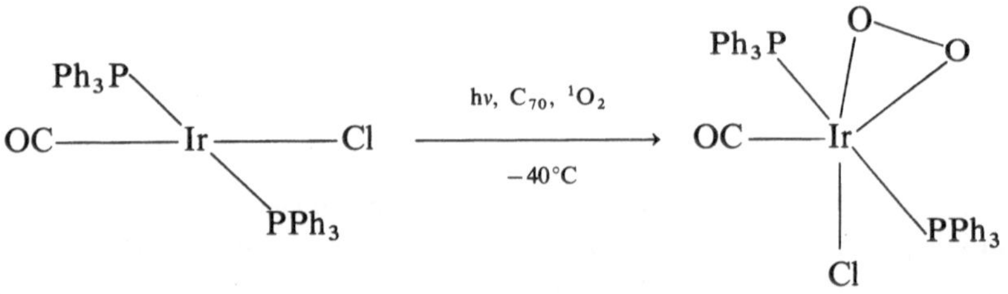

Because of their very low fluorescence yields, the fullerenes are also well suited to photophysical studies where even weak fluorescence from a sensitizer would interfere. For example, C_{60} and C_{70} have been used to sensitize the formation of singlet oxygen under conditions where two molecules of singlet oxygen collectively transfer energy to a fluorescer; the detection of the fluorescence from the energy acceptor would be difficult with most other sensitizers because of competing sensitizer fluorescence [49].

$$^3C_{70} \xrightarrow[O_2]{} {}^1O_2 + C_{70}$$

$$2{}^1O_2 \xrightarrow[Fl]{} 2O_2 + {}^1Fl \longrightarrow h\nu'$$

4 Electron-Transfer Reactions

4.1 Reduction

Electrochemical studies have shown that C_{60} is easily reduced ($E_{1/2} = -0.21$ and 0.33 V vs Ag/AgCl in tetrahydrofuran and benzonitrile, respectively [52] and -0.42 V vs SCE in benzonitrile [53, 54]). Up to six electrons can be added reversibly [55]. Several authors have shown that the fullerenes form charge-transfer complexes with amines [33, 56–59]. Wudl et al. have shown that C_{60} reacts chemically with amines, giving various substitution products [60, 61]. Since the reduction potential of $^3C_{60}$ should be higher than that of the ground state by the amount of the triplet energy [62, 63], its first reduction potential should be near 1.14 V vs SCE in benzonitrile [64]. The triplet should therefore be easily reduced by electron transfer from electron donors of lower oxidation potential.

$$D + {}^3C_{60} \longrightarrow D^{\cdot+} + C_{60}^{\cdot-}$$

Quenching of $^3C_{60}$ by electron donors occurs efficiently, and the mechanism is primarily electron transfer, as shown by the formation of the well-known transient absorptions (350 to 800 nm) of aromatic amine radical cations [64]. Because of the broad visible absorption of these radical cations, it is difficult to confidently assign visible absorptions to the C_{60} radical anion. However, with an infrared-sensitive germanium detector, a prominent transient with maxima at 950 and 1075 nm appears assigned to the C_{60} radical anion, [64] in good agreement with simultaneous and later measurements of others [17, 56, 65–67].

The rate constants for quenching (k_q) of triplet C_{60} by electron donors (amines and pyrene) were determined from the decay kinetics of $^3C_{60}$ transient absorption at 740 and 680 nm. In each case, the quenching followed Stern-Volmer kinetics [64]. A plot of log k_q vs the free energy for the electron transfer from the donor to C_{60} triplet (ΔG_{ET}) (Fig. 1) gives a fairly good correlation with rate constants calculated for electron transfer using the semiempirical Weller equation [62] (solid line). The general shape of the plot is similar to the Weller line, which approaches the diffusion rate constant of $\sim 2 \times 10^{10}$ M^{-1}s^{-1} in acetonitrile when electron-transfer is exergonic. However, the diffusion rate in the more viscous benzonitrile should be $\sim 5.6 \times 10^9$ M^{-1}s^{-1}, in excellent agreement with the maximum rate constants observed. Most of the scatter in this plot is from errors in the oxidation potentials of the donors, many of which were measured in solvents other than benzonitrile.

The EPR of the photochemically-generated C_{60} radical anion has been reported. In this study, no added donor was used [68]. It is not clear what the source of the electron in this experiment is; the authors suggest it derives from a second molecule of C_{60}.

Relatively little has been done with C_{70}. However, Verhoeven et al. showed that electron donors quench its fluorescence, and there is evidence for formation

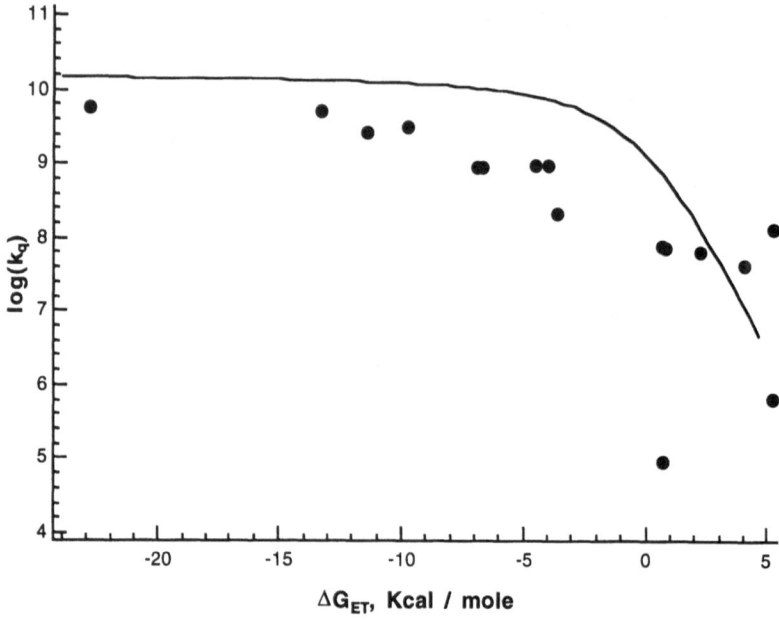

Fig. 1. Plot of log k_q vs ΔG_{ET} in benzonitrile. The *solid line* was calculated from the equation of Rehm and Weller for the less-viscous acetonitrile [57]

of ground-state charge-transfer complex with electron-rich aromatics [35]. Ground-state charge-transfer complexes formed between both C_{60} and C_{70} give ion-pairs on photoexcitation that undergo fast geminate recombination [69].

Both C_{60} and C_{70} embedded in a lipid membrane can act as efficient electron acceptors from excited electron donors adsorbed on the surface of the membrane [70]. Very interestingly, C_{60} and C_{70} can act not only as photo-sensitizers for electron transfer from a donor molecule but also as mediators for electron transfer across a membrane with very high efficiencies [71].

4.2 Oxidation

Although C_{60} is easily reduced, it is very difficult to oxidize [46, 53, 54, 72]. The only definitive electrochemical oxidation of C_{60} occurs at a potential of +1.76 V vs SCE in benzonitrile, and is irreversible [54]. The radical cation was reported to be produced by γ-irradiation at 77 K in a glass, and to absorb near 980 nm [65, 66]. Attempts to generate the radical cation ($C_{60}^{\cdot+}$) by electron transfer to singlet excited dicyanoanthracene, which has a reduction potential near +2.0 V [73] were unsuccessful. This method has been used to generate, for example, *trans*-stilbene radical cation [73, 74]. The ion pair probably does not dissociate rapidly in the relatively non-polar benzonitrile, and the lifetime of ^{1}DCA is so short that only a few per cent can be quenched at C_{60} saturation.

$$^1DCA + C_{60} \longrightarrow DCA^{\bullet-} + C_{60}^{\bullet+}$$

The excited singlet state of N-methylacridinium hexafluorophosphate (MA^+) has a reduction potential of 2.31 V vs SCE, [75, 76] sufficient to oxidize C_{60}, and also has a longer singlet lifetime than ^{1}DCA. Reduction of the cationic sensitizer produces a neutral radical and a radical cation rather than an ion pair; so that there should be no coulombic attraction to hinder the dissociation of the electron-transfer pair [76].

$$^1MA^+ + C_{60} \longrightarrow MA^{\bullet} + C_{60}^{\bullet+}$$

Irradiation of a solution containing MA^+ (2×10^{-4} M) and C_{60} in benzonitrile produced weak transient absorption at 980 nm, [77] assigned to the C_{60} radical cation because of its similarity to the spectrum reported by Kato et al. [65, 66]. The spectrum is nearly identical to that of the radical cation reported recently from a pulse-radiolysis experiment [78] and in an argon matrix [79]. From the risetime of the signal and the C_{60} concentration, the rate constant for electron transfer is diffusion-controlled.

We used cosensitization [80, 81] with biphenyl (BP, E(BP$^+$/BP) = 1.96 V vs SCE) to increase the quantum yield of $C_{60}^{\bullet+}$ formation [77]. In this process, the excited singlet (^{1}MA) abstracts an electron from a donor such as biphenyl (BP)

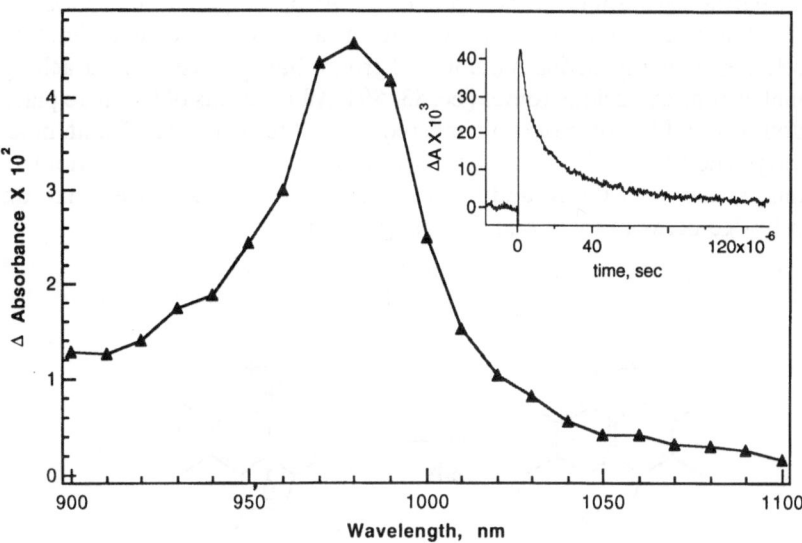

Fig. 2. Transient absorption spectrum (λ_{exc} = 420 nm) from MA^+ and C_{60} (2×10^{-4} M) containing biphenyl (0.2 M) in CH_2Cl_2 Inset: transient decay (λ_{obs} = 980 nm, average of 10 shots)

to give the BP cation radical, which has a much longer lifetime than ^{1}MA and a sufficiently high oxidation potential to oxidize C_{60}.

$$^1MA^+ + BP \longrightarrow MA^{\cdot} + BP^{\cdot+}$$

$$BP^{\cdot+} + C_{60} \longrightarrow BP + C_{60}^{\cdot+}$$

The near-IR transient absorption spectrum produced in the indirect electron-transfer experiment is shown in Fig. 2. The shape of the spectrum is very similar to that in the absence of BP, but about ten times as strong. The risetime suggests diffusion-controlled quenching of $BP^{\cdot+}$ by C_{60}.

A recent study of C_{60} in pulse radiolysis reported a spectral feature at 650 nm, assigned to the radical cation [82]. This study apparently did not use a near-IR sensitive detector, so that the strong 980 nm absorption was not observed, and it is possible that this 650 nm absorption is caused by products of radical addition to C_{60}. Reaction of triplet C_{60} with strong electron acceptors produces an exciplex and the free C_{60} radical cation in benzonitrile [24].

5 Functional Adducts

5.1 General

A great deal about the fullerenes is poorly understood, particularly how their unusual electronic properties are related to their unique structure, and the effects of partial saturation on their properties. Also, the parent molecules are poorly soluble in nonpolar solvents, and not at all in polar solvents, limiting their usefulness in many media. A number of groups have succeeded in attaching functional groups by various routes [60, 83–89]. All reactions of C_{60} take place at the double bond between the 6-membered rings, (the "pyracylene" unit, note: not pyracyclene), shown below to give dihydrofullerenes (see introduction) as the initial product [90]. (Some adducts lose nitrogen to form diradicals that can close to bridge between the 5 and 6-rings) [91, 92].

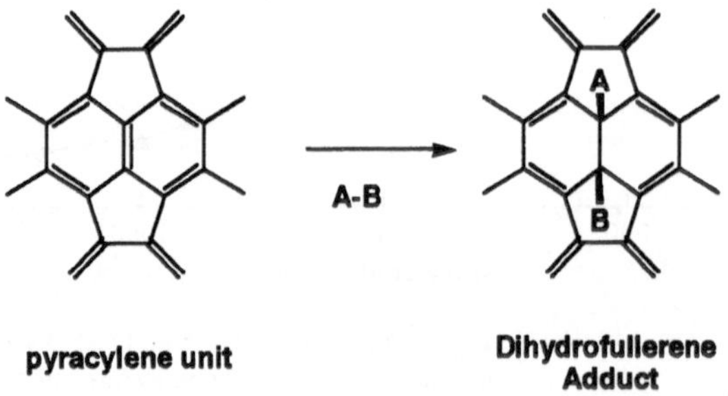

pyracylene unit   Dihydrofullerene
                              Adduct

The strong electron-withdrawing properties of the fullerene molecule are apparently also characteristic of the dihydrofullerenes. Several dihydrofullerenes have multiple reduction potentials, with the first near -1.096 V vs ferrocene [93, 94].

5.2 Diels Alder Routes to Dihydrofullerenes

Several authors have recently demonstrated a simple, high-yield route to dihydro-fullerenes via Diels-Alder addition of electron-rich dienes to C_{60} [87, 95–98]. This is a very versatile route which allows easy variation of functional groups and considerable control of the solubility characteristics of the adducts. For example, using 2-silyloxybutadiene derivatives, the cyclohexanone adduct shown below can be made after hydrolysis of the primary adduct. The corresponding alcohol is then made by reduction of the ketone, and is easily converted to amino acid derivatives which allow solubility in water [87].

A. Cyclohexanone Adduct   B. Cyclohexanol

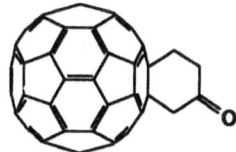
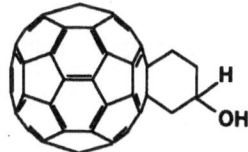

We have made preliminary photophysical studies of both the alcohol and the ketone [87, 99]. These compounds, like other dihydrofullerenes, have only indistinct, weak visible absorption, but strong UV bands. They also all have a weak long-wavelength absorption near 705 nm ($\varepsilon_{max} \sim 250$–$400$ $M^{-1} cm^{-1}$) [87]. This absorption, along with a band at ~ 430 nm [88, 95–97, 100, 101], is diagnostic for the dihydrofullerene structure [87].

Pulsed laser irradiation of both the alcohol and the ketone gave strong transient absorption spectra. Both spectra resemble that of the C_{60} triplet [99]. That of the alcohol ($\tau_T = 21$ µs) is shown in Fig. 3.

Both compounds gave singlet oxygen in high yields. The yield of singlet oxygen (which is a lower limit for the quantum yield of formation of the triplet) from the alcohol is 0.84 ± 0.1 at 532 nm and 0.72 ± 0.1 at 355 nm. Efficient triplet formation and singlet oxygen production has also been reported from an iridium derivative of C_{60} [102]. Further characterization of these compounds and their excited species and those of related derivatives is underway.

Very recently, a dihydrofullerene carboxylic acid has been shown to photo-chemically kill certain cells and cause guanosine-specific damage to DNA. Tests suggested that this action was caused by singlet oxygen [103]. This is only one of the recent reports of interesting biological activity of dihydrofullerenes [104], a field which is poised to explode.

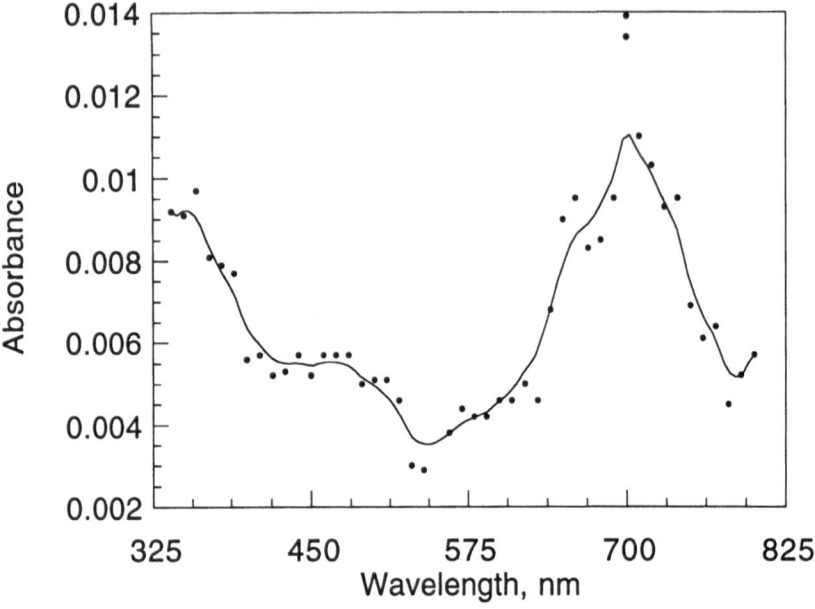

Fig. 3. Triplet-triplet absorption spectrum of cyclohexanol adduct, 3.0×10^{-5} M in benzonitrile

6 Photochemistry

There have been only a few cases of photochemical reaction of the fullerenes reported. They are known to be unstable to light, particularly in the solid state, but in most cases products have not yet been characterized [105]. There is some disagreement about the rate of this reaction, which probably depends on the purity and the physical state [10]. Photopolymerization of C_{60} has reported, and is believed to proceed via [2 + 2] cycloaddition to itself [106–108]. As mentioned above, long irradiation under oxygen gives $C_{60}O$ [50]. This reaction has recently been studied and also occurs with C_{70}; $C_{60}O$ is photochemically unstable, and disappears at a higher rate than the fullerenes, so that its yield by this route is limited [109]. It has been reported that the cage is opened on long irradiation under oxygen, but products have not been characterized [110].

6.1 Radical Addition

Although reaction of photochemically-generated species such as radicals with fullerenes does not involve excited fullerenes, this subject will be briefly mentioned here for completeness. Radicals (mostly photochemically generated) add readily to C_{60} [111–114]. The resulting adduct radicals absorb near 440 nm, with a shoulder near 640 nm [112]. They also absorb near 900 nm [115]. Their ESR spectra have been reported, and there is a suggestion of reversible dimerization [116].

6.2 Electron Transfer

As mentioned above, triplet C_{60} is readily photoreduced by amines and other donors to C_{60} radical anion and the donor radical cations [64]. We expected this reaction to lead to adducts with covalent bonds. Such adducts are formed with some amines in ground state chemistry [33, 60, 83], but the photochemical process should be more selective and easily controlled, since only one-electron reduction is possible in the photochemical process. C_{60} in the S_1 state has been suggested to produce an exciplex with triethylamine which seems to react with ground-state C_{60} to give a stable product [117]. The reduction potential of the triplet is high enough that electron-transfer from many donors such as electron-rich aromatics and alkenes should be possible.

$$^3C_{60} + D \longrightarrow C_{60}^{-\bullet} + D^{+\bullet} \longrightarrow C_{60}D$$

Adduct

Photochemical reaction of C_{60} with N,N-diethylpropynylamine, a good electron donor, gives a single [2 + 2] cycloadduct, the cyclobutenamine shown below, in >50% yield [118]. The product was characterized spectroscopically. It has not yet been demonstrated that the reaction goes via electron transfer.

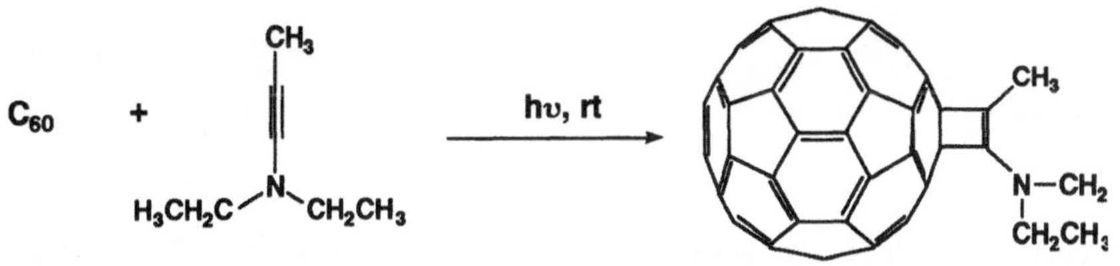

The enamine is unique in that it also has a photosensitizer (the dihydrofullerene chromophore) in the same molecule, and brief exposure to air and room light leads to cleavage of the enamine double bond, producing the ketoamide shown below. The well-known photooxidative cleavage of enamines proceeds via an intermediate 1,2-dioxetane [119, 120]. Although many 1,2-dioxetanes are relatively stable, those from enamines are not, and cleave to ketone and amide fragments, below $-40\,°C$ in most cases. The ketoamide was characterized by FAB ms (m/e = 863), IR, and ^1H and ^{13}C NMR (carbonyls at 170 ppm for the amide and 204 ppm for the ketone and overall C_s symmetry for the fullerene carbons).

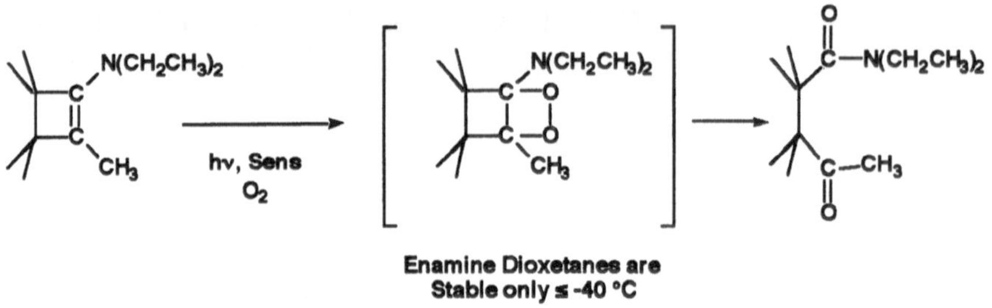

Christopher S. Foote

**Enamine Dioxetanes are
Stable only ≤ -40 °C**

Interestingly, Akasaka et al. have recently shown that 1,1,2,2-Tetramesityl-disilirane adds to excited C_{60} to give an adduct whose structure is believed to be that shown below (R = mesityl) [21]. Although it has not been demonstrated, it is believed that an electron or charge-transfer process also occurs in this case to give an exciplex (or radical ion pair) that can collapse to the observed product.

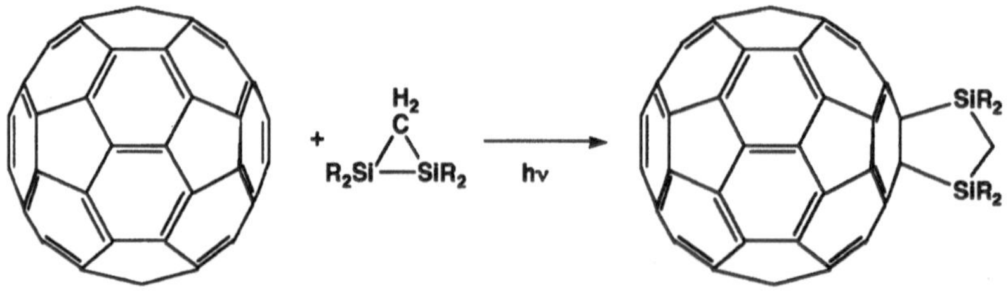

Acknowledgments. Work in this group was funded by NSF and NIH grants and carried out by undergraduate and graduate students and postdoctoral associates who are credited in the references.

References

1. Foote CS (1993) In: Prassides K (ed) Physics and Chemistry of the Fullerenes. NATO Advanced Research Workshop, Aghia Pelaghia, Crete, Greece. Kluwer, Dordrecht (in press)
2. Krätschmer W, Lamb LD, Fostiropoulos K, Huffman DR (1990) Nature 347: 354
3. Krätschmer W, Fostiropoulos K, Huffman DR (1990) Chem Phys Lett 170: 167
4. Taylor R, Hare JP, Abdul-Sada AK, Kroto HW (1990) J Chem Soc Chem Comm 1423
5. Ajie H, Alvarez MM, Anz SA, Beck RD, Diederich F, Fostiropoulos K, Huffman DR, Krätschmer W, Rubin Y, Schriver KE, Sensharma D, Whetten RL (1990) J Phys Chem 94: 8630
6. Kroto HW, Allaf AW, Balm SP (1991) Chem Revs 91: 1213
7. Taylor R, Walton D (1993) Nature 363: 685
8. Hebard AF (1993) Ann Rev Mater Sci 23: 159
9. Arbogast JW, Darmanyan AO, Foote CS, Rubin Y, Diederich FN, Alvarez MM, Anz SJ, Whetten RL (1991) J Phys Chem 95: 11
10. Arbogast J, Foote CS (1991) J Am Chem Soc 113: 8886
11. Ebbesen TW, Tanigaki K, Kuroshima S (1991) Chem Phys Lett 181: 501
12. Tanigaki K, Ebbesen TW, Kuroshima S (1991) Chem Phys Lett 185: 189

13. Sension RJ, Phillips CM, Szarka AZ, Romanow WJ, McGhie AR, McCauley Jr JP, Smith III AB, Hochstrasser RM (1991) J Phys Chem 95: 6075
14. Hung RR, Grabowski JJ (1991) J Phys Chem 95: 6073
15. Kajii Y, Nakagawa T, Suzuki S, Achiba Y, Obi K, Shibuya K (1991) Chem Phys Lett 181: 100
16. Terazima M, Hirota N, Shinohara H, Saito Y (1991) J Phys Chem 95: 6490
17. Biczok L, Linschitz H, Walter RI (1992) Chem Phys Lett 195: 339
18. Leach S, Vervloet M, Desprès A, Bréheret E, Hare JP, Dennis TJ, Kroto HW, Taylor R, Walton DRM (1992) Chem Phys 160: 451
19. Dimitrijevic NM, Kamat PV (1992) J Phys Chem 96: 4811
20. Gevaert M, Kamat PV (1992) J Phys Chem 96: 9883
21. Hung RR, Grabowski JJ (1992) Chem Phys Lett 192: 249
22. Palit DK, Sapre AV, Mittal JP (1992) Indian J Chem, Sect A 31: F46
23. Zeng Y, Biczok, L, Linschitz H (1992) J Phys Chem 96: 5237
24. Nadtochenko VA, Denisov NN, Rubtsov IV, Lobach AS, Moravskii AP (1993) Chem Phys Lett 208: 431
25. Terazima M, Sakurada K, Hirota N, Shinohara H, Saito Y (1993) J Phys Chem 97: 5447
26. Bensasson RV, Hill T, Lambert C, Land EJ, Leach S, Truscott TG (1993) Chem Phys Lett 201: 326
27. Bensasson RV, Hill T, Lambert C, Land EJ, Leach S, Truscott TG (1993) Chem Phys Lett 206: 197
28. Sun YP, Bunker CE (1993) J Phys Chem 97: 6770
29. Sun YP, Wang P, Hamilton NB (1993) J Am Chem Soc 115: 6378
30. Lee M, Song O-K, Seo J-C, Kim D, Suh YD, Jin SM, Kim SK (1992) Chem Phys Lett 196: 325
31. Kim D, Lee M, Suh YD, Kim SK (1992) J Am Chem Soc 114: 4429
32. Sibley SP, Argentine SM, Francis AH (1992) Chem Phys Lett 188: 187
33. Wang Y (1992) J Phys Chem 96: 764
34. Williams RM, Verhoeven JW (1993) Spectrochim. Acta Part A (in press) I thank Prof. Verhoeven for communicating this result
35. Verhoeven JW, Scherer T, Heymann D (1991) Recl Trav Chim Pay-Bas 110: 349
36. Williams RM, Verhoeven JW (1992) Chem Phys Lett 194: 446
37. Wasielewski MR, O'Neil MP, Lykke KR, Pellin MJ, Gruen DM (1991) J Am Chem Soc 113: 2774
38. Palit DK, Sapre AV, Mittal JP, Rao CNR (1992) Chem Phys Lett 195: 1
39. Andersson T, Nilsson K, Sundahl M, Westman G, Wennerström O (1992) J Chem Soc, Chem Comm 604
40. McLean AJ, McGarvey DJ, Truscott TG, Lambert CR, Land EJ (1990) J Chem Soc, Faraday Trans 86: 3075
41. Closs GL, Gautam P, Zhang D, Krusic PJ, Hill SA, Wasserman E (1992) J Phys Chem 96: 5228
42. Zhang D, Norris JR, Krusic PJ, Wasserman E, Chen CC, Lieber CM (1993) J Phys Chem 97: 5886
43. Steren CA, Levsten PR, van Willigen H, Linschitz H, Biczok L (1993) Chem Phys Lett 204: 23
44. Goudsmit G-H, Paul H (1993) Chem Phys Lett 208: 73
45. Samanta A, Kamat PV (1992) Chem Phys Lett 199: 635
46. Allemand P-M, Srdanov G, Koch A, Khemani K, Wudl F (1991) J Am Chem Soc 113: 2780
47. Ogilby PR, Foote CS (1983) J Am Chem Soc 105: 3423
48. Black G, Dunkle E, Dorko EA, Schlie LA (1993) J Photochem Photobiol A: Chem 70: 147
49. Krasnovsky Jr AA, Foote CS (1993) J Am Chem Soc 115: 6013
50. Creegan KM, Robbins JL, Robbins WK, Millar JM, Sherwood RD, Tindall PJ, Cox DM, McCauley Jr JP, Jones DR, Gallagher TT, Smith III AM (1992) J Am Chem Soc 114: 1103
51. Selke M, Foote CS (1993) J Amer Chem Soc 115: 1166
52. Allemand P-M, Koch A, Wudl F, Rubin Y, Diederich F, Alvarez MM, Anz SJ, Whetten RL (1990) J Am Chem Soc 113: 1050
53. Dubois D, Kadish KM, Flanagan S, Wilson LJ (1991) J Am Chem Soc 113: 7773
54. Dubois D, Kadish KM, Flanagan S, Haufler RE, Chibante LPF, Wilson LJ (1991) J Am Chem Soc 113: 4364
55. Xie Q, Pérez-Cordero E, Echegoyen L (1992) J Am Chem Soc 114: 3978
56. Sension RJ, Szarka AZ, Smith GR, Hochstrasser RM (1991) Chem Phys Lett 185: 179
57. Seshadri R, Govindaraj A, Nagarajan R, Pradeep T, Rao CNR (1992) Tetrahedron Lett 33: 2069
58. Seshadri R, Rao CNR (1993) Chem Phys Lett 205: 395

59. Seshadri R, D'Souza F, Krishnan V, Rao CNR (1993) Chem Lett 217
60. Wudl F (1992) Acc Chem Res 25: 157
61. Hirsch A, Li Q, Wudl F (1991) Angew Chem Int Ed Engl 30: 1309
62. Rehm D, Weller A (1970) Z Phys Chem N F 69: 183
63. Mattes SL, Farid S (1983) In: Padwa A (ed) Organic photochemistry. M. Dekker, New York p 233
64. Arbogast JW, Foote CS, Kao M (1992) J Am Chem Soc 114: 2277
65. Kato T, Kodama T, Shida T, Nakagawa T, Matsui Y, Suzuki S, Shiromaru H, Yamauchi K, Achiba Y (1991) Chem Phys Lett 180: 446
66. Kato T, Kodama T, Oyama M, Okazaki S, Shida T, Nakagawa T, Matsui Y, Suzuki S, Shiromaru H, Yamauchi K, Achiba Y (1992) Chem Phys Lett 186: 35
67. Greaney MA, Gorun SM (1991) J Phys Chem 95: 7142
68. Ruebsam M, Dinse K-P, Plueschau M, Fink J, Kraetschmer W, Fostiropoulos K, Taliani C (1992) J Am Chem Soc 114: 10059
69. Palit DK, Ghosh HN, Pal H, Sapre AV, Mittal JP, Seshadri R, Rao CNR (1992) Chem Phys Lett 198: 113
70. Hwang KC, Mauzerall D (1992) J Am Chem Soc 114: 9705
71. Hwang KC, Mauzerall D (1993) Nature 361: 138
72. Jehoulet C, Bard AJ (1991) J Am Chem Soc 113: 5456
73. Eriksen J, Foote CS (1980) J Am Chem Soc 102: 6083
74. Gould IR, Ege D, Moser JE, Faird S (1990) J Am Chem Soc 112: 4290
75. Gould IR, Moser JE, Armitage B, Farid S, Goodman JL, Herman MS (1989) J Am Chem Soc 111: 1917
76. Todd WP, Dinnocenzo JP, Farid S, Goodman JL, Gould IR (1991) J Am Chem Soc 113: 3601
77. Nonell S, Arbogast JW, Foote CS (1992) J Phys Chem 96: 4169
78. Guldi DM, Hungerbühler H, Janata E, Asmus K-D (1993) J Chem Soc Chem Comm 84
79. Gasyna Z, Andrews L, Schatz PN (1992) J Phys Chem 96: 1525
80. Schaap AP, Lopez L, Anderson SD, Gagnon SD (1982) Tetrahedron Lett 23: 5493
81. Spada LT, Foote CS (1980) J Am Chem Soc 102: 391
82. Hou H-Q, Luo C, Liu Z-X, Mao D-M, Qin Q-Z (1993) Chem Phys Lett 203: 555
83. Wudl F, Hirsch A, Khemani KC, Suzuki T, Allemand PM Koch A, Eckert H, Srdanov G, Webb HM (1992) ACS Symposium Ser. 481: 161
84. Tebbe FN, Harlow RL, Chase DB, Thorn DL, Campbell Jr GC, Calabrese JC, Herron N, Young Jr RJ, Wasserman E (1992) Science 256: 822
85. Taylor R, Langley J, Meidine MF, Parsons JP, Abdul-Sada AK, Dennis TJ, Hare JP, Kroto HW, Walton DRM (1992) J Chem Soc Chem Comm 667
86. Vasella A, Uhlmann P, Waldraff CAA, Diederich F, Thilgen C (1992) Angew Chem Int Ed 31: 1388
87. An Y-Z, Anderson JL, Rubin Y (1993) J Org Chem 58: 4799
88. Prato M, Suzuki T, Foroudian H, Li Q, Khemani K, Wudl F (1993) J Am Chem Soc 115: 1594
89. Prato M, Maggini M, Scorrano G, Lucchini V (1993) J Org Chem 58: 3613
90. Balch AL, Catalano VJ, Lee JW (1991) Inorg Chem 30: 3980
91. Isaacs L, Wehrsig A, Diederich F (1993) Helv Chim Acta 76: 1231
92. Prato M, Li QC, Wudl F, Lucchini V (1993) J Am Chem Soc 115: 1148
93. Shi S, Khemani KC, Li Q, Wudl F (1992) J Am Chem Soc 114: 10656
94. Paddon-Row MN private communication
95. Rubin Y, Khan S, Freedberg DI, Yeretzian C (1993) J Am Chem Soc 115: 344
96. Khan SI, Oliver AM, Paddon-Row MN, Rubin Y (1993) J Am Chem Soc 115: 4919
97. Kräutler B, Puchberger M (1993) Helv Chim Acta 76: 1626
98. Belik P, Gügel A, Spickermann J, Müllen K (1993) Angew Chem Int Ed Engl 32: 78
99. Anderson J, An Y-Z, Rubin Y, Foote CS (1993) in preparation
100. Suzuki T, Li Q, Khemani KC, Wudl F, Almarsson Ö (1991) Science 254: 1186
101. Henderson CC, Cahill PA (1993) Science 259: 1885
102. Zhu Y, Koefod RS, Devadoss C, Shapley JR, Schuster GB (1992) Inorg Chem 31: 3505
103. Tokuyama H, Yamago S, Nakamura E, Shiraki T, Sugiura Y (1993) J Am Chem Soc 115: 7918
104. Sijbesma R, Srdanov G, Wudl F, Castoro JA, Wilkins C, Friedman SH, DeCamp DL, Kenyon GL (1993) J Am Chem Soc 115: 6510
105. Taylor R, Parsons JP, Avent AG, Rannard SP, Dennis TJ, Hare JP, Kroto HW, Walton DRM (1991) Nature 351: 277

106. Rao A, Zhou P, Wang K-A, Hager GT, Holden JM, Wang Y, Lee W-T, Bi X-X, Eklund PC, Cornett DS, Duncan MA, Amster IJ (1993) Science 259: 955
107. Wang Y, Holden JM, Dong Z-H, X-X Bi, Eklund PC (1993) Chem Phys Lett 211: 341
108. Zhou P, Dong Z-H, Rao AM, Eklund PC (1993) J Chem Phys 211: 337
109. Heymann D, Chibante LPF (1993) Chem Phys Lett 207: 339
110. Taliani C, G R, Zamboni R, Danieli R, Rossini S, Denisov VN, Burlakov VM, Negri F, Orlandi G, Zerbetto F (1993) J Chem Soc Chem Comm 220
111. Dimitrijevic NM (1992) Chem Phys Lett 194: 457
112. Dimitrijevic NM, Kamat PV, Fessenden RW (1993) J Phys Chem 97: 615
113. Krusic PJ, Wasserman E, Keizer PN, Morton JR, Preston KF (1991) Science 254: 1183
114. McEwen CN, McKay RG, Larsen BS (1992) J Am Chem Soc 114: 4412
115. Guldi DM, Hungerbühler H, Janata E, Asmus K-D (1993) J Phys Chem (Submitted) I thank Prof. Asmus for communicating this result
116. Morton JR, Preston KF, Krusic PJ, Hill SA, Wasserman E (1992) J Am Chem Soc 114: 5454
117. Kajii Y, Takeda K, Shibuya K (1993) Chem Phys Lett 204: 283
118. Zhang X, Foote CS (1993) J Am Chem Soc 115: 11924
119. Foote CS, Dzakpasu AA, Lin JW-P (1975) Tetrahedron Lett 1247
120. Foote CS, Lin JW-P (1968) Tetrahedron Lett 3267
121. Akasaka T, Ando W (1993) J Am Chem Soc (115: 10366) I thank Dr. Akasaka for communicating this result

Author Index Volumes 151-169

The volume numbers are printed in italics

Adam, W. and Hadjiarapoglou, L.: Dioxiranes: Oxidation Chemistry Made Easy. *164*, 45-62 (1993).

Albini, A., Fasani, E. and Mella M.: PET-Reactions of Aromatic Compounds. *168*, 143-173 (1993).

Allamandola, L.J.: Benzenoid Hydrocarbons in Space: The Evidence and Implications *153*, 1-26 (1990).

Astruc, D.: The Use of π-Organoiron Sandwiches in Aromatic Chemistry. *160*, 47-96 (1991).

Balzani, V., Barigelletti, F., De Cola, L.: Metal Complexes as Light Absorption and Light Emission Sensitizers. *158*, 31-71 (1990).

Baker, B.J. and Kerr, R.G.: Biosynthesis of Marine Sterols. *167*, 1-32 (1993).

Barigelletti, F., see Balzani, V.: *158*, 31-71 (1990).

Baumgarten, M., and Müllen, K.: Radical Ions: Where Organic Chemistry Meets Materials Sciences. *169*, 1-104 (1994).

Bignozzi, C.A., see Scandola, F.: *158*, 73-149 (1990).

Billing, R., Rehorek, D., Hennig, H.: Photoinduced Electron Transfer in Ion Pairs. *158*, 151-199 (1990).

Bissell, R.A., de Silva, A.P., Gunaratne, H.Q.N., Lynch, P.L.M., Maguire, G.E.M., McCoy, C.P. and Sandanayake, K.R.A.S.: Fluorescent PET (Photoinduced Electron Transfer) Sensors. *168*, 223-264 (1993).

Bley, K., Gruber, B., Knauer, M., Stein, N. and Ugi, I.: New Elements in the Representation of the Logical Structure of Chemistry by Qualitative Mathematical Models and Corresponding Data Structures. *166*, 199-233 (1993).

Brunvoll, J., see Chen, R.S.: *153*, 227-254 (1990).

Brunvoll, J., Cyvin, B.N., and Cyvin, S.J.: Benzenoid Chemical Isomers and Their Enumeration. *162*, 181-221 (1992).

Brunvoll, J., see Cyvin, B.N.: *162*, 65-180 (1992).

Brunvoll, J., see Cyvin, S.J.: *166*, 65-119 (1993).

Bundle, D.R.: Synthesis of Oligosaccharides Related to Bacterial O-Antigens. *154*, 1-37 (1990).

Burrell, A.K., see Sessler, J.L.: *161*, 177-274 (1991).

Caffrey, M.: Structural, Mesomorphic and Time-Resolved Studies of Biological Liquid Crystals and Lipid Membranes Using Synchrotron X-Radiation. *151*, 75-109 (1989).